FISHERIES ADMINISTRATION
&
DEVELOPMENT

THE AUTHOR

Dr. Amita Saxena Professor in College of Fisheries Sciences at G.B. Pant University of Agriculture and Technology, Pantnagar, Uttarakhand India has been awarded as an outstanding woman of 20th century award (America). She formulated feed for air breathing and ornamental fishes/ organisms/other aquatic. Her debut in pearl culture in Uttrakhand (Hill State) is praise worthy and remarkable. She is crowned with young scientist award, several fellowship/gold medals and Bharat Jyoti/Vijay Shri award. She transferred the fisheries technologies to the farmers through visits, lectures, radio and TV talks. Nowadays she is busy in improving the rural economy of poor people through fisheries and other small scale ventures.

FISHERIES ADMINISTRATION & DEVELOPMENT

AMITA SAXENA
Professor Fisheries
GBP UNIVERSITY AG. & TECH.
PANTNAGAR 263145
INDIA

2014
Daya Publishing House®
A Division of
Astral International Pvt. Ltd.
New Delhi-110 002

ISBN 9789351302469

Published by : **Daya Publishing House®**
A Division of
Astral International Pvt. Ltd.
– ISO 9001:2008 Certified Company –
4760-61/23, Ansari Road, Darya Ganj
New Delhi-110 002
Ph. 011-43549197, 23278134
E-mail: info@astralint.com
Website: www.astralint.com

Laser Typesetting : **Classic Computer Services**, Delhi - 110 035

Printed at : **Thomson Press India Limited**

PRINTED IN INDIA

Preface

The basic concept of the book is to inform about Fishery Administration, Acts, Policies, Plans of Government of India, What are the trades, WTO, Marketing and Cooperatives. Here, the chapters are planned in such a way that readers, students, teachers, scientists should have the knowledge, so that they work accordingly for their own, state and country's bright future in terms of employment, success and for better economy. Few chapters like Indian Foreign Trade, Transport and Communication and Public Finances/Loans are described in details and in broader sense for the better understanding and to help common people.

Author expresses her gratitude to Hon'ble DG, ICAR, Hon'ble Vice Chancellor, Dean CFSc., Professors, colleagues, friends, students etc. for encouragement and constant help to fulfill this work. The literature from Government of India Planning Commission, Ministries Agriculture, Commerce and Food Processing, FAO are duly acknowledged with full gratitudes.

Author is also thankful to publisher for publishing this book.

Contents

Chapter 1

FISHERY ADMINISTRATION AND POLICIES

ADMINISTRATION

Administration is derived from two Latin words–*ad* + *ministrave* which means too look after or to take care of or to manage attains or problems. This administration is very common. General administration is a part of modern-society and there are many problems and aspects of society like health, education, law and order each part is well administered. General administration is universal. Every part of our life is administered state takes care of all the human beings. Thus modern state is administrative state as it administers every aspect of life. Plans are made and implied by administration *e.g., IRDP,* Integrated Rural Development programme made by the *administration of India.*

In India administration behaves uniting factor as there are so many diversity and all are to be centrified.

General Administration

General administration is defined as the "It is the organisation and direction of human being and material resources to achieve certain ends." This definition is given by **Presthus.**

or

The administration is the direction, coordination and control of many persons to achieve some purpose or objective.

Scope of General Administration

General administration covers mainly two things *i.e.,* one is private administration and another is the public administration. Private administration refers to the private organisations. Public administration takes care of public matters *i.e.,* matter belonging to the people of state. All the activities of Govt. are public administration.

Govt. has three paths :

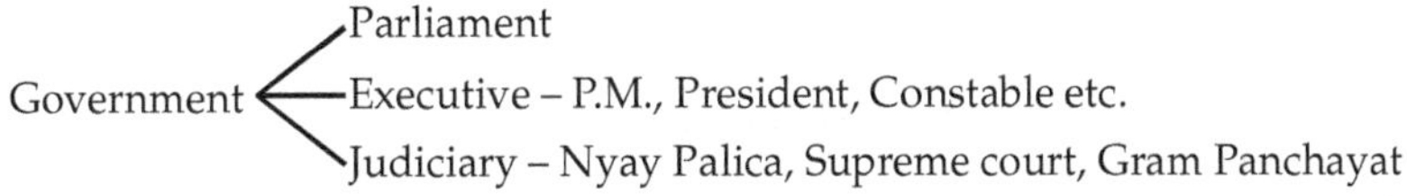

Public administration covers two types of topics :

1. **General Topics :** Organisation, its leader, financial position of organisation. Discuss problems regarding organisation, its leader, its policy, its personnel, its financial resources etc.

2. **Specific Topics :** Like administration of agriculture, administration of science and technology, administration of defence, and fisheries administration etc.

Fisheries Administration

Fisheries administration is a part of general administration. It covers so many thing but fishery administration is related to certain things. To look after the problems and matters related to fish and to take care of them is fishery administration.

Every administration presuppose four things *i.e.,* organisation, policy, personnel and resources and each administration has its own objective.

The Fishery administrations have, therefore :

1. Fish organisation and institutions.
2. Fish policy and planning.
3. Fish personnel
4. Fish financial resources.

Objective of Fishery administration is to organize and direct the fish personnel and resources so as to develop and increase the fish life quantitatively and qualitatively in Inland and offshore waters.

Scope of Fishery Administration

1. Conceptual clarification related to fishery administration *vis-a-vis* general administration or public administration.
2. *Financial Administration :* General principles, budget and planning and fisheries finance for fisheries development from central and state government, World Bank, NABARD *i.e.,* National Agricultural Bank for Rural Development, ADB *i.e.,* Asian Development Bank and UNDP etc. Institutional financing problems of finance and its recoverage.
3. *Fishery Legislations :* Indian Fisheries acts and other state law and rules. International laws of Fisheries, International Law of Oceans etc.
4. *Fishery Organisations* : Fisheries departments in central and state governments, Fish Corporations, Fish Research institutes under the ICAR, FAO etc.
5. *Fisheries Policy :* Fish development programmes in five year plans, their implementation and problems.
6. *Fishery Cooperations :* General principles of cooperation, cooperation, law in India. Fish cooperation indebtness of fisherman, rate of fish cooperatives. In fisheries-economy of a fisherman.

Public Financial Administration

It refers to the management of the finance of the Govt. including preparation of the budget method of administering various revenue resources, custody of the public funds, procedure of spending money, keeping up the financial records etc.

"Finance is the shadow of administration".

Kautilya in Arthashastra, "all undertaking depends upon the finance. Hence foremost attention shall be paid to the *treasury.* This finance is important for administration of Govt. and finance policy should be sound. This financial administration refers to the management of government finance including the preparation of budget etc. The financial administration is important for all the things."

F.A. Negro says that the financial administration is the great importance today because of the tremendous increase in the amount of money expended for government services. In the administration of finance four important agencies are involved, these are following :

1. The Executive
2. The Legislature
3. The Finance ministry
4. Audit department

The Executive

It's role in financial administration. The executive branch is Govt. or council of ministery. It is the prime duty of government to administer the state to develop it, to take care of it etc. Govt. imagines for the next year that how much money is to be utilized next year.

Annual Financial Statement

This statement is written in article 112 of Indian Constitution. The president of India shall cause to be laid before both the houses of the Indian Parliament the annual financial statement.

"The crown demands money" A.F.S. in Britain.

Loksabha has financial power to pass the budget. No money can spend from public treasury or collected from the people without the authority of law—*Article 265-266.* There are some countries like U.S.A. and Australia where parliament authorizes public loans. In India the Govt. itself can take loans. In each parliament there is a system of accountability through the controller and Auditor-general of India, (CAG) in article 148 it has power to keep accounts.

There are 2 important committees in the parliament of India namely :

1. **Estimation Committees :** This committee is to suggest economy in the Govt. expenditure. It is the permanent committee and it has few members.
2. **Public Accounts Committee (PAC) :** The committee examines the appropriation of expenditure and to report to the parliament on financial irregularities.
3. **Finance Ministry :**
   - It is very important in financial administration, it prepares the budget.
   - It manages the government finances.
   - It controls the expenditure
4. **Audit Department :** There is a Controller and Auditor general of India according to article 148. The main function of CAG is to audit the expenditure and to know to legality and desirability of the expenditure. The CAG just like as a search light.

Budget

Budget is the principle tool of financial administration and it is one important instrument through which the legislature controls the executives.

Budget derived from the **French word –** Bouqette means Bag, having financial documents but now it has lost its original meaning.

"**Budget** is a **document** presented by a **government** containing an estimate of proposed expenditure for a given period and proposed means of financing them for the approval of legislature."

This budget is the central tool of financial direction and control. It is a plan of action. It reflects the programme of the government for the coming one year. Thus, it is better to say that the budget is something more than a financial statement. It coordinates the activities of the government. The budget is prepared on the basis of certain principles.

According to the Mr. Herald De Smith there are 8 principles which guide to the budget preparation :

1. Executive Programming
2. Executive Responsibility
3. Reporting
4. Adequate Tools
5. Multiplicity of Procedures
6. Flexibility in Timing
7. Executive Direction
8. Budget Organisation

In India budget is prepared like this it start in the July through the financial year starts from 1 April to next year. All the ministers and departments are supplied with forms in July-August to prepare there estimates in advance. The preliminary estimates prepared in the 3 parts *i.e.,* **Part 1, Part 2A and Part 2B.** In October all these 3 parts of the form are presented to the finance ministry. The finance ministry compels and complete by the end of December. By the end of January the finance ministry *i.e.,* in the position to have a rough budget. In January, the **Finance-Minister** examines the budget proposal in consultation with the **Prime-Minister** and formulates a budget policy in regard to taxation proposals then the budget is considered by the cabinate, everything is done in strict secrecy. After it is being discussed in the cabinate some times in February. The finance-minister takes the approval of president of **India Act-112** to present the budget before Lok-Sabha. Budget in the parliament undergoes the following stages:

1. The President in speech to the budget session makes a reference that the annual financial statement, will be present in the house.
2. Stage is its general discussion. There is no discussion of budget on the day when it is presented to the house, after few days the Lok-sabha discuss the budget in full. Generally there is no detailed discussion. The budget is prepared on **109 demands** out of which **103** are civil demands and 6 are defence demands. All these demands are to be discussed in 26 days *i.e.,* before 31 March. Thus there is very little time.
3. State is the voting on demands.
4. Government introduces **"appropriation bill".** When it is passed and passed to Rajya Sabha for 14 days and then president signs it.

Second is the "Finance bill". It consists of all the proposals of new taxation in the budget. This is passed in Lok-Sabha then to Rajya Sabha for 14 days then President it signed.

Third is the vote on accounts. Financial year end on 31 March and starts on 1 April and discussion goes upto 30th April. So the expenditure for 1–30 April is done by vote on account.

Fourth is the demand for the supplementary grants is put before the Parliament. **In article 115** there is a provision for supplement grants.

After the Budget in Passed

The role of Finance-ministry in budgetary control. The control of finance ministry is continuous before and after the approval of budget.

Budget control is necessary before any money is spent. Parliament grants money to entire *money* government not to any particular ministry. The Finance-ministry manages the finances and must be convinced about the desirability of expenditure must have the prior approval of finance ministry then the ministers are powerful enough to sanction expenditure of any scheme upto Rs. 50 lakhs and can make purchases upto 25 lakhs. Any expenditure crossing this limit will have to get the finance ministry. After the budget passing the control of Finance-ministry is more dominating. It matches the progress of expenditures in the department, expanding money through monthly expenditure statement, periodical report and warnings. It also sees that the on spent is returned to the Finance-ministry before the close of financial year. It gives financial advise and guidance to the ministers. Therefore, in brief the Finance-ministry **control is to ensure.**

1. That no wastage of resources occurs.
2. The public money is not wasted.
3. The money spent gives adequate results.

Parliamentary Control of the Budget

1. Parliament should satisfy itself that the appropriations have been utilized economically within the framework of grants and for the approved proposes.
2. Parliament should have undertake a detailed examination of the annual budget estimates to suggest possible economies in the implementation of planes and programmes. There are three committees formed by the parliament :

1. Public Accounts Committee (PAC)
2. Estimates Committee
3. Public Undertaking Committee

1. PAC

The PAC of India modelled on the PAC of British-Parliament, which started in 1861. In India PAC is functioning from 1950, it consists of 22 members from both the houses of Indian-Parliament, on the basis of 2:1 ratio. They are elected for 2 years poll. The functions of PAC are:

1. To examine in the light of the report of the comptroller and Auditor-General (CAG). The statement of accounts showing the expenditure and income of the state corporations, trading and manufacturing units, projects etc.
2. It examines the statement of accounts showing the income and expenditure of autonomous and semi-autonomous bodies, the audit of which has been done by the (CAG).
3. Considers the report of CAG where the president may have required him to conduction audit of any stores and stocks.

2. Estimates Committee

Estimate committee is functioning since '1950' in this committee there are 30 members, all the members are elected from Lok sabha only. Its functions are :

1. It suggests economies, improvement in organisation and efficiency in administration.
2. It suggests alternative policies to bring about efficiency and economy in the administration.
3. It examines whether the money is well distributed within the limits of the policy and

it suggests the format and manner in which the estimates shall be presented to the Parliament of India.

Fisheries Legislation's in India

In India for the first time the law was formed in 1897. So, the Indian fisheries act was passed by the Governor–in–Council. The main purpose was to protect fisheries resources. Sir, Arther Cotton an Engineer who was to make dams, reported that aquatic life is influenced. He observed this up to 1860. In 1870 and after Day reported to the governor–general about destruction of fry, finger lings and boarders at the site of dam construction. Indian Fisheries Act (1897) was made by the Governor–general in council. It's important sections are :

Section (4)

If any person in India uses dynamite or any other explosive substance in any water with the intention of Catching or destroying fish in that water he shall be punished with imprisonment for 2 months or a fine of Rs. 200.

Section (5)

If any person puts any poison, lime or any other dangerous material in to any water then he will be punished for 2 months or a fine of Rs. 200.

Section (6)

1. The local Government may make further rules for the purposes mentioned below applicable to any public water and the local Govt. may apply those rules to private waters :
   (*a*) The erection and use of fixed engines in the water.
   (*b*) The construction of weirs.
   (*c*) The dimension and kind of nets to be used and the modes of their using
2. Those rules may prohibit all Fishing in any specialized water for a period of not exceeding more then 2 years.

The local government may direct that a fine of Rs. 100, for breaking any such rule and an additional fine of Rs. 10 day for breaking any rule there after and provide for the seizure, for feature and removed of the fixed engines and nets used and also removal, if seizure and fish capture.

Section (7)

A police officer or any other person specially empowered can arrest a person without warrant from Magistrate under the sections 4, 5, 6 above provided the name and address of the person arrested are not known or the person decline to give his name and address. The person will continue to be arrested until his name and address are known.

Features of Act

1. This act provided for restrictions on destruction of fish by dynamiting or other explosive in Inland waters or coastal waters.
2. It provided restruction on destruction of Fish by poisoning.
3. It provided for protection of fish in specific waters by provincial Govt. and local government.

4. It restricted the construction and use of engine weirs.
5. It also fixed the kind and size of nets and mode of using them.
6. It prohibited fishing in specialized waters for a period of 2 years.
7. Fines and imprisonment and arrest.

Comments on Act (1897)

It does not checks some of the activity related to fishing *e.g.* It doesn't check fishing activity in marine waters.

Natural resources may be destroyed which may affect the aquatic life due to natural factors like growth, mortality, death and climate etc. There is a need to be restricted the number of boats to be used in fishing and the size of fish to be caught so in futures the following suggestions are made to be incorporated in future laws.

1. Restriction on the Size of Fish
   (*a*) A minimum size of fish is prohibited in such a way that it gets a chance of breeding at least once in its life time.
   (*b*) Capture of spawn, fry, finger lings has to be prohibited.
   (*c*) Capture of fish during breeding season is to be prohibited.
2. It is with regard to construction of weirs, construction of dams, weires across a river affect the movement of fish definitely the dams and weires have their affects, which may be environmental and obstructional dams and wires are to be constructed when they are absolutely needed.
3. This is regard to Industrial wastes. Pollution of river and sea water through the discharge of Industrial wastes. Effluents and municipal sewages are to be checked.
4. With regard to licencing and leasing of Fishing rights is not to be given indiscriminately to all the people everywhere.
5. Regulation with regard to marketing of fish.

Laws of Oceans

Huge water areas—surrounded by land. Once upon a time all the seas and oceans were not related to any particular nation and were common property of all the states and nations. Territorial states have boundaries meeting with sea. Land locked countries are like — Nepal, Bhutan etc.

That part of water which is in possession of a private state is called territorial water (water adjacent to territory of states).

In older times 12 Notical miles areas in sea were in possession of a country, now 12 miles limit has been extended. Sea is important for every country.

1. The weather and climate of a state depends on the sea.
2. Sea is a medium of International trade and commerce.
3. It is important for interstate communications through under sea cables.
4. Important for protection of national Frontiers.
5. It is important from economic points of view like – Food, Industries, Petrol and Wealth (petroleum, copper), copper is so much to last 6000 years. Nickel can last 1 lakh 50 thousand years.

Gold, Silver, Zinc are available in millions of tonnes which cost can be estimated millions of dollars.

So, now 12 Notical miles is expanded by the year 1979. 75 countries of the world including India had claimed 200 miles economic zones as their territorial waters.

Exclusive Economic Zone of a Country

India – 200 miles (Approximately)

U.S.A. – 200 miles (Approximately)

Demarcation in sea is impossible for the shake of nations security, economic benefit, to make up the shortage of land base resources. India has given importance to sea. In 1976 by Amending the Article 297 of Indian Constitution. India has proclaimed 200 miles EEZ, including to its 12 miles EEZ. So now Indian sea water is 2 million square k.m., out of which on eastern coast there is 0.86 million square km and on the eastern coast there is 1.16 million square km of water. This 2 million square km is about 61% of total land surface of India and to maintain this huge sea water has created a coast–guard service. Andaman and Lakshadweep Islands have some economic zone as those of main land.

Median line is the boundary line between India and its neighbouring states. In 1958 in Geneva a conference was held, 'U.N. Conference' on law of the seas. In this convention the sovereignty was extended to 12 miles only but this convention did not give sovereign rights to states over the resources available on sea bed and just to check the free access of all the nations and to check the depletion of Fish resources. The limits was extended to 200 miles, many states particularly those nations which technologically advanced like – Japan & U.S.S.R. The extension reduced the disputes will regard to fishing between nations.

Further in 1945 FAO established a Fisheries division before that things regard to fisheries were mentioned as agreements (treaties) between the states for exploration.

Category I : Italy and Yugoslavia made a treaty over the adriatic sea in 1942.

Category II : International Council for exploration of seas in 1902.

In 1945 the president of U.S.A. Trumann declared the conservation and exploitation of fishing resources of high seas contiguous to the coasts of U.S.A. British made on the utilization of fisheries resources were made reclamation *e.g.,* U.S.A. and U.S.S.R. entered into treaty 1957 on sea fishing. In 1953 International pacific commission was established – It's main purpose was to protect Haliput.

In 1949 Inter America Tropical Tuna commission.

In 1949 Pacific Salmon commission was appointed.

Research Treaty – International council for exploration of sea 1942.

Indo-Pacific fisheries council was established in 1948.

Upto 1975 in the world there were 22 major fisheries regulated bodies created under various treaties in addition to many other bilateral treaties. When economic zone extended to 200 miles the above Called treaties were superfluous.

Main Law from 1958

International laws of seas 1958 and alter U.N. made a conference. U.N. conference was held in Geneva in 1958, then in 1960 from 24 February to 27th April 1958 the Conference continued in which 95 countries participated. 86 member states + 2 guest states *i.e.,* Ethyopia + Sudan + FRG (West German), Holy sea (Vetican city), Republic of Korea, Moracoo, Sun Marino, Swiss, Republic of Vietnam.

The conference adapted four Conventions :

1. Convention on the territorial sea and contiguous zone.
2. Convention on fishing and conservation of living resources on highseas.
3. Convention on high seas.
4. Convention on Continental self.

1. Convention I

This convention consists of 32 articles in 3 parts. Part 1 is of territorial sea. **Section I** is of part one is on the sovereign rights of states.

Section II

Section II is on the limit of the territorial sea and this limit convention extend max. upto 6 miles.

Section III

Section III is on the right of innocent passage consisting of rules applicable to all ships like–Merchant ships or Govt. ships other than war ships.

Part 2 of Convention

Part 2 of convention is on the contiguous zone, which extends up to six miles maximum.

Part 3 of Convention

Part 3 of convention I is consisting of articles to complete formalities. With regard to communist countries they claims 12 miles. India, Iran, Israil, Cylone, Columbia, Italy and Moracco etc. They use only 6 miles as territorial sea. Austria, Balzium, Brazil, Canada, Cuba, France, Japan, U.K. and U.S.A. claims only 3 miles territorial sea. Panama Claims 300 miles. Thus there is no uniformity in regard to territorial sea. (Water).

No International practice on limits of territorial sea. U.S.S.R and Communist countries claims 12 miles. India, Israil, Italy, Columbia, Moracco, Ceylon, Claimed 6 miles territorial sea. Austria, Balgium, Brazil, Canada, Cuba, France, Japan and U.S.A. Claims 3 miles so, no uniform practice.

Conclusion is that International law does not permit an extension of territorial sea beyond 12 miles. The lack of informity with regard to the territorial sea is due to the importance of fish and fish products. In the national economies of the states *e.g.,* Island extremely dependent on fish. It wants to exploit fish and fisheries products in the sea for its economic advantages. Thus it wants extension of territorial sea. U.K. which is also on fish products does not want to give Island any extension of territorial sea.

Canada wanted 6 miles territorial sea and another 6 miles as fishing zone. India claims 12 miles territorial sea including fishing zone.

In India — 6 Miles territorial zone

— 6 Miles fishing zone

U.S.S.R. agreed to Indian proposals. None of the claim was passed by 2/3rd majority in UN conference.

Article 24 of convention (1) says that the contiguous zone is never to extend beyond the 12 miles. Within this contiguous zone each state is powerful enough to prevent infringement of its customs, Fiscal, immigration and sanitary regulations and punish the offender for the violation of a above regulations.

2. Convention III

Second Convention is an fishing and conservation of living resources on high seas that water beyond the 12 miles territorial sea and contiguous water. One of the most notable achievement of the Geneva Conference *Code regulating the natural resources of the sea.* This convention is written in 22 articles.

Article I

This recognizes the rights of all states to permit their citizens and nationals in fishing on the high sea. Subject of source to restructions created by international treaties. This article further derives that if for the sake of the conservation of natural living resources. It is necessary for different states may cooperate with each other intaking appropriate measures.

Article II

Defines the conservation of living resources in the high seas. It says that the conservation of living resources is the aggregate of measures or steps rendering the possible optimum sustainable yield from the resources so as to secure a maximum supply of food and other marine products.

Article VI

Coastal states under this convention occupy a special status as to the seas near to them. They entitled to research and other benefits.

Article VIII

States having special interest in conservation of a particular sea area although not actually for fish in that area may request the states which are interested in fishing in that sea to adapt a programme. If necessary such things should be divided by arbitration.

Article IX

Defines arbitration commission consisting of 5 states.

3. Convention on High Seas

This convention is written in 37 Articles.

Article 1 : Define high seas. High sea is all parts of the sea that are not included in the territorial sea or in the internal waters of a state.

Article 2 : High seas are open to all nations. In the high seas all nations are free to navigate, to fish to spread pipe lines, cables and to flyover.

Article 3 : Provided high seas facility to the land locked state. There should be a corridor given to them.

Article 4 : All state have rights to sail ships on high seas.

Article 8 : Provided immunity of warships from the authority of any state other than flag state (state whose flag is printed on ships).

Article 9 : Authority of flag state only exist on the non-commercial ships.

Article 12 : Every ship is to render assistance to a person in distress in the high sea or a ship in collision in high sea.

Article 13 : Prohibits slave trades in ships on high seas.

Article 14 : Stops piracy.

Article 24 & 25 : Prevents sea pollution by oil in **Art. 24.** In Art. 25 provision for preventing pollution by dumping of wastes.

4. Convention on Continental Shelf

This convention is written 15 Articles.

Article 1 : Defines continental shelf like this :

"The sea bed and the soil near to the coast belt but outside the territorial sea to a depth of 200 M and beyond that limits is known as continental shelf."

Article 2 : The coast state have authority over this continental shelf and it can exploit natural resources existing there.

Article 4 : The coastal state may or may not lay cables and pipelines on continental shelf.

Article 5 : The coastal state must not may interfere with navigation, fishing or conservation of living resources of the sea while exploiting its natural resources and conducting any oceanographic research. In 1973, 1979, 1980 conference was held before 1973 upto 1964 all of 4 but 1 convention were implemented. That was convention on fishing and conservation of living resources and confusion arising has not settled till 1966 *e.g.* Ditt. states were claiming different fishing zones, *e.g.*, Iceland claimed 50 miles fishing zones which was challenged by Britain.

Canada claimed 100 miles of pollution zone and U.S.A. supported that. In 1973 U.N. conference on the law of seas took place at caracase in Venezuela (Lattin America). This was the biggest conference. In this convention there are 143 countries are participated and more than 2000 delegates attended this. This convention (conference) discuss 3 important things–

1. Open sea area.
2. Territorial sea and 200 miles EEZ.
3. Scientific pollution and Research.

Conference could not arrive any conclusion and agreement was made.

In 1979 U.N. Conference on the law of seas held at the Geneva. This was able to reach some agreements like this :

1. 12 Miles territorial sea limits was agreed by 100 states at least. There are 14 states claimed 200 miles EEZ. 20 states wanted foreign war ships to take their permission before entering coastal waters. There are 32 states claimed security zone.

In 1982, the conference was again held at New York. In this 151 countries were participated.

In 1982 UNCLOS following resolutions were passed. The conference adopted following resolutions :

1. It fixed 12 miles territorial sea and 200 miles exclusive economic zone for the coastal states.
2. It permits right of innocent passage through territorial waters and narrow straits. Article 16 strait in world today.
3. It provides an international sea bed authority. It wanted to lay mines in open seas, its initial capital 1000 miles dollars, through contribution by member states, taxes and loans.
4. The conference listed 8 states which has conducted high sea research and to take the help of these 8 states. These states are – India, France, Japan, USSR, West Germany, Canada, U.K. and U.S.A. Total 320 Articles were written, out of 151, 130 states voted for it the resolution. There are 4 states namely U.S.A., Turkey, Venezuela, Israil voted against. 17 states of European economic community (EEC) and Soviet Block remained absent from voting.

This treaty was sign in Dec. 1982 but it faced difficulties because of the following reasons:

1. U.S.A. voted against it, so the treaty will not be effective.
2. USSR and EEC countries remained absent, so it will be difficult to establish the sea bed authority.
3. To collect 1000 million dollars is be coming difficult.

Conclusion

The treaty of 1982 is much better than the previous treaties made through U.N. conference.

Social Economic Conditions of Fisherman

There are a number of people engaged in Fishing. In Hindu cast system a group of people is referred as Fisher cast. In India fishery is adopted as generation to generation activity. These peoples are socially and economically backward people. They do not get any education or training on fishing what they know about fishery is by parents, and so skillfully, they are engaged in profession so that it seems they have learned trained. Most of them lives on the banks of Inland rivers in an unhygienic condition, some of them also live on Coastal belt. They lives in villages mostly in huts, which do not have modern facilities like – Road, Electricity etc. In 1977 there were 2500 fishery villages in which 55 million fisherman lived. They are mostly illiterate, ignorant and poor. They are engaged in fishing following traditional methods, while fishing has not been mechanized or industrialized. These peoples are very large in number are not united. They are different from each other, on the basis of their sub-castes. Social custom, language provincial differences etc. so that there is no unity among them and as a result, they are exploited by the middle man, money lenders. The middle man is not actually the producer of fish, he is the real beneficiary of the fish production because he gets a good share of price:

1. The middle man is linked between fisherman and customers.
2. He finances the fisherman and exploits them because of he is the money lender.
3. He surveys the fish market.
4. Middle man some times has their own boats and nets, which they loan to the fisherman.

Working condition of Fisherman is never attractive what he owns *i.e.,* really very less. Average income of a fisherman family is very low.

They spend more for earning, food, transport, Medicine and other expenditures. In 1972 at Bombay about 75% of fisherman were indebted, to money lenders, Govt. Cooperative banks etc. and the Rs. 429 were the amount of average loan/head. For the low income sickness, building houses, fishing profession, very few amount *i.e.,* Rs. 22 were used in fishery and rest of used in other things. Working condition were not favourable to income. Sea fishing is not undertaken when climate is unfavourable. Size and capacity of boat also regulates the quantity of fish Catch. They work day and night and to various risks and they get very less. To save them or to improving their conditions 2 steps have been taken :

1. Education and training of fisherman.
2. Establishment of Fishery co-operatives.

Education and training does not help the ordinary fisherman directly. They are primarily meant for bringing out fishery graduates and to modernise fisheries through application of science and technology. India has a number of fishery educational and training centres. *e.g.,*

1. Central Institute of Fishery Education (CIFE) Bombay.
2. Central Institute of Fisheries Nautical Engg. and Training centre.
3. College of Fisheries (Pantnagar, Mangalore), Nellore, Panangarh, Tuticorin, Lambuchera, Nagpur, Udaipur, Ludhiana, etc. (Total 17)
4. Industrial fisheries courses
5. Indian Fisheries Training Centre (Barrackpore)
6. Regional Training Centre for Inland fisheries co-operatives (Agra).
7. Central Fisheries Extension Training Centre (Hyderabad).

States like — Kerala, M.P., Tamilnadu, Karnataka, Maharashtra, A.P., West Bengal, Goa, U.P. Gujarat. Lakshadweep have established many Fisheries Research and training centres, besides the FAO and ICAR are helping in education for producing fisheries personnel.

Co–operative's and their Role

Social economic condition of fisherman is inferior. There income in comparison to other industrial worker is low. They get low share of price because of the exploitation of fisherman. The fisherman do not have sufficient financial resources, therefore they borrow money from other money lenders. In brief in India fishing is substance fishing. Something is to be done to overcome. These socio economic problems of the Fisherman. One suggestion is to establishment of Fisheries co-operatives. Fisheries co-operatives are existing in foreign countries also. Which aim to improve the condition of fisherman, develop fisheries and modernize fishing. There are may be differences in the co–operatives from country to country, but cooperative is the basic idea existing everywhere *e.g.,* Japan. There are a number of good fisheries co-operatives, which have benefited the fisherman. In 1964 there were 5682 fisheries co-operatives which are grouped into 29 societies out of these 5682, 999 were inland fisheries co-operatives and rest others are coastal co-operatives. The functions of these co-operatives are following :

1. To provide loans
2. To provide banking facilities
3. To supply necessary goods for fishing.
4. Transporting facility, storing and selling the fish catches and products.
5. Conservation of fisheries resources.
6. Engaged in activity for preventing sea disaster and promoting fishing back.
7. Providing facilities for welfare of fisherman.
8. Provided education to fisherman and informations related to fisheries.
9. Collective bargaining on behalf of fisherman etc.

In Japan 1948 the fisheries co-operation association law was passed. This was amended in 1975.

Italy

In 1963 there were 460 fisheries – cooperatives out of which only 52 are inland fisheries co-operatives. There is a national consortium of fisherman's co-operatives in Italy. All the 460 co-operatives are affiliated to this national consortium and 80% of the fisherman are member of this consortium. There were about 54 fish markets and controlled by the co-operatives and through these 65% of this entire fish catch were marked.

Norway

There is a Raw Fish Act 1951 controlled the fisheries co-operatives in Norway. All fisherman are compulsory members of it. All co-operatives are marine. In 1964 Norwagian fishermen's union having 45,000 fisherman members was formed. This union takes care of economic conditions of fisherman. There is a national fisheries bank of Norway which supplies bulk of the finance to the co-operatives. The bank of Norway also gives financial assistance to these co-operatives.

Denmark

The fisheries co-operatives in Denmark are involved in the number of fisheries activity including processing, transporting and marketing of fish catch. These co-operatives are voluntary and they are not controlled by any act. The Royal Denish fisheries bank gives financial loans to fisherman.

Canada

The fisheries co-operatives are becoming important gradually, under the fisheries improvement loan act. The fisherman are given loans on the basis of personnel credit worthiness, Not on the basis of any mortgage of property.

U.S.A. and U.K.

In U.S.A. there are trade union banks which provide loans. In U.K. there is a sea fish industry authority which gives loans to the fisherman.

The description given above we can conclude that all the co-operatives attain to strengthen the economic independence of fisherman to raise fish production, to organise systematically the marketing and distribution of fishes and to ensure a fair price for the producers.

India

There are a number of co-operatives (fisheries). In 1975–76 there were about 4135 co-operatives. They were existing at different levels such as – villages, Tahsil and Distts and States etc. There are six state level co-operatives. There are 72 district/division level co-operatives and 4060 primary village level co-operative societies in 1976. District level co-operatives are largest in Tamilnadu having 19 co-operatives, and lowest in Karnataka having 4 co-operatives. These co-operatives are of different types :

1. Fish marketing cooperative society
2. Credit and supply co-operatives
3. Transport societies
4. Vessel ownership societies.

1. Fish Marketing Co-operative

The objective is to provide better economic return to the fisherman by marketing his fish produce. He is to get a reasonable price for his fish produce. The co-operatives to eliminate the fisherman in the distribution of fish. The qualities of this co-operatives are – Elimination of the middle man. The middle man really pay a very small price to fisherman. In the absence of marketing co-operatives the middle man and private merchants monopolize and dominate the fish trade which possess loss to the poor fisherman, thus these marketing co-operatives are useful in offering a better price, stabilizing price, processing and storing the fish. In Karnataka and Gujarat these marketing co-operatives are functioning well.

2. Credit & Supply Co-operative

The objectives of these co-operatives are following :

(*A*) To provide loans to members for purchasing necessary fishing equipments *e.g.* – boats, nets, paints, sail cloth and ropes etc. Easy instalments with less interests. The loans may be short-term or long-term.

(*B*) These co-operatives supply fishing equipments at reasonable prices. The credit and loan supply cooperative should be learnt to the marketing co-operatives to give greater economic benefits to the fisherman. The credit and loan should be given adequately and in time. The aim of these co-operatives should be free the fisherman from the middle man and these co-operatives should also tries for the mechanization.

3. Transport Societies

Separate transport societies should be established to transport fish from the production centres to the marketing places particularly to urban areas. In Maharashtra (Bombay) there are fish transport societies existing. In Hongkong the fish transport society existing in large number.

4. Vessel Owner–Ship Societies

Fishing boats and trollies they require large amount of money to be owned by a fisherman. A single fisherman may not be able to invest such a large amount of money, so a number of fisherman join together to purchase boats and other necessary things.

In (Kerala) **"Matsya Utpadak Societies"** have been formed. In the Karnataka same type of societies are known as **Ram paani operation.** Under such societies vessel, boat etc. are jointly owned by a group of fisherman.

Working of Fishing Co-operatives

1. They lack funds.
2. They do not have trained personnel.
3. The Fisherman members do not co-operate with each other and most of the time they don't supply fish to the co-operatives.
4. These co-operatives do not have proper processing facilities.
5. The members do not pay back the loans *e.g.,* in 1975–76. There are 2168 co-operatives in losses and the outstanding loans had gone up by 433% over the figure of 1964–65.
6. There are also administrative difficulties in these co-operatives. As a whole these fisheries co-operatives are not functioning satisfactory. There are some suggestions for the improvement of fisheries co-operatives.

Suggestions

1. The fisherman must supply atleast 50% of their fish Catch to these co-operatives so that the role of middleman is less important.
2. The cooperative should be established very near to the fish landing centres so that they get the fish transport the fish easily.
3. Gram panchayats and other Govt. organisations should lease their ponds tanks water areas to these co-operatives for a longer period.
4. The credit and loans given should be purpose oriented and on the basis of credit worthiness not on the basis of attaching or mortgaging any property or tanking security.
5. Nationalized banks should try to give loans to the fisherman.

Administrative Set-Up

Fish farmers development Agency is a project under world bank in Orissa, West Bengal, Bihar, U.P. and M.P. These five states have vast fishery resources 16 Lakh hectare stangent water area. 1.62 lakh hectare area in U.P.

There are 3 Extension officer and 1 Engineer under the **Chief Executive Officer.**

Planning

How to face the whole system and go for bright future.

Fisheries Planning

Natural resources of fisheries, utilization of these in states, blocks, Distts. and in country in a judicious way determine the prosperity and well being of its nationals. For a welfare state all the resources have to be utilized in such a way that they do not leave to environmental upset. Land and water are the basic resources.

Planning in Inland States

There are 25 states, with reference to Inland states like – U.P. but we are concerned to lotic and lentic environment like – Rivers, Streams, Irrigational Canals, Reservoirs, Lakes, swamps and village tanks. In present days the fish was available from natural resources *i.e.,* capture Fisheries and little amount from culture fisheries. From beginning fishing business was confined to special group like fisherman.

After independence a large number of manmade reservoirs for different purpose were constructed on river valley system. This actually attacked with fisheries of natural resources. Government of India adapted a policy of national planning to regulate the development of various sectors which were evaluated on the priority basis. Fisheries had lowest priority. The fisheries sector outlay was very small and what was allotted was not utilised fully. Out of these fisheries sector, maximum share was given to marine fisheries.

	Total Outlay (In Million Rs.)	*Fisheries (Sector)*	*Amount (Utilised)*	*% Total utilization its fisheries*
Plan I	2013	5.14 Million (.255%)	2.8 Million	54.4%
Plan II	4800	10.73 Million (.223%)	9.0 Million	83.9%
Plan III	7500	29.0 (.386%)	28.29	99.9%
Plan IV	15902	83.31 (.522%)	134.98	

Overall National Economy : There are 35–40 sectors. The export earning from marine products increased from 3.67 (1953) in crores to 265 in 1980. Now it is increasing in a good speed.

How planning Process takes place. Planning process has become more sophisticated. New concepts and Technology come in importance. Planning is necessary to have the development of the country, along the derived national goals and proper utilization of limited funds available for a under–developed country like–India. The planning fulfill the social and national needs open new revenue of employment for millions. Lesser the poverty and give return on investment.

Planning is done within the policy framework laid down by the Parliament and embodies in general, social, political, legal, financial and technical consideration. We are not financially but technically also poor and hence lot of efforts have to grow in improving the technical base.

Fisheries sector as compared to Agriculture, the technological basic was very poor in 50's (41–49) much of the improvement in this field came during 70's early. The process of planning has to be dynamic. It demands constant review of guidance, rules and new technologies to see if goals are being achieved or not. For planning in any sector correct basis data on the resources of and the infrastructure of and various linkages with other sectors is the primary need. Planning based on wrong promises will always flop.

It is necessary to drop details in each subsectors in various steps. Identification of projects, why we want and aims to plan, what are objectives and what results do we expect, there above are 3 main questions.

Project Preparation

The planner should ensure that the project is :

1. Within the framework of national policy, framework may vary according to the need of locality.
2. It must be technically sound.
3. Financially and economically viable.
4. Administratively workable.
5. Socially acceptable.
6. **Fits priority list** of the overall plan.
    1. Technology should be aware of to whom, who is going to incorporate that technical there. Extechnologies outdated should be avoided. All the thing framing the project must be technically sound.
    2. Financial position should be strong. Economically viable means, no loss is same return.
    3. In administration set-up is leader like. If two persons have equal power they will never work together.
    4. Socially acceptable *e.g.,* Pig cum fish culture is not acceptable in Rampur and Moradabad. Similarly fish meal having blood meal is not acceptable at several places. Another example of common cap has best growth but in market, it does not yield good price.
    5. For example, in 20 point programme, last point has least priority, 17th priority to fisheries.

The following items, should also be included :

(*A*) It is for particular geographical area or for any particular groups of population.

(*B*) The level of financing required both national finance and foreign exchange world bank will now finance for reservoir development.

(*C*) The main issue that will have to be shorted out before implementation generally legal, administrative and technical problems are to be included in this.

(*D*) Set-up guide lines and schedule for implementation of project with time frame and PERI *i.e.*, Programme evaluation and Review technique. The *other stage of* project preparation is to work *out further details* :

1. Elaboration of various micro-components.
2. Critical assessment of organization and management policy, personnel involvement. The later includes the expertise available or its absence.
3. Cost and benefit of the project.

When the above document is ready it is subject to appraisal and evaluation to competent authority, various aspects in the overall from work of planning policy.

At present so much efforts as stated above is not put in projects–preparation or generally the Government needs information on 3 points for financial sanction :

(*A*) Objective of the scheme.

(*B*) *Justification* – why the project is necessary and why the inputs asked for sanction.

(C) Description of the activities and work plan for budgetary provision.

Project evaluation and its appraisal has became more sophisticated technical needs lots of details and correct analysis. One should give the various technologies available and why a particular one is selected to achieve the goal. All the projects must have economically sound such as investment and spread over a long time, interval rate of return or cost benefit ratio in worked out. How the income to be generated will be distributed at various levels such as primary producer, middleman and retail seller. Thus the social aspect of development is emphasized. In fact the idea should withdrawn a long term perspective plan under various sub projects and programmes where they are correlated in a scientific way such that from one project you go to the second and so on the technocrats should know the methodology of appraisal and not leave solely on the "Chotta Babu".

For appraisal the project should give in a sum vary form :

1. The policy and philosophy which form the basis of the projects.
2. Goals to be achieved.
3. **Objectives :** These are the expression of goals at lower level and two are inter mixed.
4. **Targets :** These indicate specific quantified result that have to be achieved to reach the goal as a result of quantified input both input and output should be realistic once the document of main projects and its appraisal is ready the implementation phase starts, it implies.

(A) Monitary of Execution

It shows how that the programme is managed executed. It consists of accounting auditing and report. It comprises planned input and output targets with the real levels of achievements. It determines what has happened or why not happening. Monitary depends upon the level of management authority concerned with the task.

(B) Process of Evaluation

Monitary brings out the evaluation of on going scheme if need by mid course correction be introduced. This result of evaluation should be reported up to the level of decision maker and policy makers. In many cases that excost evaluation is done. It is to provide necessary feed back to planning agency for future project and also for analysis and systematic planning. The values of evaluation is influenced by the details will be given in the project report.

Review of 6th Plan (1980–85)

Fish production increased from 2.14 million tonnes in 1979–80 to 2.85 million tons in 1984–85, showing a growth rate of 3.1% per annum. Generally speaking fish production from Inland water is showing better performance than the capture Fisheries in marine sector because of introduction of scientific fish farming in tanks and ponds through FFDA's. Culture cum capture fishery in small and medium reservoir as a substantial increase in fish seed production through about by the adaptation of Commercial hatchery system during 6th plan.

The Fisheries has ample scope for development besides producing nutritive food at village level, they provide opportunities for self employment in rural areas. During 6th plan the emphasis in Inland fisheries was to increase additional fish production by fish farming and as a result of which increasing in Inland fish production from 848 thousand tonnes in 1970–80 to 1100 thousand tonnes in 1984–85.

In the marine sector the progress has been slow during the 6th plan period. One of the main reason was that against the target of introduction of 200 deep sea fishing vessel to exploit EEZ resources, only 75 vessels could be introduced. However, during the period fishing industry should considerable interest in catering fishering vessels and their number rose to 90. About 2/3rd of marine production even now is contributed by 154000 non mechanized traditional boats operating in the narrow coastal belt and the remaining 1/3rd of marine production comes from 20,000 mechanized boats. The contribution of large trawless exploiting EEZ so far has been only 1% of the total marine production.

For exploiting fisheries resources in EEZ during 6th plan period. It has decided to strengthen the "exploratory fishery" project now remained as *F.S.I. i.e., Fishery survey of India*–The F.S.I. completed the survey of 34000 square km monitored up to the end of 1983–84. As against the target of 5 major harbours fixed for 6th plan to provide landing and birthing facilities for fishing boats. There are four fishery harbours at Madras, Raychowk, Cochin and Vishakhapatnam. In addition birthing facilities were provided at 59 small landing centres and at 9 minor fishing harbours. Currently construction work is in progress at 38 sites for providing landing and birthing facilities for fishing vessels.

During the 6th plan period the work on Integrated Fisheries Project *i.e., I.F.P.* and at the central Institute of Fisheries nautical and Engineering training at Cochin was review for re-organisation and strengthening. The work of IFP in developing new projects for consumers as well as undertaking the marketing of Canned products of non–conventional varieties of fish was encouraging. The scope of coverage of Central Institute of Coastal Engineering for fisheries (CICEF) was enlarged to cover survey, design and preparation of project reports on brackish water fish forms, in addition to the on going work of society and designing of minor and major harbours.

During the 6th plan work on the new scheme of development of Inland fisheries statistics and techno, socio, economic survey of fisherman community was initiated.

A group accident insurance scheme for fisherman in Inland, marine and estuarine sectors was also introduced. Further a national welfare fund was created for under taking welfare activity for benefiting of fisherman in the country.

Marine Fisheries

India has vast potential for fishing resources comprising 2.20 million km^2 EEZ, 7517 km of coastal line. 29000 kms of rivers, 1.7 million hect of reservior, 0.902 million hectare of brackish water area and 0.753 million hectare of tanks and ponds both for Inland and marine fish production. The main thrust will be an exploitation of EEZ by promotion of investment in deep sea fishing specially to harvest resources beyond 40 fathoms. For coastal fishing besides introducing new motorised and mechanized crafts attempts will be made to expand diversified coastal fishing. The new gears and improve design of boats will also be introduced for ensuring better return of 1.8 million fisherman operating in coastal belts. The alternative raw material like fibre reinforced plastic (FRP) and ferro cement for fishing crafts will increasing be used as a substitute for scare and costly used timbers in fishing crafts. Strict enforcement of maritime zones of India *i.e.,* regulation of foreign fishing vessels. Upto 1982 conflicts were between mechanized boat operating and traditional fisherman, and then taken into consideration.

For the operation of both mechanized and deep sea fishing vessels adequate landing and birthing facilities will be provided by completing on going coast of major and minor harbours. Priority will be given to small landing centres for use by traditional fisherman operating 154000 crafts and contributing as much as 2/3rd of the total marine production insurance cover to fishing crafts and gear will be introduced and welfare scheme for fishermen will be implemented. Product development for domestic and expert markets from unconventional fish of low value landed by mechanical boats and trawlers will be taken up. In addition handling and processing facility will be strengthen to save quality fish landing of large quantity during good fishing season. Post harvest technologies in preserving processing and marketing of fish will get due attention to ensure proper and fresh supply to consumer in coastal and Inland areas.

Fish marketing in cooperative sector will be encouraged to ensure better returns to the primary producers and fair price to consumers. Emphasis will be given to set up hygienic markets.

Out of total 2457 coastal villages in the country, few selected villages will be grouped for setting fisheries. Industrial states – these states besides having got sea and shore facilities will have facilities for processing and preservation.

FSI will be strengthen and reorganised for the expeditious survey of the fishing resources of EEZ. In keeping with the growing need for trained manpower for fishery industry (CIFNET) will be upgraded and strengthen. The integrated fish projects will be restructured with emphasis on project ... with sub units on west and east coast, CICEF will intensity its efforts to accelerate brackish water survey and design of the fish farms.

The main emphasis will be on Integrated fish farming in tanks and ponds through FFDA and to introduce prawn farming in brackish water area by establishing area development prawn farming states (PFS). These culturable resources will be the only source of additional Inland fish production during 7th plan period. This approach will also creates a cader of trained fish farmers in the country to undertake scientific fish farming and provide self employment for them in rural areas. Minor and major irrigation reserviors will be developed for Capture cum culture fisheries through co-operatives societies for fisherman. Judicious stocking and management of reserviors will be undertaken by the co-operatives of corporations to raise the level of fish production from reservior. Further sewage fish culture, air breathing fish culture and reclamation of bheels for fish culture will also be get attention during the plan period.

Fish seed is the basic input for fish farming in tanks and ponds and culture cum capture fisheries in reserviors. W. Bengal has achieved a breakthrough the fish seed production in the country. It is recommended that attempts should be made in all the states to follow W. Bengal to become self sufficient in fish seed production. The national fish seed production programme will be strengthen to support state Govt. in this effort.

Conservation of fish and fisheries in lakes, reserviors, rivers and game fishery waters will have taken up to ensure sustained yield from Capture fishery. The declining trend in some of important fish sps. of fish will be studied and remedial measures taken to conservation dangered sps. The deep sea fishing to exploit EEZ diversified fishing and introduction of new mechanized fishing vessel will contribute, attribute additional fish production in marine sectors whereas fish farms in tanks, ponds and culture cum capture fisheries in reservior will added to Inland fish production conservation in coastal and capture in Inland waters will maintain fishery. The necessary infrastructure and marketing to ensure the transport of fish to consumers.

Targets : To be assigned, which is feasible and to be achieved.

Selected Development Programme

Through fish development is a fish state subject. Fisheries and fishing beyond territorial waters and Fisheries research and education are on union list hence matters related to fish development particularly in Inland fisheries development within a territorial waters or largely within the preview of State Government. Only deep sea fishing come directly on the previews of central Government. The fisheries division in union ministry of Agriculture and rural development exercises a coordinating role and assist in various programmes being implemented by state Govt. for the development of both marine and Inland fisheries will achieve national target lay out in 7th plan. Fisheries organisation under union ministry require strengthening manpower development for implementation of various schemes proposed for 7th plan. Call for the urgent consideration in marine sector, main thrust is on exploitation of EEZ by introduction of deep sea fishing trawlers, construction of Indigenous trawllers and charting of foreign vessels. The development of fishing boats, terminals marketing facilities and other shore establishments are proposed to be proved or strengthen to facilitate the operation of fishing vessels.

Similarly in Inland sector special emphasis during 7th plan will be on introducing high yielding farming techniques in tanks and ponds through FFDA. Production of quality fish seed on commercial fish scale following the example – *of West Bengal* is envisaged for additional fish production by culture.

Attempt will also be made to establish prawn hatcheries in all maritime states. All these activities in Inland sector will need specially trained manpower to transfer techniques to fish farmers for high production from brackish water farms. So special emphasis is needed on development of manpower to implement programmes of Inland fisheries with implementation of plan programmes production has increased from 0.75 million tonnes in 1950–51 to 2.85 million tonnes in 1984–85 showing significant over successive five year plan to produce more nutritive food in the country and to generate employment in rural areas for the weaker section of society accelerated growth in fisheries sector is envisaged during 7th plan period. The level of production should be reach a level of 4 million tonnes by 1989–90. The progress was quite appreciable.

FFDA

Although it should be considered the aspects of learning now being done through FFDA on a large scale. In FFDA another schemes often loan and subsidy is provided, the basic tenants of such development forming should be clearly understood on the following points :

1. The loan should be production oriented and should result in increase production.
2. The loan should be incremental *e.g.,* income generating in relation to first point.
3. Should create additional employment potential.
4. Since loans are production oriented and income generated they should be self liquidating. If any of the above purpose are not tell filled then the grant of loan is detected.

The State Fisheries Department should ensure that these objectives are achieved in view of the scares financial resources, if these are not achieved then planner could divert to finance to such other sectors where they are assure the of all the four objectives.

To make the achievement possible the following loaning system to be adopted :

1. There should be efforts towards sophisticating the levels of technology *i.e.,* technologies develop as a result of research should be passed on ultimate user *i.e.,* The fish farmer in such a manner that will increase his optional efficiency.
2. The loaning institution is the bank or department should give immediate supervision not only for financial part but also provide technical input that goes in it.
3. The bankers should develop capabilities either by themselves Collaboration with development department to provide sophisticated packages of practices including both production and market, generally post harvest technologies is not provided nor any guidance in marketing and this results the loss of profits to farmers.

In field of fishery the state co-operative banks and land developments have practically–no–role. Therefore, there is no primary agriculture society or farmer service society in fisheries sector in earlier days. The efforts has been made to provide such sectors in fisheries for better results. In 1969 there were only 8262 branches of banks in rural area surveying 70,000 rural populations/branch. In ten years period in 1979 there branches increase to 30202 and cover per branches 17000 rural population. Since last 7 years there has been further tremendous movement and banks have developed their own technical persons.

Apart from drying of individual project efforts for loaning a distt. banking plan is including technical visibility report organisational plan and transfer of technology with extension services should be included to spell out the total development and financial and marketing aspect. Pointing definite accomptability of the state, the beneficially should be a willing partners and committed to implementation of the projects, then only the high production and income to be achieved.

Monitory and evaluation of this important and has to be concurrent operation along with project implementation. The mythology of monitoring should form part of the district level plan, there should be prompt quardation between various agencies in distt. and block level.

All the grass root level workers should be given a short-term training so they are fully trained in the operation of FFDA.

Role of Banks in Fishery–Development

Introduction of lead bank scheme in 1969 gave impetus to the efforts of the banks to penetrate, rural, semiurban and backward areas. Now-a-days commercial banks, regional rural banks, farmers service societies and the few main agencies which helps in training and which helps in the project and such as poultry, duckery etc. by great scope of increasing production both in Inland and marine fisheries. The viable projects should have the following required.

1. Development of Inland fishery.
2. Construction of new ponds.
3. Purchase of spawn, fry and finger lings, feeding material and boxes etc.
4. Purchase of non mechanized fishing boat and different type of gears and other allied equipments.

Banks are generally giving advances to individual association of persons to cooperative societies who are known to have under non specific training and process necessary infrastructure for the successful important of fish farming. Now the time has changed we need certain good projects related to ornamental molluscan fisheries.

In India fishery management can really a very full if a plan of face-to-face is adopted. Banks generally advance to small and margin farmers.

Banks examines the following by fish farming :

1. Area of the water of pond
2. Kind of fish
3. Annual expenditure
4. Annual production
5. Lost of fish products
6. Net annual income aspects

Dewatering and cleaning of the ponds, bundh repairs besides this removal of weeds, labour charges and purchasing of MOC for pond preparation.

The fish farming repayment of In there loans Government provide subsidy. The upper limit is 10,000 and 2500 *i.e.,* free amount given by the Government and remaining paid by the banks.

ECONOMIC SURVEY 2009-10

- Economic survey optimistic on growth
- Current fuel prices not fiscally sustainable
- Expenditure control needed to control deficit
- Power, coal seeing a revival
- Infra capacity need to be accelerated
- Expect revenues of Rs 820 to 850 Crs in FY10
- Timely off-loading of food stocks need urgent attention
- Rationalise port service charges
- Lower peak custom duty to 7.5% from 10%
- Monetary measures must ensure credit growth
- Lower than budgeted non taxed revenue expected
- Medium term prospects of Indian economy really strong

- Major decline in consumption expenditure growth in FY10
- To give higher share to states despite fiscal strain
- States to get 1.5% more profit for people.
- To compensate states on revenue loss when GST rolled out
- Local bodies to also get share of central taxes
- Accepted major recommendations of 13th Financial Panel
- Food inflation is at present hovering close to 18%
- Moots direct food subsidy via food coupons to households
- Survey favours making available food in open market
- Survey favours monthly ration coupons usable anywhere for poor
- Fiscal deficit may be cut to 5.5%
- Gas output up 52.8 per cent to 50.2 billion cubic meters with RIL starting production
- Trade gap narrowed to USD 76.24 bn in April-December
- India 10th largest gold holding nation at 557.7 tonnes
- Large decline in customs, excise expected
- Tighten FRBM clauses for future relaxations
- Overall revenue transfers to state at 39.5%
- Services sector growth rate at 8.7%
- Virtually every second Indian has access to phone
- Fundamental policy changes needed for trade
- Wants credit available at reasonable rates on time for private sector to invest in agriculture
- Small shortfall in revenue receipt in FY10
- Overall farm GDP decline of 0.2% in FY10
- Investment growth still below GDP growth rate
- India not immune to global prices
- Auction for 3G spectrum to provide existing and foreign players to bring in new technology and innovations
- India world's 2nd largest wireless network with 525.1 million mobile users
- Slowdown in infrastructure that began in 2007, arrested
- Government initiates steps to boost private investment in agriculture
- Credit needed for private investment in agriculture
- Imports in April-December 2009 down 23.6 per cent
- Exports in April-December 2009 down 20.3 per cent
- Gross fiscal deficit pegged at 6.5 pc of GDP in 2009-10
- Favours making available food in open market
- Hype on Kharif crop failure helped hoarding

- Poor rainfall stopped prices falling
- Delay in releasing imported sugar pushed up prices
- Farm and allied sector production falls 0.2 per cent in 2009-10
- Rising food inflation a major concern
- Growth in private investments
- Growth has to quicken
- Watch and withdraw stimulus
- Risks of double dip recession seen
- Economy likely to grow up to 8.75%
- Inflation may spread, warns survey
- V-shaped recovery seen
- Economic recovery weak
- India's GDP growth rate in 2009-10 was 7.2%
- India's GDP to return to 9% in 2011-12
- High double digit food inflation in 2009-10
- Possible spurt in global commodity
- India can become world's fastest growing economy in 4 years
- Hike in fuel prices will impact inflation
- He tables economic survey in Parliament
- The fiscal deficit for the current fiscal year is expected to shoot up to 6.8% of GDP
- The economic survey is expected to project around 7.5% growth rate for the current fiscal year.

An Idea of Food Grains

Here some basic food commodities are given, the readers can achieve and draw better idea from agriculture sector. What was the earlier condition. Somehow fisherman also go for integration of fish with food grains. These description is also important for food security (students/readers must know it).

Elements of Food Security

- Availability of food at national level :
  - through domestic production and/or through imports (question of import capacity and Foreign Exchange Reserves)
- Access to food at HH level : issue of purchasing power.
- Access at intra – HH level : more an issue of gender discrimination and sociology at HH level
- Absorption of food : clean drinking water, environmental hygiene and primary health care.

DOMESTIC PRODUCTION AND AVAILABILITY OF FOOD GRAINS

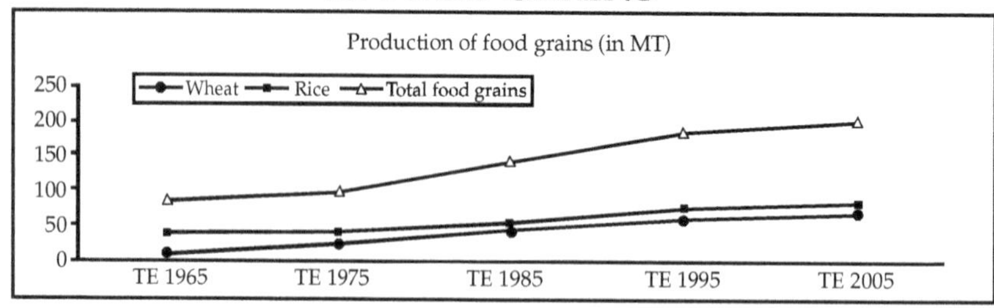

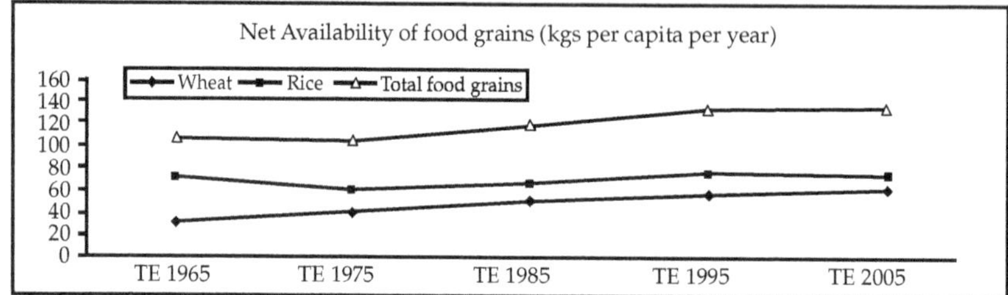

Source : Agriculture Statistics at a glance, 2005

Household Food Security : Poverty Headcount

	1994	2000
Poverty headcount, national (% of population)	36	28.6
Poverty headcount, rural (% of population)	37.3	30.2
Poverty headcount, urban (% of population)	32.4	24.7

Source : World Bank, 2005

INTERNATIONAL FOOD POLICY RESEARCH INSTITUTE

Implications of Food Grain Policy

Distribution Inefficiency

- **PDS Vs other Safety Nets**

Programs	Total program costs (billion Rs.)	Cost of transferring 1 Rupee to the poor	
		Actual	If targeting was perfect
Public Distribution Systems (PDS)*	73.16	6.68	1.52
Andhra Pradesh Rice Scheme (APRS)	10.63	6.43	1.59
SJRY (public works)**	1.18	2.28	–
Maharashtra Employment Guarantee Scheme		1.85	–
ICDS (Integrated Child Development Servicees)	8.8	1.44	–

* Due to the unavailability of state level data, total spending on PDS represents only the subsidies by the central government. This implies that the estimates of income transfer costs would have been larger state subsidies had been added

** Jawahar Gram Samridhi Yojana (JGSY) is the transformed version of the forma Jawahar Rozgar Yojana (JRY)

Source : Rashid et al. (2005), IFPRI

INTERNATIONAL FOOD POLICY RESEARCH INSTITUTE

PDS not Serving Poor States....

Share of consumption from PDS in total consumption (%)

	Rural				Urban			
	Rice		Wheat		Rice		Wheat	
	1993–4	1999–0	1993–4	1999–0	1993–4	1999–0	1993–4	1999–0
1	2	3	4	5	6	7	8	9
Andhra Pradesh	22.4	19.7	25.0	8.3	15.0	12.3	41.8	44.3
Assam ⟵	3.3	6.1	12.9	6.6	10.3	5.4	3.8	5.0
Bihar ⟵	0.3	0.9	1.8	2.9	0.2	1.1	2.4	2.7
Gujarat	25.9	19.6	14.4	15.3	19.5	10.2	7.8	5.7
Haryana	4.1	1.0	0.8	0.8	3.9	0.0	0.8	0.8
Himachal Pradesh 34.2	36.5	22.7	19.8	35.1	17.8	18.2	11.7	
J & K	4.3	27.1	5.1	15.3	27.0	47.0	50.1	30.7
Kanataka	13.3	22.9	46.9	41.1	19.8	14.1	29.1	26.5
Kerala	45.6	46.6	55.4	62.9	46.5	43.8	60.3	61.4
Madhya Pradesh	3.4	4.0	2.6	2.6	5.7	2.7	3.8	2.1
Maharashtra	13.4	15.3	16.3	19.8	15.9	7.2	8.3	7.1
Orissa ⟵	0.8	11.2	8.8	9.3	0.5	11.5	24.7	17.6
Punjab	1.4	1.3	0.6	1.1	0.9	0.8	1.4	
Rajasthan	8.7	4.3	11.5	2.0	5.2	0.0	8.6	1.7
Tamil Nadu	18.1	32.2	70.8	70.0	17.1	24.7	62.3	60.7
Uttar Pradesh ⟵	3.3	2.6	2.0	2.0	3.1	2.8	2.2	2.5
West Bengal ⟵	0.7	2.0	12.4	14.4	8.6	2.9	33.7	15.5

Source Estimated Seat NSSO 1993–94 and 1999–2000 consamet expenditure data
Access indicates the percentage of how shields reporting purchase from PDS

Implications of Food Grain Policy

Implicit taxation of Wheat and Rice for most of the years in 1980s and early 1990s (Under Importable Hypothesis)

Figure 4.1 (*a*) Nominal Protection co-efficients for Wheat in India 1965–2005

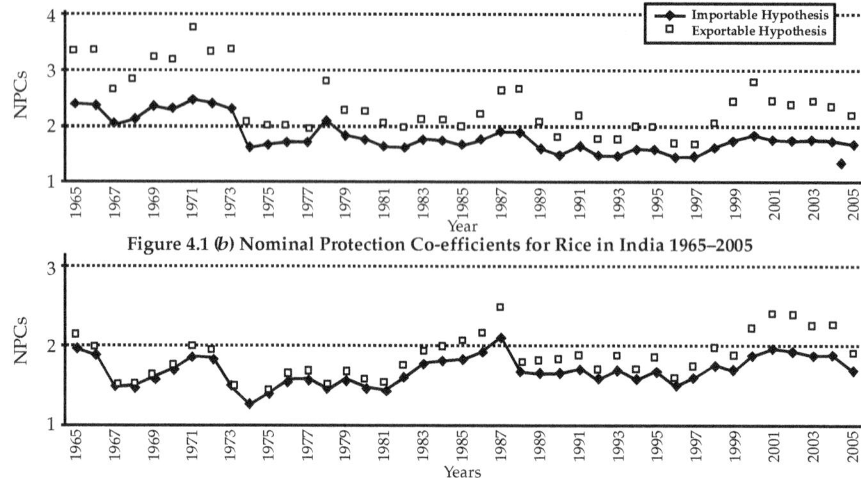

Figure 4.1 (*b*) Nominal Protection Co-efficients for Rice in India 1965–2005

Stocking

- Introduce warehousing receipt system;
- Private sector stocking to be encouraged;
- Abolish Stocking Limits Permanently;
- Strengthen Futures trading;
- Tendering for delivery for PDS.

Distribution

- TPDS
- With spread of Employment Scheme, PDS; can be gradually phased out.

Courtesy : International Food Policy Research Institute.

Base Year of Existing Price Indices in India

	Price Index	*Base Year*	*No. of included Goods*
1.	Wholesale Price Index (Effective from 1.4.2000)	1993–94	435
2.	Consumer Price Index for Industrial Workers	1982	260
3.	Industrial Production Index	1993–94	543
4.	Consumer Price Index for Urban Non-manual Employees	1984–85	180
5.	Consumer Price Index for Agriculture Labourers	1986–87	60

The early part of the year, the 52-weeks average inflation rate of 6.5 percent of Jan. 29, 2005 was higher than 5.5 percent registered a year ago.

During the post–reforms period, inflation has had a distinct decelerating trend as liberalisation of both internal and external trade and continual reduction and rationalisation of taxes has led to greater competition and cost–efficiency. Average annual WPI inflation decelerated from 10.6 percent in the first half of 1990s to 4.1 percent during 2001–02 to 2003–2004.

STEPS TAKEN FOR PREVENTING INFLATION IN INDIA

Various steps have been taken on both demand and supply side to control the inflation rate in the economy. During recent years, the Government has taken a number of steps in this direction which are as follows :

(*a*) Steps Related to Supply Side

(*i*) Open market sale of wheat and rice by Indian Food Corporation.

(*ii*) Wheat and wheat products were brought under the provisions of licensing and storage limit for preventing black marketing of these products

(*iii*) Import of wheat to maintain the buffer stocks in the economy.

(*iv*) Import of edible oils on 20% import duty and Pulses on 5% import duty under OGL.

(*v*) Import of 2 lakh tonnes Palmolin oil for the sale under public distribution system.

(*vi*) To ensure the sufficient supply of sugar, edible oil and pulses, liberalised imports were permitted.

(*b*) Steps Related to Demand Side

(*i*) To curtail fiscal deficit upto 5% of GDP.

(*ii*) To put a check on money supply increase.

Deflation

According to Crowther, "Deflation is that state in which the value of money rises and the price of goods and services falls."

The state of deflation may appear in the economy due to following reasons :

(*i*) When the Government withdraws money from circulation.

(*ii*) When Government imposes heavy direct taxes or takes heavy loans from the public (voluntary or compulsory or both).

(*iii*) When the Central Bank sells the securities in open market (which reduces the quantity of money in circulation).

(*iv*) When Central Bank controls the credit money and adopts various measures such as increase in CRR, credit rationing and direct action.

(*v*) When the Central Bank increases the Bank rate (which curtails the quantity of credit in the economy).

(*vi*) When state of over-production (excess supply over demand) takes place in the economy.

Measures of Checking Deflation

(*i*) To increase money supply.

(*ii*) To promote credit creation by the banks.

(*iii*) Curtailment in taxes so as to increase the purchasing power of the people.

(*iv*) To increase the public expenditure and to increase the employment opportunities in the economy.

(*v*) To increase the money supply in circulation by repayment of old public debts.

(*vi*) To provide economic subsidy by the Government to the industrial sector of the economy.

Money Stock Measures in India

During planning era, money supply has been increasing continuously with he rise in prices, though the increase rate in money supply has varied from year-to-year.

On the recommendations of the second working group of money supply. RBI introduced a series of money stock measures in India since 1970–71 which are as follows —

1. $M_1 \rightarrow$ Money with the Public (currency notes and coins) + Demand deposits of banks (on current and saving bank accounts) + Other demand deposits with RBI.
2. $M_2 \rightarrow M_1$ + Saving bank deposits with Post-offices.
3. $M_3 \rightarrow M_1$ + Term deposits with the bank.
4. $M_4 \rightarrow M_3$ + All deposits of post-offices.

M_1 measure represents the most liquid form of money among four money stock measures adopted by RBI. As we proceed from M_1 to M_4 the liquidity gets reduced. In other words, M_4 possesses the lowest liquidity among all these measures. The decline in liquidity indicates the shifting from 'medium of exchange' to 'store of value'. All these four money stock measures are not of equal importance. Their relative importance varies from the point of view of monetary policy.

Generally, in developed countries, the bank deposits are the most important component in money supply, while due to less banking habits in under–developed countries people want to keep their money in the most liquid from, *i.e.*, currency.

M_3 is the most important component among all money stock measures which is generally termed as **'Broad money'.**

Money Stock Measures Changes Suggested by Working Group

The working group under the chairmanship of Dr. Y.B. Reddy, Deputy Governor RBI, has suggested major changes in money stock measures (M_1, M_2, M_3 etc.) This working group was constituted in December 1997 which submitted its recommendations to RBI on June 23, 1998. The group suggested four new money measures (M_0, M_1, M_2, M_3) and three liquidity measures (L_1, L_2, L_3). Besides, the group also recommended the publishing of ***Financial Sector Survey*** *after every three months to capture the dynamic linkages between banks and rest of the organised financial sector.*

A Monetary Aggregates

M_0 = Currency in Circulation + Bankers Deposits with the RBI + 'Other' Deposits with the RBI.

M_1 = Currency with the RBI + Demand Deposits with the Banking System + 'Other' Deposits with the RBI = Currency with the Public + Current Deposits with the Banking system + Demand Liabilities portion of Savings Deposits with the Banking System + 'Other' Deposits with the RBI;

M_2 = M_1 + Time liabilities portion of Savings Deposits with the Banking System + Certificates of Deposit issued by Banks + Term Deposits (excluding FCNR (B) deposits) with a contractual maturity of up to and including one year with the Banking System = Currency with the Public + Current Deposits with the Banking System + Term Deposits (excluding FCNR (B) deposits) with a contractual maturity upto and including one year with the Banking System; and

M_3 = M_2 + Term Deposits (excluding FCNR (B) Deposits) with a contractual maturity of over one year with the Banking System + Call borrowings from 'Non–Depository' Financial Corporations by the Banking System.

B. Liquidity Aggregates

L_1 = M_3 + all deposits with the Post Office Savings Banks (excluding National Savings Certificates).

L_2 = L_1 + Term Deposits with Term Lending Institutions and Refinancing Institutions (FIs) + Term Borrowing by FIs + Certificates of Deposit issued by FIs; and

L_3 = L_2 + Public Deposits of Non–Banking Financial Companies.

Cheap Money Policy and Dear Money Policy

Cheap money policy is that monetary policy in which loans and advances are made available on low interest rate and easy terms to industries, businessmen and consumers. Cheap money policy increases the inflation rate in the economy and it is generally adopted to get rid of deflationary tendencies in the economy.

On the other hand, dear money policy is adopted to squeeze the credit utilisation facilities in the economy. Under dear money policy, interest rate is increased which helps in controlling inflation in the economy.

Development and Employment Programmes in India

An important objective of development planning in India has been to provide for increasing employment opportunities not only to meet the backlog of the unemployed but also the new additions to the labour force. The increasing diversification of the economy together with acceleration in economic growth has resulted in structural changes in the nature of the job market. Economic reforms in the areas of abolishing quantitative restrictions (QRs), reducing tariffs, reforming labour laws and abolishing SSI reservations have aimed at fostering labour–intensive production in India.

As per the results of the 55th round (1999–2000) of survey conducted by National Sample Survey Organisation (NSSO), the rate of growth of employment, on Current Daily Status (CDS) basis, declined from 2.7 percent per annum in 1983–1994 to 1.07 percent per annum in 1994–2000. The decline in the overall growth rate of employment in 1994–2000, was largely attributable to a near stagnation of employment in agriculture. As a result, the share of agriculture in total employment dropped from 60 percent in 1993–94 to 57 percent in 1999–2000. On the other hand, employment growth in all the sub-sectors within services (except community, social and personal services) exceeded 5 percent per annum.

The Approach Paper to the Mid-Term Appraisal (MTA) of Tenth Five Year Plan has reiterated that employment growth should exceed growth of labour force to reduce the backlog of unemployment. Employment strategies advocated in the Approach Paper include.

- Special emphasis to promote public investment in rural areas for absorbing unemployed labour force for asset creation.
- Identification of reforms in the financial sector to achieve investment targets in the Small and Medium Enterprises (SME) sector.
- Large–scale employment creation in the construction sector, especially for the unskilled and semi–skilled.
- Necessary support to services sectors to fulfil their true growth and employment potentials and greater focus on agro–processing and rural services.

Employment in the Organized Sector

Organized sector employment accounts for only a small proportion (7 to 8 percent) of the total workforce, and on March 31, 2003 was 27 million of which 69 percent was in the public sector. Employment in the organized sector declined by 0.8 percent in 2003 due to a decline of the public sector employment by one percent in 2003.

As per the data available from the 947 employment exchanges as on September 2004. 4.08 crore job-seekers were registered with the employment exchanges. 70 percent of which were educated (10th standard and above) and 26 percent were women job seekers. The maximum number of job seekers awaiting employment was in West Bengal (69.3 lakh). The placement was maximum in Gujarat, whereas the registration was maximum in Maharashtra. The placement effected by the employment exchanges at the All–India Level during January–September, 2004 was of the order of 1.03 lakh as against 2.12 lakh vacancies notified during this period.

Total Employment and Organised Sector Employment

Sector	*Employment (Million)*				*Growth Rate (percent per annum)*	
	1983	*1988*	*1994*	*1999–00*	*1983–94*	*1994–2000*
Total Population	718.21	790.00	895.05	1004.10	2.12	1.93
Total Labour Force	308.64	333.49	381.94	406.05	2.05	1.03
Total Employment	302.75	324.29	374.45	397.00	2.04	0.98
Org. Sector Employment	24.01	25.71	27.37	28.11	1.20	0.53
Public Sector	16.46	18.32	19.44	19.41	1.52	–0.03
Private Sector	7.55	7.39	7.93	8.70	0.45	1.87

Notes : 1. Total employment figures are on Usual States (UPSS) basis.
2. The Organised sector employment figures are as reported in the Employment Market Information System of Ministry of Labour and Pertain to 1st March of 1983, 1988, 1994 and 1999.
3. The rate of growth of total employment and organised sector employment are compound rates of growth.

Source : Planning Commission

TASK FORCE ON EMPLOYMENT OPPORTUNITIES

Strategy for Employment generation to focus on intervention in five major areas :

- *Accelerating the rate of growth of GDP, with a particular emphasis on sectors likely to ensure the spread of income to the low income segments of the labour force.*
- *Pursuing appropriate sectoral policies in individual sectors which are particularly important for employment generation. These sector level policies must be broadly consistent with the overall objective of accelerating GDP growth.*
- *Implementing focussed special programmes for creating additional employment and enhancing income generation from existing activities aimed at helping vulnerable groups that may not be sufficiently benefited by the more general growth promoting policies.*
- *Pursuing suitable policies for education and skill development, which would upgrade the quality of the labour force and make it capable of supporting a growth process which generates high quality jobs.*
- *Ensuring that the policy and legal environment governing the labour market encourages labour absorption, especially in the organized sector.*

Women Employment

Employment of women in the organised sector (both public and private) at end–March, 2003 at *4.97 million constituted 18.4 percent of the total organised sector employment compared to 18.1 percent in 2002.* As regards sectoral distribution, community, social and personnel services sectors employed 56.6 percent of women workers followed by manufacturing (20.3 percent), agriculture and allied occupations (9.4 percent) and finance, insurance, real estate and business (5.5 percent). States with higher female literacy rates have higher proportion of women in organized sector employment.

Employment in Organised Sector

(In Lakh)

*Year**	*Public Sector*	*Private Sector*	*Total*
1990	187.72	75.82	263.53
1991	190.57	76.76	267.33
1992	192.10	78.46	270.56
1993	193.26	78.51	271.77
1994	194.95	79.30	273.75
1995	194.66	80.59	275.25
1996	194.29	85.12	279.41
1997	195.59	86.86	282.45
1998	194.18	87.48	281.66
1999	194.15	86.98	281.13
2000	193.14	86.46	279.60
2001	191.38	86.52	277.89
2002	187.73	84.32	272.06
2003	185.8	84.21	270.01

* As on End of March

Rural Unemployment and Poverty Elimination Programme

According to a general survey done by the National Sample Survey Organisation (NSSO), *62% of total unemployment exists in rural sector and only 38% in urban sector of our country.* In the beginning of planning in the country the Government did not pay *heed* to the solution of unemployment problem, but it was given serious attention during the 4th plan. Eighth Five Year Plan allocated Rs. 30,000 crore for rural development and it was increased to Rs. 42874 crore for the 9th Five Year Plan. 10th Plan also has sufficient allocations for rural development.

The programmes that were especially adopted under the Fourth plan included Small Farmer Development Programme (**SFDA**). Marginal Farmer and Agriculture Labour Agency (**MFALA**), Drought–Prone Area Programme (**DPAP**), and Crash Scheme for Rural Employment. In the Fifth **Plan Food for Work Programme and Minimum Needs Programme** were launched. All these Programmes were aimed at the poorest people of the rural areas. These programmes were designed to provide financial support and secondly, to create direct employment opportunities for poorest farmers and labourers in various public works projects. During the ruling period of Janta Dal. Antyodaya Programme was started in 1977–78 to give opportunities of productive employment to the maximum number of people in the society. So that they may come out from the vicious circle of poverty.

Major on-going Programmes for Poverty Alleviation and Employment Generation

1. *National food for work programme*
2. Pradhan Mantri Gramodaya Yojana (PMGY)
3. Swarnajayanti Gram Swarojgar Yojana (SGSY).
4. Sampoorna Gramin Rojgar Yojana (SGRY).
5. Rural Housing Schemes
    - Indira Awas Yojana (IAY)
    - Credit Cum Subsidy Schemes
    - Samagra Awas Yojna
    - Innovative Scheme of Rural Housing & Habitat Development & Rural Building Centres (RBCs).
6. Rural Employment Generation Programme (REGP).
7. Prime Minister's Rojgar Yojana (PMRY)
8. Pradhanmantri Gram Sadak Yojana (PMGSY)
9. Drought Prone Area Programme (DPAP)
10. Desert Development Programme (DDP).
11. Integrated Wastelands Development Programme (IWDP)
12. Antyodaya Anna Yojana (AAY)
13. Swarna Jayanti Shahari Rozgar Yojana (SJSRY)
14. Valmiki Ambedkar Awas Yojana (VAMBAY).

During the Sixth plan, in 1980, the Government started the **National Rural Employment Programme (NREP)** in place of Rural Labour Force Programme such as Crash Plan and Food for Work Plan. The main aim of this programme was to increase the beneficial employment opportunities, to construct stable community property and to uplift the food standard of the rural poor. In order to remove the unemployment among the rural youth 'TRYSEM' (**Training to Rural Youth for Self–Employment**) plan was started in 1979. In **1983, Rural Landless Employment Guarantee Programme** (RLEGP) was started to remove the rural poverty and unemployment especially at that time when no work is available. *At the end of the Seventh plan. The Government started the Jawahar Rojgar Yojana in 1989,* which was made more extensive by combining NREP and RLEGP. Since *2 October, 1993, the Government implemented the Employment Assurance Scheme (EAS).* On 1 January, 1996, a new self-employment Programme was started in the rural areas in which provision was made to provide 50% subsidy (*maximum Rs. 7,500*) for self-employment to unemployed youth having education upto 8th standard.

Since Feb. 1, 1997, the Government introduced Ganga Kalyan Yojana as a sub-scheme of IRDP but was given the status of an independent scheme *w.e.f.* April 1, 1997. In *1999–2000 budget,* the Government introduced *Annapurna Yojana* for providing 10 kg. free foodgrains to eligible old people.

Pradhan Mantri Gramodaya Yojana (PMGY)

PMGY as introduced in 2000–01 with the objective of focussing on village level development in five critical areas *i.e.,* health, primary education, drinking water, housing and rural roads, with the overall objective of improving the quality of life of people in the rural areas.

(*i*) **Pradhan Mantri Gram Sadak Yojana (PMGSY) :** PMGSY was launched on **25th December, 2000** with the objective of providing road connectivity through good all-weather roads to all rural habitations with a population of more than 1000 persons by the year 2003 and those with a population of more than 500 persons by the year 2007. In respect of the Hill States (North–East Sikkim, Himachal Pradesh, Jammu and Kashmir, Uttaranchal) and the desert areas the objectives would be to connect habitations with a population of 250 persons and above.

As on 1 April, 2000, about 3.30 lakh habitations out of 8.25 lakh habitations in the country were not connected by all–weather roads. The PMGSY seeks to provide connectivity to about 1.60 lakh unconnected habitations besides upgradation to prescribed standards of some existing roads, at an estimated cost of Rs. 60,000 crore by the end of the Tenth Plan. For the year 2000–01, 2001–02 and 2002–03 road words of Rs. 9,296.20 crore have been cleared and are in various stages of execution. A cumulative expenditure of Rs. 4,175 crore was incurred till March 2003.

(*ii*) **Pradhan Mantri Gramodaya Yojana (Gramin Awas) :** This scheme was launched with effect from April 1, 2000. Based on the pattern of the Indira Awas Yojana, the scheme is being implemented in the rural areas throughout the country with the objective of sustainable habitat development at the village level and to meet the growing housing needs of the rural poor.

(*iii*) **Pradhan Mantri Gramodaya Yojana—Rural Drinking Water Project :** Under this programme, a minimum 25 percent of the total allocation is to be utilised by the respective States/UTs on projects/schemes for water conservation, water harvesting, water recharge and sustainability of the drinking water sources in respect of areas under Desert Development Programme/Drought Prone Areas Programme.

Swarnajayanti Gram Swarojgar Yojana (SGSY)

Integrated Rural Development Programme (IRDP) and allied programmes such as Training of Rural youth for self employment (TRYSEM), DWCRA, Million Wells Scheme (MSW), SITRA and Ganga Kalyan Yojana have been restructured into a self–employment programme called the Swarnajayanti Gram Swarojgar Yojana (SGSY) from April 1, 1999 with the following objectives:

(*i*) focussed approach to poverty alleviation;

(*ii*) capitalising advantages of group lending; and

(*iii*) overcoming the problems associated with multiplicity of programmes. The SGSY is conceived as a holistic programme of micro enterprises covering all aspects of self-employment which includes organising rural poor into Self–Help Groups (SHGs). It integrates various agencies—District Rural Development Agencies, banks, line departments, Panchayati Raj Institutions, non–governmental organisations and other semi–government organisations. The objective of SGSY is to bring the assisted poor family above the poverty line in three years by providing them income generating assets through a mix of bank credit and government subsidy and to ensure that an assisted family has a monthly net income of at least Rs. 2000. Subsidy under SGSY is uniform at 30 percent of the project cost subject to a maximum of Rs. 7500. In respect of Scheduled Castes and Scheduled Tribes, it is 50 percent subject to a ceiling of Rs. 1.25 lakh. There is no monetary limit on subsidy for irrigation projects. SGSY is funded by the centre and states in the ratio of 75 : 25.

Below the poverty line families in rural areas constitute the target group of the SGSY. Within the target group special Safeguards have been provided to vulnerable sections, by way of reserving 50% benefits for SC/ST. 40% for women and 3% for disable persons.

The scheme of SGSY covers all aspects of self–employment such as organisation of the poor into self-help groups, training, credit, technology infrastructure and marketing.

National Food for Work Programme

In line with the NCMP (National Common Minimum Programmes) National Food for Work Programme was launched on November 14, 2004 in 150 most backward districts of the country with the objective to intensify the generation of supplementary wage employment. The programme is open to all rural poor who are in need of wage employment and desire to do manual unskilled work. It is implemented as a 100 percent centrally sponsored scheme and the food grains are provided to States free of cost. However, the transportation cost, handling charges and taxes on foodgrains are the responsibility of the States. The collector is the nodal officer at the district level and has the overall responsibility of planning, implementation, coordination, monitoring and supervision. For 2004–2005, Rs. 2020 crore have been allocated for the programme in addition to 20 lakh tones of foodgrains.

Sampoorna Gramain Rojgar Yojana (SGRY)

The on-going schemes—the Employment Assurance Scheme (EAS) and the Jawahar Gram Samridhi Yojana (*JGSY*) were merged into the Sampoorna Gramin Rojgar Yojana (SGRY) on September 25, 2001. The objective of the programme is to provide additional wage employment in rural areas and also to provide food security, alongside creation of durable community, social and economic assets and infrastructure development in these areas. The programme in self-targeting in nature with special emphasis on woman, scheduled castes, scheduled tribes and parents of children withdrawn from hazardous occupation.

It is executed through two streams, *viz.*, (*i*) First Stream replaces the EAS, which is implemented at the District and intermediate *Panchayat* level 50 percent of the fund is earmarked out of the total funds available under SGRY and distributed between the *Zilla Parishad* (20 percent) and the intermediate level *Panchayat* or *Panchayat Samitis* (30 percent); (*ii*) Second Stream replaces JGSY and is implemented at the village *Panchayat* level. 50 percent of the SGRY fund is earmarked for this Stream. The fund is released to the Village Panchayats through the DRDAs/*Zilla Parishads.*

The *Gram Panchayats* can take up any work with the approval of the *Gram Sabha* as per their felt need and within available funds. 50 percent of the funds earmarked to the *Gram Panchayats* are to be utilised for infrastructure development works in SC/ST localities.

The scheme has an annual outlay of Rs. 10000 crore including foodgrains component of Rs. 5000 crore.

The scheme envisages providing 100 crore man days of wage employment every year.

This centrally sponsored scheme is implemented on a cost sharing basis between the centre and the state in the ratio of 75 : 25. In case of Union Territories, 100% expenditure is met by the centre.

Indira Avas Yojana (IAY)

The Indira Avas Yojana was started in 1985–86 as a sub–plan of RLEGP. The objective of the scheme was to construct houses for the poorest people of schedule caste/schedule tribe communities and the freed-bonded labour. The houses under the scheme were provided to them free of cost. In 1989–90, after merging the RLEGP with Jawahar Rozgar Yojana, this plan was also made a part of Jawahar Rozgar Yojana (JRY). In 1996, it was again separated from JRY and was given an independent status. The main points of this scheme are as follows :

1. Since 1993–94, the benefit of Indira Awas Yojana is being provided even to those rural poor of non–schedule caste/schedule tribe who are living below the poverty line.
2. A minimum of 60% of funds are to be utilized for construction of houses for the SC/ST people.
3. From 1995–96, IAY benefits have been extended to widows or next of kin of defence personnel killed in action.
4. Benefits have also been extended to ex-servicemen and retired members of para military forces as long a they fulfil the normal eligibility conditions of IAY.
5. 3% of funds are reserved for the disabled persons living below the poverty line in rural areas.
6. Under the plan, the allocation of the house is done in the name of the female member of the benefited family or in the joint names of husband and wife.
7. Sanitary latrine and smokeless chulha are integral to an IAY house.
8. Assistance for constructions of new houses is provided at the rate of Rs. 20000 and Rs. 22000 per unit in the plains and hilly/difficult areas respectively.

National Rural Employment Guarantee Bill, 2004 (Salient Features)

- *State Governments to provide at least 100 days of guaranteed wage employment in every financial year to every household whose adult members volunteer to do unskilled manual work.*
- *Sampoorna Grameen Rozgar Yojana (SGRY) and National Food for Work Programme to be subsumed within the Scheme once the Act is in force.*
- *Until such time as a wage rate is fixed by the Central Government, the minimum wage for agricultural labourers shall be applicable for the scheme.*
- *An applicant not provided employment within fifteen days, to be entitled to a daily unemployment allowance as specified by the State Government subject to its economic capacity, provided such rate is not less than a quarter of the wage rate for the first thirty days during the financial year and not less than a half of the wage rate for the remaining period of the financial year.*
- *Central Employment Guarantee Council to be constituted to discharge various functions and duties assigned to the Council. Every State Government to also constitute a State Council for this purpose.*
- *Panchayat at the district level to constitute a Sanding Committee of its members to supervise, monitor and oversee the implementation of the Scheme within the district.*
- *For every Block, State Governments to appoint a Programme Officer for implementing the Scheme.*
- *Gram Panchayat to be responsible for identification of the projects as per the recommendations of the Gram Sabha and for executing and supervising such works.*
- *Central Government to establish a National Employment Guarantee Fund. State Governments to establish State Employment Guarantee Funds for implementation of the Scheme.*
- *The Scheme to be self–selecting in the sense that those among the poor who need work at the minimum wage would report for work under the scheme.*

Samagra Avas Yojana

This has been launched as a comprehensive housing scheme in 1999–2000 (*w.e.f.* April 1, 1999) on pilot project basis in one block in each of 25 Districts of 24 States and in one Union Territory with a view to ensuring integrated provision of shelter, sanitation and drinking water. The underlying philosophy is to provide for convergence of the existing housing, sanitation and water supply schemes with a special emphasis on technology transfer, human resource development and habitat improvement with people's participation. Intended beneficiaries under the scheme are the rural poor, preferably those below the poverty line.

Pradhan Mantri Gram Sadak Yojana (PMGSY)

Pradhan Mantri Gram Sadak Yojana, has been launched on Dec. 25, 2000 with a plan to cover 1.6 lakh of the 2.68 lakh unconnected villages over seven years.

The PMGSY, once fully rolled out intends to provide new connectivity to about 1.6 lakh habitations with a population of 500 or more persons in the plains and those with populations of 250 or more in hills, deserts and tribal areas at an expenditure of Rs. 60,000 crore.

The objective of the scheme is to ensure that the unconnected village is connected with an all–weather road to the nearest connected habitation or to an existing all-weather road so as to enable the habitants of the village to access medical services, educational and marketing facilities.

PMGSY is a 100% centrally sponsored programme, which means that State Governments do not have to come up with counter party funding or bear interest on any money allocated by the centre for the project. The onus of roll-out of the programme is with the State administration.

The high specification laid down for executing the projects have made it different to earlier programmes of the same nature.

Drought-Prone Area Programme (DPAP)

This National Programme was launched in 1973–74 in some selected Drought Prone Areas of the country. The main objective of this plan was to reestablish the environmental balance in these areas by promoting the balanced development of land, water and other natural resources. For this programme, the arrangement of the finance is done by the Centre and the State concerned in the ratio of 50 : 50. Presently 947 blocks of 155 districts in 13 states are covered under the programme. Currently 8355 watershed projects are at various stages of implementation covering an area of more than 41 lakh hectres. **This programme is being carried on by the Rural Development Department.**

Desert Development Programme (DDP)

The Desert Development Programme was started in 1977–78 in some selected districts to check the formation of deserts, to end the drought effects in the deserts, to reestablish the ecological balance in the affected areas and to increase the land productivity and water resources in these areas. This programme is being implemented totally on the basis of Union support but the division of the funds in the hot arid areas is done between the Union and the States on the basis of 75 : 25. In 1995–96, 27.50 lakh rupees were allocated for every one thousand square kilometre hot desert area, but for any district the maximum allocation could be Rs. 8.50 crore only. In the same way, for cold desert areas in Himachal Pradesh Rs. 2 to 3 crore per district and in Jammu and Kashmir Rs. 3 crore per district were allocated. Presently 3844 watershed projects covering 19 lakh hectares are under implementation in 227 blocks of 36 districts in 7 states. This programme is being run by the Rural Development Department.

Swarna Jayanti Shahari Rozgar Yojana (SJSRY)

In the Golden Jubilee year of Independence a new scheme named Swarna Jayanti Shahari Rozgar Yojana (SJSRY) was introduced which became operational since December 1, 1997. SJSRY provides gainful employment to the urban unemployed and under-employed poor through encouraging the setting up of self–employment ventures or provisions of wage employment. The beneficiaries are identified by the urban local bodies on the basis of house-to-house survey. Under the scheme women are to be assisted to the extent of not less than 30%, disable at 3% and SC/ST at least to the extent of the proportion of their strength in the local population. It is funded on a 75 : 25 basis by the Centre and the States. The three urban schemes already in operation were merged with this new scheme. These were :

(*i*) Nehru Rozgar Yojana (NRY)

(*ii*) Urban Basic Service for Poor (UBSP)

(*iii*) Prime Minister's Integrated Urban Poverty Eradication Programme (PMIUPEP).

Universal Health Insurance Scheme

- *Inaugurated by Prime Minister on July 14, 2003.*
- *Devised by the public sector general insurance companies scheme to improve the access to the health care of the population in general and poorer section of the society in general.*
- *Reimbursement of medical expenses upto Rs. 30,000 towards hospitalisation floated amongst the entire family, death cover due to an accident for Rs. 25,000 to the head of family and compensation due to loss of earning of the earning member @ Rs. 50 per day upto a maximum of 15 days.*
- *Government contribution of Rs. 100 in annual premium to families below the poverty line.*

SJSRY is a programme which is applicable to all urban towns in India. The programme is implemented on a whole town basis with special emphasis on urban poor clusters.

Among other components the scheme has two sub-schemes where bank credits involved, namely **Urban Self Employment Programme (USEP) and Development of Women and Children in Urban Areas (DWCUA).**

Antyodaya Anna Yojana (AAY)

AAY launched in December 2000 provides foodgrains at a highly subsidized rate of Rs. 2.00 per kg for wheat and Rs. 3.00 per kg for rice to the poor families under the Targeted Public Distribution System (TPDS). The scale of issue, which was initially 25 kg per family per month, was increased to 35 kg per family per month from April 1, 2002. The scheme initially for one crore families was expanded in June 2003 by adding another 50 lakh BPL families. During 2003–04, under the AAY, against an allocation of 45.56 lakh tonnes of foodgrains, 41.65 tonnes were lifter by the State/UT Governments. Budget 2004–05 expanded the scheme further from August 1, 2004 by adding another 50 lakh BPL families. With this increase, 2 crore families have been covered under the AAY.

Valmiki Ambedkar Awas Yojana (VAMBAY)

The VAMBAY launched in December 2001 facilities the construction and upgradation of dwelling units for the slum dwellers and provides a healthy and enabling urban environment through community toilets under Nirmal Bharat Abhiyan, a component of the scheme. The Central Government provides a subsidy of 50 percent, the balance 50 percent being arranged by the State Government. Since its inception and upto December 31, 2004, Rs. 753 crore have been released as Government of India subsidy for the construction/upgradation of 3,50,084 dwelling units and 49,312 toilet seats under the scheme. For the year 2004–05, out of he tentative Central Fund allocation of Rs. 280–58 crore, upto December 31, 2004, an amount of Rs. 223.66 crore has been released covering 1,06,136 dwelling units and 20,139 toilet seats.

Development of Women and Children in Rural Areas (DWCRA)

This scheme is not in existence at present. Since April 1, 1999 DWCRA has been merged with newly introduced scheme namely Swarna Jayanti Gram Swarozgar Yojana.

Development of Women and Children in Rural Areas Programme **(DWCRA)** was started in September 1982 in the form of a sub-plan of integrated Rural Development Programme. At present this DWCRA programme is not in existence because it is merged with newly introduced programme **Swarna Jayanti Gram Swarojgar Yojana** *w.e.f.* April 1, 1999. The main aim of this programme was to provide proper self–employment opportunities to the women of those rural families who are living below the poverty line, so that their social and economic standard could be improved. Under this programme, the policy of making a group of 10–15 women was adopted corresponding to the local resources, their own choices and skills to complete the economic activities.

Integrated Rural Development Programme (IRDP)

This scheme is not in existence at present. Since April 1, 1999, IRDP with five other schemes—TRYSEM, DWCRA, SITRA, Ganga Kalyan Yojana & MWS has been merged in newly introduced Scheme namely 'Swarna Jayanti Gram Swarozgar Yojana.'

The Integrated Rural Development Programme was initially launched in 2300 blocks but was extended all over the country as a major poverty eradicating programme, on 2nd October, 1980. At present this programme is not in existence because IRDP with five other schemes (TRYSEM, DWCRA Ganga Kalyan Yojana, SITRA and MWS) has been merged in newly introduced scheme namely **Swarna Jayanti Gram Swarojgar Yojana** *w.e.f.* April 1, 1999. The main objective of this programme was to make the rural poor families economically independent so that they are able to cross the poverty line. In order to achieve this objective, the targeted groups were given productive assets for promoting self–employment. These assets were acquired by means of financial support. This financial support was given through the subsidies provided by the Government and in the form of periodical loans given by the financial institutions (Commercial Banks, Cooperative Banks and Regional Rural Banks etc.) This programme was financially nurtured in 50 : 50 ratio by the Centre and the State Governments.

Supply of Improved Tool Kits to Rural Artisians (SITRA)

This scheme is not in existence at present. Since April 1, 1999 SITRA has been merged with newly introduced scheme namely Swarna Jayanti Gram Swarozgar Yojana.

To supply modern tools to the rural artisans is also a part of IRDP. Since July 1992, this scheme was started in some of the selected districts in the form of Central supported plan but later on this scheme was extended to all over the country. At present this scheme is not in existence because it has been merged with **Swarna Jayanti Gram, Swarojgar Yojana** ***w.e.f.*** April 1, 1999. The main objective of this plan was to make the rural artisans (except weavers, tailors, embroiders and tobacco workers etc.) technically capable of improving the quality of their products and increasing their production and income with the help of modern tools.

Training to Rural Youth for Self–Employment (TRYSEM)

This scheme is not in existence at present. Since April 1, 1990, TRYSEM has been merged with newly introduced programme namely Swarna Jayanti Gram Swarozgar Yojana.

Training to Rural Youth for Self–Employment (TRYSEM) is an integral part of Integrated Rural Development Programme (IRDP). This centrally sponsored programme was started on 15 August, 1979. The main objective of this programme was to impart technical and business expertise to those rural youth who belong to the families living below the poverty line so that they may become self-employed. Under the TRYSEM programme, technical training was given to those youth of 18–35 age group who were living below the poverty line. This training was provided as per requirements of the youth and of local area. This TRYSEM programme is not in existence at present because it has been merged with newly introduced programme namely **Swarna Jayanti Gram Swarojgar Yojana** *w.e.f.* April 1, 1999.

Krishi Vigyan Kendra

Indian Council of Agricultural Research (ICAR) discharges the frontline extension education responsibilities through a network of 261 Krishi Vigyan Kendras (Farm Science Centres) in the country. During 1996–97, the Government planned to establish 119 more Krishi Vigyan Kendras. After this, during the Ninth plan, it is proposed to establish 120 more Krishi, Vigyan Kendras. According to the agriculture department, these Vigyan Kendras will organise Employment-oriented Training Programmes for the farmers (including the women and rural youth). These centres will spread the new agriculture techniques among the farmers. These centres will mainly be run by the distinguished voluntary institutions which are constantly working for the development of agriculture and village society, by the State Agriculture Universities and by the State Governments.

Rashtriya Mahila Kosh (National Women Fund)

On 30th March 1993 Rashtriya Mahila Kosh was established to meet the loan requirements of the poor women. This fund was established in the form of a society under the Society Registration Act 1860 to facilitate credit support to poor women for their socio economic upliftment. The support is extended through NGOs, women Development corporations, women co-operative societies, Self Help Groups (SHGs) and suitable state Government agencies. This fund of Rs. 31 crore which has since been raised to Rs. 100 crore with effect from December 12, 2001.

Various Women-oriented Schemes of State Governments

1. **Panchdhara Yojana :** *Madhya Pradesh Government's Scheme was launched on 1 November, 1991 for rural and tribal women which includes following five Schemes :*

   (i) ***Vatsalya Yojana*** *: For health care and facilities at the time of delivery of a child.*

   (ii) ***Gramya Yojana*** *: For providing working capital to rural women for small trade.*

   (iii) ***Aayushmati Yojana*** *: Govt. subsidy is provided for medical treatment to the poorest women.*

   (iv) ***Social Security Pension Scheme*** *: For orphan widows.*

   (v) ***Kalpavraksha Scheme*** *: For providing employment to SC/ST women in tribal areas.*

2. **Apni Beti Apna Dhan Yojana :** *Haryana Government's Scheme was launched on 2 October, 1994 under which an investment of Rs. 2500 is made by the Government in Indira Vikas Patra for newly born female child in SC/ST families which becomes Rs. 25000 after 18 years and given to the concerned girl.*

3. **Kunwar Bainu Mamerun Scheme :** *Gujarat Govt.'s Scheme was launched in 1995 in which an assistance of Rs. 5000 is provided to families having annual income of less than Rs. 7500 at the time of their daughter's marriage.*

4. **Kamdhenu Yojana :** *Maharashtra Government's Scheme provides the financial opportunities of self-employment to disabled, divorced women.*

5. **Girl Child Protection Scheme :** *Andhra Pradesh Government's Scheme aims at protecting the interest of girl child in the society.*

Jawahar Rozgar Yojana (JRY)

Jawahar Rozgar Yojana at present is not in existence because since April 1, 1999 its has been replaced by Jawahar Gram Samridhi Yojana.

In the last year of the Seventh plan, a broad rural employment programme named Jawahar Rojgar Yojana (JRY) was started on 1 April, 1989 by dissolving the National Rural Employment Programme (NREP) and Rural Landless Employment Guarantee Programme (RLEGP).

Since 1993–94, the JRY was implemented in the following three streams :

First Stream

Comprising of general works under JRY and two sub-schemes namely, the **Indira Awas Yojana (IAY) and Million Wells Scheme (MWS).**

Second Stream

Second steam of JRY (also called as the intensified JRY) is being implemented in 120 identified districts with additional allocation.

Third Stream

Third stream of JRY consists of Special and Innovative projects.

JRY was one of the major wage employment programmes. It was implemented in all the villages of the country through the Panchayati Raj institutions. It contributed to a great extent in creating durable rural infrastructure which has a critical importance in the development of village economy thereby improving the standard of living of the rural poor.

Jawahar Gram Samridhi Yojana (JGSY)

At present this scheme is merge with Sampoorna Gramin Rojgar Yojana w.e.f. September 25, 2001.

Jawahar Gram Samridhi Yojana (JGSY) was launched *w.e.f.* 1st April, 1999 to ensure development of rural infrastructure at the village level by restructuring the erstwhile Jawahar Rojgar Yojana (JRY). The **Primary objective** of JGSY was creation of demand driven community village infrastructure including durable assets at the village level and assets to enable the rural poor to increase the opportunities for sustained employment. The **secondary objective** was generation of wage employment for the unemployed poor in the rural areas.

Main features of JGSY were as follows :

1. Main emphasis on creation of rural infrastructure at the village level.
2. Implementation of the scheme entirely by village panchayats.
3. Empowerment to Gram Sabha for approval of Schemes/works.
4. 22.5% of JGSY funds for individual beneficiary schemes for SCs/STs.
5. 3% of annual allocation for creation barrier free infrastructure for the disabled.
6. DRDA/Zila Parishad is responsible for overall guidance, co-ordination, supervision, monitoring and periodical reporting.

Million Wells Scheme (MWS)

This scheme is not in existence at present. Since April 1, 1999 MWS has been merged with newly introduced scheme namely Swarna Jayanti Gram Swarozgar Yojana.

During 1988–89, Million Wells Scheme was started as a sub-plan of NREP/RLEGP in order to provide open irrigation wells free of cost to the poor belonging to the schedule caste/ schedule tribe, to marginal and small farmers and to freed bonded labours. Since April 1, 1989 this plan was continued under Jawahar Rozgar Yojana, but during 1995–96 it was given an independent status. MWS was funded by the centre and states in the ratio of 80 : 20. It also lost its existence at present at it has been merged with Swarna Jayanti Gram Swarojgar Yojana, *w.e.f.* April 1, 1999.

Agro–Service Centres

In the forth plan, assistance was provided to the unemployed graduates and diploma holders for opening Agro-Service Centres. The main objectives of this plan are as follows —

1. To provide self–employment opportunities to the technical labour.
2. To provide maintenance and repairing facilities for agricultural machinery and tools to farmers at their farm places only.
3. To establish convenient centres for spare parts, fuel, lubricating oil, and other engineering inputs.
4. To provide inputs like, fertilizer, pesticides etc.

Marginal Farmers Development Agencies (MFDA)

One of the objectives of Fourth Five Year Plan (1969–74) was to provide loans to the small farmers so that they may use the modern techniques and adopt the intensive farming. On the directions of Planning Commission, such agencies were established for recognising the small farmers and presenting various plans to the banks for solving the financial problems of the small farmers.

Employment Guarantee Scheme of Maharashtra

In 1972–73, the Government of Maharashtra started the Employment Guarantee Scheme. It was the first plan of such type in which the **'Right to work'** ensured in the Constitution was accepted. If any person demands work it is responsibility of the Government to provide job to the person under this scheme.

The main objectives of this plan are as follows :

1. To provide profitable and productive employment to a person in the accepted village projects which will increase the productivity in the economy.
2. To implement labour intensive productive projects such as those relating to small irrigation, water and land protection, canals, land development and afforestation.
3. To enforce these schemes through departments itself instead of contractors so that 60% of total expenditure is made on labour and remaining 40% expenditure is made on material, capital instruments, supervision and administrative services.

Rural Health Infrastructure

In rural areas, services are provided through a network of integrated health and family welfare delivery system. As on 31 March, 2004 an extensive network of 3,043 Community Health Centres, 22,842 Primary Health Centres and 1,37,311 Sub-centres were in existence to provide primary health care at grass root level. Further 8,669 new Sub-centres were sanctioned during the year 2003–04. One Sub-centre manned by one female and a male multipurpose worker covers a population of 5,000 in plain areas and 3,000 in hilly, tribal and backward/difficult terrain areas. One Lady Health Visitor supervises six Sub-centres. One Primary Health Centre covers a population of 30,000 in the plain areas and 20,000 in tribal and difficult terrain areas. One Community Health Centre covers 80,000 to 1,20,000 population. It has 30 indoor beds, well equipped laboratory and X–ray facility.

Intensified Jawahar Rozgar Yojana (The Second Stream of JRY)

From 1993-94, the Intensified Jawahar Rozgar Yojana is being implemented in those 120 backward districts of 12 states of the country which are badly affected with unemployment and under-unemployment problems. These states are Andhra Pradesh, Bihar, Gujarat, Jammu & Kashmir, Karnataka, Madhya Pradesh, Maharashtra, Orissa, Rajasthan, Tamil Nadu, Uttar Pradesh and West Bengal. **Since January 1, 1996 this plan was merged with Employment Assurance Scheme (EAS).** The main points of this plan were as under—

1. Under this programme, those works are given priority which provide ample employment opportunities, like creating small irrigation facility on barren lands, forestry etc.
2. This plan also includes various activities which create rural infrastructure including primary education institutions.

Innovative and Special Employment Scheme (The Third Stream of JRY)

The third stream of Jawahar Rozgar Yojana which is known as the Innovative and Special Employment Scheme is being implemented since the year 1993–94. In this scheme, special and modern projects are included whose objective is to stop the migration of labour, to encourage the female employment and to develop the productivity in the desert areas. Following are some of the main points of this scheme :

1. Under this plan, the schemes like **Operation Black Board** were also given place which fulfil the main objective of Jawahar Rozgar Yojana. The construction of the school building and of class rooms has been given priority. Operation Black Board scheme was started in 1987.
2. All the projects coming under the third stream are given acceptance by a Screening Committee which is constituted under the chairmanship of the Secretaries of Rural Employment and Poverty Eradicating Departments of the Central Government.

Employment Assurance Scheme (EAS)

At present this programme is not in existence as it is merged in Sampoorna (Gramin Rozgar Yojana since September 25, 2001).

The Employment Assurance Scheme was started from 2nd October, 1993 in 1,778 development blocks in the rural areas of 261 districts. During 1994–95, this scheme was also implemented in 697 development blocks of Drought-Prone Area Programme (DPAP) and Desert Development Programme (DDP). In this way, by December 31, 1995, this programme had been extended to 2,475 development blocks of the country. On January 1, 1996 after merging the second stream of Jawahar Rozgar Yojana (Intensified JRY) in Employment Assurance Scheme, it became operative in 3206 blocks of the country. During 1997–98, this plan was extended to 1123 new blocks and it covered all 5448 rural blocks of the country. The main points of this scheme were as under :

1. The main objective of this scheme was to provide profitable employment of not less than 100 days to every desirous villager of ages between 18 years and 60 years during the lean agricultural season. The secondary objective of the scheme was to create economic infrastructure and community projects for creating sufficient employment and development activities.
2. This programme was reorganised from April 1, 1999 and it was made the single wage employment programme and implemented as a centrally sponsored Scheme on a **cost sharing ratio of 75 : 25.** (Earlier this ratio was 80 : 20 for the centre and state

governments). The Central assistance was provided directly to the District Rural Development Agency or to Zila Parishad.

3. **EAS** was a Demand Driven Programme. That is why, under it no physical target was prescribed.

Rural Landless Employment Guarantee Programme (RLEGP)

Since 1989–90, this programme was merged with Jawahar Rozgar Yojana.

The Rural Landless Employment Guarantee Programme was started in the rural areas on 15th August, 1983, with the objective of creating employment, constructing the productive projects and improving the rural life, but the guarantee part of this programme could however not be implemented due to the lack of resources. The total expenditure of this programme is financed by the Central Government. The resources are allotted to the State/Union Territories on the basis of determined standards, in which 50% weightage is given to the cultivators and marginal farmers on the basis of their number of the remaining 50% weightage is given on the basis of poverty. Under this programme, the wages of the labourers are given according to the Minimum Wages Act. Some part of the wages is given in the form of foodgrains on reduced prices. A Condition was made in the programme that the labour cost amount of any project should not be less than 50% of the total expenditure. Under this programme, contractors were not permitted. Out of the total amount of expenditure 10% had been planned for schedule castes/tribes. Under this programme, the funds for Social Forestry, Indira Awas Yojana and Million Wells Scheme were also allocated.

The Sangam Yojana for the Handicapped

*The Sangam Yojana declared on 15 August, 1996 is one of the various plans related to Social Welfare. Under it, those handicapped who are living in the rural areas are organised in a group. Every organised group named '**Sangam**' is given an assistance of Rs. 15,000 for performing their economic activities.*

CAPART

Council for Advancement of People's Action and Rural Technology (CAPART) was formed on September 1, 1986. CAPART is a registered body under the Ministry of Rural Development. The head office of CAPART is at New Delhi. CAPART has nine Regional committees/Centres at Jaipur, Lucknow, Ahmedabad Bhubaneshwar, Patna, Chandigarh, Hyderabad, Guwahati and Dhanwad. The Regional Communities are empowered to sanction projects proposals to voluntary agencies upto an outlay of Rs. 20 lakh in their respective regions. Its main objective is to encourage and assist the voluntary activities for implementing projects for rural prosperity. Since inception and upto June 2003 CAPART has sanctioned 21621 projects involving an amount of Rs. 701.63 crore and has released Rs. 545.85 crore. Some important features related to it are as under :

1. CAPART extends assistance to Jawahar Rozgar Yojana, Organisation of the Beneficiaries of Poverty Eradicating Programme, Integrated Rural Development Programme. Development of Women and Children in Rural Areas and Other related Organisation.
2. The Rural Development Department of the Indian Government provides the required funds to CAPART.

Rural Electrification in India

Rural electrification involves supply of energy for two types of programmes : (*a*) production oriented activities like minor irrigation, rural industries etc. and (*b*) electrification of villages. Rural Electrification Corporation (*REC*) was established in July 1969 to finance various projects of rural electrification. REC has been providing assistance to the State Electricity Boards for taking up system improvement projects for strengthening of transmission and distribution system and small generation power projects like wind energy and hydel projects. At the time of establishment of Corporation in 1969 only 13% villages were electrified. This number went upto 83.7% in March 2004. Cumulatively, 305064 villages have been electrified and 8207482 pumpset had been energised upto March 2004 out of the total estimated potential energisation of 19.5 million pumpsets. Under Kutir Jyoti Programme over 48.5 lakh single point connections were released at a cost of about Rs. 317 crore to the rural households of families below poverty line by March 2002.

Kutir–Jyoti Programme

The Indian Government started a 'Kutir-Jyoti Programme' in 1988–89 for improving the living standards of the scheduled caste and tribal families, including the rural families who live below the poverty line. Under this programme, a government assistance of Rs. 400 is provided to the families who are living below the poverty line (including Dalits and Adivasi) for providing single point electricity connection in their houses. Under Kutir–Jyoti Programme over 48.5 Lakh single point connections were released and Rs. 317 crore disbursed to the households of rural families below poverty line by March 2002.

Remote Village Electrification

The Ministry of Non–conventional Energy Sources is implementing a programme since 2001–02 for the electrification of remote census villages and all unelectrified remote hamlets through renewable energy means.

More than 24,500 villages/hamlets (which are not likely to be electrified through grid–extension by 2012) were tentatively identified for this purpose. As on 31 March 2004, 1,563 remote villages and 316 remote hamlets were electrified and projects are under implementation in 1,517 remote villages and 721 hamlets. The target for the Tenth Plan period is electrification of 5,000 such villages.

The projects are implemented through State Nodal Agencies for Renewable Energy, Power Departments, Electricity Boards, Corporate Entities for power generation, transmission and distribution setup by the Central or State Governments, Non Governmental Organisations, Cooperative Societies and similar non-profit bodies, District-level bodies, Panchayati Raj Institutions, Village Councils and Private Sector with emphasis being on provision of energy services.

Rural Water Supply and Sanitation

The important points related to this programme are as under :

1. National Drinking Water Mission (NDWM) was established in 1986.
2. In 1991, the name of this Mission was changed to **'Rajiv Gandhi National Drinking Water Mission' (RGNDWM).** The objective of this Mission is to provide safe drinking water in sufficient quantity to the whole rural population in the coming years.
3. In August 1984, this programme was transferred from urban Development Ministry to the Rural Development Department.
4. Under this programme, a provision has been made for spending 35% of the allocated fund for solving the problem of drinking water of the schedule caste/schedule tribe.
5. National Human Resource Development Programme (NHRDP) was introduced in 1994 with the aim of utilising human resources of that particular area to meet requirements of water supply and cleanliness in rural areas and rural people.

6. NHRDP was introduced with a basic aim to train at least one beneficiary (generally a woman) of lower strata in each village.
7. Individual latrine for SC/ST and freed bonded labour living below and poverty line is constructed with average cost of Rs. 2,500. 20% of this cost is shared by local panchayat or the beneficiary and the remaining amount is shared by the Government as subsidy.

Current Scenario : Rural Water Supply & Sanitation

Water is a State subject and the Schemes for providing drinking water facilities are implemented by the States. The Central Government supplements the efforts of the States by providing financial and technical support. The Tenth Plan envisages provision of safe drinking water to all rural habitations. Two major programmes are being implemented to achieve this objective. These are the Accelerated Rural Water Supply Programme (ARWSP) and the Pradhan Mantri Gramodaya Yojana – Rural Drinking Water (PMGY–RDW). As reported in the last survey with an investment of over Rs. 45,000 crore (upto March 31, 2004), considerable success has been achieved in meeting the drinking water needs of the rural population. There are more than 3.7 million hand pumps and 1.73 lakh piped water schemes installed in the rural areas. As on March 2004, 95 percent of rural habitations have been fully covered (FC) and 4.6 percent are partially covered (PC) and 0–4 percent are not covered (NC) with drinking water facilities.

National Social Assistance Programme (NSAP)

This programme has been implemented since August 15, 1995. The three main components of this programme are as under :

1. National Old Age Pension Scheme (NOAPS).
2. National Family Benefit Scheme (NFBS).
3. National Maternity Benefit Scheme (NMBS).

The main points related to the above mentioned schemes are as under :

1. Under National Old Age Pension Scheme there is a provision of giving pension of Rs. 75 per month to the applicant (female or male) who is above 65 years of age and is living below the poverty line.
2. Under the National Family Benefit Scheme, in the case of natural death of the income earning member of the family (male or female in the age group of 18 to 65 years), the suffering family will get a lump sum amount of Rs. 5,000 in the form of Survivor Benefit. In the case of accidental death, this assistance amount will be Rs. 10,000.
3. Under National Maternity Benefit Scheme, a financial assistance of Rs. 500 is given to women of poor families having age of 19 or above at the time of giving birth to first two children. This assistance is for benefiting women at pre and post maternity time.
4. This programme is 100% centrally financed programme which is implemented by the local institutions like Panchayats/Municipalities.

Kasturba Gandhi Education Scheme

Kasturba Gandhi Education Scheme was introduced in the country on August 15, 1997, on the occasion of 50 ears of Independence. This scheme will specially cover districts having lower women literacy rate. A provision of Rs. 250 crore was made in 1997–98 budget for establishing such schools for girl's education.

Urban Unemployment and Poverty Eradication Programme

The unemployment existing in the Urban areas is divided into 2 categories :

Members of Parliament Local Area Development Scheme (MPLADS)

Every Parliament member of both the houses has been given an authority to recommend various development projects for his/her Parliament Constituency to the concerned District Magistrate. Under this scheme, a ceiling of Rs. 10 lakh has been fixed for one such project. A total annual ceiling of Rs. 1 crore was fixed for each MP when the scheme was introduced in Dec. 1993, but on Dec. 23, 1998 it was increased to Rs. 2 crore. The District Magistrate has to investigate at least 10% of the total work done every year under this scheme. This scheme was introduced by the Government to enable the Members of Parliament to actively participate in the development programmes relating to their respective areas. Since the other on going schemes like IRDP, JRY etc. are tagged with specific rules and regulations and thus, leave no favour to Members of Parliament for initiating any development programme independently in the area. Realising this very practical difficulty the then Prime Minister Mr. P.V. Narsimha Rao introduced this scheme on 23rd December 1993. In the beginning, this scheme was implementation by Rural Development Ministry but since October 1994, the implementation of this scheme was transferred to the Department of Programme Implementation till March 31, 1999 a sum of Rs. 3626.38 crore has been released under this scheme. Out of this amount Rs. 2315.40 crore have been spent which 64% of the released amount.

Prime Minister's Rozgar Yojana

The scheme (PMRY) was launched on October 2, 1993 and initially was in operation in Urban areas. From April 1, 1994, the scheme is being implemented throughout the country. It was proposed to establish 7 lakh tiny units in Industry, Service and Trade areas and to create about 10 lakh employment opportunities during the Eighth plan period. During 9th plan period the scheme has been confined in the revised from by the government. Under this scheme every selected educated unemployed youth in the age group of 18–35 years and having family income below Rs. 40,000 is provided a loan of upto Rs. 1 lakh for opening his own enterprise and Rs. 2 lakhs for other activities. Projects involving two or more than two partners may be given a loan upto Rs. 10 lakhs. Under this scheme, 15% of the total project cost (maximum Rs. 15000) is given to the beneficiary as subsidy, 5% of equity is to be invested by the beneficiary himself and the remaining cost of the project is financed by the concerned bank. The entrepreneurs of these tiny units are provided adequate training and also given assistance of raw material and marketing, if required. Micro-enterprises from commercial sector should not comprise more than 30 percent. This scheme is being administered by Union Industry Ministry. SC/ST and other backward classes have been given reservation of 22.5% and 27% respectively.

Mahila Samridhi Yojana

With the objective of providing economic security to the rural women and to encourage the saving habit among them, the Mahila Samridhi Yojana was started on 2 October, 1993. Under this plan, the rural women of 18 years or above age can open their saving account in the rural post office of their own area with a minimum Rs. 4 or its multiplier. On the amount not withdrawn for 1 year, 25% of the deposited amount is given to the depositor by the government in the form of encouragement amount. Such accounts opened under the scheme opened under the scheme are provided 25% bonus with a maximum of Rs. 300 every year.

The Department of women and Child Development, the model agency for MSY, decided in April 1997 that no new MSY accounts should be opened from 1 April, 1997 onwards but the existing accounts could be maintained.

Recently it is decided to merge this scheme with **'Mahila Swayam Sidha Yojan'** a new scheme announced on July 12, 2001.

Twenty Point Programme (TPP)

Under the slogan of ***'Garibi Hatao'*** *Twenty Point Programme was started in 1975. This programme was reconstituted twice in the years 1982 and 1986. The reconstituted Twenty Point Programme 1986 is in operation since April 1, 1987. The twenty points of this programme are as follows —*

1. *Attack on Rural Poverty*
2. *Strategy for Agriculture dependent on Rains.*
3. *Better Utilisation of Irrigation Water.*
4. *Bigger Harvests.*
5. *Enforcement of Land Reforms.*
6. *Special Programmes for Rural Labour*
7. *Clean Drinking Water.*
8. *Health for all.*
9. *Two–Child Norm.*
10. *Expansion of Education*
11. *Justice for Schedule castes and Schedule tribes.*
12. *Equality for Women*
13. *New Opportunities for the Youth.*
14. *Residential houses for People.*
15. *Improvement of the Urban Slums.*
16. *New Strategy for Forestry.*
17. *Protection of Environment.*
18. *Concern for the Consumer.*
19. *Energy for the Villages.*
20. *A Responsible Administration.*

Jan Arogya Bima Policy

Jan Arogya Bima Policy is primarily meant for the large segment of the population who can not afford the high cost of medical treatment. The limit of cover per person is Rs. 5,000 per annum. The premium for an adult individual is Rs. 70 per annum up to 45 years of age. Persons beyond 45 years and up to 70 years are also covered with slightly higher premium. Two dependent children below the age of 25 are also covered for the same compensation amount of Rs. 5,000 per annum per child but at a concessional premium of Rs. 50 per person. The cover provides for reimbursement of medical expenses.

Raj Rajeshwari Mahila Kalyan Yojana

A new policy offering security to women in the age group of 10 to 75 years irrespectively of their income, occupation or vocation was introduced with effect from 19th October, 1998. For a premium of Rs. 15 per annum, the policy provides a cover of Rs. 25,000 for permanent total disablement of the insured women. The policy also provides a cover of Rs. 25,000 for the death of her husband. For the death of an unmarried women, the policy provides a cover of Rs. 25,000 which will be payable to her nominee/legal hair.

Bhagyashree Child Welfare Bima Yojana

A new policy covering a girl child was introduced *w.e.f.* 19th October, 1998. It covers one girl child in a family upto the age of 18 years whose parent's age does not exceed 60 years. The premium per girl child is Rs. 15 per annum. In case of death of both or either of the parents, an amount of Rs. 25,000 would be deposited in the name of the girl child with a financial institution. Fixed annual disbursements to the girl child upto the age of 18 years would be made from the amount to her credit and the balance amount to her credit would be disbursed to her on attaining the age of 18 years. In the event of death of the girl before attaining the age of 18 years, due to accident and surgical operations, the balance amount, standing to the credit of the girl child, will be paid to the surviving parent/guardian.

Indira Mahila Yojana

From August 20, 1995 the Union Government launched this plan in 200 development blocks in the country, which was to be extended to the remaining segments after a short interval. The main objective of this plan is to create awareness among the women and to provide the income resources to them. This plan will establish co-ordination among the various plans related to women so that the available funds can be properly used for the welfare of women. Under the Indira Mahila Yojana. Women groups are formed in the villages and urban slums which work with the supports of the Indira Mahila Kendras established at Anganwadi level. Road construction, rural electrification, increasing the non-traditional energy sources, social forestry, education and health Programmes have also been included in Indira Mahila Yojana.

Recently it is decided to merge this scheme with **'Mahila Swayam Sidha Yojana'** a new scheme announced on July 12, 2001.

Balwadi Nutrition Programme

The Balwadi Nutrition Programme was started in the year 1970–71 with an objective to provide full nutrition, entertainment facilities and informal school for providing early education to the children of 3–5 years age. There are 5,053 Balwadis on the village/tribal and urban slums of the country, in which 2.25 lakh children are getting benefit. This programme is being implemented by five voluntary organisation of the National level, to whom the Government provides financial assistance.

Vande Matram Scheme

Maternal mortality in India is estimated to be 407 in 1998 (SRS–RGI). Most of the maternal deaths are due to complications of pregnancy like haemorrhage, anaemia, toxaemia, obstructed labour, unsafe abortions and post-mortum sepsis. These complications if identified and treated well in time, could save the lives of these women.

Vande Matram Scheme which was launched on 9 February, 2004 is a major initiative in public–private partnership with the Federation of Obstetric and Gynaecological Society of India (FOGSI). FOGSI have volunteered to participate in the national endeavour for reducing maternal mortality and improving health of women particularly during pregnancy and child birth. The scheme envisages provision of free outpatient services including antenatal check-up to all pregnant women and family planning counselling to new mothers regularly by the government and private doctors at their facilities on a fixed date.

Mid–Day Meal Scheme for School Children

A few State Governments in the country have implemented Mid–Day Meal Scheme for Children. Since August 15, 1995, the Central Government also decided to implement this scheme to promote the primary education in the country. Under this scheme, children studying in schools run by local bodies and Government aided private primary schools are provide mid–day meal free of cost. If school does not possess the cooking facility, every eligible student is given three kgs of foodgrains per month. Only those students are eligible for the benefit who attend at least 80% of the total school days. The total expenditure of the scheme has been estimated to be Rs. 2080.9 crore. In the very first year of the Mid–Day Meal Scheme 3.4 crore children were provided meal assistance. The aim of the scheme is to provide incentive to 11 crore children of 1–4 classes for attending school regularly for 3 years and also to provide nutritious food to them.

Scheme of Providing Loan for House Construction in the Rural Areas

In the budget of 1997–98, the Central Finance Minister had declared that from August 15, 1997 a special scheme for house construction would be started in the rural areas. In this scheme which is prepared by the National Housing Bank, a loan of Rs. 2 lakh can be provided for constructing a house on one's own land or for repairing an old house on the condition that the recipient of loan must contribute one-third amount of the total cost by his own sources.

Ganga Kalyan Yojana

The Ganga Kalyan Yojana was another scheme sponsored by the Centre. This Yojana was started from Feb. 1, 1997 in all the districts of the country. The objective of this Yojana was to assist the farmers by means of subsidy, maintenance support and loan related arrangement for undertaking minor irrigation schemes covering both surface and ground water. The farmers were provided Rs. 5,000 per hectare. The benefit of this yojana was provided only to the people of targeted group (that is small and marginal farmers below the poverty line). The assistance provided to these people was the mixture of government subsidy and term loans by financial institution. The expenditure incurred on this scheme was divided between the Central Government and the State Government's in the ratio of 80 : 20. Since **April 1, 1999 Ganga Kalyan Yojana has been merged with Swarna Jayanti Gram Swarozgar Yojana.**

Annapurna Yojana

Annapurna Yojana was proposed in 1999–2000 budget proposals. This scheme was inaugurated by Prime Minister on 19th March. 1999 in Sikhera village of Ghaziabad district.

Initially this scheme provide 10 kg foodgrains to senior citizens who were eligible for old age pension but could not get it due to one reason or the other.

Later on this scheme has been extended to cover those people also who get old age pensions. This extension has come into force w.e.f. January 14, 2001. About 68 lakh additional people will be benefited under this extended scheme.

Foodgrain are provided to the beneficiaries at subsidised rates of Rs. 2 per kg of wheat and Rs. 3 per kg of rice. The scheme is operational in 25 states and 5 union territories. More than 6.08 lakh families have been identified and the benefits of the scheme are passing on to them.

Shiksha Sahayog Yojana

The scheme was launched on 31 December 2001, with the object to lessen the burden of parents in meeting the educational expenses of their children. It provides scholarships to students of parents living below or marginally above poverty–line and who are covered under Jan Shree Bima Yojana and are studying in 9 to 12 standard.

A scholarship amount of Rs. 300 per quarter per child is paid for a maximum period of four years and for maximum two children of a member covered under Jan Shree Bima Yojana.

No premium is charged for this benefit. As on 31 March 2004 scholarship were disbursed to 160473 beneficiaries.

Food for Work Programme

This programme was initially launched *w.e.f.* February 2001 for five months and was further extended. The programme aims at augmenting food security through wage employment in the drought affected rural areas in eight States *i.e.,* Gujarat, Chhattisgarh, Himachal Pradesh, Madhya Pradesh, Maharashtra, Orissa, Rajasthan and Uttaranchal. The Centre makes available appropriate quantity of foodgrains free of cost to each of the drought affected States as an additionality under the programme. Wages by the State government can be paid partly in kind (upto 5 kgs. of foodgrains per man day) and partly in cash. The workers are paid the balance of wages in cash, such that they are assured of the notified Minimum Wages. This Programme was extended upto March 31, 2002 in respect of notified "natural calamity affected districts."

Krishi Sramik Samajik Suraksha Yojana (KSSSY–2001)

KSSSY–2001 has been launched in the country since July 1, 2001. Initially this scheme has been introduced in 50 districts of the country. Labour Ministry with Life Insurance Corporation of India have jointly launched this scheme so as to provide social security to agricultural labourers. Persons between age of 15–50 can join the scheme minimum membership of the group at commencement should be 20. More than 30 crores of agricultural labourers will get life and accident insurance as well as pension benefits. The beneficiary under this scheme contributes one rupee per day (*i.e.,* Rs. 365 per year) while the government contributes two rupees per day (*i.e.,* Rs. 730 per year) for every beneficiary.

A sum of Rs. 20,000 will be given in case of natural death while Rs. 50,000 will be given in accidental cases if the age of insured labour is below 60 years. On partial permanent disability due to accident, the amount payable is Rs. 25,000. Besides, a lump sum survival benefit is paid to the member at the end of every 10 year after entry into the scheme. Pension is paid to members on reaching age 60.

Social Security Pilot Scheme for Labours of Unorganised Sector

On January 23, 2004 (*i.e.,* on birthday of Netaji Subhash Chandra Bose) the government has inaugurated a new pilot scheme named a Social Security Pilot Scheme for the welfare of labours of unorganised sector. This scheme was approved by he cabinet on January 7, 2004. This pilot scheme will be introduced in 50 selected district and run through Employees Provident Fund organisation. This scheme will ensure various facilities of family pension, insurance and medical to labours of unorganised sector.

This pilot scheme is a part of proposed unorganised Sector Workers Bill. About 37 crore workers will be benefited after the approval of this bill.

Under this scheme labour of age group 18–35 years has to contribute Rs. 50 per month while Rs. 100 per labour per month will be contributed by the employer. For the age group of 35–50. These amounts will be Rs. 100 lakhs and Rs. 200 respectively.

Khadi and Village Industries Development Scheme

The Khadi and Village Industries provide important opportunities of non-farming employment to the rural people. In the budget for 1995–96, a proposal was made for beginning a new scheme under which the Khadi and Village Industries Commission (KVIC) was to be provided Rs. 1000 crore on a consortium basis through banking system. KVIC grants loans to

Khadi and Gramodyog units directly or through state level khadi and village level boards (KVIB). Centre and the State Governments provide guarantees on loans given by commercial banks to Khadi and Village Industries Board.

Jan Shree Bima Yojana

On August 10, 2000, the government has launched a new scheme for the people living below the poverty line. The scheme is named as Jan Shree Bima Yojana. This scheme was proposed in budget proposals of 2000–2001 by Finance Minister. The scheme has replace Social Security Group Insurance Scheme (SSGIS) and Rural Group Life Insurance Scheme (RGLIS). This newly launched scheme will cover the people of 18 to 60 years both from rural as well as urban sector.

Under this scheme beneficiaries will have life security of Rs. 20000 (in cases of natural death) and Rs. 50000 (in case of accidental death) by paying an annual premium of Rs. 200 only. Beneficiaries living below the poverty line will have to pay only Rs. 100 as premium and the remaining amount of Rs. 100 will be paid from the Social Security Fund of LIC.

National Programme for Education of Girls at Elementary Level (NPEGEL)

The objective of the scheme is to provide additional support to education of girls at the elementary level through the following additional initiatives : (*i*) to develop a school, as a model girl–child friendly school, to the cluster level, (*ii*) to provide additional incentives such as stationary, slates, work books, and uniforms and to meet any other locally–felt need within the existing ceiling of Rs. 150 per child per annum, (*iii*) additional interventions like award to schools/teachers, student evaluation, remedial teaching, bridge courses alternative schools, learning through open schools, teacher training and child care centres at the cluster level within a ceiling of Rs. 60,000 per annum. (*iv*) mobilization and community monitoring with a ceiling of Rs. 95,000 per cluster over a five–year period; (*v*) development of materials, and (*vi*) planning, training and management support.

Various Development and Employment Programme in India—At a Glance

S.No.	*Programme/Plan/Institution*	*Year of Beginning*	*Objective/Description*
1.	Community Development Programme (CDP)	1952	Over-all development of rural areas with people's participation.
2.	Intensive Agriculture Development Programme (IADP)	1960–61	To provide loan, seeds, fertilizer tools to the farmers
3.	Intensive Agriculture Area Programme (IAAP)	1964–65	To develop the special harvests
4.	Credit Authorisation Scheme (CAS)	November 1965	A scheme of Qualitative Credit Control of Reserve Bank
5.	High Yielding Variety Programme (HYVP)	1966–67	To increase productivity of foodgrains by adopting latest varieties of inputs for crops.
6.	Indian Tourism Development Corporation (ITDC)	October 1966	To arrange for the construction of Hotels and Guest houses at various places of the country.
7.	Green Revolution	1966–67	To increase the foodgrains, specially wheat production
8.	Nationalisation of 14 Banks	July 1969	To provide loans for agriculture, rural development and other priority sectors.
9.	Rural Electrification Corporation	July 1969	Electrification in rural areas
10.	Housing and Urban Development Corporation	April 1970	Loans for the development of housing and provision of resources for technical assistance.

S.No.	Programme/Plan/Institution	Year of Beginning	Objective/Description
11.	Scheme of Discriminatory Interest Rate	April 1972	To provide loan to the weaker section of the society at the concessional interest rate of 4%
12.	Employment Guarantee Scheme of Maharashtra	1972–73	To assist the economically weaker sections of the rural society
13.	Accelerated Rural Water Supply Programme (ARWSP)	1972–73	For providing drinking water in the villages
14.	Drought-Prone Area Programme (DPAP)	1973	To try an expedient for protection from drought by achieving environmental balance and by developing the ground water
15.	Crash Scheme for Rural Employment (CSRE)	1972–73	For rural employment
16.	Marginal Farmer and Agriculture Labour Agency (MFALA)	1973–74	For technical and financial assistance to marginal and small farmers and agricultural labour
17.	Small Farmer Development Agency (SFDA)	1974–75	For technical and financial assistance to small farmers.
18.	Command Area Development Programme (CADP)	1974–75	To ensure better and rapid utilisation of irrigation capacities of medium and large projects
19.	Twenty Point Programme (TPP)	1975	Poverty eradication and raising the standard of living
20.	National Institution of Rural Development	1977	Training, investigation and advisory organisation for rural development
21.	Desert Development Programme (DDP)	1977–78	For controlling the desert expansion and maintaining environmental balance
22.	Food for Work Programme	1977–78	Providing foodgrains to labour for the works of development
23.	Antyodaya Yojana	1977–78	To make the poorest families of the village economically independent (only in Rajasthan State).
24.	Training Rural Youth for Self-Employment (TRYSEM)	August 15, 1979	Programme of training rural youth for self-employment
25.	Integrated Rural Development Programme (IRDP)	October 2, 1979	All-round development of the rural poor through a programme of asset endowment for self-employment
26.	National Rural Employment Programme (NREP)	1980	To provide profitable employment opportunities to the rural poor.
27.	Development of Women and Children in Rural Areas (DWCRA)	September, 1982	To provide suitable opportunities of self-employment to the women belonging to the rural families who are living below the poverty line.
28.	Rural Landless Employment Guarantee Programme (RLEGP)	August 15, 1983	For providing employment to landless farmers and labourers.
29.	Self-Employment to the Educated Unemployed Youth (SEEUY)	1983–84	To provide financial and technical assistance for self-employment
30.	Farmer Agriculture Service Centre's (FASC's)	1983–84	To popularise the use of improved agricultural instruments and tool kits.
31.	National Fund for Rural Development (NFRD)	February 1984	To grant 100% tax rebate to donors and also to provide financial assistance for rural development projects

S.No.	*Programme/Plan/Institution*	*Year of Beginning*	*Objective/Description*
32.	Industrial Reconstruction Bank of India	March 1985	To provide financial assistance to sick and closed industrial units for their reconstruction
33.	Comprehensive Crop Insurance Scheme	April 1, 1985	For insurance of agricultural crops
34.	Council for Advancement of People's Action and Rural Technology (CAPART)	September 1, 1986	To provide assistance for rural prosperity
35.	Self–Employment Programme for the Urban Poor (SEPUP)	September 1986	To provide self-employment to urban poor through provision of subsidy and bank credit.
36.	Service Area Account (SAA)	February 1988	A new credit policy for rural areas
37.	Formation of Securities and Exchange Board of India (SEBI)	April 1988	To safeguard the interest of investors in capital market and to regulate share market
38.	Tourism Finance Corporation of India (TFCI)	1989	To arrange the finance for the schemes related to tourism
39.	Jawahar Rozgar Yojana	April 1989	For providing employment to rural unemployed
40.	Nehru Rozgar Yojana	October 1989	For providing employment to urban unemployed
41.	Agriculture and Rural Debt Relief Scheme (ARDRS)	1990	To exempt bank loans upto Rs. 10,000 of rural artisans and weavers
42.	Scheme of Urban Micro Enterprises (SUME)	1990	To assist the urban poor people for small enterprise
43.	Scheme of Urban Wage Employment (SUWE)	1990	To provide wages employment after arranging the basic facilities for poor people in the urban areas where population is less than one lakh
44.	Scheme of Housing and Shelter Upgradation (SHASU)	1990	To provide employment by the means of shelter upgradation in the urban areas where population is between 1 to 20 lakhs
45.	National Housing Bank Voluntary Deposit Scheme	1991	To utilise black money for constructing low cost housing for the poor.
46.	National Renewal Fund (NRF)	February 1992	To protect the interest of the employees of Public Sector
47.	Supply of Improved Toolkits to Rural Artisans	July 1992	To supply modern toolkits to the rural craftsmen except the weavers, tailors, embroiders and tobacco labourers who are living below the poverty line
48.	Employment Assurance Scheme (EAS)	October 2, 1993	To provide employment of at least 100 days in a year in villages.
49.	Members of Parliament Local Area Development Scheme (MPLADS)	December 23, 1993	To sanction Rs. 1. crore per year to every Member of Parliament for various development works in their respective areas through DM of the district
50.	Scheme of Infrastructural Development in Mega Cities (SIDMC)	1993–94	To provide capital through special institutions for water supply, seewage, drainage, urban transportation, land development and improvement of slum projects undertaken in Mumbai, Kolkata, Bangalore, Chennai and Hyderabad.
51.	Scheme of Integrated Development of Small and Medium Towns	Sixth Five Year Plan	To provide resources and create employment in small and medium towns for prohibiting the migration of population from rural areas to big cities.

S.No.	Programme/Plan/Institution	Year of Beginning	Objective/Description
52.	District Rural Development Agency (DRDA)	1993	To provide financial assistance for rural development
53.	Mahila Samridhi Yojana	2 October, 1993	To encourage the rural women to deposit in Post Office Saving Account
54.	Child Labour Eradication Scheme	August 15, 1994	To shift child labour from hazardous industries to schools.
55.	Prime Minister's Integrated Urban Poverty Eradication programme (PMIUPEP)	November 18, 1995	To attack urban poverty in an integrated manner in 345 town having population between 50,000 to 1 lakh
56.	Group Life Insurance Scheme in Rural Areas	1995–96	To provide insurance facilities to rural people on low premium
57.	National Social Assistance Programme	1995	To assist people living below the poverty line
58.	Ganga Kalyan Yojana	1997–98	To provide financial assistance to farmers for exploring and developing group and surface water resources.
59.	Kasturba Gandhi Education Scheme	August 15, 1997	To establish girls schools in districts having low female literacy rate
60.	Swarna Jayanti Shahari Rozgar Yojana (SJSRY)	December, 1997	To provide gainful employment to urban unemployed and under employed poor through self-employment or wage employment.
61.	Bhagya Shree Bal Kalyan Policy	Oct. 19, 1998	To uplift the girls conditions.
62.	Rajrajeshwari Mahila Kalyan Yojana	Oct. 19, 1998	To provide insurance protection to women.
63.	Annapurna Yojana	March 1999	To provide 10 kg. foodgrains to senior citizens (who did not get pension).
64.	Swarna Jayanti Gram Swarozgar Yojana	April 1999	For eliminating Rural poverty and unemployment and promoting self-employment.
65.	Samagra Awas Yojana	1999–2000	For providing shelter sanitation and drinking water
66.	Jawahar Gram Samridhi Yojana (JGSY)	April 1999	Creation of demand driven community village infrastructure
67.	Jan Shree Bima Yojana	Aug. 10, 2000	Providing Insurance Security to people living poverty line
68.	Pradhan Mantri Gramodaya Yojana	2000	To fulfil basic requirement in rural areas
69.	Antyodaya Anna Yojana	Dec. 25, 2000	To provide food security to poor.
70.	Ashraya Bima Yojana	June 2001	To provide compensation to labourers who have lost their employment
71.	Pradhan Mantri Gram Sadak Yojana (PMSGY)	Dec. 25, 2000	To line all villages with Pacca Road
72.	Khetihar Mazdoor Bima Yojana	2001–2002	Insurance of Landless Agricultural workers
73.	Shiksha Sahyog Yojana	2001–2002	Education of Children Below Poverty Line
74.	Sampurna Gramin Rojgar Yojana	Sept. 25, 2001	Providing employment and food security.
75.	Jai Prakash Narain Rojgar Guarantee Yojana	Proposed in 2002–03 Budget	Employment Guarantee in most poor distts.
76.	Valmiki Ambedkar Awas Yojana (VAMBAY)	Dec. 2001	Constructing Slum houses in urban areas
77.	National Slum Development Programme	Aug. 1996	Development of Urban Slums.
78.	Social Security Pilot Scheme	Jan. 23, 2004	Scheme for labours of unorganised sector for providing family pension, insurance and medical.

S.No. Programme/Plan/Institution	Year of Beginning	Objective/Description
79. Vande Matram Scheme	Feb. 9, 2004	Major initiative in public–private partnership during pregnancy check-up
80. National Food for Work Programme	Nov. 14, 2004	Programme to intensify the generation of supplementary wage employment.

Selected Programmes at Elementary Education State

- ***District Primary Education Programme :*** *Launched in 1994 in 42 district, this is a Centrally sponsored scheme aiming at providing access to primary education for all children, reducing primary dropout rates to less than 10 percent increasing learning achievement of primary school students by at least 25 percent and reducing gender and social groups to less than 5 percent. During 2001–02 the programme has been further expanded to 23 districts bringing the total coverage to 271 districts spread over 18 States.*
- ***National Programme of Nutritional Support to Primary Education (Mid-Day meal Scheme) :*** *Launched on 15th August, 1995 with the objective to boost the Universalisation of Primary Education by impacting upon enrolment, attendance, retention and nutritional needs to children studying in classes I—V. Under this programme, more than 10 crore children are being targeted for coverage. At present, Gujarat, Kerala, Tamil Nadu, Madhya Pradesh, Chhattisgarh, Orissa, Karnataka and Delhi are providing cooked meals. The remaining State/UTs are distributing foodgrains (wheat/rice).*
- ***Lok Jumbish :*** *This project in 1992 and has completed in two phases upto June 1999. This programme is being implemented in Rajasthan and has show a positive impact of micro-planning and school mapping process through community support. Year 2001–02 is the third year of phase III of the project. The project is spread over 13 districts of the State, covering 101 blocks.*
- ***Pradhan Mantri Gramodaya Yojana (PMGY) :*** *This programme was launched during 2000–01 and envisages Additional Central Assistance (ACA) for basic minimum services in certain priority areas. The scheme has six components covering elementary education, primary health, rural shelter, rural drinking water, nutrition and rural electrification. A minimum of 10 percent of ACA for all components except nutrition (for which it is 15 percent) has been fixed. The allocation for the remaining 35 percent of ACA would be decided by the States and UTs among the components of the scheme, as per their priorities. Funds for elementary education sector under PMGY are utilized to further the goal of Universalisation of Elementary Education.*

Chapter 2

FISHERIES ACTS, LAWS AND ISSUES

COASTAL AQUACULTURE AUTHORITY ACT, 2005

The following Act of Parliament receive the assent of the President on tie 23rd June, 2005, and has been published for general information by the Ministry of Law and Justice (Legislative Department).

The Coastal Aquaculture Authority Act, 2005
No. 24 of 2005 (23rd June, 2005)

"An Act to provide for the establishment of a Costal Aquaculture Authority of regulating the activities connected with coastal aquaculture in the coastal areas and for matters connected therewith or, incidental thereto."

But it enacted by Parliament in the Fifty-sixth Year of the Republic of India as follows;-

CHAPTER I : PRELIMINARY

1. (1) This. Act may be called to Coastal Aquaculture Authority Act 2005.
   (2) Provisions of section 27 shall come into force at once and the remaining provisions of this Act shall come into force on such date as the Central Government may, by notification in the Official Gazette appoint.
2. (1) In this Act, unless the context otherwise requires,-
   (*a*) Authority means the Coastal Aquaculture Authority established under sub-section (I) of section 4;
   (*b*) "Chairperson" means the Chairperson of the Authority;
   (*c*) "Coastal aquaculture" means culturing, under conditions is ponds, pens, enclosures or otherwise, in coastal areas, of shrimp, prawn, fish or any other aquatic life in saline or brackish water, but does not include fresh water aquaculture;
   (*d*) "Coastal area" means the area declared as the Coastal Regulation Zone, for the time being, in the notification of the Government of India in the Ministry of Environment and Forests (Department of Environment, Forests and Wildlife No. S.O.114 (E), dated the 19th February, 1991 and includes such other area as the Central Government may, by notification in the Official Gazette, specify;
   (*e*) "Member" means the member of the Authority appointed under sub-section (3) of section 4 and includes the Chairperson and the member-secretary;
   (*f*) "Prescribed" means prescribed by rules made under the Act.
   (*g*) "Regulations" means the regulations made by the Authority under this Act.

(2) Words and expressions used herein and not defined but defined in the Environment (Protection) Act, 1986 shall have the meanings respectively assigned to them in that Act.

CHAPTER II : GENERAL POWERS OF CENTRAL GOVERNMENT

3. The Central Government shall take all such measures as it deems necessary or expedient for regulation of coastal aquaculture by prescribing guidelines, or ensure that coastal aquaculture does not cause any detriment to the coastal environment and the concept of responsible coastal aquaculture activities to protect the livelihood of various sections of the people living in the coastal areas.

CHAPTER III : THE COASTAL AQUACULTURE AUTHORITY

4. (1) With effect from such date as the Central Government may, by notification in the Official Gazette, appoint in this behalf, there shall be established for the purposes of this Act an Authority to be called the Coastal Aquaculture Authority.

(2) The head office of the Authority shall be at such place as the Central Government may decide.

(3) Authority shall consist of the following members who shall be appointed by the Central Government, namely:-

(*a*) The Chairperson who is, or has been, a Judge of a High Court;

(*b*) One member who is an expert in the field of coastal aquaculture;

(*c*) One member who is an expert in the field of coastal ecology nominated by the Department of Ocean Development of the Central Government;

(*d*) One member who is an expert in the field of environment protection or pollution control nominated by the Department of Ocean Development of the Central Government;

(*e*) One member to represent the Ministry of Agriculture of the Central Government;

(*f*) One member to represent the Ministry of Commerce of the Central Government;

(*g*) Four members to represent the coastal States on rotation basis;

(*h*) One member secretary.

(4) The term of office of the Chairperson and every other member shall be three years.

(5) The salaries and allowances payable to, and the other terms and conditions of service of, the members shall be such as may be prescribed.

5. A person shall be disqualified for being appointed as a member if he-

(*a*) Has been conceited and sentence to imprisonment for an offence which, in the opinion of the Central Government, involves moral turpitude; or

(*b*) Is an undischarged insolvent; or

(*c*) Is of unsound mind and stands so declared by a competent court, or

(*d*) Has, in the opinion of the Central Government, such financial or other interest in the Authority as is likely to affect prejudicially the discharge by him of his functions as a member.

6. Subject to sub-section (5) of section 4, any person ceasing to be a member shall be eligible for reappointment as such member for not more than two consecutive terms.

7.

(1) The Authority shall meet at such times and places and shall observe such rules of procedure in regard to the transaction of business at its meeting (including the quorum thereat) as may be specified by regulations.

(2) If for any reason the Chairperson is unable to attend to any meeting of the Authority any other member chosen by the members present at the meeting shall preside at the meeting.

(3) All questions which come up before any meeting of the Authority shall be decided by a majority of votes of the members present and voting and in the event of an equality of votes, the Chairperson or in his absence the person presiding shall have and exercise a second or casting vote.

8. No act or proceeding of the authority shall be invalidated merely be reason of -

(*a*) Any vacancy in, or any defect in the constitution of, the Authority;

OR

(*b*) Any defect in the appointment of a person acting as a member of the Authority; or

(*c*) Any irregularity in the procedure adopted by the Authority not affecting the merits of the case.

9.

(1) For the purpose of discharging its functions, the Authority shall appoint such number of officers and other employees as it may consider necessary on such terms and conditions as may be specified by the regulations.

(2) The Authority may appoint, from time-to-time, any person as adviser or consultant as it may consider necessary on such terms and conditions as may be specified by the regulations.

10. All orders, decisions and other instruments of the Authority shall be authenticated under the signature of the chairperson or nay other member or any officer of the Authority authorised by the Chairperson in this behalf.

CHAPTER IV : POWERS AND FUNCTIONS OF AUTHORITY

11. (1) Subject to any guidelines issued by the Central Government under section 3 the Authority shall exercise the following powers and perform the following function namely:-

(*a*) To make regulations for the construction and operation of aquaculture farms within the coastal areas;

(*b*) To inspect coastal aquaculture farms with a view to ascertaining their environmental impact caused by coastal aquaculture;

(*c*) To register coastal aquaculture farms;

(*d*) To order removal or demolition of any coastal aquaculture farms which is causing pollution after hearing the occupier of the farm; and

(*e*) To perform such other function as may be prescribed.

(2) Where the Authority orders removal or demolition of any coastal of sub-section (1) the workers of the said farm shall be paid such compensation as may be settled between the workers and the management through an authority consisting of one person only to be appointed by the Authority and such authority may exercise such powers of a District Magistrate for such purpose, as may be prescribed.

12. Subject to any rule made in this behalf, any person generally or specially authorized by the Authority in this behalf, may wherever it is necessary to do so for any purpose of this Act, at all reasonable times, enter on any costal aquaculture land, pond pen or enclosure and -

    (*a*) Make any inspection, survey, measurement, valuation or inquiry;

    (*b*) Remove or demolish any structure therein; and

    (*c*) Do such other acts or things as may be prescribed;

    Provided that no such person shall enter on any coastal aquaculture land, pond, pen, or enclosure without giving the occupier of such aquaculture land, pond or enclosure at least twenty-four hours' notice in writing of his intention to do so.

13. (1) Save as otherwise provided in this section, no person shall carry, or caused to be carried on, coastal aquaculture in coastal area or traditional coastal aquaculture farm which lies within the Coastal Regulation Zone referred to in subsection (9) and is not used for coastal aquaculture purpose on the appointed day unless he has registered his with the Authority under sub-section (5) or in pursuance of sub-section (9), as the case may be.

    (2) Notwithstanding anything contained in sub-section (1) a person engaged in coastal aquaculture, immediately before the appointed day, may continue to carry on such activity without such registration for a period of three months from that day and if he make an application for such registration under sub-section (4) within the said period of three months, till the communication to him of the disposing of such application by the Authority.

    (3) The registration made under subsection (5) or in pursuance of subsection (9)-

    (*a*) Shall be valid for a period of five years;

    (*b*) May be renewed from time-to-time for a like period; and

    (*c*) Shall be in such form and shall be subject to such conditions as may be specified by the regulations.

    (4) A person who intends to carry on coastal aquaculture shall make an application for registration of this farm before the Authority in such form accompanied with such fees as may fee prescribed for the purpose of registration under subsection (5).

    (5) On receipt of an application for registration of a farm under subsection (4), the Authority shall consider the application in the prescribed manner and after consider the application either register the farm or reject the application:

    Provided that the Authority shall not reject the application without recording the reason for such rejection.

    (6) The Authority shall, after registering a farm under sub-section (5), issue a certificate of registration-in the prescribed form to the person who has made the application for such registration.

    (7) In the case of a farm comprising more than two hectares of water spread area, no application for registration to commence any activity connected with coastal aquaculture shall be considered under sub-section (5) unless the Authority after making such inquiry as it think fit, is satisfied that registration of such farm shall not be detrimental to the coastal environment.

    (8) Notwithstanding anything contained in this section–

(*a*) No coastal aquaculture shall be carried on within two hundred meters from High Tide Lines; and

(*b*) No coastal aquaculture shall be carried on in creeks, rivers and backwaters within the Coastal Regulation Zone declared for the time being muter the Environment (Protection) Act, 1986:

(9) Provided that nothing in this sub-section shall apply in the case of a coastal aquaculture farm which is in existence on the appointed day and to the non-commercial and experimental coastal aquaculture farms operated or proposed to be operated by any research institute of the Government or funded by the Government;

Provided further that the Authority may, for the purposes of providing exemption under the first proviso, review from time to time the existence and activities of the coastal aquaculture and the provisions of this section, shall apply on coastal aquaculture farms in view of such review.

Explanation - For the purpose of this sub-section, "High Tide Line" means the line on the land up to which the highest water line reaches during the spring tide.

(*d*) Notwithstanding anything contained in this section any traditional Coastal aquaculture farm which lies within the Coastal Regulation Zone declared by the notification of the Government of India in the Ministry of Environment and Forests (Department of Environment Forest and wildlife) No, S.O. 114 (E) dated the 19th February, 1991 and is not used for coastal aquaculture purposes on the appointed day shall be registered under sub-section (5) by producing before the Authority, by this person who is the owner of such farm, documentary proof of such ownership failing which such farm shall not be roistered under sub-section (5) and if such person after such registration does not utilise farm, within one year, registration shall be by cancelled the Authority.

(10) A person, who intends to renew the registration of a farm–made under sub-section (5) or in pursuance of sub-section (9) may make an application within two months before the expiry of such registration to the Authority in the prescribed form accompanied with the prescribed fees and the Authority shall, after receiving such application, renew the registration and for such purpose make an entry with its seal on the registration certificate relating to such form issued under sub-section (6).

(11) The Authority may refuse to renew the registration of a farm under sub-section (10) if the Authority is satisfied that the person to whom such registration is made has failed to utilise such form for coastal aquaculture purposes or without any reasonable cause has violated any provision of this Act or the rules and regulations made thereunder or any direction or order made by the Authority in pursuance of section 11:

Provided that such refusal to renew the registration shall not be made without providing such person an opportunity of being heard.

Explanation 1. For the purposes of this section, 'appointed day' means the date of establishment of the Authority.

Explanation 2. For the removal of doubts it is hereby declared that the expression "renew the registration" used is sub-sections (10) and (11) shall be construed to included further renewal of the registration.

14. If any person carries on coastal aquaculture or traditional coastal aquaculture or causes the coastal aquaculture or traditional coastal aquaculture to be carried on in contravention of sub-section (1) of section 13, he shall be punishable with imprisonment for a term which may extend to three years or with fine which may extend to one lakh rupees, or with both.
15. No court shall take cognizance of an offence under section 14 without a written complaint filed by an officer of the Authority authorised in this behalf by it.

CHAPTER V : FINANCE ACCOUNTS AND AUDIT

The Central Government may after due appropriation made by Parliament, by law, in this behalf pay to the Authority in each financial year such sums as may be considered necessary for the performance of functions of the Authority under this Act.

(1) The Authority shall have its own fund and all sums which may from time-to-time, be paid to it by the Central Government and all the receipts of the Authority (Including any sum which any State Government or any other authority or person may hand over to the Authority) shall be credited to the fund and ail payments by the Authority shall be made therefrom.

(2) All moneys belonging to the fund shall be deposited in such banks or invested in such manner as may subject to the approval of the Central Government, be decided by the Authority.

(3) The Authority may spend such sums as it thinks fit for performing its functions under this Act, and such sums shall be treated as expenditure payable out of die fund of the Authority.

Ministry of Law and Justice (Legislative Department)
The Produce Cess Laws (Abolition) Act, 2006

No. 46 of 2006

Dated : September 25, 2006

An act to repeal the Agricultural Produce Cess Act, 1940 and the Produce Cess Act, 1966.

BE it enactedby Parliament in the Fifty-seventh Year of the Republic of India as follows :

1. This Act may be called the Produce Cess law (Abolition) Act, 2006. — Short title
2. The Argicultural Produce Cess Act, 1940 is hereby repealed. — Repeal of Act 27 of 1940
3. The Produce Cess Act, 1966 is hereby repealed — Repeal of Act 15 of 1966
4. (1) The repeal by this Act of any enactment shall not, — Saving
   (a) affect any other enactment in which the repealed enactment has been applied incorporated or referred to;
   (b) affect the validity, invalidity, effect or consequences of anything already done or suffered, or any right, title, obligation or liability already acquired, accrued or incurred or any remedy or proceeding in respect thereof, or any release or discharge of or from any debt, penalty, obligation, liability, chaim or demand, or any indemnity already granted, or the proof of any past act or thing;
   (c) affect any principle or rule of law, or established jurisdiction, form or course of pleading, practice or procedure, or existing usage, custom.

India has a traditional base in aquaculture. Mixed farming in impoundments is in vogue. Growth of commercial aquaculture in the country and the present status is mainly due to the efforts of the Marine Products Export Development Authority with its promotional programmes through the Offices established in all the maritime states. The Aquaculture Section of MPEDA started functioning since. 1978. Since 1982 selective stocking and supplementary feeding were popularized in the traditional farming areas. In the year 1985 MPEDA formulated financial assistance schemes for promoting aquaculture. One of the most popular schemes of MPEDA is that of 25% financial assistance for Development of new farms subject to a maximum of Rs 2.5 lakhs per beneficiary. Though not many people have availed subsidy assistance, the combined effort of financial assistance and technical assistance of MPEDA resulted in development of more than 190000 ha of farms. In the year 1988 MPEDA promoted two hatcheries namely the Orissa Shrimp Seed Production and Research Centre (OSSPARC) and Andhra Pradesh Shrimp Seed Production, Supply and Research Center (TASPSRS) for production of P. monodon seeds required for stocking the scientific farms. As a result of these development today we have more than 361 hatcheries involved in production of almost 15 billion seeds of PL 15 and above. This includes the hatcheries involved in production of *Macrobrachium rosenbergii.* Maximum number of hatcheries were developed before the year 1996. As technology for production of shrimp was not popular during 90s, MPEDA through TASPARS took up a demonstration production successfully and produced 4.2 tonnes of black tiger shrimp/hectare/crop. In 1993-94 farming the brackish water shrimp in direct sea water of higher salinity was taken up by entrepreneurs. As a result maximum area came under farming of black tiger shrimp in 90s. MPEDA has been involved in every aspect of shrimp farming with the entrepreneurs including disease management and intervened in the management of shrimp diseases in 1994 and contained the disease. Subsequently to ensure quality seeds for aquaculture, MPEDA introduced financial assistance schemes for PCR laboratories. For sustainable and environment friendly aquaculture MPEDA introduced schemes for effluent treatment systems in shrimp farms. As a result today (2005-06) 191074 ha have been developed out of which about 140500 ha have been under production producing a total quantity of about 143000 tonnes of tiger shrimp a productivity of around 1000 kgs/ha annum. Scampi culture also picked up recently in the country. In the year 2005-06, it is estimated that 51 584 ha has been developed under farming out of which about 43500 ha was production producing about 43000 tonnes of scampi with a productivity of around 1000 kgs/ha/annum.

Table 1 : Status of shrimp and scampi culture in various maritime states of India (as on 31.3.2006)

State	*Area under (Ha)*	*% of Total Area*	*Production (MT)*	*% of Total Production*	*Productivity (HT/Ha/ Yr.)*
Andhra Pradesh	91268	49.56	107772	57.94	1.18
West Bengal	54758	29.74	46087	24.78	0.84
Orissa	11560	6.27	10419	5.60	0.90
Tamil Nadu	5452	3.0	7641	4.10	1.40
East Coast Total	163038	88.54	171919	92.43	1.05
Kerala	15122	8.21	7247	3.89	0.47
Karnataka	3406	1.85	1886	1.01	0.55
Gujarat	1337	0.72	3362	1.80	2.51
Maharashtra	881	0.48	917	0.49	1.04
Goa	331	0.18	659	0.35	1.99
West Coast Total	21077	11.44	14071	7.56	0.68
Total	**184115**	**100.00**	**185990**	**100.00**	**1.01**

As the supplementary feed in a major input in farming, maximum feed mills were also established around 1993. There are 28 feed mills as on March, 2006 with an annual production capacity of 150000 tonnes of palletized feed. The average annual production capacity of majority of feed mills in 5000 tonnes. It is estimated that only above 60% of the installed capacity is utilized.

The details of state-wise area under shrimp and scampi farming, % of total area, total production, % of total production and productivity are given in Table 1.

Though the potential area available on east and west coast is almost equal, the development on the east coast is more to the tune of above 25% of the available potential area while in the west coast the development is only above 5% of the area and therefore MPEDA is concentrating on developing the undeveloped area vast area available in the state of Gujarat. The attempts are yielding results.

Table 2 : Details of hatcheries, feed mills, PCR labs and LCMSMS (as on 31/3/2006).

State	*No*	*Hatcheries Capacity in Million*	*No*	*Feed mills Capacity in LMT*	*PCR Labs*	*LCMS MS Labs*
Gujarat	2	45	Nil	Nil	1	1
Maharashtra	8	245	Nil	Nil	2	1
Karnataka	14	321	Nil	Nil	4	1
Kerala	29	537	1	0.20	11	2
Tamil Nadu	83	3078	1	0.40	21	1
Andhra Pradesh	199	9735	25	0.85	41	3
Orissa	15	475	Nil	Nil	5	1
West Bengal	11	166	1	0.05	2	1
Total	**361**	**14700**	**28**	**1.50**	**87**	**11**

Details of State-wise facilities of hatcheries, feed mills, PCR labs developed are given in the Table 2.

Table 3 : National Residue Control Plan for 2006

Programme	*Batches*	*Persons*
HACCP Principles & Traceability System	39	1053
Sanitation Control Procedures	01	21
Verification & Audit	04	61
Sanitation Control Procedures (AFDO)	01	40
FAO/INFOFISH Training on Audit	03	120
Lead Auditor Training (QAS)	02	18
Audit Training (NMFS)	01	51
HACCP Training in Aquaculture	07	210
Training on HACCP in Aquaculture	01	58
Total	**59**	**1632**

The areas which needs to be addressed for increasing the production of shrimp in India are, containing recurrence of sporadic shrimp disease, reduction in the cost of shrine/scampi production, providing adequate credit facility for new farm development, providing insurance cover for shrimp/scami farms, revival of abandoned shrimp farms by meeting the immediate requirements etc.

Processing and Quality Control

For export processing, adequate number of seafood processing plants have been established in the country. As on 31.03.2006, 371 seafood processing plants were in operation with registration form MPEDA. The seafood processing plants are of world class and are implementing HACCP. Out of the 370 seafood processing plants, 170 are having approval for export of fish and fishery products to the European Union. As Hazard Analysis and Critical Control Point (HACCP) system of quality assurance is mandatory for export to the major markets, MPEDA took special efforts to ensure adequate trained personnel for seafood HACCP implementation. The details of HACCP training conducted and the number of persons trained are given in Table 3.

Apart from meeting the standards required for establishments, in order to the export to the European Union, India has to monitor the residue level of stilbenes, stilbene derivatives and esters, steroids, compounds in Annexure IV of 2377/90, Antibacterial substances, Antihelmintics, Organochlorine compounds including Pc Bs, Chemical elements, Mycotoxin, Dyes, as per the directive 96/23/EC. The National Residue Control Plan (NRCP) covers aquaculture products, feed and hatchery water. Samples are taken from farms, hatcheries, processing factories and feed mills. During 2006 it is proposed to analyze 1777 samples for carious parameters. This involves huge investment on sophisticated equipments, employing a good number of technically qualified personnel and huge expenditure on chemicals, glass ware etc. to analyze various parameters more particularly chloramthenicol and nitrofurans. 11 High Performance Liquid Chromatograph with Mass Spectrometer (HPLC MS) have been installed all along the coastal states.

The issues in processing and quality control are huge expenditure for up-gradation of seafood processing establishments, huge investment on analysis on antibiotics, various efforts to counter rejection of seafood from EU on account of bacterial inhibitors, countering micro-biological issues in seafood in EU as there is no harmonization of bacteriological standards among the member countries of EU, muddy and mouldy smell in Japan etc.

Marine Products Export and Issues

The export of marine products has been showing a increase over the period from 1999-2000 to 2005-06 but for some decline in one or two years. India was the 3rd largest supplier of shrimp to Japan, 4th largest exporter of shrimp to USA and the 5th top most supplier of shrimp to Europe, topmost supplier of Cephalopods to Europe during 2004-05. India contributes to 2.58% share of World seafood export trade, stands 16th position in world seafood export. India is the 3th largest fish producing nation next to China and Peru. The major markets for Indian marine products are EU, USA and Japan as seen form Table 4.

Shrimp is the major item contributing 65% of marine products export earnings in the country. It is evident from the Figure - 1 till 1987-88 shrimp export was about 55000 tonnes. From 1988-89 culture shrimp also contribute for export. Share of wild caught remained almost stagnant. Export grew steadily. Quantity of shrimp export was about 140000 tonnes. Cultured shrimp contribution in export was 60% by quantity and 80% by value. This substantial increase has been possible due to the better price for culture shrimp because of the uniform size.

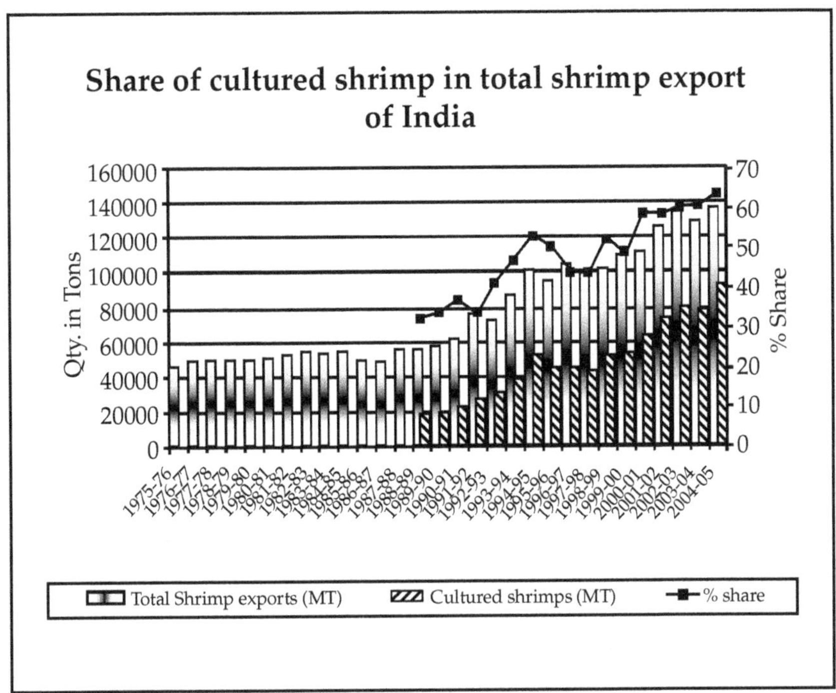

Table 4 : Major Markets for Indian Marine Products 2005-06

Country	*Qty (MTS)*	*Share%*	*Value in million US$*	*Share %*
EU	123736	27	456	29.27
USA	54236	12	368	23.63
Japan	55364	12	255	16.33
China	112050	24	162	10.40
SE Asia	55043	12	126	8.06
M.East	20621	5	66	4.21
Others	36921	8	126	8.10
Total	457490	100	1559	100

India has been able to increase its share in the global fish export from 1.7% in the year 1978 to 2.07% in 2003 as shown in the Table 5.

Some of the important issues in the export trade which need to be addressed with capacity building are bio-terrorism and response act of 2002 of USA, countering the (of weighted average anti-dumping duty of 10.17% with duty of 4.94%, 9.71; 15.36%) for the three mandatory respondents. The additional bond insisted by customs in USA, issues of ecolabeling traceability and risk assessment.

Table 5 : Share of India in global fish export (value in billion US$)

Year	*Global fish export*	*Indian fish export*	*Share %*
1970	2.94	0.05	1.70
1975	6.22	0.12	1.93
1980	15.49	0.28	1.81
1985	17.25	0.31	1.80
1990	35.75	0.48	1.34
1995	52.03	1.08	2.08
1997	53.42	1.29	2.41
1998	51.16	1.17	2.29
1999	52.83	1.11	2.10
2000	55.29	1.43	2.59
2001	56.19	1.26	2.24
2002	58.24	1.42	2.24
2003	63.28	1.31	2.07

A SHIFT FROM RIGHTS TO RESPONSIBILITIES IN THE EEZ

Should Coastal States Re-Evaluate their Commitment to the 1982 United Nations Convention on the Law of the Sea?

Many coastal Sates rely on the resources of their seas, the adjoining seas and the high seas to supply the basic needs of their people. The historic use of the seas and the effect of zonal claims on these uses are a major for these.

Continental shelf claims have tended to dominate the attention of the coastal States to the detriment of the responsibilities expected of coastal States who have ratified the Law of the Sea Convention (LOSC). A look at the rights and the responsibilities of a coastal State in its exclusive economic zone (EEZ) is pertinent.

Exclusive Economic Zone

The EEZ shall not extend beyond 200 nautical miles from the baselines from which the breadth of the territorial sea is measured. The benefits to the coastal State of this zone are related to the exploration, preservation and exploitation of all resources, whether mobile or sedentary, renewable or nonrenewable. Most of the right reserved for other States are concerned with either navigation or communications.

For a coastal State to be able to successfully administer an EEZ, the extent of the zone should be established. The outer limited of the zone are calculated to an maximum of 200 nautical miles form the baselines of the coastal State. It is not practical to claim these zones and expect a master of a vessel to know where the outer limits of the zones are if the coordinates of the limits of the EEZ are not provided by the coastal State. The extent of the EEZ should be given due publicity if the coastal State is to be able to exercise its rights and fulfill its responsibilities, and the depiction any boundary should be such that it is practical for any user.

Under the LOSC, a baseline could be one of the following: the low water line or the normal baseline, straight baseline, closing line of bays or achipelagic baseline.

Straight line may be used to close bays and to replace the low water line as the normal baseline. Lateral and medianline boundaries, between littoral or opposite States, should also be finalised.

Enforceable Laws

Laws that are enforceable by the coastal State include those governing pollution, seaworthiness of vessels, local crimes and civil liabilities. To undertake an assessment of the responsibilities within an EEZ, it is necessary to consider some responsibilities in other zones that impact an EEZ.

Resources

Coastal States have the rights to natural resources in the EEZ. The resources that can be found in the zones are varied and include fish, gas, oil and other hydrocarbons.

However, both the sea and the seabed contain far more than the obvious, and mud, clays, oozes, manganese nodules, polymetallic sulphides and energy sources are factors that should be considered. For a coastal State to be aware of these assets, significant research has to be undertaken by a coastal State or on its behalf.

After territorial integrity, the second most important consideration of a coastal State is usually the exploitation and possible depletion of the resources of the adjoining seas. This includes the technological developments that have made it possible for greater use to be made of the resources found in the sea, whether close to a coast or in and under the high seas.

The coastal State is obligated to promote the utilisation of living resources. This infers that the optimum harvest should be made. And, if for some reason, the coastal State is unwilling or unable to achieve this optimum level of harvesting, it should then make some of these stocks available to landlocked States in the region to harvest.

Marine Scientific Research

Marine scientific research is not defined in the LOSC, but can be accepted to mean any type of research survey or investigation undertaken in marine areas. Marine scientific research is necessary for the exploration, exploitation and protection of resources and the environment, particularly the environment in which the resources are found. The research will include bathymetry, oceanology, oceanography, marine geoscience and precise positioning accuracy.

The determination of the existence and extent of resources could be expensive, so the research to undertake and establish this information would have to be prioritised.

In the EEZ, coastal Sates have sovereign rights for the purpose of exploring and exploiting, conserving and managing the natural resources, whether living or non-living, and any related marine scientific research. The LOSC has a number of general provision which apply. They include that marine scientific research shall be conducted exclusively for peaceful purposes, shall be conducted with appropriate scientific methods and means compatible with this convention, shall not unjustifiably interfere with other legitimate uses of the sea compatible with this convention and shall be duly respected in the course of such use, and shall be conducted in compliance with all relevant regulations adopted in conformity with this convention, including those for the protection and preservation of the marine environment.

Coastal States are also expected to consent to marine scientific research by other States and international organisations provided, certain conditions are met. These include that the research is not related to the exploration or exploitation of the living or non-living resources in the region, and that the research does not involve drilling on the continental shelf, the use of explosives, the use of harmful substances or the construction, operation or use of artificial islands, installations or structures. Information related to projects to be undertaken in an area have to be provided to the coastal State. The information relates to the nature and objective of the project; the methods and means, including the names, tonnage, type and class of vessels and equipment to be used; the precise geographical area; the dates of first arrival or deployment of vessels or equipment and the final departure date; the sponsoring institute (its director and person in charge's names); and the extent to which the coastal State may participate in the project.

If any of this information is inaccurate or not forthcoming, or if there is any other outstanding information, it could be grounds to deny consent. Other provisions include that the coastal State may designate specific areas where exploration will shortly commence and, therefore, decline consent. The coastal State may also undertake projects with international organisations.

Navigation and Communications

The traditional rights of innocent passage and freedom of navigation have has been a marked increase in the rights of a coastal State in that it is now able to exert its jurisdiction beyond the 12-nautical mile territorial sea for the protection of its environment and its resources.

The provision of maritime safety information and the requirement that the coastal State has to manage activities, such as the laying of pipelines or the erection of structures on the seabed in all zones, is an important responsibility of a coastal State. The coastal State has the sole right to construct or authorise the construction of installations and artificial islands in its EEZ. It also has exclures jurisdiction over its own structures and jurisdiction for customs, fiscal health, safety and immigration over all structures. Due notice be given of the construction of such installations, and they should have a permanent means of warning mariners of their presence. Safety zones may also be declared around the structures. While shipping is expected to take the necessary steps to avoid the structures., it is beholden on the coastal State that, where possible these structures are not erected in recognised international sea lanes.

The LOSC provides the right to lay and maintain cables and pipelines in the EEZ. This empowers the coastal State to prevent and control pollution from these pipelines without impeding the laying there of. In the EEZ, other States laying cables and pipelines may not infringe on the rights of the coastal State.

Archipelagic States are obligated to provide acceptable sea lanes through their archipelagic waters. If an archipelagic State is to enjoy the privileges of its status, it will be obliged to ensure that it is turn provide the required support to the vessels passing through its waters.

Freedom of Navigation

Freedom of navigation is defined as the high seas being open to all nations, and that no State may validly purport to subject any part of them to its passage outside of territorial waters and has always been regarded as a right in international customary law. It imposes no restriction on the vessel except that it should not be engaged in either piracy or slave trading, and the vessel may not participate in research, exploration or exploitation in the EEZ of a coastal State without that State's permission.

The sea is being used, to an ever-greater extent, for world transportation, and there is the ever-increasing treat of pollution. Pollution is a major peacetime marine hazard. It can take the form of oil pollution from a maritime casualty, from land-based sources, from nuclear or noxious waster disposal or transportation, or the dumping of material damaging to the environment.

The LOSC defines pollutions of the marine environment and dumping as the introduction by man, directly or indirect, of substances or energy into the marine environment, including estuaries, which result, or is likely to result in such deleterious effects as harm to living resources and marine life. Other definitions included are hazards to human health, hindrance to marine activities, including fishing and other legitimate uses of the sea, impairment of quality for use of sea water and reduction of amenities.

Coastal States have traditionally used the seas and oceans for the disposal of wastes. Little thought was given to the possibility that the oceans were incapable of accommodating the waste. The LOSC requires that a coastal State prevents, reduces and controls pollution of the marine environment from land-based pollution, dumping by both local and foreign parties or pollution from the atmosphere. It is also possible for pollution to occur from other sources, such as activities on the seabed, marine scientific research, the introduction of foreign species, seabed mining and the manner of mining waste disposal.

The coastal State must cooperate with the competent international organisations to establish international rules and regulations to prevent collisions at sea and other damage to vessels that could result in pollution. Routing schemes designed to minimise the threat of collisions are encouraged in the LOSC. Collisions or standings could cause oil, noxious or harmful substances to be lost into the sea. Coastal States are also required to prevent pollution from vessels flying their flag in other States' waters in accordance with international conventions.

Sources of Marine Pollution

The pollution that should be prevented in accordance with the LOSC includes the release of toxic, harmful or noxious waste; the pollution from vessels, deliberately or as a result of accidents and collisions; the pollution from installations, structures and devices used for exploration or exploitation; and the pollution from any other installation, structure or device capable of polluting the marine environment.

Pollution from submarine cables and pipelines usually occurs as a result of an accident. A pipeline can be damaged by trawling, anchoring of vessels and oil rigs, seabed mining and other seabed activities. Pipelines are subject to wear and tear, and poor maintenance could also result in pollution occurring.

The coastal State should adopt laws and regulations within its area of jurisdiction to prevent the pollution from artificial islands, any installation or structure erected to explore or exploit marine resources and any other structure erected in the EEZ or on the continental shelf. Dumping is regarded internationally as being different to the pollution that is caused by shipping, as it is an act for which a trip by a vessel may have been specifically arranged. Of particular concern to coastal States is the disposal, by dumping, of radioactive waster and matter. While States have adopted legislation to prevent this type of dumping, and to control the passage of vessels carrying hazardous material, the major concern is still the threat of accidents to vessels transporting the material.

The LOSC article 216 makes provisions for the enforcement of laws relating to dumping in the various zones and areas. It states that the coastal State will be responsible for enforcement in its zones and on its continental shelf, that the flag State will be responsible for enforcement on the aircraft and vessels of its registry, and that any State, in regard to the loading of wastes and other matter in its territory and at its installations, is not obliged to institute proceedings when another State has done so in accordance with article 211.

In the event of a maritime casualty occurring, even beyond the territorial waters of the State, a State may take whatever actions are necessary to prevent the pollution of its zones or coast line. The State is expected to conduct the necessary inspections while the vessels are in port, and to prevent their sailing if it is considered that a risk of pollution exists. Coastal States may require a vessel in transit to provide information on its identity, port of registry and its last and next part of call if there is evidence that it has been responsible for pollution in that State's territorial waters or EEZ. If the vessel refuses to provide the information, the coastal State has the right to inspect the vessel and, if there is evidence to substantiate the claim, the vessel may be detained and proceedings initiated against it. A coastal State may adopt laws and regulations to prevent, reduce and control pollution of the marine environment through the air space under their sovereignty.

Where a vessel is considered by the competent officers of a State to be unseaworthy and a possible pollution threat to the interests of the coastal State, it may prevent the vessel from sailing to any destination except a convenient shipyard or repair yard. Where the necessary repairs have been affected, the vessel should then be allowed to proceed. It is possible, therefore, for some coastal States to enjoy the benefits of the resources of a large maritime area, which they may have claimed as an EEZ. At the same time, The coastal State has to assume all the responsibilities related to the area claimed and do everything in its power to protect the resources both for itself and for the international community.

Marine Food Fishes of Andaman and Nicobar Islands

The Andaman and Nicobar Islands, comprise over 500 islands, islets and rocky outcrops of various sizes in Bay of Bengal. The Islands are situated between 6° 45′ In latitude and 92° 10′ and 94° 15′ E longitude in a North - South direction. The islands have a land area of 8,293 sq. km. They have a total coastline of 1,962 km which is about one-fourth of the coastline the India mainland. Of the 2 million sq. km. of the exclusive economic zone of India, the area around the Andaman and Nicobar Islands comprises 30%. Being oceanic islands, the continental shelf around them is narrow with an area of about 16,000 sq. km. only. The limited shelf area is comensated by extension of marine habitats into numerous bays, creeks and inlets on the landward side of the islands.

From the fish and fisheries view point, Andaman and Nicobar Islands are the most interesting and fascinating owing to the diversity of Icthyofauna occurring in the varied marine habitats, such as mangroves, creeks, rocky and extensive sand beaches, muddy shores, coral reefs etc.

Marine food fishes are abundantly available in the offshore, inshore and mangrove areas of Andaman and Nicobar Islands.

About 21,550 islanders are currently involved in fisheries. The overall annual fishery yield from waters around the islands for 2001 was estimated by the Department of Fisheries of A & N to be in the order of 28,000 tonnes. There is a general opinion among all stakeholders that the fisheries resource of A & N waters are greatly under-exploited. This is reflected in the weight of individual fish caught. Catches of groupers each weighing 40-50 kg is quire normal Important commercial fish groups associated with these waters include shark and rays, sardine, grouper, snapper, emperor, croaker, sweetlips, mackerel, mullet and tuna. Only artisanal and present and there is almost no commercial exploitation of finish resources by Indian effort in the seas around Andaman and Nicobar Islands.

India ranks seventh among the major fishing nation of the world and it contributes about 45% of the fish landings of the Indian Ocean. The Indian inshore fishes form a part of the Indo-Pacific region. This region covers a vast expanse extending from East Africa and the Red Sea to Northern Australia, Southern Japan, the Hawaiian Islands and the islands of Polynesia. The dispersal of marine forms largely depends upon oceanic temperature, salinity and currents, besides other factors, such as availability of expanses of deep waters, coastline configuration, submarine contours etc. The marine fisheries of these islands are of utmost national importance in augmenting the country's food resources and for fetching a considerable amount of foreign exchange through export of fishes in whole or value-added form to several countries. Both pelagic and demersal fisheries, consisting of a rich variety of fishes, are harvested at present. The level of exploitation is however believed to be no where near maximum sustainability yield.

Oceans cover three-quarters of the earth's surface. Fishes occupy every livable habitat in the sea and adjacent waters: coral reefs, backwaters, mangrove creeks, mud flats, rocy shores, kelp beds and ocean depths. The intertidal or littoral zone of the marine environment is delimited by the tidemarks of low and high water. The littoral zone is broad and provides a distinct habitat characterised by wide fluctuation in temperature and salinity. Non-commercial fishes like gobys, blennys and juveniles of commercial fishes are often found in tide pools, which serve as 'nursey areas' providing food and shelter to these vulnerable juveniles. Beyond the littoral region, the ocean is divided into benthic and pelagic realms. Fishes that live on or near the bottom pollute the benthic habitat. The shallowest part of the benthic realm is the sub littoral or continental shelf zone, extending from the littoral zone out to a depth of 200 meters. High productivity and consequently dense population of fishes characterise the continental shelf zone. Typical species of this sub littoral zone consist of groupers (Serrandiate), croakers (Sciaenidae), stingrays (Dasyaidae), lizardfishes (Synodontidae), rabbit fishes (Siganidae), goatifishes (Mullidae) and flounders (Pleurotectidae). Beyond 200 meters, the bottom usually drops more steeply and there are fewer non-commercial fishes in this zone compared to the sub littoral zone.

The pelagic realm is divided into a neritic zone over the continental shelf and an oceanic zone beyond the continental shelf. The neritic zone is the most productive of all the marine habitats. Much of the fishery resources are harvested form this rich zone. Typical neritic fishes are the snappers (Lutjanidae), kingfishes (Carangidae), sardines (Clupeidae), anchovies.

The trader/importer/processor/restaurants make more money in B grade fish then A grade. This is because they get B grade products at 10 to 12 cents lower prices than A grade. The final fish dish served to the consumer makes no difference between A grade and B grade. For example, let us take the case of Ribbonfish, the largest item of finfish exported to China in terms of quantity. When the fishes are caught in the nets and hauled into the vessel, they struggle for survival and in the process hurt each other with their powerful snouts. So there will be scratches on the body and even burst bellies through which the intestine protrudes. The real attraction for the ribbonfish is its shining silver colour, tasty meat and very cheap price. To meet the very high demand of fish at affordable for the people of china, the country has to import huge quantities of Ribbonfish from various sources. And India is the largest supplier of this much needed fish. That being the situation, the Chinese traders know that Ribbonfish will not be available with 100% silver intact unless it is caught by hook. But the fish caught by hook constitutes only a very negligible portion of the total exports from India. But it is business. They take advantage of the grading system of the fish. If India can produce to eat Chinese special dishes in pouches with Ribbonfish, the packers will get higher prices and our foreign exchange earnings will go fantastically up.

Many reputed Chinese companies are interested in joint venture projects with Indian packers for making such traditional Chinese special dishes from cheap fishes like Ribbonfish/Croaker etc. Interested packers/exporters are advised to explore and exploit the new emerging business-opportunities in China. And within the country there are lot of demand of fish/prawn dishes and products.

PRODUCTION, QUALITY CONTROL AND EXPORT OF FISH AND FISHERY PRODUCTS

In the year 2003 the world fish production was in the order of 132.52 million tonnes. Out of which culture production was 42.30 million tonnes contributing 31.91% of production. It is estimated that about 75% of the production goes for human consumption and 25% goes for non food purposes. The global annual per capita fish consumption (1999-2001 average) was 16.1kg. In India the total fish production was 5.9 million tons out of which 2.2 million tonnes comes from culture. A major portion of culture production is Indian major carps consumed in the domestic market. For export purpose black tiger shrimp and scampi are farmed. Out of the total production 4.87 million tonnes is consumed in domestic market, only a small portion is exported. Estimates show that 0.4 million tonnes of fish is used for non food purpose mainly due to poor handing. The annual per capita fish consumption in India is 4.8 kgs (1999-201 average).

Similar to the trend in global fish production, in India also the sources of production are capture fisheries and culture fisheries. Aquaculture is considered as the only alternative to capture fisheries to sustain fish production. It is known that production form capture fisheries is stagnating over the years. Aquaculture production is contributing substantially to the economy of several developing countries like India, Thailand, Indonesia, etc. Shrimp export alone contributes to around US$ 1 billion in India out of the total export earnings of 1559 million US$.

In the farming sector about 60000 ha is under traditional farming located in the states of West Bengal, Karnataka and Kerala. The remaining 130000 ha are developed scientific farms located in the maritime states like AP, Orissa, Tamil Nadu, Kerala, Karnataka, Maharashtra and Gujarat. In the traditional farms generally there is no selective stocking of seeds. Whatever seeds enter are allowed to grow in the impoundments without artificial feeding. In the scientific farming areas, the species cultured are black tiger shrimp *Penaeus monodon* and scampi, *Macrobrachium rosenbergii* mainly for export purpose. In the scientific farms hatchery produced seeds are stocked and properly managed with good aquaculture practices to produce quality shrimp/scampi.

Capture Fisheries and Issues

A major share of fish for domestic consumption and export is contributed by the mechanized fishing vessels numbering over 53000 operating all along the west and east coast of India. These boats are mostly 32 ft overall length (OAL). Majority of them are trawlers. The remaining more than 100000 non mechanized vessels also contribute to the production from sea. Freezer vessels are a few and there are no factory vessels in the country.

While production from capture fisheries is stagnation there is a scope for increasing capture fisheries by diversified crafts and gears for resource specific fishing such as tuna fishing. While India is known for a good resource of tuna it is not exploited due to lack of adequate fishing vessels, lack of knowledge on tuna capture, lack of technology for sashimi grade tuna processing. These areas need capacity building for increasing production in India.

ADB GRANT FOR FISHING COMMUNITIES IN KERALA, TAMIL NADU

The Manila-based Asian Development Bank (ADB) has approved a $ 5-million grant (approximately Rs. 23.22 crore) to help fishing communities and others affected by the December 2004 tsunami in Kerala and Tamil Nadu restore and diversify their livelihood.

The grant project, from the Japan Fund for Poverty Reduction announced by the Government of Japan, will pilot new ways to generate income in the fishing communities in selected districts that have less risk, are more productive and less vulnerable to natural disasters.

According to a press release issued here from the ADB Media Centre in Manila on Friday, the project will help in setting up a fully serviced fishing village complex in these districts. It will help about 1,000 people and will serve as a model for replication in other districts. The fishing village will support a range of fishing-related activities like fish curing, vending, marketing and trading, and even ice production and supply.

The village will also nurture fish processing and marketing through special outlets, as well as new low-energy fish processing techniques and coastal market infrastructure for hygienic fish marketing. In addition, cooperative retail outlets will be established. The project will restore 1,000 acres of small farmlands, establish small-scale milk dairies with market links, develop organic vegetable farms, establish agro-processing cooperatives, and provide training to improve production process of traditional products.

Environment friendly fish breeding activities, like inland aquaculture farms for breeding high value fish and cage farming in the backwaters, will also be supported to provide long-term jobs for about 1,000 people and increase fish production. "To accelerate economic recovery in the affected areas, there is a need to restore, initiate, and facilitate the development and implementation of sustainable livelihood activities," said A.K. Jorgensen, an ADB Principal Urban Specialist. Fishing is the dominant industry along the coast, so its destruction by the tsunami adversely affected all communities in the area. Many people had fisheries-relate jobs or business that suffered, while the whole local economy experienced a serious slump. The Government, NGOs, and beneficiaries will contribute $700,000 equivalent toward the project's total cost of $5.7 million. The Disaster Management and Mitigation Department in T.N. and the Department of Disaster Management in Kerala are the executing agencies of the project, which will be carried out over three years.

'Developed Countries Must Take Initiative to Push WTO Talks'

India on Tuesday countered allegations that it was responsible for a collapse of the ministerial level meeting to put the stalled WTO talks back on track, with commerce and industry minister Kamal Nath saying that developed countries needed to take the initiative to move the talks forward.

Making it clear that the onus was on developed countries to push the talks forward, Nath ruled out the possibility of India softening its stance to accommodate the demands of US for steep tariff reduction for farm products without adequate cut in agricultural subsidies.

"It is very clear that the developmental content of this round, which was its mandate, has to be redeemed. Rushing towards completion of the round does not mean that we sacrifice on the content," he told a press conference here on Tuesday.

The statement comes a day ahead of Nath's visit to Geneva for consultations with WTO director-general Pascal Lamy, besides bilateral meetings with others members to break the deadlock of current round of talks.

Rehabiliation Package for Seafood Industry - Time Frame to be Formed for Clearing Bad Debts

The members/loanees are required to submit a reasonable, One Time Settlement proposal to the bankers within a given time frame. It was suggested that the members may submit the proposals by the end of September on which banks have to take a decision by the end of December. It was also suggested that the Associations/MPEDA may use their good offices in cases wherever negotiations are needed between banks and members.

However, willful defaulters/cases of fraudulent availment of loans would be deals as per the policy of the banks. It was also suggested that banks may keep in view the report of the committee while taking decisions on one time proposals submitted by individual members.

Vizhinjan Harbour Project to be Completed in Two Years

Minister Shri S. Sharma Fisheries of Kerala said the rehabilitation of fishermen displaced for the project would be carried out is collaboration with other agencies. The Government would hold discussions with local communities to sort out resettlement problems. A special project would be drawn up to relocate families and create hygienic conditions at the Vizlingam Fishing organizations and political scheme.

Market opportunities for Canadian salmon in Korea. Consumption patterns are changing rapidly for certain food products among the younger Koreans; however the consumption of seafood and fish has not changed to any noticeable level. In fact capita consumption of fish is increasing.

The above mentioned Canadian fish species are very popular among importers as Canada has been able to maintain a very positive image in Korea with respect to fish and seafood quality. Canadian suppliers should continue to maintain the current quality standards at reasonable prices. The growth of an affluent middle class in Korea has recently led to a burgeoning market for some seafood products including lobster and hokigai. With the rapid growth of Hotel, Restaurant and Institution (HRI) business in Korea, the demand for frozen low priced finfish species should be strong in the future. The seafood consumption is increasing. Koreans consumed 4.2 million tons of seafood in 2005 (approx. 49 kg per person) - among the highest in the world. The fishing industry is unable to meet domestic demand, and consequently has led to the development of fish farming and import of large amount of fish and fish products from other countries Import of fishery products have increased, this is mainly due to the decreasing fish stock in Korea's coastal waters and the growing domestic demand.

Total seafood imports in 2005 was 1.1 million tons valued at US$ 2 billion, mainly from China, Russia, Japan, USA, Vietnam and Thailand. Korea is an emerging market for Norwegian seafood exporters to Asia. Total Norwegian seafood export to Korea in 2005 was 14.100 tons at a value of US$ 25 million. Norway is the 9th largest exporter of fishery products to Korea. Salmon and Mackerel are the major fish species imported from Norway, hereof salmon import from Norway was 4,869 tons valued at US$ 20 million in 2005. Particularly, Norwegian salmon has a dominant position in the market, taking 99% market share for fresh salmon and about 58% of frozen salmon. Frozen salmon is mainly used as raw material for the smoking industry in Korea.

Norwegian seafood has a very good reputation in Korea related to quality and hygiene standards. This could contribute to further expansion of seafood export to Korea for new species and products, such as cod, king snail, among others. It is the most opportune time for the Indian exporters of marine products to evolve a proper marketing strategy to enter the ever growing, fish hungry Korean market. Although Indian products like frozen shrimp [both IQF/ block frozen and cooked], ribonfish, blue swimming crab/three-spot crab [both whole and half cut], surimin are already popular in Korea, the country's market share is very negligeable compared to the wide reputation and market share enjoyed by Norway, Canada, the US, Russia and China. Indian seafood packers can easily forge strategic business partnership with reputed Korean traders/processors and greatly increase the value and volume of export to that country. High class products that meet the standards of the sophisticated Korean consumers command high prices and Indian exporters can reap a rich harvest.

In recent years new types of O_2 absorbers (scavengers) have been developed like the one that emit the same volume of CO_2 as that of O_2 absorbed. Even microwavable O_2 as that of O_2 absorbed. Even microwavable O_2 absorbers are available. Other intelligent packaging system includes Time Temperature Indicators (TTI), Radio Frequency Identification (RFID) and anti-microbial plastic technologies.

For fish products development of gas indicators like CO2 and O2 sensitive label which can change colour at set concentration of gas is an encouraging development. These intelligent packaging techniques, which communicate the quality of the packed food, are viewed upon as an effective way to ensure quality and safety of food to the user.

Packing Materials for MAP

In the area of packaging material also rapid advances have been made to meet the requirements of modern food industry. As a result of continuous research world wide, packaging material with improved quality and new packaging materials developed with optimal barrier properties.

While selecting packaging materials for MAP the following six characteristics have to be fulfilled.

1. Puncture resistance.
2. Sealing reliability.
3. Anti fogging properties.
4. Cabondioxide permeability.
5. Oxygen permeability.
6. Water transmission rate.

Though a number of packaging material are available for MAP the following four are considered most suited:

1. Poly Vinyl Chloride (PVC)
2. Poly Ethylene Terephthalate (PET)
3. Poly Propylene (PP) and
4. Poly Ethylene (PE)

MAP Techniques

There are two different techniques to create modified atmosphere in a package

1. Gas flushing.
2. Compensated Vacuum

In the former method air inside the package is replaced by a continuous gas stream so that the air in the pack surrounding the product in diluted. The greatest advantage of gas flushing is the speed of the process.

In compensated Vacuum method first the air inside the pack is removed by vacuuming and then breaking the vacuum with the desired gas mixtures. Since this is a two-step process, it is slow and time consuming.

Future Prospects of MAP Products

'Convenience food' and 'ready meal solutions' are the new mantra in the modern food processing industry. With emphasis on convenience, taste, quality and safety, demand for MAP products of microwaving king are going to increase dramatically in the coming years. The MAP products offer enhanced shelf-life, improved hygiene, reducted in-store labour, cleaner operation, product quality and safety.

Consumer attitudes have undergone a sea change in the past two decades Food products are now being perceived as 'meal solutions' rather than 'food items'.

WTO POLICY INDIA MOVES WTO PANEL

The bond and an antidumping duty have brought exports down to $252mn in 2006 from $485mn in 2005.

India has moved the World Trade Organization (STO) disputes panel against a directive by the US customs border protection that seeks a customs bond on shrimp exports to the US.

The bond a cash guarantee given to the US customs border protection for an amount calculated at 100% of the duty payable on total exports during the previous one year, is over and above the antidumping duty of 10.17% imposed on Indian shrimp.

According to the US customs, the bond is to make sure there is enough money in case there is an increase in the duty, which is reviewed periodically. The first of these reviews is under process and a preliminary determination under this review has already raised it to 10.54%.

The bond, coupled with the duty, appears to have taken a toll on Indian shrimp exports to the US, which are down to $252 million (Rs. 1,058.40 core) in 2006 from $485 million in 2005. Even the number of exporters to the US came down from 228, at the start of the duty four years ago, to 74 as on 31 January 2007, the start of second administrative review of the duty.

A.J. Tharakan, national president of the Seafood Exporters Association of India (SEAl), which had taken up the issue before the US court of international trade, confirmed that the government recently filed the papers before the WTO disputes panel. Senior Supreme Court counsel K. Venugopal is to represent India in Geneva when the case comes up for hearing in a month or two.

India had raised the issue before WTO in June 2006 and in October had requested establishment of a panel to look into the matter. Though the US had objected to the formation of a panel, a renewed request saw the WTO constitute a panel on 26 January.

It was against this backdrop that India made its latest move, said Tharakan. Brazil, China, the EU, Japan and Thailand had reserved their right to participate in the panel proceedings as third parties.

The bond is called a "continuous" one since it was valid for multiple transactions for a term, generally a year. India claimed in its submission that the bonds were a barrier to international trade.

The matter had also been taken up before the US Court of International Trade which last month had allowed SEAl's petition. The association had argued that the bond requirement was a violation of international trade practices and not in accordance with the law, said Tharakan.

IRAN BECOMES AQUACULTURE REFERENCE CENTER

Based on the approval of the Network of Aquaculture Centers in Asia-Pacific's (NACA) Governing Council, Iran was selected for the first time as NACA reference center.

Head of the Research Institute of Iran's Fishery, Abbas-Ali Motallebi, told *IRNA* on Saturday two research centers in Iran were selected as NACA reference centers after the comprehensive plan the institute submitted to the 17th governing council meeting of NACA in Thailand.

The two research centers in Iran include the one in Tonkabon for research on cold water fish and another one in Rasth for research on estrogen.

Research plans and projects in the fields of ecology, aquaculture and nutrition, health and diseases, genetic study of Caspian Sea salmon and other species of rare indigenous cold water fish are drawn up at these two research centers, he said,

"Promotion of Iran's cold-water aquaculture industry as well as its unique climatic and ecological conditions among NACA member states account for its more active cooperation with regional and international bodies," Motallebi said, "Besides, the decisive role of the Research Center of Iran's Fishery in policy making and giving proposals on research and executive plans in the field account for its growing reputation on national, regional and international levels."

SHRIMP BONDS ARE UNFAIR, RULES US COURT

India's seafood export industry has received a shot in the aim with the US Court of International Trade's (CIT) preliminary ruling that the additional bonding requirements imposed by the country's Customs and Border Protection (CBP) for shrimp are against the law.

CIT adds that these rules have not been applied fairly, and have ended up imposing an excessive burden on international trade. The problems with the CBP had forced India and other affected countries to take up the matter with the World Trade Organisation (WTO), whose dispute settlement body had appointed a panel to look into the matter.

There is additional support for the Indian cause from the US Government Accountability Office (GAO), an independent, non-partisan body responsible for federal government oversight. In a recent report, the GAO says that the bonds have been inconsistently applied and might impose excessive burden on global trade.

The CIT ruling has been welcomed in Kochi by seafood exporters who feel it will lend support to India's fight at the WTO level against the continuous bonds. The country's marine exports to the USA have been seriously hit by the 10.17% antidumping duty and Customs bond.

'This preliminary ruling will make it much easier to export seafood to the USA, since a bond will not be required now,' says Seafood Exporters' Association of India (SEAI) president AJ Tharakan. 'SEAI has also filed a suit with the CIT, which is yet to be considered. 'The present ruling may positively influence our case.'

The preliminary ruling of CIT was based on a plea by the US National Fisheries Institute NFI), a union of those in the seafood trade, which challenged CBP's requirement that companies importing shrimp subject to anti-dumping duty orders should obtain a continuous entry bond.

It was in August 2004 that CBP implemented the rule asking importers of agriculture or aquaculture products, subject to duties, to post bonds equal to the value of shipments brought in during the course of the previous 12 months.

In December 2005, NFl and 27 of its members moved the CIT, seeking an injunction against the CBP order.

The bond amount, matching antidumping duty cash deposits, was to be over and above the duty paid. This, NFl contended, was effectively a demand for double the amount of security required under the antidumping legislation.

The institute had argued that this was contrary to the law, and that shrimp importers had been uniquely and unfairly singled out for this additional bonding requirement.

The GAO report, which comments on the unfair treatment, also calls for a thorough examination of the impact of CBP policies on the entire imports business in the USA. A final decision on whether the CBP's actions are 'arbitrary and capricious' will be taken later this year.

It is interesting to note that the anti-dumping duty on six shrimp-exporting countries - India, Thailand, Ecuador, China, Brazil and Vietnam-does not appear to have reversed growth in US shrimp imports, although they have certainly hit Indian exports hard.

Statistics show that US shrimp imports in October 2006 alone were a record 63,504 tonnes, 13% higher than in the same month of the previous year. Incidentally, while all shrimp varieties were up, only black tiger shrimp, exported mainly from India, declined by 26%.

Indian exports to the USA in the first half (April to September) of fiscal 2006-07 showed a 23.4% drop. Value realisation in rupee terms plummeted by 16.7%, from Rs. 8.96 billion ($201 million) during the same period last year, to Rs. 7.68 billion ($173 million).

Source : *Seafood International.*

FPI SALE BECOMES MORE LIKELY

At least four offers have been tendered to buy some or all the assets of Fishery Products International Ltd (FPI), the Newfoundland-based seafood company with a division in the USA.

Two other offers came from Newfoundland companies that had previously tried to purchase some FPI assets, such as closed plants on the Burin Peninsula-the Barry Group and Ocean Choice International Inc.

The fourth and latest offer has come from High Liner, formerly National Sea Products, the Lunenberg, Nova Scotia, company that is a rival for some of FPI's business. High Linear operates a plant in Portsmouth, NH, not far from FPI's plant in Danvers, Mass.

Failing processors in 1984 and is still governed by the FPI Act, a law passed to ensure the assets of the company do not leave provincial control.

Newfoundland and Labrador. Fisheries minister Tom Rideout says the government will look most favourably on a 'package deal' not a sell-off of individual assets. He indicated little enthusiasm for the internal group's offer.

John Risley, a member of the FPI board of directors, owns nearly 15% of FPI stock and is the chairman of Clearwater Seafoods, a Nova Scotiabased company.

After the High Liner offer was made, Mr. Risley called it 'a good company' and added that 'the seafood industry is badly in need of consolidation.

FPI has been in negotiations on and off for two years with members of the Fish, Food and Allied Workers union, trying to get plant workers to accept a cut in pay before the company would reopen plants on the Burin peninsula.

The company has also been attempting to restructure its newfoundland fish processing operations to increase efficiency and cut losses-reportedly as high as C$1 million ($852,000) per month. Its 1700 plant workers and 200 trawler operators have been without an agreement since March 31, 2005.

Source : *Seafood International*

Australia ranks as Thailand's sixth-largest shrimp export market. Last year's volume totalled 17,840 tonnes, accounting for 5.1 per cent of the Kingdom's total shrimp export volume, while value increased 2.6 percent to Bt 2.39 billion.

Thailand's largest export market for shrimp is the US, followed by the EU, Japan, Canada and South Korea.

Yuthasak Supasorn, executive director of the institute, said rising exports of shrimp products had prompted Thai farmers to shift from black-tiger shrimp farming to vannamai shrimp of white shrimp, which give a higher yield. So far, vannamai shrimp accounts for 90 per cent of the country's total shrimp farms.

However, white shrimp has a lower export price than black-tiger shrimp, which will directly affect the country's total export value even though export volume has increased.

PRAWN PRODUCTION TO TOUCH RS. 10,000 CR. BY 2012 (A TARGET)

The federation's president IPR Mohan Raju and secretary V Balasubramaniam told that prawn culture in the country, with the exit of all the major corporate players, is entirely a small scale enterprise. More than 1.5 lakh farmers are growing prawns in 1.6 lakh hectares of brackish water areas both on the east and west coasts of the country. The production is about 1.35 lakh tonne and the entire lot is exported.

PFFI estimates that over 12 lakh hectares of fallow saline water area is suitable for prawn culture in the country, (6.5 lakh hectares on the east coast and 5.5 lakh hectares on the west). Of this only 1.5 lakh hectares are being utilised. The federation plans to spur development of at least one lakh hectare more in five years. The investment needed would be Rs. 2,500 crore.

Investments would be needed to increase quality seed output to 10 billion (Rs. 150 crore), and feed production to 2.5 lakh tonne (Rs. 36 crore). The new development and investments would create other one lakh new direct jobs in the rural areas.

There is good scope for developing fresh water prawn culture in over 65,000 hectares with Andhra Pradesh in the lead (25,000 hectares) followed by Maharashtra (19,000 hectares).

EATING SUSTAINABLE SEAFOOD – THREE TIPS TO STEER CLEAR OF FISHERIES COLLAPSE

The world's fish populations are increasingly endangered from over fishing, pollution, and over consumption. A study late last year reported that major fish species, including tuna, scallops, lobster, and flounder, could be effectively extinct by the middle of the century. But fishery collapses are not inevitable, say Brian Halweil, senior researcher at the World watch Institute.

Here are Brian's tips on how to continue enjoying the health benefits of seafood while avoiding fishery depletions and the toxins present in many fish :

1. Eat less of the big fish such as salmon, tuna, swordfish and sharks. These are among the most vulnerable populations, and also the fish that live the longest, have the most fat, and accumulate the most toxins over their lifespan.
2. Eat lower on the marine food chain, including smaller species such as clams, oysters, molluscs, anchovies, and sardines. Smaller species are less endangered because they are more abundant, reproduce faster, and feed lower on the food chain (so they don't consume other fish themselves). They also have less fat and don't accumulate as many toxins as the larger, longer-lived fish species.
3. Keep in mind how fish are caught Some trawling nets are so large they could pull a 747 jet off the ocean floor. Instead, choose fish caught by line, pot, ornet (or other artisanal methods) and avoid trawl-caught fish.

Whether you have health concerns or are concerned about fish populations in general, it's a good idea to eat lower on the marine food chain.

Source : *Worldwatch Institute.*

U.S. PROPOSED CUTS IN WORLD FISHERIES SUBSIDIES

The United States has sent to the World Trade Organization (WTO) a proposal to cut national fisheries subsidies that the United States says contribute to overfishing.

The proposal calls for 'a broad prohibition on subsidies to the harvesting of marine wild capture fisheries' and 'effective disciplines for programs that are not included in the prohibition.' The subsidies that would be banned support activities that contribute to the overcapacity oft he world's fishing fleets, the depletion of many commercial fish stocks, endanger fragile marine ecosystems and distort trade, according to the Office of the U.S. Trade Representative (USTR).

Under the U.S. proposal, such activities as capturing young fish to be raised in pens or famrs, or harvesting unpinned ocean fish to use as feed would be prohibited.

FRESH FISH, SEAFOOD AMONG FASTEST-GROWING FOOD SEGMENTS

AC Nielsen looked at food and beverage segments selling over $1 billion (C771.5 million) to determine which were showing the most growth. Of those, drinkable yoghurt topped the list with 18 percent growth, but right behind was fresh and seafood, with 12 percent growth. "The changing of our world population affects our food and beverage products", the report stages. "As populations age in developed markets, consumers have become more concerned with eating healthier."

Across the globe, fresh, healthy and convenient products showed the bigges jumps.

In Europe, for example, the top growth categories were fresh herbs and spices, up 15 percent); drinkable yogurt, up 14 percent; dairy substitute drinks, up 14 percent; sports and energy drinks, up 13 percent; and fresh fish, shellfish and seafood.

In North America, the top growth category was frozen meal starters, up 95 percent, followed by sports and energy drinks at 51 percent.

In the Asia-Pacific region, frozen meal starters were the top growth segment, up 48 percent, followed by fresh herbs and spices at 47 percent.

In Latin America, frozen meat substitutes were the top category, up 66 percent, with dairy substitute drinks at 40 percent the second-largest.

"Around the world, consumers are balancing health and nutrition concerns with a desire for convenience and value," said Jane Perrin, senior vice president and managing director for AC Nielsen Global Services.

The United Kingdom and United States in particular showed increased consumption of fish and seafood, the report said.

Source : *Infofish.*

TIME TO MONITOR SEAFOOD DEMAND

Increasing consumer demand for seafood is a trend continuing to gain stam in 2007. In certain areas of the world, and for certain species, growth in seafood consumption is near explosive and is contributing to higher seafood prices.

Vietnam increased seafood exports more than 8 percent in January. U.K. seafood consumption rose more than 4 percent in value in 2006. Unilever expects its rate of growth to improve between 3 percent and 5 percent this year, and Norwegian seafood exports are increasing for all species. Growth in exports of farmed salmon for the first five weeks of this year is around 15 percent.

In Norway we have a rather odd phenomenon : While demand for Norwegian farmed salmon is almost outof control, the majority of analysts are more concerned with the fluctuations in salmon prices. Greater emphasis should be given to demand when the salmon industries being analyzed.

The seafood industry has yet to create an instrument that can determine how demand for seafood will develop.

In the Salmon industry, there are good systems and models to assess how many salmon are in sea cages and how much salmon can be expected in the market in a year or two. This is often the background for how share rates are set for stock-listed salmon companies, and subsequently, that is why few, pay much attention to demand.

Growth so far this year is more than 15 percent higher than last year, and is a distinct indication of how growth will develop the rest of the Year. That's completely independent of the fact bird flu was discovered on a turkey farm in the United Kingdom.

Supposing the surge in demand we have see so far this year continues until the end of December, and current prices remain at the same level, we would be talking about a new record year that would overshadow 2006, itself a record year.

There are several factors indicate a tremendous surge in the market for seafood, and growth for salmon will remain at sustainable level.

Seafood has become an alternative to meat because it is regarded as wholesome. There has been a change in direction toward eating fresh fish, but quotas for wild fish species are under pressure.

As the omega-3 wave gains momentum, it will result in added demand for the actual source of this miraculous remedy, seafood.

Regardless, one thing is certain : The industry would be well-advised to develop an analysis system that gives a clearer idea of how much seafood people will eat in the future.

SEAFOOD PACKAGING TO ENSURE FRESH PURCHASES

Dublin based researchers have come up with an innovative solution to fish spoils and designed seafood packaging that contains a sensor which changes colour when seafood has spoiled.

The sensor responds to the presence of basic volatiles responsible for the characteristic rotten fish odor. Its pH-sensitive dye changes from yellow to an orange/red colour that lets customers know if pre-packed fish products are spoiled. The device will also enable suppliers to quickly assess the freshness of their stock and offer a further guarantee of freshness to the consumer.

The innovative solution has been tested in collaboration with the country's biggest seafood producer Oceanpath Ltd, Howth, and supermarket Super Quinn with very positive results.

Source : *Barkeeper.*

POLICY AND ACTS

Seafood Export to Double by 2010

Export of seafood from India is likely to be worth over $4 billion by 2010 from current export estimates of $2 billion, provided the capacity of its fleet of fish catching vessels are expanded by equipping them with more accurate remote sensing tools as also substantially enhancing fiscal assistance to seafood exporters through Marine Products Export Development Authority (MPEDA).

It has been projected that India's seafood exports, which stagnated at $1.6 billion in 2005-06 and moved up to $2 billion in first nine months of current fiscal, have the potential to accelerate faster in view of their growing demand in trading blocs such as the European Union, the United States, Canada and West Asia.

The current fleet of fish catching vessels in the country is less than 50,000 in number with the capacity of each remaining at about 1.5 tonnes.

This capacity needs to be expanded to 5 tonnes to increase fish acreage with the latest remote sensing equipment to exploit the deep shipping potential of India's marine products.

Export to Canada, Tunisia, Puerto rico, Russia, Lithuania, Fiji Island and Bangladesh showed a positive growth, whereas export to Mexico, Cyprus, Australia and the Maldives showed a negative trend. India's major export items include frozen fish, cuttlefish, squid and dried items.

States such as Andhra Pradesh, Tamil Nadu, Kerala, Maharashtra, West Bengal, Gujarat and Orissa have huge marine products potential that needs to be harnessed in a manner that can enhance India's export potential provided all possible incentives and encouragement in terms of policies and finance is given to exporters, according to the paper.

–Seafood Journal

SHRIMP ANTI-DUMPING DUTY

In a bizarre twist to the second administrative review of shrimp anti-dumping duties, yet another organisation in the US, the Louisiana Shrimp Association, has sought review of Indian firms.

While the Southern Shrimp Alliance, the petitioners who originally initiated the anti-dumping duties on Indian shrimp has requested for review of over 250 exporters, the majority of them not exporters to the US, Louisiana Shrimp Association has cast a wider net, seeking review of over 300 companies.

Analysts here see the move of the SSA to have a large number of companies under review mainly to create a panic in the industry here and havoc in the US department of commerce.

The move appears thwarted with the entry of LSA, clearly indicating that any settlement with SSA would not be binding and the DoC would have to consider LSA request too. Extract extra money.

Source : *Financial Express*

NATIONAL FISHERIES INSTITUTE AND U.S. TUNA FOUNDATION TO MERGE

National Fisheries institute (NFI) and U.S. Tuna Foundation (USTF) announced the merger of the two organizations. The combined organization will strengthen the seafood community's ability to educate Americans about the health benefits of seafood. It will operate as a single organization.

'Scientific evidence continues to mount that American families should eat more seafood', said NFI President John Connelly. 'We must educate the public about the importance of increasing fish consumption, including canned and pouch tuna. NFI members welcome Bumble Bee Foods LLC, Chicken of the Sea International, and StarKist Seafood, which sell more than 90 percent of shelfstable tuna enjoyed in this country. By joining forces, the seafood community will amplify our common message for the American consumer : Seafood is essential to a healthy diet.'

USTF President Anne Forristall Luke said, 'USTF members share NFI's goal to help Americans eat more fish by supplying them with affordable, delicious and healthy proteins. As one of the most popular fish enjoyed in the U.S., tuna contributes to growth of the seafool category. The proteins, vitamins and omega-3 fatty acids packed in tuna contribute to heart health and brain development throughout life. Tuna companies look forward to working with our partners in the seafood community to educate the public about the value of our foods to every person's healthy diety.'

NFI will establish a Tuna Council, comparised initially of Bumble Bee, Chicken of the Sea and StarKist. The Tuna Council will focus its energies on issues specific to tuna and be funded through a special assessment of the participating companies.

Connelly will lead the combined organization at NFI's offices in McLean, Virginia. Forristall Luke will help lead the transition and implementation of the merger. NFI and USTF expect the integration to be completed in spring 2007.

RIGHT POLICY PURSUED : HILSHA OUTPUT CAN BE RAISED BY 40 PC

The squeezed 13 percent contribution of hilsha, the national fish of Bangladesh, to total annual fisheries output can be enhanced to 40 percent shortly if a right policy is pursued.

Fisheries officials stressed the urgency of framing hilsha-procurement policy at a time when the government carried on drives in the coastal belts against the annihilation of hilsha fry, commonly known as 'jatka.'

For lack of such policy measures, production of hilsha, country's one of most cherished treasures for its unique taste and economic value, has been on a steady decline mainly for indiscriminate fishing of the under-sized hilsha.

Fisheries Department sources said production will definitely.

PLAN TO PROMOTE POKKALI CULTIVATION

In an attempt to revive the fast-depleting Pokkali cultivation in the district, the district administration, along with the Pokkali Land Development Authority (PLDA), is chalkingout schemes to encourage the traditional, organic mode of paddy cultivation.

The PLDA was aiming to increase the extent of paddy fields cultivated this season in the district, according to its member-secretary Zeenath Akbar. The Pokkali mode of cultivation, which was labour-intensive, had been suffering greatly due to the shortage of workers as well as due to the reluctance of the farmers to cultivate the paddy fields. The cultivation method, which alternated paddy cultivation with prawn farming in the waterlogged fields lying below sea level, had of late been disturbed as the landowners started extending the duration of prawn farming.

UN AGENCY LEADS FIGHT FOR BINDING INTER NATIONAL TREATY TO FIGHT ILLEGAL FISHING

Stepping up the fight against illegal fishing, which is depleting world stocks, the United Nations Food and Agriculture Organization is leading efforts for the adoption by 2009 of a final draft for a legally binding international agreement establishing control measures in ports where fish is landed, transhipped or processed.

The proposed agreement will be based on a voluntary FAO model scheme which outlines recommended "Port State" control measures, which include running background checks on boats prior to granting docking privileges and undertaking inspections in port to check documentation, cargos and equipment. These are widely viewed as one of the best ways to fight illegal, unreported or unregulated fishing.

Fishing without permission, catching protected species, using outlawed types of gear or disregarding catch quotas are among the most common offences.

At a meeting in Rome last week of FAO's Committee on Fisheries (COFI), 131 governments and the European Commission agreed to start a process leading to such a pact. Additional consultations will be held in this year and next to produce a draft version for final approval at COFI's next meeting in 2009.

Illegal, unreported or unregulated fishing undermines good management of world fisheries, has negative impacts on fish stocks, including those upon which poor fishers depend, and implies significant costs both in terms of lost fishing revenue and money spent combating it.

FAO's model scheme recommends training inspectors to increase their effectiveness and improving international information-sharing about vessels with a history of such activities to help authorities turn away repeat offenders.

In meeting also entrusted FAO with drafting technical guidelines on recommended best practices in deep sea fisheries; producing guidelines on the use of protected areas for better management; undertaking a study on the probable impacts of climate change in order to evaluate management and policy responses; and convening a conference on the needs of small-scale fisheries that employ some 34 million people in the developing world.

In addition to delegations from FAO Members, 41 intergovernmental organizations and 29 non-governmental organizations also participate in COFI.

PRAWN FARMERS' BODY TO INTRODUCE EXOTIC SPECIES

To promote and voice the concerns of prawn farmers, an organisation called the Prawn Farmers Federation of India (PFFI) has been formed. It would strive to bring into its fold over a lakh small farmers, their families and several lakh direct and indirect employees.

In a press release, Mr. V. Balasubramaniam, Interim Secretary, PFFI and Joint Secretary, Tamil Nadu Coastal Aqua Federation, said during discussions on issues to be addressed, it was unanimously resolved that the Federation will give top priority to the issue of introduction of exotic species.

Representatives of the federation along with representatives from the All-India Prawn Hatcheries Owners Association and the Sea Food Exporters Association will make a representation to the Union Ministry of Agriculture, to ban the introduction of the exotic P. Vannamei species of prawn in India since it will destroy the thriving black tiger prawn culture.

India is the leading producer of black tiger prawns in the world at present.

India has over 1.5 lakh hectares under prawn cultivation with around 1.2 to 1.4 million hectares potential brackish water area available. More than 91 per cent of the one lakh plus farmers have small scale of less than 2 hectares.

About six per cent of the farmers' own land between 2-5 hectares and the remaining three percent own land over 5 hectares. Total production is around 1.35 lakh tonnes with an average production of less than 1,000 kg per hectare.

Andhra Pradesh is the hub of prawn farming activity in India, while Tamil Nadu is a major player, the release added.

TECHNOLOGICAL CHANGES AND ITS IMPLICATIONS ON THE EMPOWERMENT OF FISHERWOMEN

Agriculture, animal husbandry and fisheries provide maximum employment in the primary sector and form the major source of income and livelihood security of about 65-67 percent of population in India. With the introduction of liberalization policy and thrust for adoption of improved technologies, there has been a spectacular increase in production in all segments and the country has witnessed rapid structural change.

When opportunities and resources are given, women can take an efficient, dynamic and active role for the development of nation. In fisheries women has a pivotal role and are actively involved both in pre-harvest, and post harvest sectors. As in any other filed in this sector also rapid changes have taken place for the last three decades. Technological changes make work easier, faster, challenging less threatening and reduce drudgery. Fisher women spend long hours every day for tedious and mostly unpaid labour-intensive and time-consuming activities. Consequently, there is a real need to develop appropriate technologies for women to reduce their workload in non-remunerative activities as well as to increase and improve the quality of their income-generating works. The active participation of women in fisheries needs to be recognized, as they in fact help to ensure the distributive justice among rural poor and economic stability by contributing sustainability in post harvest fisheries sector. Hence Empowerment of fisherwomen should be given top priority in all fisheries development programmes.

Roles of Fisherwomen in India

Out of the 1.2 million fisherfolk in post harvest sector, women occupy a considerable proportion of more than 0.5 million (Sathiadhas, 1998). Women constitute about 25% of the labour force engaged in pre harvest activities, 60% in export marketing and 40% in internal marketing. Their role in household management is far higher than that of women in other sectors. Majority of workers in the pre-processing and processing plants of shrimp are women. Women also occupy a very good proportion of the work force in export-oriented processing of cuttlefish, lobsters and finfish varieties. Unlike other segments, in processing industry women are required to learn and practice new technology day-to-day, and adapt to changes quickly as per the market requirement.

In Tamil Nadu, women are involved in traditional activities like fish curing, marketing, net making, prawn seed collection and innovative activities like seaweed collection. In Andhra Pradesh, main occupation of women include collecting fish and molluscan shells in addition to their contribution in fish drying, curing, marketing, shrimp processing and net making. The women from other communities are also involved in activities like fish drying and curing in West Bengal. Women play a major role in fish marketing in the state of Maharashtra, while they have a predominant role in handling and processing activities in Gujarat. In Lakshadweep, particularly Minicoy, the major fishery products are produced mainly by women. However, the over all structural changes and technological development and adoption in the marine fisheries sector have dislodged a good proportion of women from the sectors like fish drying, curing, dry fish trade and net making. (Sathiadhas *et al.*, 2005).

Indirect Contribution of Fisherwomen in Marine Sector

In a fisher's family, the responsibility of household management including food, childcare, education, health, sanitation, financial management and getting and repaying the debts are mostly on the shoulders of women members. The burden of her responsibility doubles during the off-season. Mechanisation and consequent increase in number of fishing days has aggravated her problems.

The rise in the cost of living and uncertainty of fish catch from wild has forced fisher folk to look for alternative sources of income. With the advent of technological advances in the field of aquaculture, the new and profitable occupational patterns of fisher folk may be directed towards small-scale aquaculture that can be easily taken up by fisherwomen. For women, projects with special emphasis on fisheries could be started to develop their skills, knowledge, habits and attitude towards adoption of new and economically viable technologies.

Problems Faced by Fisher Women

Women in fisheries sector have to play a multifaceted role and there are numerous problems associated with them. Women face increasing work-loads as men migrate to other areas in search of job. Women's operations are generally small-scale and their incomes are low compared to their male counterparts.

Increase in education status of women can improve productivity and it in turn will improve the household health and nutrition. Women usually work 16-18 hours a day.

INDIA NOT BOUND BY ANY DEADLINE FOR DOHA ROUND CONCLUSION : KAMALNATH

The Commerce and industry minister Kamal Nath says any breakthrough will materialise only if rich countries come up with an offer.

The minister said India has not committed itself to any deadline regarding the conclusion of the Doha Round.

A breakthrough before June-end, stated by World Trade Organisation's (WTO's) director-general Pascal Lamy, will materialise only if developed countries come up with an offer, taking into account livelihood concerns of developing countries, Nath said. WTO talks should not be driven by the expiry of the Trade Promotion Authority of the US on June 30 and the expiry of German presidency of the EU, Nath added.

"If the contents of the offer by rich nations are so good that it is acceptable to India and other developing nations, then we can conclude the Doha Round even next week. But if it is not, then June 30 is no deadline," the commerce minister said.

Echoing the views of Carnegie Endowment's director, trade, equity and development project, Sandra Poleski, Nath said the extension of the Trade Promotion Authority in the US and French polls were their internal matters.

The chief roadblock was that rich nations have not yet made concrete offers on tariffs and farm subsidies, Nath said, adding, India would rather be satisfied with a "no deal" than the consequences of a bad deal. He also stressed India would put its foot down on any deal which would lower its economic growth rate from 9%.

Pointing out that even the Uruguay Round of WTO lasted eight years from 1986-94, Polaski said, "If a decision is not reached even after June 30, the mandate should be renewed to take up the issues more seriously. But if Pascal Lamy decides to suspend talks after June 30, it would be a self inflicted wound on the WTO," she added.

After June 2007, all member nations should discuss the possibility of adopting an approach that is less crisis-driven but more problem-solving, she added.

SEA TURTLES AFFECTED

In India mechanised bottom trawling that involves towing trawl nets along the sea floor is aimed at shrimps due to their economic importance.

"We need to look at other fish resources apart from shrimp, if fishing has to continue as an important source of livelihood," says Dr. Devadasan, Director of CIFT in Cochin. Besides the overexploitation of the target species, the process of shrimp trawling adversely affects the lives of sea turtles, and several other juvenile species of fishes. Studies conducted by CIFT and other fisheries organisations in India and the world over have seen than the shrimp trawl is a non-selective fishing gear.

The non-targeted species go by the name of by catch. An estimated average of 27 million tonnes of by-catch is discarded annually by the world's marine fishing fleets.

In India, studies show that an estimated quantity of 1,30,000 tonnes of by-catch get discarded annually along the east of India.

Although some by-catch products such as catfish, snappers, eels, pomfrets and mackerel have their own lucrative markets the issue of by-catch poses a problem because they comprise mostly of juveniles and young ones. Trawl fisheries in different parts of the world are now required to use By-Catch Reduction Devices (BRD). These devices help exclude the non-targeted species.

Marine Fisheries Act at a Glance

(*a*) The Indian Fisheries Act of No. 4 of 1897 GOI.

(*b*) The Indian Fisheries Act as adopted and applied by State of Saurastra, 1897.

(*c*) The Mysore Game and Fish Preservation Act, 2 of 1901, Government of Mysore.

(*d*) The Game and Fish Protection Regulation Act 12 of 1914, Government of Travancore (1914) (modified 1921).

(*e*) Cochin Fisheries Act 3 of 1917 (modified 1921), Government of Cochin

(*f*) Andaman and Nicobar Islands Fisheries Regulation 1 of 1938.

(*g*) The United Provinces Fisheries Act 45 of 1948.

(*h*) Government of Travancore-Cochin Fisheries Act 34 of 1950.

(*i*) The Maharashtra Fisheries Act 1960 (modified 1962), Government of Maharashtra.

(*j*) The Indian Fisheries (Pondicherry Amendment) Act 18 of 1975.

(*k*) The Indian Wildlife Act 1972. 2 Ib-The territorial waters, continental shelf, EEZ and other Maritime Zones Act, 1972.

(*l*) The Marine Products Export Development Authority Act, 1972

(*m*) The Maritime Zones of India (Regulation of fishing by foreign vessels) Act, 1981

(*n*) The Kerala Marine Fishing Regulation Act and Rules 1980 (Act 10 of 1981)

(*o*) The Goa Marine Fishing Regulation Act, 1980

(*p*) The Maharashtra Marine Fishing Regulation Act 1981, Government of Maharashtra

(*q*) The Orissa Marine Fishing Regulation Act 1981 (Orissa Act 10 of 1982) and the Orissa Marine Fishing Regulation Rules 1983

(*r*) The Tamil Nadu Marine Fishing Regulation Rules 1983

(*s*) The Karnataka Marine Fishing Regulation Act, 1986

(*t*) The The Andhra Pradesh Marine Fishing Regulation Act, 1994

(*u*) The Gujarat Fisheries Act, 2003

(*v*) Andaman and Nicobar Marine Fishing Regulation Act, 2003 (*w*) Lakshadweep Marine Fishing Regulation—Rules, 2004

The marine fishing regulation Acts (a-w above) have been formulated following a model bill circulated by the Government of India to all Maritime state Governments for regulation of exploitation of marine fisheries resources in territorial waters of India. These Acts demarcate fishing zones in territorial waters for fishing by non-mechanized and mechanized fishing vessels. The distance from the shore earmarked for each category varies from state to state. In general, 5 to 10 km is reserved for operation by artisanal (non-mechanized) vessels.

Kerala and Goa were the first to enact the Marine Fisheries Act in 1980, as Andaman and Nicobar islands enacted the Act in 2003 while Lakshadweep did so in 2004.

Andaman and Nicobar islands

(*i*) Vessels up to 30 HP only are allowed to operate up to 10 km

(*ii*) Vessels above 30 HP are allowed to operate beyond 10 km

(*iii*) Every year 15 April to 31 May shall be the closed season for bottom trawlers and vessels engaged in shark fishing

(*iv*) Every year, 1 May to 30 September shall be the closed season for fishing sea shells

(*v*) Fishing nets below 20 mm mesh size are prohibited

(*vi*) Trawl nets of standard mesh size fitted with turtle excluder device alone are permitted

(*vii*) Only gill nets, shore seines and dragnets with mesh size above 25 mm are allowed to operate

Lakshadweep

(*i*) Use of purse seine, ring seine, pelagic, mid water and bottom trawl of less than 20 mm mesh size is prohibited except live bait net

(*ii*) Use of draft gill net of less than 50 mm mesh size and shore seine of less than 20 mm mesh size is prohibited

Fisheries-related policies

Recognizing the importance of coastal ecosystems and the country's reliance on these natural resources, several regulations and notifications have been promulgated by the central and state Governments. The important ones are :

(*i*) Indian Ports Act (1963)

(*ii*) Wildlife Act (1972)

(*iii*) Water Act (1974)

(*iv*) Environment (Protection) Act (1986)

(*v*) General standards for discharge of wastewaters in marine coastal areas (1993)

(*vi*) Notifications declaring certain coastal areas as a marine sanctuary or marine national park

(*vii*) Notification declaring coastal stretches as Coastal Regulatory Zone (CRZ) and regulating the activities in the CRZ (1991, 1994, 1996)

(*viii*) Environment impact assessment notification (1994).

These Acts are meant for regulating fishing in public and private waters through a system of licensing by Governments Authorities. The Ministry of Environment and Forests (MoEF), Government of India, issued the Coastal Zone Regulation (CRZ) notification in February, 1991 under the Environment (Protection) Act, 1986 indicating that the coastal stretches influenced by tidal action up to 500 m from the High Tide Line (HTL) shall be treated as Coastal Regulation Zone where setting up of new industries (except directly relating to water front) are prohibited. Hatcheries have been mentioned excluded.

To protect the corals and coral reefs, there is provision under the powers conferred under Section 10(3) of the Mines and Minerals (Regulations and Development) Act 1957.

Government of India targetted certain policies in previous five years plans but desirable success has not been achieved. The target must be fulfilled in next five year plan.

Chapter 3

INDIA'S FOREIGN TRADE

After independence, Indian foreign trade has made cumulative progress both qualitatively and quantitatively. Though the size of foreign trade and its value both have increased during post-independence era, this increase in foreign trade cannot be said satisfactory because Indian share in total foreign trade of the world has remained remarkable low. In 1950, the Indian share in the total world trade was 1.78%, which came down to 0.6% in 1995. According to the Economic Survey 2004–05 this share percentage of 0.6% continued in years 1997, 1998 and 1999. Since 1970, this share has remained around 0.6% but very slightly increased to 0.8% in 2003. Currently India is the 31st leading exporter and 24th leading importer in world merchandise trade which clearly indicates that Indian has failed to increase its share in the total world trade. Trade policy (2004–09) has set a target of achieving 1.5% share in global trade by 2009.

Table 1 : Indian Share in World Trade

(in %)

Year	*Export*	*Import*	*Trade*
1950	1.85	1.71	1.78
1960	1.03	1.69	1.36
1970	0.64	0.65	0.65
1980	0.42	0.72	0.57
1990	0.52	0.66	0.59
1991	0.50	0.56	0.53
1992	0.53	0.61	0.57
1993	0.58	0.60	0.59
1994	0.60	0.63	0.61
1995	0.60	0.60	0.60
1996	0.60	0.60	0.60
1997	0.60	0.60	0.60
1998	0.60	0.60	0.60
1999	0.60	0.80	0.70
2002	0.70	0.80	0.80

India's trade links with all the regions of the world have increased over the years. In view of the current wave of world-wide globalisation, India has taken major initiatives to diversity its exports as also their destinations. Indian exports cover over 7500 commodities to about 190 countries while imports from about 140 countries account for over 6000 commodities. Table 2 indicates the volume of Indian trade in post-independent era.

It is a remarkable fact that during whole planning period our balance of trade has remained unfavourable. Our imports have exceeded exports, showing a trade deficit.

Only two financial years *i.e.,* 1972–73 and 1976–77 were exceptional in showing favourable balance of trade worth Rs. 104 crore and Rs. 68 crore respectively. The deficit in balance of trade in our country has been generally increasing, even through our foreign trade has been getting much more broad based. The Government has introduced a number of measures for reducing deficit in the balance of trade. The main objective was to control imports on the one hand and to promote exports on the other. The basic reason of increasing deficit in balance of trade in India has been the high import bill of petroleum products. Since July 1991, the government adopted the policy of economic liberalisation and a series of economic reforms were adopted in the country.

Selected Indicators of External Sector

(April-Sept.)

		1990-91	*1998-99*	*1999-00*	*2000-01*	*2001-02*	*2002-03*	*2003-04*	*2003-04*	*2004-05*
1.	Growth of Exports BOP (%)	9.0	–3.9	9.5	21.1	–1.6	20.3	20.4	9.0	23.2
2.	Growth of Imports BOP (%)	14.4	–7.1	16.5	4.6	–2.8	14.5	24.4	23.6	39.0
3.	Exports/Imports BOP (%)	66.2	72.1	67.8	78.5	79.4	83.4	80.7	74.9	66.4
4.	Import Cover of FER (No. of months)	2.5	8.2	8.2	8.8	11.5	14.2	16.9	14.8	13.8
5.	External Assistance (net)/TC%	26.2	10.2	8.2	4.8	13.4	–29.4	–13.1	–1.8	3.6
6.	ECB (Net)/TC(%)	26.8	55.5	3.1	50.6	–19.0	–15.9	–7.3	1.4	20.9
7.	Non-Resident deposits/TC(%)	18.3	12.2	14.2	27.2	33.0	27.99	17.5	18.5	–12.3
	As % of GDPmp									
8.	Exports	5.8	8.3	8.4	9.9	9.4	10.6	10.8		
9.	Imports	8.8	11.5	12.4	12.7	11.8	12.7	13.3		
10.	Trade balance	–3.0	3.2	–4.0	–2.7	–2.4	–2.1	–2.5		
11.	Invisible balance	–0.1	2.2	2.9	2.2	3.1	3.3	4.3		
12.	Current Account Balance	–3.1	–1.0	–1.0	–0.5	0.7	1.2	1.8		
13.	External Debt	28.7	23.6	22.1	22.6	21.2	20.3	17.8		

Notes : (*i*) TC : Total capital flows (net).

(*ii*) ECB : External Commercial Borrowing.

(*iii*) FER : Foreign Exchange Reserves, including gold, SDRs, and IMF reserve tranche.

(*iv*) GDPmp : Gross domestic product at current market prices.

(*v*) As total capital flows are netted after taking into account some capital outflows, the ratios against item no. 5, 6 and 7 may, in some years, add up to more than 100 percent.

(*vi*) Rupee equivalents of BOP components are used to arrive to GDP ratios. All other percentages shown in the upper panel of the table are based on US dollar values.

Table 2 : Foreign Trade in India

(in crore Rs.)

Year	*Import*	*Export*	*Trade Deficit*
1950–51	608	606	2
1960–61	1122	642	480
1970–71	1634	1535	99
1980–81	12549	6711	5838
1990–91	43198	32553	10645
2000–2001	230873	203571	27302
2001–2002	245199	209018	36181
2002–2003	297206	255137	42069
2003–2004	359108	293367	65714
2004–2005 (April–Dec.)	333907	242435	91472

Table 2A : Export Targets for Next Five Years

(in billion dollars)

Year	*Export Target*
2004–05	75
2005–06	88
2006–07	104
2007–08	125
2008–09	150

Devaluation of rupee in 1991, and the convertibility of Indian rupee in trade account and current account during 1993–94 and 1994–95 respectively improved the balance of trade position in 1993–94. But the deficit again increased during the subsequent years.

Table 3 : Foreign Trade in India

(in Million $)

Year	*Export*	*Import*	*Trade Deficit*
1997–98	35006	41484	6478
1998–99	33218	42389	9171
1999–2000	36822	49671	12849
2000–2001	44560	50536	5976
2001–2002	43827	51413	7586
2002–2003	52719	61412	8693
2003–2004	63843	78149	14306
2004–2005 (April–Feb.)	69798	93628	23830

Table 4 : Trends of India's Foreign Trade

Quantity : Thousand tonne
Value : Rs. crore & US $ Million

			2002–03			2003–04		
			Qty.	*Rs. Cr.*	*$ million*	*Qty.*	*Rs. Cr.*	*$ million*
1		*2*	*24*	*25*	*26*	*27*	*28*	*29*
I.		**Agricultural and allied products : of which**	—	33691	6962	—	36247	7888
I.	1.	Coffee	184.9	994	205	188	1086	236
I.	2.	Tea and mate	184.4	1663	344	178	1637	356
I.	3.	Oil cakes	177.6	1847	382	325	3348	729
I.	4.	Tobacco	100.5	1022	211	121	1096	238
I.	5.	Cashew Kernels	129.4	2053	424	99.7	1700	370
I.	6.	Spices	278.0	1659	343	267.0	1544	336
I.	7.	Sugar and molasses	1870.2	1814	375	1299.2	1235	269
I.	8.	Raw cotton	11.7	50	10	179.6	942	205
I.	9.	Rice	4967.8	5831	1205	3412.1	4708	1025
I.	10.	Fish and fish preparations	—	6928	1432	—	6106	1329
I.	11.	Meat and meat preparations	—	1377	284	—	1714	373
I.	12.	Fruits, vegetables and pulses (excl. cashew kernels, processed fruits & juices)	—	1692	350	—	2358	513
I.	13.	Miscellaneous processed foods (incl. processed fruits and juices)	—	1484	307	—	1403	305
II.		**Ores and Minerals (excl. coal) of which**	—	7591	1568	—	8876	1932
II.	1.	**Mica**	33.8	41	8	—	106	23
II.	2.	Iron ore (million tonne)	57093.3	4200	868	—	5173	1126
III.		**Manufactured goods of which**	—	198760	41070	—	228246	49671
III.	1.	Textile fabrics & manufactures (excl. carpets hand-made) of which	—	43753	9041	—	44234	9626
III.	1.1.	Cotton yarn, fabrics, made–ups etc.	—	16217	3351	—	15600	3395
III.	1.2.	Readymade garments of all textile materials	—	27536	5690	—	28634	6231
III.	2.	Coir yarn and manufactures	—	355	73	—	357	78
III.	3.	Jute manufactures incl. twist & yarn	—	907	187	—	1113	242
III.	4.	Leather & leather manufacturers incl. leather footwear, leather travel goods & lather garments.	—	9844	1848	—	9938	2163
III.	5.	Handicrafts (incl. carpets handmade) of which	—	5742	1186	—	4867	1059
III.	5.1.	Gems and jewellery	—	43806	9052	—	48586	10573

1		2	24	25	26	27	28	29
III.	6.	Chemicals and allied products	—	28456	5880	—	34915	7598
III.	7.	Machinery, transport & metal manufactures including iron and steel	—	43474	8983	—	56615	12321
IV.		**Mineral fuels and lubricants (incl. coal)**	—	13102	2707	—	17159	3734
V.		**Others**	—	1993	412	—	2839	618
VI		**Total**	—	**255137**	**52719**	—	**293367**	**62843**

India Improves its Share in World Trade

According to WTO report, world export growth rate during 2002 has declined to 4% but Indian exports registered. The third highest growth rate of 14% in the world after China (22%) and Czech Republic (15%). Report adds that Indian ranked 30th in merchandise exports and improved its rank to 24th in imports.

Table 5 : Commodity Composition of Exports

(April–Oct. 2004)

	Commodity Group	*Percentage share*			
		April–March		*April–October*	
		2002–03	*2003–04*	*2003–04*	*2004–05*
I.	Primary products	16.6	15.5	14.2	14.1
	Agriculture & allied	12.8	11.8	11.0	9.5
	Ores & minerals	3.8	3.7	3.2	4.7
II.	Manufactured Goods	76.6	76.0	77.4	73.7
	Textiles incl. RMG	21.1	19.0	19.0	16.3
	Gems and jewellery	17.2	16.6	18.5	17.5
	Engineering goods	17.2	19.4	19.0	20.1
	Chemicals & related product	14.2	14.8	14.6	14.2
	Leather and manufactures	3.5	3.4	3.5	3.0
III.	Petroleum, crude and prdts.	4.9	5.6	5.9	8.6
IV.	Others	1.9	2.9	2.6	3.5
	Total Exports (I + II + III + IV)	100.0	100.0	100.0	100.0

Source : DGCI & S. Kolkata.

India's exports have been witnessing robust growth and displaying a tendency of moving to a higher growth trajectory since 2002–03. The sharp recovery witnessed in 2002–03 was further consolidated in 2003–04, with exports registering a growth rate (in US dollar value and on customers basis) of 21.1 percent on top of a rise of 20.3 percent in the preceding fiscal. Volume increase was the main contributor to this strengthening of export performance. Net terms of trade, which had increased on an average by 1.5 percent per annum in the 1990s, have witnessed a continuous decline since 1999–00. This deterioration in prices of exports relative to imports has been significant in the last two years and seems to have been affected, *interalia,* by the resurgence in international crude oil prices. However, given the strong growth in exports in volume terms, the income terms of trade, which measure the import purchasing power of exports, has consistently improved during the 1990s (except 1996–97). In the recent past, between 2000–01 and 2002–03, this capacity to import on the basis of exports increased by 10.0 percent per year. It reflects the growing competitiveness of Indian exports, with volumes increasing with decline in relative unit prices.

Table 5A : Imports of Principal Commodities

(April–Oct. 2004)

Commodity Group	*Percentage share*			
	April–March		*April–October*	
	2002–03	*2003–04*	*2003–04*	*2004–05*
1. POL	28.7	26.3	26.4	30.4
2. Pearl, Precious & semi precious stones	9.9	9.1	8.6	8.0
3. Capital goods	12.1	13.3	11.2	10.0
4. Electronic goods	9.1	9.6	9.6	9.3
5. Gold and silver	7.0	8.8	9.6	8.7
6. Chemicals*	6.9	7.4	7.4	7.2
7. Edible oils	3.0	3.3	4.0	2.5
8. Coke, Coal and Briquettes	2.0	1.8	2.0	2.8
9. Metalferrous ores and Metal scrap	1.7	1.7	1.8	2.2
10. Professional instruments and optical goods	1.8	1.6	1.6	1.4
11. Others	17.8	17.1	17.8	17.5
Total Imports	100.0	100.0	100.0	100.0

**Organic and Inorganic.*

Despite some slowdown in the second half, export growth continues to be buoyant in the current financial year. Exports registered an increase of 27.03 percent in US dollar terms in April–February 2004–05, substantially higher than the annual target of 16 percent for 2004–05. Imports went up by 36.33% during the first eleven months of year 2004–05. Hence the trade deficit during April 2004 to February 2005 stood at $ 23830 million. Under its newly announced Foreign Trade Policy 2004–09, Government, encouraged by a 20 percent plus growth rate in three of the last four years, has fixed an ambitious target of US$ 150 billion for exports by the year 2008–09, implying an annual growth rate in US dollar terms of around 20 percent, thus doubling the share of India in global exports to 1.5 percent.

Composition of India's Foreign Trade

Indian foreign trade registered a number of structural changes during the planning period. The percentage of non-traditional goods in total exports has continuously increased the exports of chemical and engineering goods have shown a high growth rate. During past few years, handmade good (including gems and jewellery) have become one of the important export commodities. India is making exports of a few traditional items including tea, coffee, rice, pulses, spices, tobacco, jute, iron ore etc.

Besides, the imports of petroleum products, capital goods, carbon chemical and compounds, medical and pharmaceutical products are also import by Indian. Pearls, gems and stones are also imported on a large scale but after their processing these are exported from the country. Other imports include edible oils, fertilizers, non-ferrous metals, paper and paper broads, pulp and waste paper etc.

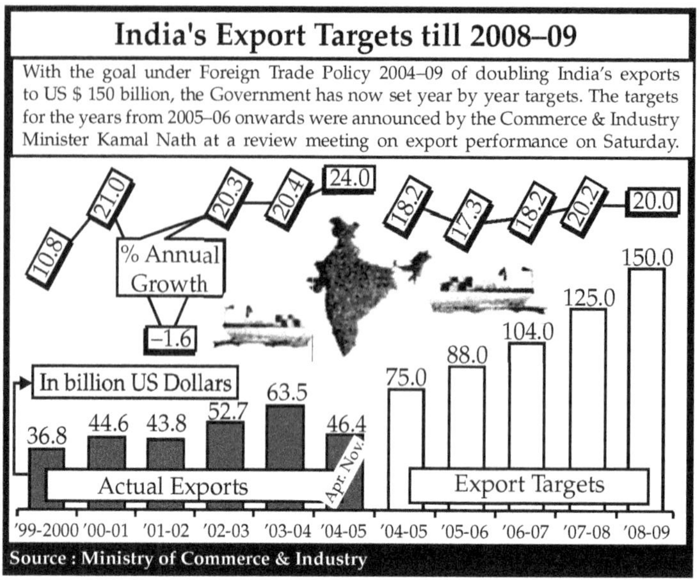

Indian exports cover a wide range of agricultural and industrial products as also various handicrafts, readymade garments and leather manufactures etc. Project exports which include consultancy, civil construction and turn-key contracts have also made a significant progress in the recent years. Recently electronic hardware and software exports have increased in a significant way mainly to the advanced countries. Imports have also increased substantially, bulk of which comprise item like petroleum products, fertilizers etc. precious stones for export production and capital goods, raw materials, consumables and intermediates for industrial production and technological upgradation.

Direction of Foreign Trade

The maximum foreign trade in India is done with United States of America, though Indo–American trade relations have been adversely affected due to imposition of American trade laws like Super-301 and Special-301 during past 3–4 years. The Government of India has, however, not made any change in its trade policy with USA. According to the data available during 2003–04, India had the export worth Rs. 52798 crore with USA while total imports from USA were estimated to be about 23136 crore.

India is having the maximum trade with OECD countries which EEC registers the maximum trade with India. Table 7, indicates the direction of Indian foreign trade.

The direction of Indian trade registered a change during recent past years. Indian trade has been partially shifted from West–Europe to East–Asia and OECD countries. The high growth rate in Japan and ASEAN countries gave a high demand and favourable market to Indian exports. This has been one of the major reasons responsible for increasing Indian exports to East-Asian region of the world. Contrary to it, the countries of West–Europe (France, Spain, Portugal etc.) registered the phase of depression in their economies. That is why Indian exports to these countries got reduced over the past recent years.

USA is India's largest trading partner accounting for nearly 18% of our exports and 6.4% of the imports. The share of non-traditional items and value added products have been increasing in our exports to USA while one imports comprised mostly on engineering products and chemical products.

Table 6 : India's Major Trading Partners, 2002–2004
[Percentage share in total trade (exports + imports)]

	Country			*April-October*	
		2002–03	*2003–04*	*2003–04*	*2004–05*
1.	USA	13.4	11.6	12.2	11.1
2.	UK	4.6	4.4	4.4	3.7
3.	Belgium	4.7	4.1	4.1	3.8
4.	Germany	4.0	3.9	3.8	3.5
5.	Japan	3.2	3.1	3.2	2.6
6.	Switzerland	2.4	2.7	3.3	3.0
7.	Hong Kong	3.1	3.4	3.6	2.8
8.	UAE	3.8	5.1	4.2	5.5
9.	China	4.2	5.0	4.3	5.6
10.	Singapore	2.5	3.0	2.5	3.3
11.	Malaysia	1.9	2.1	2.1	1.9
	Total (1 to 11)	**47.9**	**48.2**	**47.6**	**46.8**

Source : DGCI and S. Kolkata

Till 1991, 18% of the total Indian exports were exported to USSR and other socialist countries of East-Europe. Due to division of USSR and structural changes in economies of East-Europe this export percentage got reduced to 1.2% only during 2003–04. However, exports to Russia have shown an increasing trend during past two years.

Export growth to OECD region and other developing country regions slowed down in 2003–04 to 12.3 percent and 28.3 percent, respectively, from 22.0 percent and 32.0 percent respectively in 2002–03. Overall export to Latin America region remained subdued as exports to Brazil and Mexico, major destinations for India, declined in 2003–04. Exports to Eastern Europe witnessed a turnaround, mainly due to higher exports to Hungary and Romania, with the region retaining its share in total exports. In 2003–04, while the high international crude petroleum prices resulted in a rise in the share of the OPEC region in India's imports, the consequent gains in terms of trade for the OPEC also led to an increase in the region's share in India's exports. Exports to Asian countries maintained their rising profile with robust export growth of 26.1 percent to the ASEAN region in 2003–04. Imports sourced from the ASEAN region also grew by 44.3 percent taking the two-way trade to 9.3 percent of India's total external trade in 2003–04. Export growth to OECD region, developing country region, OPEC and Eastern Europe accelerated in April–October 2004, and the share of these regions in total Indian exports were 46.2 percent, 33.0 percent, 15.2 percent and 1.4 percent, respectively.

The sourcing of imports in 2003–04 showed higher share from regions like OPEC, Eastern Europe and other developing countries (especially from Asia), with the share of OECD region remaining broadly unchanged. Given the robust growth in imports in 2004–05, share of imports from OPEC and Eastern Europe increased to 8.3 percent and 1.7 percent respectively, while the share of OECD region and other developing countries moderated to 33.9 percent and 19.8 percent respectively in April–October 2004.

Trade with the ASEAN continues to be buoyant in the current year with exports registering a growth of 50.0 percent and imports a rise of 21.1 percent in April–October 2004, mainly because of higher trade with Indonesia, Malaysia, Thailand and Singapore.

Table 7 : Foreign Trade Direction

		Foreign Trade	Export				Import			
			Rs. Crore		% Share		Rs. Crore		% Share	
			2002–03	2003–04	2002–03	2003–04	2002–03	2003–04	2002–03	2003–04
		1	2	3	4	5	6	7	8	9
I.		**OECD of which**	**127679**	**136151**	**50.0**	**46.4**	**112766**	**135889**	**37.9**	**37.8**
I.	1.	EU of which	54173	61816	21.2	21.1	56434	62248	19.0	17.3
I.	1.1	Belgium	8042	8298	3.2	2.8	17964	18270	6.0	5.1
I.	1.2	France	5198	5886	2.0	2.0	5295	5010	1.8	1.4
I.	1.3	Germany	10195 @	11693 @	4.0 @	4.0 @	11637 @	13411 @	3.9 @	3.7 @
I.	1.4	Netherlands	5071	5924	2.0	2.0	1867	2461	0.6	0.7
I.	1.5	U.K.	12081	13892	4.7	4.7	13439	14862	4.5	4.1
I.	2	North–America	56110	56306	22.0	19.2	24245	26471	8.2	7.4
I.	2.1	Canada	3379	3507	1.3	1.2	2741	3336	0.9	0.9
I.	2.2	USA	52730	52798	20.7	18.0	21505	23136	7.2	6.4
I.	3.	Other OECD of which	11789	10934	4.6	3.7	15726	24794	5.3	6.9
I.	3.1	Australia	2440	2685	1.0	0.9	6469	12174	2.2	3.4
I.	3.2	Japan	9021	7854	3.5	2.7	8887	12258	3.0	3.4
II.		**OPEC of which**	**33462**	**43971**	**13.1**	**15.0**	**16950**	**25905**	**5.7**	**7.2**
II.	1.	Iran	3169	4219	1.2	1.4	1250	1226	0.4	0.3
II.	2.	Iraq	1040	345	0.4	0.1	0.1	1	Neg.	Neg.
II.	3.	Kuwait	1213	1466	0.5	0.5	869	655	0.3	0.2
II.	4.	Saudi Arabia	4553	5162	1.8	1.8	2443	3390	0.8	0.9
III.		**Eastern–Europe of which**	**4639**	**5139**	**1.8**	**1.8**	**3837**	**5673**	**1.3**	**1.6**
III.	1.	GDR*	*	*	*	*	*	*	*	*
III.	2.	Romania	133	220	0.1	0.1	221	330	0.1	0.1
III.	3.	Russia @@	3407	3280	1.3	1.1	2868	4410	1.0	1.2
IV.		**Other LDCs** of which**	**78558**	**95674**	**30.8**	**32.6**	**58172**	**72037**	**19.6**	**20.1**
IV.	1.	Africa	8261	9665	3.2	3.3	5292	4703	1.8	1.3
IV.	2.	Asia	64534	81010	25.3	27.6	47658	61830	16.0	17.2
IV.	3.	**Latin–America and Carribean**	5763	4999	2.3	1.7	5222	5504	1.8	1.5
V.		**Others**	**10800**	**12435**	**4.2**	**4.2**	**105480**	**119504**	**35.5**	**33.3**
VI.		**Total**	**255137**	**293367**	**100.0**	**100.0**	**297206**	**359108**	**100.0**	**100.0**

@ Figures for unified Germany Neg. — Negligible

* Including under F.R.G. (Item I. 1.3 above) with the reunification of Germany.

** Excluding members of OPEC.

@@ Refers to former USSR before 1992–93.

Trade with SAARC region countries was also buoyant with exports to the region growing by 47.8 percent and imports sources from it rising by 24.8 percent in 2003–04.

China has emerged in 2003–04 as India's third highest trading partner, after the US and UAE, overtaking countries like UK and Belgium.

Trade Policy

Over the years, trade policy has undergone fundamental shifts to correct the earlier anti-export bias through the withdrawal of quantitative restrictions (QRs), reduction and rationalization of tariffs, liberalization in the trade and payments regime and improved access to export incentives, besides a realistic and market based exchange rate. The focus of these reforms has been on liberalization, openness, transparency and globalization with a basic thrust on outward orientation focusing on export promotion activity and improving competitiveness of Indian industry to meet global market requirements. In early 2002, the Government presented a Medium Term Export Strategy (MTES) for 2002–07 providing a vision for creating a stable policy environment with indicative sectorwise targets, with a mission to achieve one percent of global trade by 2007. The Export and Import (EXIM) Policy framed for the period 2002–07 and unveiled on 31 March, 2002 also seeks to usher in an environment free of restrictions and controls. Synergy between these policies/strategies is expected to realize India's strong export potential and enhance the overall competitiveness of is exports.

Trade policy reforms in the recent past, with their focus on liberalization, openness, transparency an globalization, have provided an export friendly environment with simplified procedures for trade facilitation. Such continued trade promotion and trade facilitation efforts of the Government have also aided the current strengthening of export growth.

Sanctioned and Established Agricultural Export Zones

Agri-Produce	*State*	*Area/Dist.*	*Proposed Investment (Rs. crore)*	*Present States*
Wheat	M.P.	Ujjain, Indore, Bhopal	86.4	Functioning
Apple	J & K	Srinagar, Baramula, Annatnag, Kupwara	85.0	Not-functioning
Mango & Grapes	Andhra Pradesh	Rangareddy, Medhak, Mehboob Nagar	57.0	Functioning
Apple	Himachal Pradesh	Shimla, Sirmaur, Kulloo, Mandi, Kinnaur	57.0	Not-functioning
Mango pulp, Vegetables	Andhra Pradesh	Chittoor	53.0	Not-functioning
Mango	U.P.	Lucknow, Unnao, Hardoi, Sitapur, Barabanki	44.7	Not-functioning
Potato, Onion, Lehsun	M.P.	Malwa, Ujjain, Indore, Dewas, Dhar, Shajapur, Ratlam, Neemuch, Mandasaur	50.0	Not-functioning
Basmati Rice	U.P.	Eastern U.P.	39.8	Not-functioning
Potato	W. Bengal	Hugali, Burdwan, Midnapur, Narayanpur, Hawrah	36.6	Functioning
Onion, Lehsun	Gujarat	Bhavnagar, Surendra Nagar, Amreli, Rajkot, Junagarh, Jamnagar	35.0	Not-functioning

Trade Policy Outlook (1991–92 to 2000–2001)

Trade Policy reforms over the last decade have provided an export friendly environment with simplified procedures conducive to enhancing export performance. The focus of these reforms have been on liberalisation, openness, transparency and globalisation with basic thrust on outward orientation focusing on export promotion activity moving away from quantitative restrictions and improving competitiveness of Indian industry to meet global market requirements. Over the years significant changes in the EXIM policy has helped to strengthen the export production base, remove procedural irritants, facilitate input availability besides focusing on quality and technological upgradation and improving competitiveness. Steps have also been taken to promote exports through multilateral and bilateral initiatives, identification of thrust areas and focus regions.

Agricultural Export Zones

The EXIM policy 2001 introduced the concept of Agricultural Export Zones to give primacy to promotion of agricultural exports and effect a reorganisation of our export efforts on the basis of specific products and specific geographical areas. By focusing specially on areas wherever there is a convergence of these two factors, the intention is to transform these Zones into Regional Rural Motors of the export economy. The scheme is centered around the cluster approach of identifying the potential products, the geographical region in which these products are grown and adopting an end-to-end approach of integrating the entire process right from the stage of production till it reaches the market. Such an approach would strengthen backward linkages with a market oriented approach, enhance product acceptability and its competitiveness abroad as well as in the domestic market, increase value addition to basic agricultural produce, reduce cost of production through economies of scale, provide remunerative returns to the forming community in a sustained manner, improve access to the produce/products of the Agriculture and Allied sectors in the international market, improve product quality and packaging, promote trade related research and development and increase employment opportunities. It would help in internationalisation of our agriculture by filling critical gaps including information on prices, demand quality standards, etc. and help in shifting the terms of trade in favour of agriculture.

Till March 31, 2004, the Union government had sanctioned and notified 48 Agriculture Export Zones (AEZ) which are being set up in 19 various states.

Various trade facilitation measures announced in the review of credit policy by the RBI in October 2004 including liberalization of guarantee by Authorized Dealers (ADs) for trade credit, relaxation of time limit for export realization for Export Oriented Units (EOUs), relaxation in booking of Forward contracts by exporters/importers and undertaking of a fresh survey by RBI for evaluation of the impact of the measures taken by it to reduce the transaction cost for exports. Government also announced, on August 31, 2004, a new Foreign Trade Policy for the period 2004–09, replacing the hitherto nomenclature of EXIM Policy by Foreign Trade Policy (FTP). A vigorous expert-led growth strategy of doubling India's share in global merchandise trade in the next five years, with a focus on the sectors having prospects for export expansion and potential for employment generation, constitute the main plank of the policy. These measures are expected to enhance international competitiveness and aid in further increasing the acceptability of Indian exports.

Special Economic Zones (SEZs)

A scheme for setting up Special Economic Zones (SEZs) in the country to promote exports was announced by the Government in the Export and Import Policy on 31 March, 2000. The SEZs are to provide an internationally competitive and hassle-free environment for exports and are expected to give a further boost to the country's exports.

The Policy has provided provisions for setting up SEZs in the public sector, joint sector or by State Governments. It was also announced that some of the existing Export Processing Zones would be converted into Special Economic Zones. Accordingly the Government has issued notification for conversion of all the existing Export Processing Zones (EPZs) into Special Economic Zones. In addition the Government have also granted approval for setting up of 25 Special Economic Zones in the private/joint sector/by the State Governments at

Positra, Mundra and Dahej (Gujarat), Navi Mumbai and Kopta (Maharashtra), Nanguneri (Tamil Nadu), Kulpi, Salt Lake and Calcutta Leather Complex, (West Bengal), Paradeep and Gopalpur (Orissa), Bhadohi, Kanpur, Moradabad and Greater Noida (U.P.), Kakinada and Visakhapatnam (Andhra Pradesh), Indore (Madhya Pradesh), Sitapura and Jodhpur (Rajasthan), Vallarpadam/Puthvypeen (Kerala), Hassan and Baikampady (Karnataka) and Ranchi (Jharkhand).

Some of the distinctive features of the scheme are : a designated duty free enclave to be treated as foreign territory for trade operations and duties and tariffs; SEZ units could be for manufacturing/services; no routine examination of import policy in force; SEZ units to be positive net foreign exchange earners in three years; no fixed wastage norms; duty-free goods to be utilised within the approval period of give years; performance of SEZ units to be monitored by a Committee consisting of the Development Commissioner; subcontracting of part of production and production process allowed for all sectors, including jewellery units; 100 percent foreign direct investment through automatic route in the manufacturing sector; 100 percent Income tax exemption for five years and 50 percent for two years thereafter and 50 percent of the ploughed back profit for the next three years; external commercial borrowing through automatic route, etc.

At present 711 units are in operation in eight SEZs. Exports by SEZ units during 2003–04 were of the order of Rs. 14,004 crore as compared to Rs. 10,057 crore during 2002–03.

EXIM Policy for Tenth Plan (2002–2007)

The Export-Import (EXIM) Policy for the period 2002–07 was announced on 31 March, 2002 by the NDA Government with the objective to consolidate the gains of the previous Export-Import Policies and to set for itself a challenging target to achieve one percent share in global trade by 2007 but this policy was replaced by Foreign Trade Policy announced by UPA government for the period 2004–09.

The basic tenets of the new Policy were to continue the process of liberalisation, simplification of procedure, reduction of transaction cost to our exporters and importers, and finally focusing on areas of country's core competence like agriculture, leather, textile and the small scale sector, including cottage and handicraft sectors. The process of removing Quantitative Restrictions on imports was almost complete and this Policy put the view to eliminate QRs on experts also.

The important highlights of the policy were :

- Annual target to boost exports upto 80 billion dollars.
- Indian Share in World Trade is targeted to be 1% from the present level of 0.67%.
- Elimination of Quantitative Restrictions (QRs) on exports.
- Special incentives for agricultural exports.
- Focusing sharply on SEZs.
- DEPB and EPCG schemes not only retained but also made more flexible.
- **Cottage Sector & SSI**
  - (*i*) Common service providers in industrial clusters can avail of EPCG schemes.
  - (*ii*) Access to market initiative fund.
  - (*iii*) Eligibility for export house status brought down to Rs. 5 crore.
- **Electronic Hardware**
  - (*i*) EHTP modified to enable IT sector to face zero duty regime under ITA–1.
  - (*ii*) Domestic sales to count as export obligation where import duty is zero.
- **Chemical & Petro Chem**
  - (*i*) Reimbursement of 50% of registration fees.

(*ii*) Free export of samples with out any limit.

- **Gems & Jewellery**

  (*i*) Customs duty on rough diamond imports removed. Licensing regime removed too.

  (*ii*) Value-added norms for jewellery exports cut to 7% and 3% from 10%.

- **Market Extension**

  (*i*) Focus Africa launched. First phase will cover Nigeria, South Africa, Kenya, Ethiopia, Tanzania and Ghana.

  (*ii*) Links with CIS countries to be revived.

  Kazakhistan, Kyrgyzstan, Uzbekistan, Turkmenistan, Ukrain and Azerbaijan to be targeted.

- Transport subsidy to units in Jammu & Kashmir and the North-East to offset the disadvantage of being for from ports.
- To boost relocation of industries to India, plant and machinery allowed to be imported without a licence where the depreciated value of such relocating plants exceeds Rs. 50 crore.
- SEZs have been described as the **'best of our dream projects.'** Bank branches in these zones as overseas branches free of CRR, SLR and priority sector lending requirements.

Export Promotion Measures

A Medium-Term Export Strategy which is co-terminus with the Tenth plan period (2002–07) was announced in January 2002 with the objective of enhancing India's exports. A number of programmes/schemes have been launched which include schemes like Assistance to States for Developing Export Infrastructure (ASIDE), establishing Agri Export Zones, Market Access Initiative, strengthening "Focus LAC" programme, introducing "Focus Africa" programme, etc. In the EXIM Policy, 2003–04 and EXIM Facilitation Measures announced in January 2004, besides, the focus on Service exports and policies to strengthen Special Economic Zones (SEZs), a new programme called "Focus CIS" has been introduced. Thus measures are being taken from time to item to increase India's exports.

FOREIGN TRADE POLICY 2004–09

Objectives and Strategy

The new Foreign Trade Policy (FTP) takers an integrated view of the overall development of India's foreign trade and essentially provides a roadmap for the development of this sector. It is built around two major objective of doubling India's share of global merchandise trade by 2009 and using trade policy as an effective instrument of economic growth with a thrust on employment generation. Key strategies to achieve these objectives, interalia, include : unshackling of controls and creating an atmosphere of trust and transparency; simplifying procedures and bringing down transaction costs; neutralizing incidence of all levies on inputs used in export products; facilitating development of India as a global hub for manufacturing, trading and services; identifying and nurturing special focus areas to generate additional employment opportunities, particularly in semi–urban and rural areas; facilitating technological and infrastructural upgradation of the Indian economy, especially through import of capital goods and equipment; avoiding inverted duty structure and ensuring that domestic sector are not disadvantaged in trade agreements; upgrading the infrastructure network related to the entire foreign trade chain to international standards; revitalizing the Board of Trade by redefining its role and inducting into its experts on trade policy : and activating Indian Embassies as key players in the export strategy.

Foreign Trade Policy 2004–09 : At a Glance

- *1.5% share of global trade by 2009.*
- *Special focus on five traditional exports—agriculture, handicrafts, handlooms, leather and footwear and gems and jewellery so as to make exports as employment oriented.*
- *Absolute export sector to be exempted from service tax for cutting down the export cost.*
- *Three new export promotion schemes have been introduced :*
    - *Target Plus Scheme*
    - *Vishesh Krishi Upaj Yojana*
    - *Served from India*
- *Vishesh Krishi Upaj Yojana to be introduced for boosting exports of agriculture products.*
- *Served from India scheme will boost exports of services.*
- *New 'Service Export Promotion Council' has been constituted for services.*
- *Free Trade and Warehousing Zones (FTWZs) have been planned on lines of Special Economic Zones (SEZs) so as to given boost to free trade cut in trading limits for export oriented units of Star Export House Category.*
- *Target Plus Scheme for exporters incentives.*
- *Liberalisation of EPCG Scheme.*
- *DEPB scheme to be continued but new substitute scheme to be introduced shortly.*
- *Duty free imports for services importers.*
- *Liberal import conditions for seeds.*
- *Ban on old machinery imports lifted.*

Foreign Trade Policy 2004–09
(EXIM Policy 2004–09)

The new foreign trade policy of the union government was unveiled by the Commerce Minister Mr. Kamal Nath on August 31, 2004. The new policy reflected no major departure from the five year EXIM Policy of 10th plan (*i.e.,* for the period 2002–07), even as it carried a definite mark of the common minimum programme of the ruling alliance. With foreign trade largely freed and import duties falling progressively, the policy has necessarily restricted itself to a facilitating role. Now the sphere of trade has been extended and the new foreign trade policy goes beyond the traditional focus on pure exports.

The new trade policy highlights the view, "Trade is not an end in itself, but a means to economic growth." Exports are not just for foreign exchange generation but are an engine of growth that can create incremental economic activity in the country." For establishing this view point into reality, new foreign trade policy has set a broader objective of doubling India's share (from 0.7% in 2003 to 1.5% in 2009) in global merchandise trade within next five years and giving a big push to employment generation.

The salient features of the new trade policy are :

- An express thrust of the policy is on the exports of farm produce.
- While exporters of specified farm goods would benefit from a new duty free credit entitlement upto 5% of the FOB value of exports, for all other agriculture products exporters would be allowed to import capital goods free of duty under EPCG (Export Promotion Capital Goods) Scheme.
- Policy comprises **'Special focus initiatives'** for employment intensive sectors like gems and jewellery, handlooms, handicrafts, leather and footwear.

- The threshold for becoming a status holding exporters has been brought down from Rs. 45 crore to Rs. 15 crore and export clusters of Rs. 250 crore (as against Rs. 1000 crore at present) have been qualified as 'Town of Export Excellence.'
- Small and medium exporters would be benefited from the liberalisation of import of capital goods and the additional flexibilities for the fulfilment of export obligation under EPCG scheme.
- Three new export promotion schemes—'Target Plus', 'Vishesh Krishi Upaj Yojana' and 'Served from India'—have been introduced. These schemes would benefit exporters who excel others, exporters of farm products and exporters of services respectively.

The **target plus scheme** gives incentives based on incremental export growth. The duty free entitlement for exporters of fruits, vegetables, flowers and minor forest produce would be freely transferable and can be used for import of variety of inputs and goods.

Under the **'served from India'** scheme, the policy tries to create a powerful brand for Indian Services exports in the global market.

- New Policy focusses a lot on procedural simplification and rationalisation of measures for the exporters. This in the long run would reduce transaction cost for the exporters.
- The new policy aims to create warehousing infrastructure for both exports and imports in India and wants to increase competition for the Free Trade Zones (FTZs) set up by UAE.
- For giving a boost to foreign trade, policy proposes to set up FTWZs (Free Trade and Warehousing Zones) on the lines of SEZs (Special Economic Zones). These zones would be established in the areas proximate to sea–ports, airports or dry ports so as to offer easy access by rail and road.
- Policy makes room for setting up of bio-technology parks across the country.
- The policy announcement on service tax exemption for all goods and services exported—including those from the Domestic Tariff Area (DTA)–will be available on 161 tradable services.
- Policy offers duty free import of labour intensive services. Such services are to be associated with labour intensive industries like gems and jewellery, handloom and handicraft, leather and footwear.

New Export Promotion Schemes

A new scheme to accelerate growth of exports called ***'Target Plus'*** *has been introduced. Under the scheme, exporters achieving a quantum growth in exports are entitled to duty free credit based on incremental exports substantially higher than the general actual export target fixed. Rewards are granted based on a tiered approach. For increments growth of over 20 percent, 25 percent and 100 percent, the duty free credits are 5 percent, 10 percent and 15 percent of f.o.b. value of incremental exports. Another new scheme called* **Vishesh Krishi Upaj Yojana** *has been introduced to boost exports of fruits, vegetables, flowers, minor forest produce and their value added products. Export of these products qualify for duty free credit entitlement equivalent to 5 percent of f.o.b. value of exports. The entitlement is freely transferable and can be used for import of a variety of inputs and goods. To accelerate growth in export of services so as to create a powerful and unique 'Served from India' brand instantly recognized and respected the world over, the earlier duty free export credit (DFEC) scheme for services has been revamped and re-cast into the* ***'Served from India'*** *scheme. Individual service providers who earn foreign exchange of at least Rs. 5 lakhs, and other service providers who earn foreign exchange of at least Rs. 10 lakhs are eligible for a duty credit entitlement of 10 percent of total foreign exchange earned by them. In the case of stand-alone restaurants, the entitlement is 20 percent, whereas in the case of hotels, it is 5 percent. Hotel and restaurants can use their duty credit entitlement for import of food items and alcoholic beverages. To make*

India into a global trading-hub, a new scheme to establish ***Free Trade and Warehousing Zone*** *(FTWZs) has been introduced to create trade-relate infrastructure to facilitate the import and export of goods and services with freedom to carry out trade transactions in convertible currencies. Besides permitting FDI up to 100 percent in the development and establishment of these zones, each zone would have minimum outlay of Rs. 100 crores and five lakh sq. mts. built up area. Units in the FTWZs qualify for all other benefits as applicable for SEZ units.*

Balance of Payments Position in India

Balance of payments accounts consists of two accounts–current account and capital account. Current account includes all the debit and credit entries of invisible items side-by-side with trade items, while the capital account is related with the entries of capital transactions in the country. The current account is the important one because it indicates the health of economy. Indian trade balance has been in adverse state for the last five decades which resulted the deficit in current account of the balance of payments.

India's balance of payments remained reasonably comfortable in both 2000–01 and 2001–02. The current account deficit which signifies country's overall current liabilities has come down from the level of 3.1% of GDP in 1990–91 to a current account surplus of 0.3% of GDP in 2001–02, implying a situation where the current receipts exceed current liabilities. On the capital account inflows have remained buoyant (except in 1995–96) bolstering the reserves during this period. Overall balance of payments position has, thus transformed over the past decade from a difficult one at the beginning of 1990s.

India's Preferential Trade Agreement with 'Mercosur' Countries

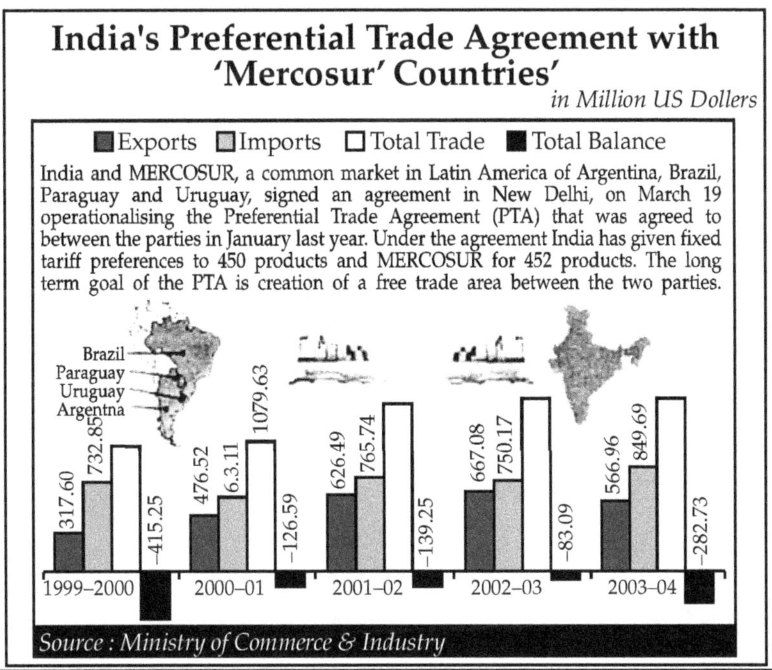

India's Preferential Trade Agreement with 'Mercosur' Countries
in Million US Dollers
Exports
Imports
Total Trade
Total Balance
India and MERCOSUR, a common market in Latin America of Argentina, Brazil, Paraguay and Uruguay, signed an agreement in New Delhi, on March 19 operationalising the Preferential Trade Agreement (PTA) that was agreed to between the parties in January last year. Under the agreement India has given fixed tariff preferences to 450 products and MERCOSUR for 452 products. The long term goal of the PTA is creation of a free trade area between the two parties.
Brazil
Paraguay
Uruguay
Argentna
317.60
732.85
–415.25
476.52
6.3.11
1079.63
–126.59
626.49
765.74
–139.25
667.08
750.17
–83.09
566.96
849.69
–282.73
1999–2000
2000–01
2001–02
2002–03
2003–04
Source : Ministry of Commerce & Industry

BOP : At a Glance

(A) Current Account

1. *Visible balance of trace merchandise*
2. *Invisible balance of trade*

(a) *Service : Tourism, Transport, Software service.*

(b) *Private transfers*

(c) *Account of Investment Income*

(B) Capital Account :

(a) *FDI : Indian and Abroad*

(b) *Portfolio Investment : Loans, Banking capital, NRI deposits.*

OTHER MEASURES TAKEN BY THE GOVERNMENT FOR IMPROVING BALANCE OF PAYMENTS

Full Convertibility of Rupee in the Current Account

In budget proposals of 1992–93, a new system named LERMS (Liberalised Exchange Rate Management System) was introduced and since March 1, 1992 double exchange rates system was adopted. Under new system, the exporters could sell 60% of their foreign exchange earnings to authorised foreign exchange dealers on open market exchange rate, while 40% sale was made compulsorily on exchange rates decided by the RBI. Besides, authorised exchange rate became applicable for providing foreign exchange to the Government for the most essential imports, while importers of other items had to manage themselves foreign exchange in the open market. This step was taken to discourage imports. LERMS showed good result and encouraged by it, the Government introduced in 1993–94, full convertibility of rupee in trade account. By adopting this step the Government abolished double exchange rate system for export and import and implemented LERMS based on open market exchange. In budget proposals of 1994–95, then Union Finance Minister, Dr. Manmohan Singh, declared the full convertibility of rupee in the current account. This full convertibility, however, did not meet the norms prescribed by the IMF under Article–VIII of the Agreement. Article–VIII does not lay any restrictions on current account transactions among the nations.

The high officials of IMF declared in New Delhi on August 5, 1994 that IMF expected the Government of India to implement three important steps before getting its affiliation in clause VIII :

1. Abolition of foreign currency (ordinary) non-repatriable deposit scheme.
2. Abolition of the ceiling for providing foreign exchange for foreign tours of foreign education.
3. To provide facility to interest repatriation on non-resident none deposits.

Full Rupee Convertibility Ruled Out

Deposit sufficient forex reserves Finance Ministry has ruled out full convertibility on capital account in near future. Finance Minister declared to put it off till the government is able to fully rein in expanding fiscal deficit and growing inflationary pressure.

On 19th August, 1994 the RBI declared certain relaxation while declaring full convertibility of Indian rupee in current account.

These relaxation are :

1. The repatriation of income earned from investments by NRIs and their overseas corporate bodies will be allowed in a phased manner over a three year period.
2. The existing uppermost ceilings for providing foreign exchange for foreign tours, education, medical treatment, gift and services was converted into indicative ceilings

beyond which foreign exchange could be obtained for bonafide current payments after making a reference to the RBI.

3. Interest repatriation facility was provided on deposits of non-resident non-repatriable (NRNR). Accounts from October 1, 1994, but the principal amount remained non-repatriable earlier.
4. No new deposits under foreign currency (ordinary) non-repatriable deposit scheme were to be accepted after August 20, 1994, but deposits accepted before October 1, 1994 under FCONR Scheme will get the facility of interest repatriation.

The process of easing the restrictions was formalised in August 1994 with India accepting Article VIII status of the IMF. There has been further relaxation of restrictions on current transactions in the following years.

Meaning of Convertibility of Money

Prior to First World War, the whole world was having gold standard under which the currency in circulation was allowed to get converted either in gold or in other currencies based on gold standard. Such currency was called convertible money. But after First World War all the countries dropped gold standard and paper currency was used which was non-convertible.

In existing money standards convertibility of money has different meaning. Presently, convertibility of money implies such a system in which country's currency becomes convertible in foreign exchange and vice–versa.

Since August 19, 1994 Indian rupees has been made fully convertible in current account transactions related to goods and services.

Committee on Full Convertibility of Rupee

Presently, Indian rupee is fully convertible in current account transactions of balance of payments. It is under consideration of the Government to make Indian rupee fully convertible in capital account. On February 8, 1997 RBI appointed a special group under the chairman ship of Mr. S.S. Tarapore for providing suggestions to make Indian rupee fully convertible in capital account. The other members of the committee were Mr. Surjeet Bhalla, Mr. M.G. Bhinde, Kirit Parekh and Mr. A.V. Rajwade.

The Committee has suggested to adopt measures in three phases so as to make Indian rupee fully convertible in capital account, by the end of 1999–2000. Inspite of Committee's recommendations, full convertibility of Indian rupee in Capital account has not been adopted by the government.

Liberalisation of Gold Import Policy

Gold Import Policy announced in 1992, gave permission to NRIs and Indian tourists coming from abroad to bring gold upto 5 kgs while coming to India. On 31 December, 1996 the Government again liberalised this gold import policy and raised this ceiling of gold import upto 10 kgs. From January 5, 1999, the gold imported/brought under the specified ceiling was initially charged import duty in foreign exchange at the rate of Rs. 400 per 10 grams which was reduced to Rs. 250 per 10 gm in Union Budget 2001–2002. This facility was made available only to those persons who return to India from their foreign tours of at least 6 months duration.

In budget proposals of 2003–04, the import duty on gold biscuits and coins was reduced from Rs. 250 per 10 grams to Rs. 100 per 10 grams.

Import Liberalisation

The Central Government in accordance with the commitments made to World Trade Organisation (WTO), announced relaxation in imports of a number of goods. Various goods were transferred from restricted list to Special Import Licence (SIL) and also goods from SIL list were shifted to Open General Licence (OGL). The Govt. has removed quantitative restrictions on imports of 714 products from 1st April, 2000, and further removed the restrictions of the remaining 715 products from April 1, 2001.

Export Promotion Capital Goods Scheme (EPCGS)

EPCG scheme has been started to permit the exporters to import capital goods on concessional imports duties. Under new export–import policy (1997–2002) exporters of goods and services can import capital goods by paying only 10% import duty. Under EPCG such importers of capital goods have to export goods of 4 times CIF value within next five years.

Improvements in Advance Licensing

A new value–aided based Advance Licensing System has been introduced in which duty free imports of raw material and components are permitted upto a certain percentage of declared export value. Physical quantities and standards have not been determined for individual imputs. Self–certified Advanced Licensing facility has been provided to export houses, commercial houses, star trading houses and super star trading houses. The export time ceiling under this advance licensing scheme in EXIM policy 1997–2002 was extended from 12 months to 18 months. Similarly, the vatidity duration for advance licensing was also extended from 12 months to 18 months.

Export Oriented Units, Export Processing Zone and Special Economic Zone Schemes

The Government has liberalistion and scheme for export oriented units and export processing zones. Agriculture, horticulture, poultry, fisheries and dairying have been included in export oriented units. Export processing zone units have also been allowed to export through trading and star trading houses and can have equipments on lease. These units have been allowed cent percent participation in foreign equities.

1. **Export Processing Zones :** Before getting converted into Special Economic Zones (SEZs), these zones were playing important role in promoting exports of the country. These zones were created to develop such an environment in the economy which may provide capability of facing international competition. The Export Processing Zones (EPZs) set up as enclaves, separated from the Domestic Tariff Area by fiscal barriers, are intended to provide a competitive duty free environment for export production.
2. **Export Oriented Units :** Since 1981, the Government introduced a complementary plan of EPZ scheme for promoting export units (making export of their cent percent production). Under this scheme the Government provides various incentives to increase the production capacity of these units so as to increase exports of the country. This scheme offers a wider source of raw materials, hinterland facilities, availability of technological skills, existence of an industrial base and the need for a large area of land for the project. The EOU have put up their own infrastructure. Exports by EOUs during 2002–03 were of the order of Rs. 22729 crore while exports during 2003-04 were estimated at Rs. 27012 crore.

   As on March 2004, 1764 export oriented units are working in he country.
3. **Export Houses, Trading Houses, and Star Trading Houses :** To increase the marketable efficiency of exporters, the Government introduced the concept of export houses, trading

houses and star trading houses. Those registered exporters who have shown good export performances over past few years have been given the status of export houses, and trading houses. Units having such classification are required to achieve the prescribed average export performance level and earning of foreign exchange. These units are provided some special facilities and benefits by the Government.

On March 31, 2002, 60 Super Star Trading Houses and Star Trading Houses, 372 Trade Houses and 1832 Export Houses were working in the country. Since April 1, 1994 a new category named **Golden Super Star Trading Houses** was added by the Government which has the highest average annual foreign exchange earning. On March 31, 2001 there were 4 Golden Super Star Trading House working in the country.

4. **Export Promotion Industrial Parks (EPIP) :** A Centrally-sponsored '**Export Promotion Industrial Park** (EPIP)' scheme has been introduced in August 1994 with a view to involving the state governments in the creation of infrastructure facilities for export oriented production. It provides for 75% (limited to Rs. 10 crore) grant to state government towards creation of such facilities. The Central Government has so far approved 25 proposals for establishments of EPIPs in the states of Punjab, Haryana, Himachal Pradesh, Rajasthan, Karnataka, Kerala, Maharashtra, Tamil Nadu, Andhra Pradesh, U.P., Gujarat, Bihar, J & K, Assam, M.P., West Bengal, Orissa, Meghalaya, Manipur, Nagaland, Mizoram and Tripura. While the EPIP in Rajasthan at Sitapura (Distt. Jaipur), Bangalore (Karnataka), Ambarnath (Distt. Thane, Maharashtra), Kakkinad (Distt, Ernakulam, Kerala), Surajpur (Distt. Gautambudh Nagar, U.P.), Gummidipoondi (Chengalpattu Distt. Tamil Nadu), Pashamylaram (Distt. Mendak, Andhra Pradesh) and Amingaon (Near Guwahati, Assam have been completed. Exports have already commenced from Karnataka, Kerala & Rajasthan EPIPs. Other parks are at various stages of implementation.

5. **Setting up of Special Economic Zones :** A new scheme for setting up Special Economic Zones (SEZs) in the county to promote exports was announced by the Government in the Export and Import Policy on 31 March, 2000. The Policy provided for setting up of SEZs in the public, private, joint sector or by State Governments. It was also announced that some of the existing Export Processing Zones would be converted into Special Economic Zones. Accordingly, the Government has issued notifications on 1 November, 2000 for conversion of the existing Export Processing Zones at Kandla (Gujarat), Santa Cruz (Maharashtra) and Cochin (Kerala) into Special Economic Zones. Notification has also been issued for conversion of the private sector EPZ at Surat (Gujarat) into Special Economic Zone at the request of the promoters.

At present there are 711 units in operation in the light functional SEZs. Exports by SEZ units during 2003–04 were of the order of Rs. 14004 crore as compared to Rs. 10057 crore during 2002–03.

Approvals have also been given for setting up of 17 special Economic zones at Positra (Gujarat), Navi Mumbai and Khopta (Maha–Mumbai) in Maharashtra, Nanguneri (Tamil Nadu), Kulpi and Salt Lake (West Bengal), Paradeep and Gopalpur (Orissa), Bhadoi, Kanpur, Greater Noida and Moradabad (U.P.), Vishakhapattanam and Kakinada (Andhra Pradesh), Indore (Madhya Pradesh), Hasan (Karnataka) and Vallarpadam/Puthuvypean (Kerala) on the basis of proposals received from the private sector/state government. The Zones are at various stages of implementation. The Special Economic Zone scheme intends to provide an internationally competitive and hassle–free environment for exports. The SEZ scheme is expected to give a further boost to the country's exports.

The government has offered more concessions for Special Economic Zones (SEZ) to bolster exports and foreign exchange in flows. All services rendered to a developer on a unit within SEZ would be exempted from Service Tax. This means that a service provider of any of the 51 odd taxable services including telephone, brokerage and insurance will not levy service tax within the zone.

SEZ Exports Present Better Result

Against the over-all export growth of 10% during the first half of current fiscal 2003–04, exports from Special Economic Zones registered a growth of 29% to $1.29 billion as against $ 1.0 billion during the same period of last fiscal.

EOUs recorded a growth of 24% during the period under review.

First Export Promotion Industrial Park in India

The first EPIP was officially inaugurated by Central Minister of State for Commerce on March 22, 1997 at Sitapura near Jaipur. This Park has been developed by RIICO (Rajasthan Industrial Investment Corporation) with the assistance of Central Government and developed in main area of 365 acre. The total cost of this Park was Rs. 47.15 crore including central assistance of Rs. 10 crore.

The Central Government has already sanctioned such type of 6 other industrial parks in other States but RIICO got the lead in establishing Sitapura Park within a record period of 16 months, which became the first EPIP in India.

State Government's proposal for establishing another EPIP in Bhiwadi in Rajasthan in under consideration of the Central Government.

Export of Services

For India, services account for 51 percent of GDP and 31 percent of total exports. The potential for growth, however, continues to be large. There was an upward shift in the trend growth of services exports (in US dollar terms) from 7.9 percent in the first half of the decade of the 1990s to 15.3 percent during 2000–01 to 2003–04. Software and other miscellaneous services (including professional, technical and business services) have emerged as the main categories in India's export of services. The relative shares of travel and transportation in India's service exports have declined over the years, while the share of software exports has gone upto 49 percent in 2003–04. The buoyant growth of professional, technical and business services has provided a cushion against the slowdown in traditional services such as travel and transportation. The share of other miscellaneous services, which was around 20 percent until 2003–04, registered a sudden rise to 37 percent in April–September 2004. The comparative advantage to India in software, telecom and other business services is well–documented in several studies.

Export of Major Services as Percent of Total Services Exports

Year	*Travel*	*Transportation*	*Software*	*Miscellaneous**
1995–96	36.9	27.4	10.2	22.9
2000–01	21.5	12.6	39.0	21.3
2001–02	18.3	12.6	44.1	20.3
2002–03	16.0	12.2	46.2	22.4
2003–04	16.5	13.1	48.9	18.7
Apr. Sept. 04	9.9	11.7	36.9	37.2

*Miscellaneous services excluding software.

Services exports grew by 20.2 percent in 2003–04. With the liberalization of exchange restrictions on current account, services imports have also increased over time. Travel payments, for example, rose on account of outward movement of workers and professionals, and a spurt in outbound tourist traffic from India and exceeded travel receipts in April–September 2004. However, overall faster growth of services receipts over payments resulted in the net surplus from services trade increasing from US $980 million in 1990–91 to US $6,591 million in 2003-04.

Export Development Centres

The Government has made a plan to invest about Rs. 50,000 crore for developing infrastructure in 23 Export Development Centres identified by Union Commerce Ministry in 1995–96. The major portion of this investment will be made by the Private Section.

1. Tirupur : Hosiery and Weaving Industry.
2. Moradabad : Brass Ware Handicraft.
3. Ludhiana : Heavy Machinery and Hosiery.
4. Surat : Gem and Jewellery.
5. Panipat : Handloom textiles.
6. Alleppi : Coconut and coir.
7. Jalandhar : Sport goods.
8. Ranipat (Amboor) : Leather.
9. Nagpur : Hand-made equipments.
10. Vishakhapattnam : Fish and Fish products.
11. Meerut : Sport goods.
12. Aligarh : Brass locks.
13. Agra : Leather shoes.
14. Khurja : Clay pots.
15. Kanchipuram : Silk
16. Selam : Hand-made items.
17. Sivakashi : Match boxes.
18. Ambala : Scientific equipments
19. Rajkot : Engine pump.
20. Wapi (Ankleshshwar) : Chemicals
21. Jamnagar : Brass spare parts.
22. Batala : Machine equipment.
23. Bhagalpur : Weaving.

Liberalisation of Foreign Investment Policy

The Government has introduced a number of policy measures to achieve an annual target of foreign investment worth $ 10 billion. These are as follows :

1. 48 high priority industries of the country have been given the self-approval facility for foreign equity investments upto 51%. Three industries related to mining and 9 other capital intensive infrastructural industries have been permitted to have foreign equities of 50% and 74% respectively.

2. NRIs have been granted investment permission, with cent percent (100%) equity for repartriability in high priority industries.
3. The condition of installing new machinery has also been removed for foreign capital investment.
4. RBI has granted permission to foreign nationals of Indian origin for acquiring housing assets without seeking prior permission of RBI.
5. FERA conditions were liberalised with effect from January 8, 1993 and later on FERA was replaced with FEMA.
6. Foreign companies have been granted permission to use their trade marks for selling their commodities in India since May 14, 1992.
7. The Government has allowed foreign institutional investors to invest in India capital market. This permission will be applicable only if they are registered with SEBI and get approval under FERA from RBI. The portfolio investment by Foreign Institutional Investors (FII) in primary and secondary markets has been increased from 24% to 40% of issued share capital of any company, subject to the approval of the Board of Directors of the concerned company. This limit is further raised to 49% in 2001–2002 budget. FIIs have been directed to allocated their total investments between equity and debentures in the ratio of 70 : 30.
8. The Government has signed the 13th April, 1992 MIGA (Multilateral Investment Guarantee Agency) convention to assure protection to foreign investors.
9. The Government has allowed all the industries except a few (only 6 industries that is electronics, aerospace and defence, industrial explosive and hazardous chemicals, medicines, Alcohol drinks, Cigarettes and Sigars) where foreign direct investment can be made without approval from Foreign Investment Promotion Board.

Foreign Investment

Aggregate foreign investment flows experienced a rapid rise during 2003–04. On a year–on–year basis, such flows (net) increased by 255 percent. While foreign direct investment (FDI) flows (net) increased only marginally (by around US$200 million), portfolio flows (net) witnessed an eleven-fold increased, accounting for nearly 77 percent of net foreign investment inflows.

During the first half of the current year, foreign investment (net) flows have been lower, compared to that in the corresponding previous period. The near–halving of foreign investment flows during the first half of the current year can be explained by a reduction in portfolio flows. Quarterwise estimates, however, points to an improvement in both the components of foreign investment flows in the second quarter of the current year. While FDI (net) flows increased from US$771 million in April-June 2004–05 to US$ 1.3 billion in July-September 2004–05, portfolio flows also picked up from only US$81 million in the first quarter to US$430 million in the second quarter. There are indications of a gain in momentum of portfolio inflows in the more recent months of the current financial year.

Foreign Direct Investment

Aggregate FDI inflows into India were somewhat lower during 2003–04 as compared to that during 2002–03. The reduction is attributable to a small decline (US$379 million) in fresh equity capital inflows in 2003–04. Reinvested earnings during 2003–04 at US$1.8 billion were more or less the same as in 2002–03. FDI flows into India, on BOP basis, after rising sharply from 1999–2000, have been showing a decline since 2001–02. FDI (net) undertaken by Indian enterprises overseas, was also lower at US$1.3 billion during 2003–04, compared to US$1.8 billion in 2002–03.

Foreign Direct Investment Approval and Inflows

Sl.No.	Financial Year	Amount in Rupees (in crore)		Amount in US$ (in million)	
		Approvals	Inflows	Approvals	Inflows
1.	1991–1992#	1,345	408	527	165
2.	1992–1993	5,546	1,094	1,976	393
3.	1993–1994	7,469	2,018	2,428	654
4.	1994–1995	9,971	4,312	3,178	1,374
5.	1995–1996	36,608	6,916	11,439	2,141
6.	1996–1997	40,206	9,654	11,484	2,770
7.	1997–1998	40,033	13,548	10,984	3,682
8.	1998–1999	30,324	12,343	7,532	3,083
9.	1999–2000	17,976	10,311	4,266	2,439
10.	2000–2001	25,207	12,645	5,754	2,908
11.	2001–2002	14,645	19,361	3,160	4,222
12.	2002–2003	7,904	14,932	1,654	3,134
13.	2003–2004	6,224	12,117	1,353	2,776
14.	2004–2005*	6,784	11,726	1,475	2,549
	Total	**250,062**	**131,385**	**67,210**	**32,290**

Aug. March: *Upto November 2004.

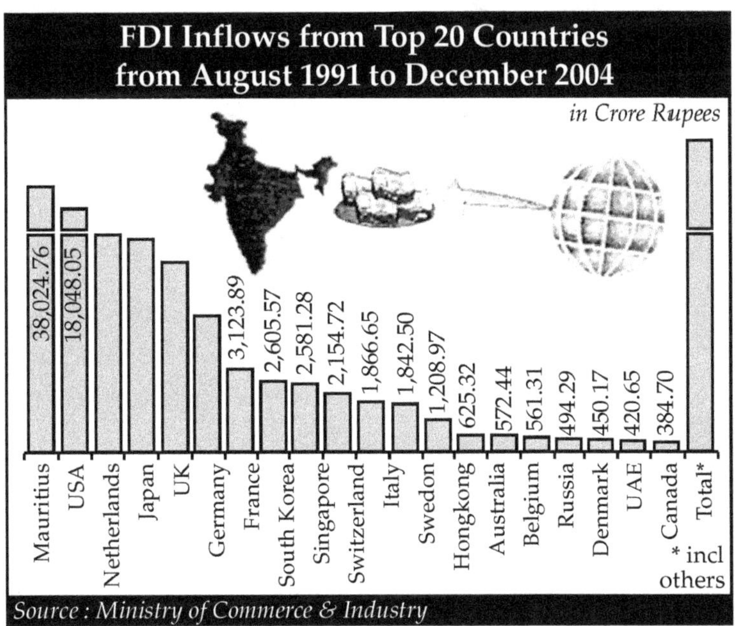

The declining trend of FDI flows into India appears to have reversed during the current year, with such flows during the first half of 2004–05 almost US$1 billion higher than that during April–September 2003–04. On a quarterly basis, FDI (net) flows into India improved from US $ 1.3 billion during. April–June 2004–05 to US$1.8 billion during July–September 2004. FDI (net) abroad has been broadly at the same level in the two quarters.

Aggregate FDI flows (net) into India during April–September 2004–05 are estimated at almost 70 percent of such flows during the whole of 2003–04, thereby indicating a turn around in the year 2004–05.

Sectorwise FDI Approvals and Inflows
(Cumulative from August 1991 to November 2004)

Rank	*Sector*	*No. of Approvals*	*Amount Approved*	*Percentage Approved Amount in Rupee Terms*	*Amount Inflows*	*Percentage of total Inflows in Rupees Terms**
1	*2*	*3*	*4*	*5*	*6*	*7*
1.	Fuel Power	371	43,687 (11,888)	17.47	—	—
	Oil Refinery	338	26,079 (7,185)	10.43		
	Total (power + oil refinery)	709	69,766 (19,073)	27.90	10,433 (2,459)	9.78
2.	Telecommunications (radio paging, cellular mobile, basic telephone services)	805	41,371 (11,441)	16.54	11,231 (2,674)	10.53
3.	Transportation industry	1,105	21,110 (5,427)	8.44	12,123 (2,909)	11.36
4.	Electrical equipments (including computer software and electronics)	4.746	18,947 (4,888)	7.58	16,093 (3,793)	15.09
5.	Metallurgical industries	428	15,412 (4,212)	6.16	2,043 (481)	1.92
6.	Services Sector (financial, non-financial and others)	1,332	16,917 (4,417)	6.77	8,752 (2,174)	8.20
7.	Chemicals (other than fertilizers)	1,074	12,618 (3,492)	5.05	6.405 (1,653)	6.00
8.	Food-processing industries	784	9,562 (2,746)	3.82	4,481 (1,128)	4.20
9.	Hotel and tourism	537	4,909 (1,353)	1.96	981 (232)	0.92
10.	Paper and pulp (including paper products)	137	3,116 (832)	1.25	1,282 (336)	1.20
11.	Other sectors (total of remaining sector excluding above)	7,078	36,334 (9,330)	14.53	57,561 (14,451)	30.80
	Total for all sectors (August 1991 to November 2004)	19,444	2,50,062 (67,211)	100.00	1,31,385 (32,290)	100.00

Regarding the share of top 10 nations in India's FDI inflows (in US$ terms) during August 1991 to November 2004, Mauritius (34.49%) stands first followed by USA (17.1%), Japan (7.33%), Netherlands (7.16%) and UK (6.56%). Among the five top states attracting major share of FDI approvals were Maharashtra (14.8%) Delhi (12.2%), Tamil Nadu (9.05%), Karnataka (7.63%) and Gujarat (4.97%).

FDI is seen as a means to supplement domestic investment for achieving a higher level of economic growth and development. FDI benefits domestic industry as well as the Indian consumers by providing opportunities for technological upgradation, access to global managerial skills and practices, optimal utilization of human and natural resources, making Indian industry internationally competitive, opening up export markets, providing backward and forward linkages and access to international quality goods and services. Most importantly, FDI is central for India's integration into global production chain, which involve production by multinational corporations spread across locations all over the world.

Policy for Foreign Direct Investment (FDI)

Promotion of FDI forms an integral part of Industrial Policy. The role of FDI in accelerating economic growth is by way of infusion of capital, technology and modern management practices. Deptt. of Industrial Policy and promotion has put in place a liberal and transparent foreign investment regime where most activities are open to foreign investment on the automatic route without any limit on the extent of foreign ownership. For the facilitating and increasing the FDI inflow in the country following agencies play an important Role—

(i) ***Foreign Investment Promotion Board (FIPB) :*** *Provides a time bound, transparent and proactives FDI regime for approval of FDI investment proposals.*

(ii) ***Foreign Investment Implementation Authority (FIFA) :*** *It interacts directly with investors alongwith concerned ministry/department of centre/states.*

There are two modalities for FDI approval–Automatic route and Government Approval Route, F.D.I. is not permitted in the areas of :

(i) Agriculture and Plantation (other than Tea plantations), (ii) Real Estate Business (other than development of integrated townships and settlements, (iii) Retail trading, (iv) Lottery business, gambling and betting, (v) Private security services, (vi) Railways, (vii) Atomic Energy.

F.D.I. policy has special incentives for foreign investors :

- *Original investment and the returns on investment are fully repatriable.*
- *Payment of lumpsum fee and royalty to foreign technology provider is permitted under the automatic route within prescribed limits.*
- *Payment of royalty on use of trade marks and brand name without transfer of technology is also permitted.*

Share of Top Five States Attracting FDI Approvals
(From August 1991 to November 2004)

Rank	*Name of State*	*Approvals*			*Amount of FDI Approved*		*% with Total FDI Approved*
		Total	*Tech.*	*Financial*	*Rupees in crore*	*US $ in billion*	
1.	Maharashtra	5,037	1,318	3,719	37,020	9,621	14.80
2.	Delhi	2,810	307	2,503	30,519	8,445	12.20
3.	Tamil Nadu	2,681	618	2,063	22,642	5,894	9.05
4.	Karnataka	2,639	502	2,137	19,075	4,833	7.63
5.	Gujarat	1,236	568	668	12,437	3,273	4.97

Major Initiatives to Attract Foreign Direct Investment (FDI)

- *In pursuance of Government's commitment to further facilitate Indian industry, Government has permitted access to FDI through automatic route, except for a small negative list. Latest revision to further liberalise the FDI regime are as under:*
- *Increase in the FDI limits in "Air Transport Services (Domestic Airlines)" up to 49 per cent through automatic route and up to 100 per cent by non-resident Indians (NRIs) through automatic routes. (No direct or indirect equity participation by foreign airline is allowed).*
- *Foreign investment in the banking sector has been further liberalised by raising FDI limit in private sector banks to 74 per cent under the automatic route including investment by FIIs. The aggregate foreign investment in a private bank from all sources will be a maximum of 74 per cent of the paid up capital of the bank and at all times, at least 16 per cent of the paid up capital held by residents except in regard to a wholly owned subsidiary of a private bank. Further, the foreign banks will be permitted to either have branches or subsidiaries, not both. Foreign banks regulated by a banking supervisory authority in the country and meeting Reserve Bank's licence criteria will be allowed to hold 100 per cent paid capital to enable them to set up wholly owned subsidiary in India.*
- *FDI ceiling in telecom sector in certain services (such as basic, public mobile radio trunked services (PMRTS), global mobile personal communication service (GMPCS) and other value added services), has been increased from 49 per cent to 74 per cent, in February 2005. The total composite foreign holding including but not limited to investment by FIIs, NRI/OCB, FCCB, ADRs GDRs, convertible preference share, proportionate, Foreign investment in India promoters investment companies including their holding companies etc., will not exceed 74 per cent.*
- *In January 2004, guidelines on equity cap on FDI, including investment by NRIs and Overseas Corporate Bodies (OCBs) were revised as under:*
  - *FDI upto 100 per cent is permitted in printing scientific and technical magazines, periodicals and journals subject to compliance with legal framework and with the prior approval of the Government.*
  - *FDI upto 100 per cent is permitted through automatic route for petroleum product marketing, subject to existing sectoral policy and regulatory framework.*
  - *FDI upto 100 per cent is permitted through automatic route in oil exploration in both small and medium sized fields subject to and under the policy of the Government on private participation in exploration of oil fields and the discovered field of national oil companies.*
  - *FDI upto 100 per cent in permitted through automatic route for petroleum products peptides pipelines subject to and under the Government policy and regulations thereof.*
  - *FDI upto 100 per cent is permitted for Natural Gas/LNG pipelines with prior Government approval.*

Foreign Exchange Reserves in India

The foreign exchange reserves of the country include three important components :

1. Foreign Exchange Assets of RBI.
2. Gold Stock of RBI.
3. SDR holdings of the Government.

The foreign exchange reserves of India consist of foreign currency assets held by the RBI, gold holding of the RBI and SDRs. India's foreign exchange reserves crossed the historic and unprecedented mark of $ 100 billion on December 19, 2003. A strong balance of payments position in recent years has led to a steady accumulation of India's foreign exchange reserves. During 2003-04. India experienced an all-time high reserve accumulation of US $ 36.9 billion (including valuation changes, gold, SDR and the reserve position at the IMF). In the current year total foreign exchange reserves stood at US $ 128.9 billion on February 4, 2005. This represents an accretion of US $ 15.9 billion in the year so far, as against an accretion to reserves of US $ 31 billion in the corresponding period of 2003-04.

In the era of liberalisation started in 1991, due to various economic reforms, particularly liberalisation of exchange rate restriction, the foreign exchange reserves grew by $ 3-4 billion a year but from 2000, these accelerated by $ 4.3 billion in 2000-01, $ 11.8 billion in 2001-02 and $ 21.3 billion in 2002-03 and $ 36.9 billion in 2003-04.

After 1991, Indian foreign exchange reserves have rapidly increased due to various reasons which are as follows:

1. Devaluation of Rupees.
2. Availability of loans from international institutions.
3. Availability of foreign exchange from NRIs under various schemes.
4. Increased foreign investment (both direct and indirect).
5. Full convertibility of Rupee on current account.
6. Absorption of dollar supplies by Central Bank
7. Easy access to external commercial borrowing.

FEMA Approved but Money Laundering Bill Sent to Select Committee

In the budget of 1997-98, the Government proposed to replace FERA-1973 by FERA (Foreign Exchange Management Act) FEMA (Foreign Exchange Management Act) 1999 was approved by both the houses of Parliament in December 1999. After the approval of the President FEMA-1999 has come into force w.e.f. June 1,2000. Under FEMA-1999 provisions related to foreign exchange have been modified and liberalised so as to simplify foreign trade and payments. FEMA-1999 will make favourable development in foreign money market.

Finance Minister Mr. Yashwat Sinha also introduced Money Laundering Bill with FEMA Bill 1999. This bill was passed by Lok Sabha on December 2, 1999 but due to objections of left parties and Congress over a few provisions of the bill. It was referred to the Select Committee of Rajya Sabha. According to this bill, acquisition. Possession or owning of money, movable/immovable assets from crime. Especially drugs and narcotics transactions would tantamount to money laundering. Money laundering has been defined as a cognisable crime under this bill.

Factors Responsible for Record addition to Foreign Exchange Reserves in March & April 2004

Foreign exchanges reserves of India has recorded unprecedented addition during March and April 2004 Factors responsible for this surge are following:

- *Sale of pre-earned foreign exchange by exporters in view of strong rupee against dollar so as to hedge against the loss due to over valuation of rupee.*
- *High investment by FIIs in ICICI Bank's issue and ONGC's issue.*
- *Increasing Transfers by NRI's*
- *ECB by Indian Companies.*

Foreign Debt Burden on India

Foreign assistance plays an important role in economic development of developing countries. Foreign assistance includes debt and aid. Debts are repaid with interest amount and hence too much dependence on foreign debts damages the economy of a developing country. Foreign debts have played an important role in economic development of India.

India's external debt stock, which was US $ 83.8 billion at end-March 1991, increased over the years to reach US $ 105.4 billion as the end of December 2003 and 111.8 billion dollar at end of March 2004.

Movement in key debt sustainability indicators point towards further consolidation of external debt during 2003-04. The total external debt to GDP ratio improved to 17.8 per cent at end-March 2004. The proportion of short-term debt in total external debt declined from 4.8 per cent at end-March 2003 to 4.3 per cent as on March 31, 2004, which however rose to 5.7 per cent at end-September 2004 with a rise in import-related trade credits. Debts service payments as a proportion of current receipts rose in 2003-04 mainly due to exceptional transactions, namely, prepayments and redemption of RIBs. Excluding these one-off transactions, debt service ratio worked out to 10.4 per cent in 2003-04.

The share of concessional debt in the total external debt stock of the country remained more or less constant between 35 per cent to 37 per cent after 2001. However, by international standards, India's share of concessional debt continues to be high, particularly among the top 15 debtor countries, India's share of concessional debt is the highest.

In terms of indebtedness classification, the World Bank, in Global Development Finance 2004, has categorized India as a *less indebted,* country for the year 2002. This categorisation is continuing since 1999. In 1998, India was considered a moderately indebted country.

International Comparison of External Debt-2002

			Debt Substainability Indicators			
S.No	*Country*	*Total External Debt*	*Debt to GNP*	*Debt Service*	*Short term debt to total external debt*	*Concessional debt to total debt*
		(US $ billion)	*(ratio as per cent)*			
1	*2*	*3*	*4*	*5*	*6*	*7*
1.	Brazil	227.9	52.5	68.9	10.3	1.4
2.	China	168.3	13.4	8.2	28.5	17.8
3.	Russian Federation	147.5	43.3	11.3	11.1	0.4
4.	Mexico	141.3	22.6	23.2	7.0	0.9
5.	Argentina	132.3	138.4	18.3	11.2	0.9
6.	Indonesia	132.2	80.3	25.0	17.6	24.0
7.	Turkey	131.6	72.7	46.8	11.5	3.5
8.	**India***	**104.4**	**20.7**	**14.9**	**4.4**	**38.4**
9.	Poland	69.5	37.2	22.5	12.8	9.5
10.	Philippines	59.3	71.4	20.2	9.4	21.1
11.	Thailand	59.2	47.6	23.1	20.1	16.6
12.	Malaysia	48.6	54.9	7.3	17.2	6.6
13.	Chile	41.9	68.1	32.8	9.0	0.7
14.	Hungary	35.0	54.4	33.9	16.2	0.3
15.	Colombia	33.9	43.3	40.2	11.2	2.7

* According to World Bank data.

Source: Global Development Finance 2004, The World Bank.

India's indebtedness position vis-a-vis other emerging economies has also improved over the years. In terms of absolute debt levels, India improved from *third* largest debtor after Brazil and Mexico in 1991 to *eighth* in 2002. In terms of external debt indicators like short-term debt to total external debt ratio and short-term debt to foreign exchange reserves ratio, India's position is quite encouraging among the top 15 debtor countries. The short-term debt to total external debt ratio for India is lowest at 4.4 per cent whereas China and Thailand have higher ratios at 28.5 per cent and 20.1 per cent respectively.

India's External Debt Outstandings

Period	*End-March*			*End-September*
	2002 R	*2003 R*	*2004 R*	*2004 QE*
	(US $ Million)			
Long-term Debt	96,089	100,344	107,060	107,105
Short-term Debt	2,745	5,009	4,770	6,485
Total External Debt	98,843	105,353	111,830	113,590
	(Rupees crore)			
Long-term Debt	4,68,932	4,77,093	4,71,157	4,93,979
Short-term Debt	13,396	23,793	20,725	29,930
Total External Debt	4,82,328	5,00,886	4,91,879	5,23,909
	(Ratio as per cent)			
External Debt to GDP	2.12	20.3	17.8	*
Short-term debt to Total External Debt	2.8	4.8	4.3	5.7
Short-term debt to Foreign Currency Assets	5.4	7.0	4.4	5.7
Debt Service to Current Receipts	13.4	15.8	18.3	*
Concessional debt to total debt	35.9	36.6	36.0	35.1

R : Revised; QE : Quick Estimates

* Not computed for the broken year.

India's External Debt – A Change in Profile

In terms of composition, India's external debt has shifted in favour of private debt over the last decade. The ratio of Government and non-Government debt, which was roughly 60:40 during 1990 to 1995 declined to 40:60 by end-September 2004. Larger accumulation of private debt was essentially under 'NRI deposits' and 'export credit & commercial borrowings'. With the increasing importance of external debt of the non-Government variety, the share of concessional debt, although stagnant in the range of US $ 38-40 billion in absolute terms, fell from 45.9 per cent in 1991 to 36 per cent in 2004. The proportion of short-term debt in total external debt has also declined over the years. Notwithstanding the increase, albeit moderate, in the absolute level of external debt, both solvency and liquidity indicators show sings of continuous improvement.

External Assistance to India

India has obtained foreign assistance from various sources to meet its requirements. Both capitalist and socialist countries and International Institutions have given assistance to India. A number of countries belonging to **'India Development Forum'** like Austria, Belgium, Canada, Denmark, France, Germany, Japan, Netherland, Norway, Sweden, Britain and USA have provided assistance to Indian as a result of the initiative taken by the World Bank (IBRD) and its associate institutions like. IDA. Among socialist countries-Bulgaria, Hungary, Poland, USSR (Now disintegrated) also gave assistance to India Australia, New Zealand, Switzerland and oil exporting countries are also in the list of countries providing external assistance to India. The lion's share in external assistance to India is received from the member countries India Development Forum. From the point of view of utilised assistance USA has provided the largest amount in the form of loans and grands to India. PL 480/665 assistance accounts for a big share in USA external assistance. Japan ranks second followed by Germany and Britain. It clearly indicates that the maximum external assistance to India has been provided by developed and highly industrialised nations of the world.

The major part of external assistance received in India has to be repaid in foreign currencies. The loan repayable in India rupee comprises of only 1% of total external assistance. Thus 99% of sanctioned external assistance has to be repaid in foreign currencies.

Compositional Share and Key Indicators of India's External Debt

Components (per cent)	*(At end-March) Share in total debt (per cent)*				*Ratios*	*Key indicators (ratio as per cent)*			
	1991	*1995*	*2000*	*2004 (Sept.)*		*1991*	*1995*	*2000*	*2004*
A. Long-term debt	**89.8**	**95.7**	**96.0**	**94.3**	1. Total external debt-to-GDP	28.7	30.8	22.1	17.8
(*i*) Multilateral	25.0	28.8	32.0	26.6	2. Total external debt-to-current receipts	328.9	235.8	145.6	95.5
(*ii*) Bilateral	16.9	20.5	18.5	14.6	3. Short-term debt-to-external debt	10.2	4.3	4.0	4.3
(*iii*) IMF	3.1	4.4	0.0	0.0	4. Short-term debt-to-GDP	2.9	1.3	0.9	0.7
(*iv*) Export credit & Commercial borrowings	17.3	19.8	27.2	24.2	5. Debt service ratio	35.3	25.9	17.1	10.4*
(*v*) NRI deposits	12.2	12.5	13.8	26.9	6. Short-term debt to forex assets	382.1	20.5	11.2	4.4
(*vi*) Rupee debt	15.3	9.7	4.5	2.0					
B. Shot-term debt	10.2	4.3	4.0	5.7					
C. Total debt (A+B)	**100.0**	**100.0**	**100.0**	**100.0**					
Memo item:									
1. Government debt	59.6	60.1	47.7	38.9					
2. Non-government debt	40.4	39.9	52.3	61.1					
3. Total debt (1+2)	**100.0**	**100.0**	**100.0**	**100.0**					

* Excluding exceptional transactions such as prepayments and redemptions of Resurgent India Bonds (RIBs)

Indian Development Forum

The World Bank constituted India Aid Club in 1958 to provide financial assistance during the periods of Second Five Year Plan (1956-61) and Third Five Year Plan (1961-66) Presently it has been renamed as **'India Development Forum'** which is still providing financial assistance for development Plans.

In Tokyo conference held on Sept. 1996 (all the previous conferences were held in Paris) India Development Forum sanctioned the assistance of $ 6.68 billion for 1996-97. This amount was increased to $ 6.74 billion for the year 1997-98. In 1997-98 the conference was held in Paris on June 24-25, 1997. The percentage of concessional loans in total sanctioned loan during 1996-97 was 33% as against 41% in 1995-96. Japan contributed the maximum share in sanctioned loans during 1996-97 and 1997-98 Japan has assured to grant $ 1.154 billion for the year 1997-98.

Table 9 : Foreign Assistance Obtained by India in Various Five Year Plans

	Loan		*Grant*		*PL 480/665*		*Total Foreign Assistance*	
	Sanction	*Utilisation*	*Sanction*	*Utilisation*	*Sanction*	*Utilisation*	*Sanction*	*Utilisation*
Upto the end of Fourth Plan	9,665.3	8,572.6	753.1	712.7	2,637.5	2,636.8	13,055.9	11,922.1
Fifth Plan (1974-79)	7,912.1	5,920.4	1,795.3	1,156.9	136.6	182.0	9,843.8	7,259.3
1979-80	1,295.1	1,048.6	564.4	304.5	-	-	1,859.5	1,353.1
Sixth Plan (1980-85)	15,197.2	9,118.5	1,564.1	1,785.4	-	-	16,761.3	10,903.9
Seventh Plan (1985-90)	42,231.5	20,119.6	2,739.5	2,580.2	-		44,971.0	22,699.8
1990-91	7,601.3	6,170.0	522.1	534.3	-	-	8,123.4	6,704.3
1991-92	11,805.8	10,695.9	901.8	919.1	-	-	12,707.6	11,615.0
Eight Plan 1992-97	62,127.2	51,813.6	8,765.2	4,848.4	-	-	70,892.4	56,662.0
Ninth Plan 1997-98	14,865.0	10,823.4	2,101.0	921.3	-	-	16,966.0	11,744.7
1998-99	8,320.8	12,343.4	209.8	895.5	-	-	8,530.6	13,238.9
1999-2000	17,703.7	13,330.7	2,615.3	1,073.9	-	-	20,319.0	14.404.6
2000-2001	17,184.1	13,527.1	940.6	727.2	-	-	18,124.7	14,254.3
2001-2002	21,630.0	16,111.7	3,465.0	1,447.6	-	-	25,095.0	17,559.3
Tenth Plan 2002-2003	19,875.7	13,898.3	1,296.1	1,835.8	-	-	21,171.8	15,734.1
2003-2004	14,754.4	15,271.0	2,350.7	2,073.4	-	-	17,105.1	17,344.4

Source: Economic Survey 2004-05.

Table 10 : Percentage of Foreign Assistance in Various Five Year Plans

Plan	*% of Foreign Sources in Total Outlay*
First	9.64
Second	22.45
Third	28.25
Three Annual Plans	36.36
Fourth	12.92
Fifth	12.80
Sixth	7.70
Seventh	9.04
Eighth	6.61
Ninth	6.96

It is slightly lower than the previous year's assistance of $ 1260 million. In 1998 Germany has been the second largest country providing bilateral assistance to India, whereas last year USA was the second largest country after Japan. In 1997-98 World Bank was the largest international institution providing multilateral assistance of $ 3 billion to India. In its annual meeting held in Paris on 23-24 May, 2000, IDF has not sanctioned any assistance to India for the period 2000-2001. If could not hold its meeting in 1998 & 1999.

The fisheries industry in India is quite huge with vast coastline. It has great potential in Inland and marine sector. It has opportunity for foreign investors. The fisheries foreign trade policy is given below as such items permitted by Government of India.

1. **Items permitted**
   (i) No Quantitative restrictions on export.
   (ii) Licence under Foreign Trade Policy not required for import of 125 species/groups of fish, crustaceans, molluscus and other aquatic invertiberates covered under FREE policy in Chapter 3 of ITC (HS) classification of Export &Import items under the EXIM policy.
   (iii) Import of five groups of live fish permitted under Restricted Policy (EXIM Code 0301)
   (iv) Import of Whale Shark (Rhincodon types) and parts and products of the species is restricted.
2. **Promotional measures**
   (i) Central assistance to States for development of critical infrastructure for export such as roads, inland container depots, container freight stations, Export Promotion Industrial Parks and for equity participation in infrastructure projects.
   (ii) Encouragements to State Governments for setting up Export Zones.
   (iii) Declaration of Towns of Export Excellence to encourage setting up of critical infrastructure for export production, encourage common service providers and facilitate availability of better technological services and integrate benefits under the other schemes of EXIM Policy for the units in such towns.
   (iv) Market Access Initiative Schemes for encouraging increased marketing efforts by exporters/Brand promotion
   (v) Schemes to promote the Concept of Total Quality Management.
3. **Import for export production**
   (i) Advance authorization for duty free import of inputs for export production.
   (ii) Duty free import authorisation (DFIA) Scheme

   Scheme DFIA is issued to allow duty free import of inputs, fuel, oil, energy sources, catalyst which are required for production of export product. DGFT, by means of Public Notice, may exclude any product(s) from purview of DFIA. This scheme is in force from 1st May, 2006.

   Entitlement Provisions of paragraph 4.1.3 (FTP)shall be applicable in case of DFIA. However, these Authorizations shall be issued only for products for which Standard Input and Output Norms (SION) have been notified. Pre-export Authorization shall be issued with actual user condition and shall be exempted from payment of basic custom duty, additional customs duty, education cess, anti-dumping duty and safeguard duty, if any. A minimum 20% value addition shall be required for issuance of such authorization.

   (iii) Manufacturer exporters, merchant exporters tied to supporting manufacturers and service providers eligible for import of capital goods at 5% Customs duty linked to fulfillment of export obligation in 8 to12 years under EPCG Scheme.
4. **EOU/EPZ/SEZ**
   (i) Scheme of 100% EOU/Export Processing Zone/Special Economic Zone for export production continues. No trading units permitted under the scheme.

5. Package for Marine Sector

a. Duty free import of specified specialized inputs/chemicals and flavoring oils as per a defined list shall be allowed to the extent of 1% of FOB value of preceding financial years export. Use of these special ingredients for seafood processing will enable us to achieve a higher value addition and enter new export markets.

b. To encourage the existing mechanized vessels and deep sea trawlers to adopt modern technology for scientific exploitation of our marine resources in an eco-friendly manner and boost marine sector exports, it is proposed to allow import of monofilament long line system for tuna fishing at a concessional rate of duty.

c. The present system of disposal of waste of perishable commodities like seafood after inspection by a customs official is very cumbersome and leads to development of unhygienic conditions. To overcome this, a self removal procedure for clearance of waste shall be applicable, subject to prescribed wastage norms.

Illustrate of Foreign Trade Policy Announced on 1.4.2007

Focus Market Scheme (FMS)

Objective is to offset high freight cost and other externalities to selected international markets with a view to enhance India's export competitiveness in these countries

Entitlement Exporters of all products through EDI enabled ports to notified countries (as in Appendix 37C of HBP v1) shall be entitled for Duty Credit scrip equivalent to 2.5% of FOB value of exports for each licensing year commencing from 1st April, 2006. However additional Markets notified in Appendix 37C of HBP v1 shall be entitled for Duty Credit scrip on exports w.e.f 1.4.2007.

APPENDIX 37C OF HANDBOOK OF PROCEDURES VOL. I

Sr.No	*Focus Market Code*	*Country Code*	*Country*
Countries in Latin American Block			
1	L 001	015	ARGENTINA
2	L 002	039	BOLIVIA
3	L 003	073	CHILE
4	L 004	109	ECUADOR
5	L 005	317	PARAGUAY
6	L 006	319	PERU
7	L 007	427	URUGUAY
8	L 008	433	VENEZUELA
Countries in African Block			
9	A 001	011	ANGOLA
10	A 002	035	BENIN
11	A 003	041	BOTSWANA
12	A 004	050	BURKINA FASO
13	A 005	053	BURUNDI
14	A 006	057	CAMEROON
15	A 007	061	CANARY IS

Sr.No	Focus Market Code	Country Code	Country
16	A 008	063	CAPE VERDE IS
17	A 009	067	CAFRI REP
18	A 010	069	CHAD
19	A 011	085	COMOROS
20	A 012	087	CONGO P REP
21	A 013	115	ETHIOPIA
22	A 014	116	ERITREA
23	A 015	117	EQUTL GUINEA
24	A 016	135	FR S ANT TR
25	A 017	141	GABON
26	A 018	143	GAMBIA
27	A 019	167	GUINEA
28	A 020	169	GUINEA BISSAU
29	A 021	199	COTE D'IVOIRE
30	A 022	227	LESOTHO
31	A 023	229	LIBERLIA
32	A 024	231	LIBYA
33	A 025	241	MADAGASSCAR
34	A 026	243	MA;AWO
35	A 027	249	MALI
36	A 028	255	MAURITANIA
37	A 029	257	MAURITIUS
38	A 030	265	MOROCCO
39	A 031	267	MOZAMBIQUE
40	A 032	269	NAMIBIA
41	A 033	289	NIGER
42	A 034	339	REUNION
43	A 035	345	RAWANDA
44	A 036	347	SAHARWI A.DM RP
45	A 037	349	SAO TOME
46	A 038	353	SENEGAL
47	A 039	355	SEYCHELLES
48	A 040	357	SIERRA LEONE
49	A 041	363	SOMALIA
50	A 042	371	ST HELENA
51	A 043	385	SWAZILAND
52	A 044	399	TOGO
53	A 045	407	TUNISIA
54	A 046	417	UGANDA
55	A 047	459	CONGO D.REP
56	A 048	461	ZAMBIA
57	A 049	463	ZIMBABWE

Note : The export performance of notified markets under Para 3.9.2.1 of Policy shall be regularly reviewed and the list of countries in Appendix 37 C may be modified from time to time, in terms of Para 3.9.5 of the Policy.

Focus Product Scheme (FPS)

Objective is to incentives export of such products, which have high employment intensity in rural and semi urban areas, so as to offset infrastructure inefficiencies and other associated costs involved in marketing of these products.

APPENDIX 37D OF HANDBOOK OF PROCEDURES VOL. I
VALUE ADDED FISH PRODUCTS

Sl.No	*Item*	*ITC(HS)*	*Description*
1	46		Shrimp – breaded, battered, marinated and other such prepared products
2	47		Shrimp pickle
3	48		Shrimp Curry
4	49		AFD Shrimp, AFD Powder
5	50		Shrimp IQF Raw
6	51		Shrimp IQF blanched/cooked
7	52		Shrimp in Tray/pouch packs
8	53		Squid – breaded, battered, marinated and other such prepared products
9	54		AFD Squid
10	55		Squid IQF raw
11	56		Squid IQF blanched/cooked
12	57		Squid in Tray/pouch packs
13	58		Cuttlefish AFD
14	59		Cuttlefish IQF Raw
15	60		Cuttlefish IQF blanched/cooked
16	61		Cuttlefish in tray/pouch packs
17	62		Cuttlefish breaded, battered, marinated and other such prepared products
18	63		Fish fillets / loins / steaks etc in tray /vacuum pouches
19	64		Braded fish fingers / fish fillets, precooked loins and other such prepared products
20	65		Fish pickle
21	66		Fish curry
22	67		Lobster cooked / half cut IQF/packed in tray / pouches
23	68		Stuffed crab
24	69		Breaded crab cakes/ Crab cake
25	70		Pasteurized crab meat
26	71		Raw crab meat/soft shell crab
27	72		Mussel/clam meat pickle
28	73		Surimi analogues
29	74		Canned Tuna

Courtesy : Ministry of Commerce and Industry Directorate of Foreign Trade.

Courtssey : www.MPEDA.Com/Trade.

Entitlement Exports of notified products (as in Appendix 37D of HBP v1) through EDI enabled ports to all countries shall be entitled for Duty Credit scrip equivalent to 1.25% of FOB value of exports for each licensing year commencing from 1st April, 2006. However, additional products notified / clarified in Appendix 37D of HBP v1 shall be entitled for Duty Credit scrip on exports w.e.f 1.4.2007.

The value of world trade in natural resources – including fisheries, fuels, forestry products and mining – reached 3.7 trillion dollars in 2008, close to a quarter of world merchandise trade.The WTO latest forecast marks a rise from the 9.5 per cent issued in March. Unless there are unanticepated negative economic impacts it will estimated 10% in second half of 2010 (Lamy Pascal DG.WTO 2010)

Impact of Recession

- India's Forx reserves USD 271,641 billion.
- Exports for October 2008 contracted by 15% on a year on - year basis
- Growth contraction has come after a robust 25% plus average export growth since 2003.
- Positive correlation between growth in exports and country's GDP.
- A decelegrating export growth has implication for India even though our economy is far more domestically driven than those of east Asia. Still the contribution of merchandise exports to GDP has risen steadily over the past six years – from about 10% of GDP in 2002–03.
- A sharp fall in export growth has lead to loses of job.
- For instance, the manufacturing sector has suffered approximately 10 million job losses.

Chapter 4

TRANSPORT AND COMMUNICATION

After forming and making a product the publicity, exposure advertisement is essential and its transport to various places is compulsory so there must be some transport system and communication.

Railways

The first train in India was started on a small rail route of 34 kilometer between Bombay and Thane on 16 April, 1853. At present, the Indian Railways consist of an extensive network spread over 63221 km. comprising Broad Gauge (46807 km) Meter Gauge (13290 km) and narrow Gange (3124 km). With such a large rail route, the Indian Railway has become the biggest railway of Asia and second in the world. Upto the end of March 2004, about 28% of the route kilometers, 36 per cent of running track kilometer and 36 per cent of total track kilometer has been electrified. About 16 lakhs people are employed in Indian Railways, which is the largest number engaged in any undertaking in the country. Railways absorb about 40% of the total central government employees.

The Indian Railways have played an integrating role in the social and economic development of the country. Railways also have an advantage of being less energy intensive and more environment friendly.

Measures to Strengthen Safety and Upgrade Railway Infrastructure

- *Special Railway Safety Fund–a non-lapsable fund–with Rs. 17,000 crore set up to wipe out the arrears in the renewal of vital safety equipment within a fixed time-frame of 5-7 years, for replacing old assets including tracks, bridges, as per recommendation of the Railway Safety, Review Committee.*
- *Provision of 25 Watt VHF set at roadside stations for communication between drivers, guards and station staff.*
- *Fund allocated from the diesel cess being used for safety-related works pertaining to level crossings and for Road over and under bridges.*
- *Track circuiting of run-through lines on all stations with permitted speed of more than 75 kmph completed.*
- *Bogie mounted brake system being introduced on wagons in place of the conventional pull/push road type air brake system. Development and indigenous manufacture of Centre Buffer Couplers.*
- *A cohesive strategy for enhanced security of rail user under formulation as per recommendations of the High-level Committee on Enhanced Security on Railways.*

Cooperation between public and Railway administration in secured through various committees including Zonal Railway Users' Consultative Committee and Divisional Railway Users' Consultative Committees. The rolling stock fleet of Indian Railways in service as on 31 March, 2003 comprised 52 steam, 4,699 diesel and 2,930 electric locomotives. Currently, the

Railways are in the process of inducting new designs of fuel efficient locomotives of higher horse power, high-speed coaches and modern bogies for freight, traffic. Modern signalling like panel interlocking, route relay inter-locking centralised traffic control, automatic signalling and multi-aspect colour light signalling, are being progressively introduced. Indian Railways made impressive progress regarding indigenous production of rolling stock and variety of other equipment over the years and are now self-sufficient in most of the items.

Indian Railways Gaugewise Route (as on Dec., 31, 2004)

Gauge	*Route km*
Broad Gauge (1.676 mm)	46,807
Metre Gauge (1.000 mm)	13,290
Narrow Gauge (762 mm and 610 mm)	3,124
Total	**63,221**

Rail Vikas Nigam Ltd. (RVNL) was set up in January 2003 as an effort to create new institutional mechanisms for implementing railway projects through a blend of budgetary support and non-budgetary initiatives. It is implementing a part of the National Rail Vikas Yojana. An outlay of Rs. 717 crores has been provided for RVNL during the year 2004-05 to execute 38 projects which form part of the Golden Quadrilateral and port connectivity routes.

Historic Events of Indian Railways

1849: *Great Indian Peninsular Railway (GIPR) incorporated. A 'guarantee system' that assured 5 per cent returns of a British railway companies in India established.*

1853: *Railways begin in India with private funds and government support.*

1868: *Calcutta and South Eastern Railways (CSER) suffers floods losses, transfers all line to government in return for capital costs, becoming the first state owned railways.*

1882: *Almost 75 railway operations, owned by the private sector and princely states and a variety of track gauges. Post of Director General of Railways (DGR) is created in the Central Public Works Deptt. (CPWD) to coordinate the network.*

1889: *State takes over Nizam Railways.*

1900: *GIPR taken over by state.*

1901: *Based on the recommendations of Sir Thomas Robertson Committee, a Railway Board is set up with three members.*

1904: *More members including in the Board.*

1905: *Powers and structure of Railway Board and formalised. It is now under Deptt. of Commerce and Industry with an independent Chairman.*

1907: *Govt. purchases all major line and releases them to Pvt. operators.*

1920: *Based on Actworth Committee recommendations govt. Takes over actual management of all railways and separates railway finances from general finances. This practice is followed even today.*

1922: *Retrenchment Committee under Lord Inchcape recommends drastic cuts in workforce and expenses.*

1922: *Railway Board reorganised, overriding power given to Chief Commissioner railways.*

1924: *Railway finances separated from general govt budget.*

1925: *First Railway Budget presented.*

1925: *EIR Co. and GIPR, the largest networks, taken over the state.*

1937: *The post of Minister for Transport and Communications created. The minister was a civil servant and dealt with the Railway Board.*

1951: *Zonal grouping of Railways begins.*

A new category of seventeen inter-city train services called Jan Shatabdi was also introduced during 2002-03. These trains have most of the characteristics of the prestigious Shatabdi Express trains and also have specially designed second class chair car accommodation. These trains are more affordable and, therefore, accessible to a much wider spectrum of the travelling public.

Railway Passenger Insurance Scheme

The Indian Railways have introduced 'Railway Passenger Insurance Scheme' *w.e.f.* August 1, 1994. Under this scheme, the victims of terrorist activities are given compensation at the same rates as rail accidents. In matters of death and permanent disability, the amount of compensation is upto rupees two lakhs whereas on being injured, this amount can range from Rs. 16,000 to Rs. 1,80,000.

This insurance scheme is introduced by the Indian Railways in collaboration with United India Insurance Company.

Indian Transport Sector : An Overview

Railways and roads are the dominant means of transport carrying more than 95 per cent of the total traffic in the country. The transport sector has expanded manifolds in the first 50 years of planned development, both in terms of spread and capacity.

***Railways**–The Indian Railways has completed 150 years of service to the nation in May 2002. Railway, with a capital base of about Rs. 55,000 crore and a network of about 63,000 route km, is the principal mode of transportation for bulk freight and long distance passenger traffic. The main thrust in the Tenth Plan is on the capacity expansion through modernisation and technological upgradation of railways system, improvement in quality of service, rationalisation of tariff in order to improve the share of rail freight traffic and to improve safety and reliability of railways services.*

***Roads**–The Tenth Plan thrust is on improving accessibility. Capacity and riding quality of roads; removing deficiencies in the existing road network; revamping maintenance practices; augmenting resources for road development through private sector participation and levy of user charges.*

***Ports**–The ports act as trans-shipment point between water transport and service transport and therefore, play a crucial role in the transportation system for facilitating international trade. There are 12 major ports and 185 minor/intermediate ports along India's 5600 km coastline. The major ports handle about 75 per cent of the port traffic of the country and remaining 25 per cent is handled by minor/State ports.*

The thrust in the Tenth Plan is on creation general and bulk cargo handling facilities with focus on container traffic and improvement in the efficiency and productivity through private sector participation by introduction of organisational charges and rationalisation of manning scales.

***Inland Water Transport (IWT)**–IWT is an efficient, environmentally clean and economical mode of transport. India is richly endowed with waterways comprising river systems and canals. It is estimated that total of 14.544 km of waterways could be used for passenger and cargo movement. However, capacity of the sector is grossly under-utilised as most navigable waterways suffer from hazards like shallow water and narrow width of channel during dry weather, silting of river beds' and erosion of banks, absence of adequate infrastructure facilities like terminals for loading and breathing and surface road links.*

The thrust in the Tenth Plan is on development of infrastructure facilities with a focus on North-East region and private sector participation so that there is a gradual shift of domestic cargo from rail and road modes to inland water transport, increasing its share from the present level of less than one per cent to at least two per cent.

***Civil Aviation**–Civil Aviation, the fastest mode of transport for movement of passenger and cargo traffic, is broadly into three distinct functional entities, viz., regulatory, cum-development, operational and infrastructure. The main objective of the Civil Aviation development in the Tenth Plan is to provide world class infrastructure facilities an efficient, safe and reliable air services to meet domestic and international travel needs of trade and tourism.*

Konkan Railway Plan

In March 1990, the Konkan Railway Plan was started to provide a link by the shortest rail route between the states of Goa, Maharashtra, Karnataka and Kerala. This includes, the 760 km distance between Apta and Managalore. The Registration of Konkan Rail Corporation was made on 26 July, 1990 under the company law, 51% share equity of the corporation belongs to Indian Railways. The Konkan Railway Project has been completed on January 26, 1998 with the total investment of Rs. 3500 crore on January 26, 1998. The rail traffic has been started between **Roha** (Maharashtra) and Mangalore (Kerala). Konkan Railways ensures maximum speed of 160 km per cent hour. Now Konkan Railways has been running well all along the West Cost of India.

Important Highlights of Konkan Railways

1.	*Total Rail Route Distance (Roha to Mangalore)*	*760 km*
2.	*No. of Long Bridges*	*179*
3.	*No. of Small Bridges*	*1819*
4.	*No of Tunnels*	*92*
5.	*Length of Longest Tunnel (Near Karwude)*	*6.5 km*
6.	*Length of Longest Bridge (Near Honawar on Sharwati River)*	*2.065 km*
7.	*Height of the Highest Bridge (Near Ratnagiri on Panwal River)*	*64 metre/210 feet*
8.	*Maximum speed*	*160 km/hour*

Progress of Loco in India

1893: *First railway foundry set up at Jamalpur.*

1895: *First locomotive built with old parts at Ajmer workshop.*

1899: *'Lady Curzon', first locomotive built in India at Ajmer.*

26th January 1950: *Chittaranjan Loco Works (CLW) set up. Builds first steam engine, 'Deshbandhu'.*

1952: *TELCO begins productions of YG locos.*

1959: *First steam loco designed by CLW.*

1961 : *CLW makes the first 1500 DC electric loco 'Lokmanya'.*

Salient Features of the National Rail Vikas Yojana

1. *Capacity bottlenecks in the critical sections of the railway network will be removed at an investment of Rs. 15,000 crore over the next five years. i.e. during the Tenth Plan period.*
   - *(i) Strengthening of the Golden Quadrilateral and its Diagonals to enable the Railways to run more long-distance mail/express trains and fright trains at a higher speed of 100 kmph, at a cost of Rs. 8,000 crore.*
   - *(ii) Strengthening of rail connectivity to ports and development of multimodal corridors to hinterland at a cost of Rs. 3,000 crore.*
   - *(iii) Construction of four mega bridges-two over the River Ganga, one over River Brahmaputra, and one over the River Kosi, at a cost of Rs. 3,500 crore.*
   - *(iv) Accelerated completion of last mile and other important projects, at a cost of Rs. 763 crore.*
2. *Completion of all viable Sanctioned Railway Projects, within the next 10 years.*
3. *'Operation Cleanlines' to significantly improve the standards of sanitation at railway stations on platforms and inside railway compartments.*

Road Transport

Roads now carry 85 per cent of passenger traffic and 70 per cent of freight traffic. While highways make up only 2 per cent of the overall road network by length, they account for around 40 per cent of this traffic. A series of initiations have been undertaken in recent years, to set the stage for a quantum leap in India's road system. These initiatives combine new institutional arrangements, high-way engineering of international standards, founded on a self-financing revenue model comprising tolls and a cess on fuel. Three initiatives in the road sector were begun in recent years: The National Highway Development Project (NHDP), Pradhan Mantri Bharat Jodo Pariyojana (PMBJP) and Pradhan Mantri Gram Sadak Yojana (PMGSY). NHDP dealt with building high quality highways. The PMBJP dealt with linking up major cities to the NHDP Highways. The PMGSY addressed rural roads.

India has one of the largest road networks in the world aggregating to about 3.3 million km. at present. The ninth plan laid emphasis on a coordinated and balanced development of road network in the country under:

(*i*) Primary road system covering National Highways.

(*ii*) Secondary and feeder road system covering State Highways and Major district roads.

(*iii*) Rural roads including village roads and other district roads.

While the Central Sector Programme pertains mainly to National Highways the responsibility for development of other categories of roads vests with the State/Union territories.

Although India has the third largest road network in the world, efficient and speedy transportation is not facilitated due to large proportion of unsurfaced roads (50 per cent) and the over dependence on National Highways. The National Highways (65569 km) account for about 2% of the total road network but carry as much as 40% of the total road traffic in the country. National and State Highways which are around 172000 km in length, occupy a pre-eminent position as nearly 60% of the freight and 87% of passenger traffic move on them.

National Highway Development Project (NHDP)

Indian roads carry 85 per cent of the passenger and 70 per cent of the freight traffic of the country. The highways, even though they make up only 2 per cent of the road network by length, carry 40 per cent of this traffic. For many years, India lagged behind many countries of the world which built expressways capable of sustained speeds of over 100 kilometre per hour (kph). In recent years, a concerted effort has been undertaken, through new institutional arrangements and improved highway engineering, founded on a revenue model comprising tolls and cess on fuel, to build roads which deliver 80 kph sustained performance. Under the National Highways Development Project (NHDP)-the largest highway project ever undertaken by the country and with the shortest timespan for completion 14,279 kilometre of National Highways are to be converted to 4/6 lanes, at a total estimated cost of Rs. 65,000 crore (at 2004 prices). The NHDP consists of the following components:

(*i*) The Golden Quadrilateral (GQ-5,846 kilometre) connecting the four major cities of Delhi, Mumbai, Chennai and Kolkata.

(*ii*) The North-South and East-West Corridors (NS-EW 7,300 Kilometre) connecting Srinagar in the North to Kanyakumari in the South and Silchar in the East to Porbandar in the West.

(*iii*) Port connectivity and other projects 1,133 kilometre.

Progress of NHDP
(As on January 31, 2005)

(Kilometres)

Length	*GQ*	*NS-EW*	*Post Connectivity & other Projects*	*Total*
Total	5,846	7,300	1,133	14,279
Completed	4,480	675	263	5,418
Under implementation	1,366	857	455	2,678
Balance length to be awarded	Nil	5,768	415	6,183
Cumulative expenditure (in Rs. crore)	20,115	2,131	1,928	24,174

As on January 31, 2005, 5,418 kilometre of NHDP has been completed, the bulk of which (4,480 km) lie on the GQ. The expenditure so far has amounted to Rs. 24,174 crore. There, are 2,678 km under construction. Contracts for 6,183 km are yet to be awarded. It is expected that the GQ would be substantially completed by December 2005, and the NS and EW corridors would be completed by December, 2007. There are constraints faced in timely completion of NHDP which include (*i*) delays in land acquisition and removal of structures, (*ii*) law and order problem in some States and (*iii*) poor performance of some contractors which the completion of more than 75 per cent of the GQ, a substantial impact upon the economy is already visible. At this stage there is a need to focus attention on corridor management and road safety and NHAI has put in place a corridor management policy.

Rural Roads

The Pradhan Mantri Gram Sadak Yojana (PMGSY) was launched in December 2000 as a 100 per cent centrally sponsored Scheme to provide rural connectivity to unconnected habitations with population of 500 persons or more (250 in case of hilly, desert and tribal areas) in the rural areas by the end of the Tenth Plan period. It is funded by the diesel cess in the Central Road Found, and through borrowing from domestic financial institutions and multilateral funding agencies. Augmenting and modernizing rural roads has been included as an item of the NCMP. Thus, the scope of PMGY has been expanded to include both construction of new links and upgradation of existing through routes associated with such link routes to form one complete subnetwork, for providing connectivity between the village and the market.

A survey undertaken to identify the "core network" as part of the PMGSY showed that over 1.70 lakh unconnected habitations needed to be taken up under the PMGSY. This would require new construction of 3.69 lakh kilometre and upgradation of 3.68 lakh kilometres of rural roads, at a total cost of Rs. 1,33,000 crore as against earlier estimates of Rs. 60,000 crore. This does not include the cost of 5-year maintenance of link routes and 10-year maintenance of through routes taken up under PMGSY.

Project proposals amounting to Rs. 14,789 crore have been cleared upto November, 2004 for 35,296 road works covering 103,010 kilometres. Rs. 10,207 crore have been released to States/UTs. In terms of outcomes, 22,930 road works adding up to 60,024 km of roads have been completed, and expenditure of Rs. 7,866 crore has been incurred by States/UTs upto October 2004. The National Rural Roads Development Agency (NRRDA), an agency of the Ministry of Rural Development registered under the Societies Registrations Act provides operational and technical support for the programme.

The World Bank supports the Rural Connectivity Programme through a series of tranches of IBRD loans/IDA credit.

State Sector Road

Since the State Highways and District and Rural Roads are under the responsibility of State Governments, these are developed and maintained by various agencies in State and Union developed in rural areas under Pradhan Mantri Gram Sadak Yojana (PMGSY). The objective of the PMGSY is to link all villages with a population of more than 500 with all-weather roads by the year 2007. The States are also assisted through financial assistance from the Central Road Fund for development of selected roads which are of Inter-State and economic importance.

This Central Road Fund has been given a status by Central Road Fund Act enacted in December 2000.

Pradhan Mantri Bharat Jodo Pariyojana

In addition to National Highway Development Projects 14,000 km-4/6 laning and development of another 10,000 kms has been launched under the Pradhan Mantri Bharat Jodo Pariyojana. The Stretches are being identified on the basis of traffic density, connectivity of State Capitals with NHDP and important centres of tourism and economic activity. The total project cost is about Rs. 40,000 crore. The projects would be awarded on Build, Operate, Transfer (BOT) basis. The project was started on Jan. 14, 2004 by taking 4 laning Pune Nasik section of NH-50. Following National highways have been taken up in the project:

- Chennai-Trivandrum-Pondicherry
- Belgaum-Panji
- Thane-Nasik-Dhule
- Pune-Dhule-Nasik
- Dhule-Nagpur-Raipur-Aurang
- Indore-Ujjain-Khalghat
- Bhopal-Jabalpur
- Agra-Bharatpur-Jaipur
- Ambala-Chandigarh-Shimla
- Chandigarh-Kirathpur
- Delhi-Dehradun
- Bagha border-Amritsar-Jalandhar
- Amritsar-Pathankot
- Slyok-Gangatok
- Guwahati-Itanagar
- Dobaka-Dimapur-Kohima
- Silchar-Imphal
- Silchar-Kohima

Under the project 17 State capital, untouched under NHDP would be connected with a NHDP highways.

Water Transport

Under Water Transport, Coastal Transport and Overseas. Transport have their own importance. In the coastal areas of the sea, navigation is comparatively cheaper. On the 6,000 km of natural peninsular coastline of India, there are 12 major and 185 minor and intermediary operable ports providing congenial and favourable conditions for the development of domestic transport infrastructure. The development and the management of major ports rests with respective Port Trusta under the central government, the state governments administer the major ports. Among the major ports, Kandala, Mumbai, Majhgaon, New Mangalore, Cochin and JNPT (Jawahar Lal Nehru Port, Mumbai) are situated on the Western Coast and Tuticorin, Chennai, Vishakhapattanam, Paradeep, Kolkata and Haldia ports are situated on the Eastern coast. Jawahar Lal Nehru port is the latest big port equipped with most modern facilities. The ports of Mumbai and Cochin and natural sea-ports whereas the Kandala sea-port is used in tides. Vishakhapattanam is the deepest port of India among the sea-ports of the eastern coast. Chennai is the oldest sea-port. Kolkata port is situated on the river bank.

In addition, a new major port at Ennore, 25 km north of Chennai has been constructed with ADB assistance and operationalised. The responsibility for development and management of major ports rests with respective Post Trusts under the Central Government, while state government administer the minor ports.

Besides there are 185 minor and intermediate ports. The minor ports are located in Gujarat (40), Maharashtra (53) Goa (5) Daman and Dui (2) Karnataka (9) Kerala (13) Lakshadweep (10) Tamil Nadu (14) Pondicherry (1) Andhra Pradesh (12) Orissa (2) West Bengal (1) and Andaman & Nichobar (23).

The capacity of the Indian Ports increased from 20 million tonnes (MT) of cargo handling in 1951 to 389.5 MT. as on 31 March, 2004. At the beginning of the Tenth Plan, the capacity of major ports was about 344 MT. It is proposed to be increased to 470 MT by the end of the Tenth Plan.

The process of phased corporatisation has been initiated for the major ports. A beginning has been made by registering Ennore Port Company Ltd., for managing Ennore Port. It has also been decided to corporatise existing major ports starting with Jawaharlal Nehru Port at Navi Mumbai.

Shipping

Shipping plays an important role in the transport of India's economy. Approximately 90 per cent of the country's trade volume (77 per cent in terms of value) is moved by sea. Presently, India has the largest merchant shipping fleet among the developing countries and ranks 19th in the world in terms of shipping tonnage. Indian shipping sector facilitates not only the transportation of national and international cargoes but also provides a variety of other services such as cargo handling services, ship building/repairing, freight forwarding, lighthouse facilities and training of maritime personnel. As on 31st July, 2004, the net operative tonnage consisted of 659 ships totallings 7.74 million Gross Registered Tonnage (GRT) and 12.84 million DWT.

India's shipping fleet as on 31 July, 2004, consist of 211 overseas vessels with 6.6 million GRT and 11.39 million DWT and 441 coastal vessels with 0.80 million GRT and 0.86 million DWT.

Inland Water Transport

India has got about 14,500 km of navigable waterways which comprise rivers, canals, backwaters, creeks, etc. At present, however, a length of 3,700 km of major rivers is navigable by mechanized crafts but the length actually utilised is only about 2,000 km. As regards canals, out of 43,000 km of navigable canals, only 900 km is suitable for navigation by mechanized crafts. About 18 million tonnes of cargo is being moved annually by Inland Water Transport (IWT).

Private Sector Participation in Ports

In order to improve efficiency, productivity and quality of services as well as to bring in competitiveness in port services, the port sector has been thrown open to private sector participation. This is in consonance with the general policy of liberalisation/globalization of economy of the Government of India. It is expected that private sector participation would result in reducing the gestation period for setting up new facilities, also bringing the latest technology and improved management techniques.

Various areas of port functioning such as leasing out existing assets of the port, construction/creation of additional assets, leasing of equipment for port handling and leasing of floating crafts from the private sector, pilotage and captive facilities for port based industries have been identified for participation/investment by the private sector.

Major Ports in India

India has about 5600 km **main coastline** *serviced by 12 major ports which are under the purview of the Central Government.*

West Coast Line	*East Coast Line*
1. Mumbai	1. Kolkata/Haldia
2. Jawaharlal Nehru at Nhava Sheva	2. Paradeep
3. Kandhla	3. Vishakhapattnam
4. Marmugao	4. Chennai
5. New Mangalore	5. Ennore
6. Cochin	6. Tuticorin

The Inland Waterways Authority of India (IWAI) came into existence on 27 October, 1986 for development, and regulation of inland waterways in the country and to act as advisor to the Central and State Governments on matters relating to Inland Water Transport. The Authority undertakes various schemes for development of IWT related infrastructure on National Waterways. The Head Office of the Authority is located at Noida. IWAI also has its regional offices at Patna, Kolkata, Guwahati and Kochi and Sub offices at Allahabad, Varanasi, Bhagalpur, Farakka and Kollam.

Tenth Plan Objective for the Port Sector

During the Tenth Five year Plan 2002-07, it is proposed to enhance capacity and improve productivity in major ports with focus on measures aimed at modernisation, rendering cost-effective services, enhancement of service quality, commercialization through corporations and increased private sector participation.

Shipping Companies

There were 140 shipping companies in the country in operation as on 31 March, 2003 including the Shipping Corporation of India, a public sector undertaking. Of these, 107 are engaged exclusively in coastal trade, 32 in overseas trade and the remaining 10 in both coastal and overseas trade. Shipping Corporation of India, Which is the biggest shipping line of the country, has a merchant fleet of 88 ships of 2.66 million GRT as on 31 March, 2003 and operated on almost all maritime routes. Its tonnage accounts for about 42% of total India tonnage.

National Waterways

The Inland Waterways Authority of India was set up on October 27, 1986. This statutory body is entrusted with the responsibility of development, maintenance and regulation of national waterways.

The Government has identified 10 important waterways for consideration to declare them as National waterways. The following have so far been declared as National Waterways and the same are being developed for navigation by Inland Waterways Authority of India–the Ganga between Allahabad and Haldia (1620 km) on October 27, 1986, the Sadia-Dhubri-stretch of river Brahamputra (891 km) on October 26, 1988 and the Kollam-Kattapuram stretch of west coast canal (168 km) alongwith Champakare canal (14 km) and Udyogmandal canal (23 km) in Kerala with effect from February 1, 1999. Techno-economic studies on many other waterways such so Godawari, Krishna, Barak Sundarbans Buckingham canal, Brahmani East-Coast canal, DVC canal etc. have been completed and found viable.

Sagar Mala Project

Sagar Mala project is a multi dimensional project to infuse a new life into some of the neglected sectors of water transport such as inland waterways and coastal shipping as well as making the large ports of world class order so as to make Indian shipping industry most competitive in the world. Rs. 1,00,000 crore project encompasses all the facts of the maritime sector, including ports, shipping, shipbuilding, inland waterways as well as maritime education and training.

It aims to fully realise the potential of the sector, which will play a major role in the accelerated development of our country and its economy. The project's main thrust is on encouraging private and foreign investment, and privatising port facilities, under the model, to keen pace with the expected growth in India's sea-borne traffic in the coming years. Sagar Mala a golden sea chain of transport along the coastline is an ambitions project convering all facets of maritime transport, including ports, shipping and landline waterways and even cruise tourism.

The main areas of the development of the project are the following:

- Making inland water transport a viable alternative to the heavily burdened rail and road transport.
- Rapid development of domestic ship-building and ship repairing industries.
- Establishment of two maritime universities on each coast so as to produce world class maritime manpower.
- Installing two training ships on each coast besides equipment and simulators in our training institutes and universities.
- Developing 14-20 minor ports exclusively for the purpose of coastal shipping.
- Infrastructure development on the existing three National Waterways.
- Enlarging the storage, loading-unloading facilities at all the major ports.
- Developing new ports with private public partnership.

Air Transport

For Civil Aviation, Air India, Indian Air Lines, Vayudoot, Pawan Hans and Private Air Services are available. Air India and Indian Air Lines were established under the Air Corporation Act, 1953. The Air India was established with the purpose of international air flights and Indian Air Lines for flights within the country. Air India has extended its Air Services in five Continents. Its main office is situated in Mumbai. For International Flights, the aerodromes of Delhi, Mumbai, Kolkata, Chennai and Thiruvananthpuram have been declared as International aerodromes. The responsibility of management and development of these aerodromes was entrusted to International Airports Authority of India. With effect from April 1, 1995, the responsibility of maintenance of International aerodromes alongwith Internal aerodromes has been entrusted to

International Airways Authority of India. Air India owns a fleet of 18 air crafts and operates 189 flights per week servicing 41 stations (28 International and 13 domestic.)

Airports Authority of India

The Airports Authority of India (AAI) was formed on 1 April, 1995 and is responsible for providing safe, efficient air traffic services and aeronautical communication services for effective control of air traffic in the Indian are space. The Authority manages 126 airports inclined 11 international, 86 domestic airports and 29 civil enclaves at defences airfields. It controls and manages the entire Indian space extending beyond the territorial limits of the country, as accepted by the International Civil Aviation Organisation (ICAO).

The Authority has a Civil Aviation Training College at Allahabad for importing training on various operational areas like Air Traffic Control, Radars, Communication etc. The authority maintains the National Institute of Aviation Management and Research (NIAMAR) at Delhi for imparting aviation management training programmes and refresher courses.

Indian Airlines is the major domestic air carrier of the country. It operates to 58 domestic stations (including 2 seasonal stations, *i.e.*, Jaisalmer and Puttaparthy) with its wholly owned subsidiary Alliance Air Indian Airlines also operates to 17 international stations, *viz.*, Bangkok, Singapore, Kuala Lampur, Yangon, Kathmandu, Colombo, Dhaka, Malee, Kuwait, Sharjah, Dubai, Fujairah, Ras-al-Khaimah, Muscat, Doha and Bahrian India Airlines has a fleet of 62 aircraft.

First Indian Airport in Private Sector

The first Indian airport in private sector is under construction at Cochin. It is being constructed by the company named 'Cochin International Airport Ltd.' This airport will also provide service for International flights.

Vayudoot was established on Jan. 20, 1981 as a Corporation. Basically, it was established to fulfil the requirements of Air Services in North-Eastern area, but by and by it has extended its services in other areas also where the services of Indian Air Lines are not available. Vayudoot had 16 aeroplanes in 1992-93. Vayudoot, too, showed net loss in 1991-92 and 1992-93 amounting to Rs. 30.6 crore and Rs. 24.2 crore respectively. After 1992-93, Vayudoot has been merged with Indian Air Lines.

Airport Privatisation

It has been decided to restructure the Airports of Airports Authority of India as and when found suitable through long term lease. This will improve the managerial efficiency bring the standard of services/facilities at par with the international standards and to attract investment from private sector. The existing airports at Delhi, Mumbai, Chennai and Kolkata are being taken up for this exercise.

The Government has also approved the proposal for setting up of new airports of International Standards at Bangalore, Hyderabad and Goa with private sector participation.

In civil Aviation sector, 49% Foreign Direct Investment (FDI) in Airlines, 74% FDI in airports development (100% FDI with permission of Government) is allowed.

Pawan Hans was established on Oct. 15, 1985 as a Corporation. Its original name was Helicopter Corporation of India which was later renamed as Pawan Hans Helicopter Ltd. (PHHL). Basically, it was established to provide helicopter services in the exploration of oil to ONGC (Oil and Natural Gas Commission). Besides, it also provides services to inaccessible areas and charter services for the promotion of tourism. It has developed the helicopter market in India and provided a thrust to the industry as a whole. It has been awarded ISO 9002 certification for its entire gamut of activities and is first aviation company in India to achieve the same. In the recent years, the permission to operate the private air taxis in the field of Air Transport has been made available by the Government, as a result of which the sole right of Public Sector in Air Transport has come to an end. Air Taxis Services can be operated as non-scheduled flights both on 'chartered' and 'non-chartered' basis to all airports in the country open to civil operations. The Air Taxi-operators are free to charge any fare on their own commercial discretion.

The Recommendations of Rakesh Mohan Committee on Infrastructure

In Oct. 1994, under the Chairmanship of Sri Rakesh Mohan, Director General of NCAER (National Council of Applied Economic Research), a 17 member committee was constituted by the Govt. The committee handed-over its report of the Finance Ministry on June 22, 1996 which was released on Jan 8, 1997. The committee in its Report, has said that it is possible to increase the investment in basic infrastructure area from the present level 5.5% of GDP to 7% by 2000-01 and 8% by 2005-06. This will increase to $ 2000-01 and further to $ 50 billion by 2005-06 in place of present annual investment of $ 17 billion. The committee has recommended wide concessions in taxes to promote private investment in basic infrastructure areas.

In order to fulfil the financial needs of basic areas, the committee has recommended the use of funds available with insurance companies, Assets Management Companies of Employee's Provident Fund. The committee has said that it is better to create small insurance companies and divide GIC and its four allied companies. The other recommendations of the committee are as under:

Telecommunication

- *To open the long distance inter-circle services to private sector by 2001.*
- *To create a company named 'India' Telecom for the Telecommunication department.*
- *To consider the increased privatisation of MTNL (Mahanagar Telephone Nigam Ltd.).*
- *To deposit all the license fees connected with telecommunication into one basic fund to that loans and equities may be provided to projects concerned with basic infrastructure.*

Electricity

- *The formula of fixation of electricity charge should be based on investment and there must be a possibility of 1% increase in levy on an average per year in addition to adjustment for inflation.*
- *In order to promote of Private sector, a minimum charge of electricity should be fixed.*
- *The State Electricity Boards should be reconstituted at an early date.*
- *The Central Electricity Tribunal should be decentralized.*

Road

- *To establish a 'Highway Development Fund' by levy of surcharge of 50 paisa on diesel per litre and Re. 1 on petrol per litre.*
- *To construct Super National Highway and by-passes etc. on an extensive scale in collaboration with private sector.*
- *To begin 'Highway Infrastructure Saving Scheme' on the tune of National Saving Scheme.*
- *To create a Road Board for the development of integrated main routes.*
- *To create a Road for the development of integrated main routes.*
- *To have 'Four Lanes' for some present National Highways.*

Sea-ports

- *At least 2 'Mega ports' should be developed in the country. There should be at least one on both the coasts.*
- *To exempt the private collaboration for sea-ports from corporation taxes.*
- *The financial arrangement for sea-ports should be done through primary market or financial institutions.*

Merger of NAA and IAAI

The two Airport Authorities–National Airports Authority (NAA) and International Airports authority of India (IAAI) have been organisation named as Airports Authority of India (AAI). This newly created AAI is presently controlling both domestic and international airports of the country. Prior to it NAA was controlling domestic airports, while IAAI was managing international airports of the country.

Postal Network

India has the largest postal network in the world. At the time of independence there were 23-344 post offices throughout the country of these 19,184 post offices were in the rural areas and 4,160 in urban areas. Today, the country has 1,55,669 post offices, of which roughly 89% were outside cities. As a result of this seven-fold growth in the postal network, today India has the largest postal network in the world.

On an average, a post office serves an area of 21.11 sq kms and a rural areas are opened subject to satisfaction of norms regarding population, income and distance stipulated by the Department. There is an element of subsidy to the extent of 85 per cent of cost in opening post offices in hilly, desert and inaccessible areas, whereas the subsidy in opening post offices in normal rural are areas is to the extent of 66.66% per cent of the cost.

New National Telecommunication Policy
(Announced on 26 March, 1999)

New Telecommunication Policy (NTP'99) approved by the Central Cabinet was announced on 26th March, 1999 by the chairman of GOT (Group on Telecommunications) and the same day it was declared by Telecommunication Minister in a news conference. The new policy has come into force from April 1, 1999 and the Telecommunication Policy of 1994 comes to an end.

Under the new policy, the participation of the Private sector is emphasized. The new participators of the Private sector will once more be allowed to operate on the basis of Revenue Sharing in addition to admission fee. The details of this will be issued by TRAI. The present operators who are operating under 'Licence Fee System' will be able to enter the new Revenue Sharing System or not is a matter to be announced later after consultation with the Attorney General.

Under the new policy, the target is fixed to make available the 'Telephone on Demand' upto the year 2002. In the country, the Teledensity is targeted upto 2005 @ 7 per thousand population and upto 2010 @ 15 per thousand population. In the Rural Sector, the Teledensity at present in 0.4 per thousand population. It has been decided to increase it up to 2010 @ 4 per thousand population. All the villages will be brought under the control of Telecommunication Services in the next year and all the distt. Headquarters will be brought under Internet upto 2000. Under the new policy, STD and ISD will be open to Private Sector *w.e.f.* 1 March, 2000 and 1 March, 2004 respectively.

In regard to Cellular Phone service, a third operator Telecommunication Deptt/Mahanagar Telephone Nigam Ltd. (MTNL) is being introduced in place of two-fold operators under the new policy. The introduction of the fourth operator will depend on the recommendations of TRAI. Thus, its has been broadly accepted to allow the entrance of more companies in place of 'Double Monopoly' (Telecommunication Deptt/MTNL). Their number will be fixed by TRAI.

Policy in the Telecom Sector

Key Policy Developments

- *National long distance service opened up for unrestricted entry.*

- *Termination of monopoly of VSIN for International Long Distance (ILD) services has been preponed to March 31, 2002 from March 31, 2004.*
- *25 new Basic Service License Agreements signed by private operators. Fourth cellular operator, one each in four metros and thirteen circles has been permitted with seventeen fresh licenses issued to private companies in September/October 2001.*
- *Wireless in Local Loop (WLL) has been introduced for providing telephone connections in urban, semi-urban and rural areas promptly.*
- *Process of disinvestments of VSNL is underway.*

Other Policy Initiatives

- *Sanchar Sagar project being executed through 'DWDM' Technology to meet the bondwith demand of IT sector.*
- *Asynchronous Transfer Mode (ATM) being introduced in 8 cities for high-speed data transmission.*
- *VSNL has developed its capacity to provide international bandwidth on demand.*
- *HTL disinvestment completed.*
- *License conditions for Global Mobile Personal Communications by Satellite (GMPCS) have been finalized on November 1, 2001.*
- *Policy of Voice mail/Audiotex service was announced in July 2001 by incorporating a new service called "Unified Messaging Service".*
- *ISPs have been permitted for setting up of Submarine cable landing station for international gateways for Internet.*
- *Thirteen ISPs have been given clearance for commissioning of international gateways for Internet using satellite medium for 29 gateways.*
- *16 companies are operating Public Mobile Radio Trunked Service (PMRTS) started in 1995 in 25 cities. Policy for PMRTS in terms of NTP' 99 has been announced on November 1, 2001 under which fresh licenses will be granted only in digital technology.*
- *National Internet Backbone (NIB) has been commissioned.*

In the new policy, the policy making of Telecommunication deptt. and work of providing licence has been separated from the work of service rendering. The matters pertaining to service will be made upto 2001 by corporisation. For the uplift of services, Deptt. of Telecom Services will be established separately. The participators who rend services will also be under the charge of TRAI.

In the new Telecommunication Policy, the powers of TRAI are made more explicit. Its jurisdiction will be restricted to 'Service Providers'. TRAI will issue instructions to all Service Providing Agencies and the personal disputes will be settled by its as well. This includes the disputes between Telecommunication Deptt. and Private operators. In this policy, if has also been made clear that the position of TRAI is consultative and the Government is not bound to consult TRAI.

According to the New Telecom Policy (NTP) 1999, the Government has opened the National Long Distance Service to private operators without any restriction on the number of operators with effect from August 13, 2000. The Government has also issued guidelines for license to Infrastructure Provider-II (IP-II) for leasing/renting out/selling end-to-end bandwidth. For Infrastructure Provider-I (IP-I) providing assets such as Dark fibres, Right of Way, Duct Space and Tower, no formal license is required. They are only required to be registered as IP-I.

Broad Extension of Telephone Network

The Telecommunication services in India have improved significantly since Independence.

With the opening of Telecom sector to private investment and establishment of an independent regulator, the matter of separation of service provision function of the Department of Telecommunications (DOT) and providing a level playing field to various service providers including the government service provider, has been achieved. From 1 October, 2000 a new Public Sector Undertaking, *viz.*, Bharat Sanchar Nigam Limited (BSNL), has been formed to take overall the service providing functions of the erstwhile Department of Telecommunication Services (DTS).

Initially, the telephone exchanges were of manual type, which were subsequently upgraded to Automatic Electro-Mechanical type. In the last one-and-a-half decade, a significant qualitative improvement has been brought about by inducting Digital Electronic Exchanges in the network on a very large scale. The number of exchanges which was around 300 then, increased to 36,772 by July 2003. Today 100 per cent telephone exchanges in the country are of electronic type.

Rapid expansion in the telecom sector has been accompanied by a simultaneous significant technological change. The total number of lines grew by about 40 per cent in 2003-04, to cross 76 million at end March 2004. At the same time, there was a continuing massive shift in the technology of access from fixed line to mobile telephony. In 2003-04, fixed lines grew by less than 3 per cent, while mobile telephones grew by 159.2 per cent. The growth of mobile phones during the year was accelerated by the introduction of the Calling Party Pays (CPP) regime also introduced in May 2003.

After the announcement of New Telecome Policy 1999, progress in telecom in India has been extremely rapid. The total number of telephone (basic and mobile) rose from 22.8 million in 1999 to 88.6 million at the end of October, 2004. During 2003-04 itself, 21.92 million telephones were added, which was equal to the total number of phones installed as of 1999. During the first seven months of the 2004-05, 12 million phones have been added. Overall, teledensity rose from just 2.32 in 1999 to 8.2 in October 2004.

The structure and composition of telecom growth has undergone a substantial change in terms of mobile versus fixed phones and public versus private participation. In 1999, both mobile phones and private sector separately accounted for 5 per cent of total number of phones. In October 2004, mobile phones accounted for 50 per cent of total phones and the private sector accounted for 44 per cent of total phones.

International Comparison of Teledensity (2003)

Countries	*Teledensity*
Australia	126.18
Bangladesh	1.56
Brazil	42.38
China	42.32
India	6.60
Indonesia	9.17
Nepal	1.70
Pakistan	4.42
Sri Lanka	9.57
U.K	143.13
U.S.A.	116.43

Source: ITU December, 2003.

Note: India's teledensity as on March 31, 2004 was 7.20.

Although India's 88.62 million strong telephone network, including mobile phones, is one of the largest in the world, with the low telephone penetration rate of about 8.20 phones per hundred population, the country offers vast scope for growth. Present projections suggest that by the end of 2007, the total number of phones could reach 250 million.

Over the recent period, public sector operators (BSNL and MTNL) have lost market share in fixed telephony from 98.65 per cent to 91.39 per cent. In the past two years, Public Sector Undertaking (PSUs) have actually seen a decline in the number of fixed lines, while such lines have grown in the private sector. At the same time, the PSUs actually gained market share in mobile telephony, going from 3.98 per cent 20.21 per cent share of the market. Overall, the share of PSUs declined from 90 per cent to 55.6 per cent

The growth of tele-density has required substantial financial investment. One important source for this investment has been FDI. From August 1991 to August 2004, 926 proposals of FDI of Rs. 41.368 crore were approved. The actual FDI inflow of approximately Rs. 5,763 crore between January 2001 and August 2004 alone was about 56 per cent of the total FDI flow in telecome since its inception in 1991. In terms of approval of FDI, the telecom sector is the second largest, after power and oil refineries.

Telephone Connections

(million lines)

	Public	*Private*	*Overall*
Telephone connections as on March 31, 2004			
Mobile	6.00 (7.84)	27.70 (36.19)	33.70 (44.03)
Fixed	40.48 (52.89)	2.36 (3.08)	42.84 (55.97)
Overall	46.48 (60.73)	30.06 (39.27)	76.54 (100)
Growth (per cent) during 2003-04			
Mobile	127.27	167.37	156.20
Fixed	–0.12	116.51	2.93
Overall	7.67	162.53	40.10

Communication Convergence Bill 2001

In pursuance of the New Telecom Policy (NTP)-1999, action was taken to prepare a new comprehensive statue to replace the Indian Telegraph Act 1885 keeping in view the rapid convergence of telecom, computers, television and electronics. Accordingly, the Communication Convergence Bill, 2001. A Bill to promote, facilitate and develop in an orderly manner the carriage and content of communications (including broad casting, telecommunication and multimedia), for the establishment of an autonomous Commission [i.e., the Communications Commission of India] to regulate carriage of all forms of communications, and for establishment of an Appellate Tribunal [i.e., the Communications Appellate Tribunal] and to provide for matter connected therewith introduced in the Lok Sabha on 31 August, 2001. The Bill was referred to the Standing Committee on Information Technology for Examination. The Committee submitted its Report on 20 November, 2002, and the observations and recommendations made by it are under consideration.

Key Policy Developments in the Telecom Sector

- *The International long distance business was opened up for unrestricted entry.*

- *The monopoly of VSNL over International Long Distance (ILD) service was ended and VSNL was privatised.*
- *Large number of villages covered through Wireless in Local loop (WLL).*
- *The National Internet Backbone (NIB) was commissioned.*
- *The USO Administrator was appointed and his office is functioning.*
- *Radio Frequency Spectrum allocation in being modernised and automated to efficiently address the dynamic needs of the liberalised sector.*

Major Initiatives for Infrastructure Policy

Infrastructure Finance

- *An Infrastructure Equity Fun of Rs. 1,000 crore is being set up to help in providing equity investment of infrastructure projects.*
- *An institutional mechanism is being set up to coordinate debt financing by financial institutions and banks of infrastructure projects larger than Rs. 250 crore. Industrial Development Finance Corporation (IDFC) will act as the coordination institution with responsibility for different sectors being shared with Industrial Development Bank of India (IDBI) and Industrial Credit and Investment Corporation of India (ICICI).*

Power

- *Electricity Act notified in June 2003.*
- *28 States signed the tripartite agreement for one-time settlement of the dues of State Electricity Boards (SEBs) to Central Public Sector Undertakings (CPSUs), and offer securiting the dues, 27 states issued bonds amounting to Rs. 28,98385 crore, August 2003 onwards.*
- *50,000 MW hydro electric initiative launched in May 2003.*

Telecom

- *Unified Access Service License regime introduced in October 2003.*
- *Telecommunication Interconnection Usage Charges (IUC) Regulation, notified on October 29, 2003.*
- *Universal Service Obligation Fund set up as a separate non-lapsable fund in January 2004.*

Roads

- *The 'Golden Quadrilateral' is in progress.*
- *Pradhan Mantri Bharat Jodo Project for development of 10,000 kms of roads connecting state capitals with National Highways launched in January 2004.*

Civil Aviation

- *International airports at Delhi, Mumbai, Chennai, and Kolkata are proposed to be upgraded to the standards of world class airports.*
- *New international airports at Bangalore, Hyderabad, and Goa are planned, with private sector participation.*

Railways

- *National Rail Vikas Yojana launched.*
- *Seventeen inter-city train services called Jan Shatabdi introduced.*
- *Rail Vikas Nigam set up in January 2003.*

Electronic Mail Service

It is a **'Store and Forward'** communication system based on computers. This system does not require the presence of both sender and receiver. This service is rapidly getting popularity because it is computer based system and provides facilities of data collection, recollection and storing. This system transfers data easily from one end to the other. This system can further be extended for use in EDI (Electronic Data Interchains), Software improvements, Store and Forward Fax Systems.

Gramin Sanchar Sewak Scheme (GSS)

It is a pilot scheme launched on 24 December, 2002 by the Prime Minister through Grameen Dak Sewak rural post offices, who are willing to work as franchisee for BSNL on the existing STD/ISD/PCO franchisee basis. In this scheme, GDSDA valunteers are called Grameen Sanchar Sewak (GSS), who carry a mobile fixed wireless terminal (FWT) with display unity in a carry bag and visit door to door to provide telephone facility to the rural population in his routine beat in the village. The FWT has been supplied by the BSNL and the scheme makes use of the existing WLL network. The pilot scheme has been implemented throughout the country except in Andaman and Nicobar, Haryana and Punjab Telecom Circles. As on 31 March, 2003, around 8,100 villages are being covered by 1,860 GSSs.

Voice Mail

Voice Mail Service provides the facilities of collecting and reobtaining voice messages through telephone. This service is useful for the persons who are often on tours and do not have telephone facility. The customer of this service has one voice mail number and a mail box. Any message can be left with the mail box which can again be collected from it, if required. This service also reduces the financial burden of exchange bill.

Voice Mail has broad scope which can include targeted Audio Tax Service.

The described above services (like road, railways, airways, ports, ship, telecommunications) may be utilized for the purpose of fisheries/agriculture development and transportation of fisheries agriculture goods from one place to another.

Chapter 5

PUBLIC FINANCE

Financial Relations Between Centre and States

India possess a federal structure in which a clear distinction is made between the union and the state functions and sources of revenue. Our constitution provides residual powers to the Centre. Article 264 and 293 explain the financial relations between the Union and State Government.

Although the states have been assigned certain taxes which are levied and collected by them. They also share in the revenue of certain union taxes and there are certain other taxes which are levied and collected by the Central Government but whole proceeds are transferred to the states.

The constitution provides residuary powers to the Centre. The Indian constitution makes a clear division of fiscal powers between the Centre and State Governments.

(*A*) The List I of Seventh Schedule of Indian Constitution enlists the union taxes which are as follows:

1. Taxes on income other than agriculture income.
2. Corporation tax.
3. Custom duties.
4. Excise duties except on alcoholic liquors and narcotics not contained in medical or toilet preparation.
5. Estate and succession duties other than on agricultural land.
6. Taxes on the capital value of assets except agricultural land of individuals and companies.
7. Rates of stamp duties on financial documents.
8. Taxes other than stamp duties on transactions in stock exchanges and future markets.
9. Taxes on sales or purchases of newspapers and on advertisements therein.
10. Taxes on railway freight and fares.
11. Terminal taxes on goods or passengers carried by railways, sea or air.
12. Taxes on the sale or purchase of goods in the course of interstate trade.

(*B*) List II of Seventh schedule enlists the taxes which are within the jurisdiction of the states:

1. Land revenue.
2. Taxes on the sale and purchase of goods, except newspapers.

3. Taxes on agricultural income.
4. Taxes on land and buildings.
5. Succession and estate duties on agricultural land.
6. Excise on alcoholic liquors and narcotics.
7. Taxes on the entry of goods into a local area.
8. Taxes on the consumption and sale of electricity.
9. Taxes on mineral rights (subject to any limitations imposed by the Parliament).
10. Taxes on vehicles, animals and boats.
11. Stamp duties except those on financial documents.
12. Taxes on goods and passengers carried by board or inland waterways.
13. Taxes on luxuries including entertainments, betting and gambling.
14. Tolls.
15. Taxes on professions, trades callings and employment.
16. Capitation taxation.
17. Taxes on advertisements other than those contained in newspapers.

(*C*) Apart from taxes levied and collected by the states, the constitution has provided for the revenues for certain taxes on the union list to be allotted, partly or wholly to the states. These provisions fall into various categories:

1. Duties which are lived by the union government but are collected and appropriated by the states. These include stamp duties, excise duties on medical preparations containing alcohol or narcotics.
2. Taxes which are levied and collected by the union, but the entire proceeds of which are assigned to the states, in proportion determined by the Parliament. These taxes include:
   (*i*) Succession and Estate duty.
   (*ii*) Terminal taxes on goods and passengers.
   (*iii*) Taxes on railway freight and fares.
   (*iv*) Taxes on transactions in stock exchanges and future markets.
   (*v*) Taxes on sale and purchase of newspapers and advertisements therein.
3. Central taxes on income and union excise duties are levied and collected by the union but are shared by it with the states in a prescribed manner.
4. Proceeds of additional excise duty on mill made textiles, sugar and tobacco which are levied by the union since 1957 in replacement of state sales taxes on these commodities, are wholly distributed among the states in a manner as to guarantee their former incomes from the displaced sales taxes.

Service Tax

Service tax was introduced on 1994-95 in a small way to operationalise the principle of neutrality of the tax system to different forms of production and in recognition of the fact that value additions, whether in manufacturing or service, should form the basis of taxation. Taxation of services faces three problems. First, a large number of service providers are in the informal sector without a regular system of accounts. Second, revenue potential of some of the services may not be commensurate with the efforts involved in the identification, assessment and enforcement. Third lack of standards in assessments, in view of the wide variations in value additions across and within sectors, may lead to disputes. A system of self-assessment is in vague since April

2001, whereby service providers, on whom the responsibility of payment of tax vests, are required to file a simple return. The rate of service tax was increased from 5 per cent in 1994-05 to 8 per cent in all the taxable services from May 14, 2003. As a major step towards integrating the tax on goods and services, Budget for 2004-05 extended the credit of service tax and excise duty across goods and services. To offset the negative revenue impact on account of such a move, the service tax rate was increased from 8 per cent to 10 per cent. Successive Budgets have new services into the tax net. In 2005-06 budget, service tax rate has been retained at 10% but 21 new services have been included under the sphere of service tax. Besides, annual turnover upto Rs. 4 lakh has been exempted from services tax.

Finance Commission

Finance Commission is constituted to define financial relations between the Centre and the States. Under the provision of Article 280 of the constitution, the President appoints a Finance Commission for the specific purpose of devolution of non-plan revenue resources. The functions of the Commission are to make recommendations to the President in respect of:

1. The distribution of net proceeds of taxes to be shared between the union and the states and the allocation of share of such proceeds among the states.
2. The principles which should govern the payment of grants-in-aid by the Centre to the States.
3. Any other matter concerning financial relations between the Centre and the States.

Finance Commission	*Year of Establishment*	*Chairman*	*Operational Duration*	*Year of Submitting Report*
I	1951	K.C. Niyogi	1952-57	1952
II	1956	K. Santhanam	1957-62	1956* and 1957
III	1960	A.K. Chanda	1962-66	1961
IV	1964	P.V. Rajamannar	1966-69	1965
V	1968	Mahaveer Tyagi	1969-74	1968* and 1969
VI	1972	Brahma Nand Reddy	1974-79	1973
VII	1977	J.M. Shellet	1979-84	1978
VIII	1983	Y.B. Chawan	1984-89	1983* and 1984
IX	1987	N.K.P. Salve	1989-95	1989
X	1992	K.C. Pant	1995-2000	Nov. 26, 1994
XI	1998	A.M. Khusro	2000-2005	Jan. 15, 2000* & 7 July, 2000 & 31 Aug. 2000.
XII	2003	C. Rangrajan	2005-10	Report submitted on Nov. 30, 2004.

* Interim Report.

In above context so far 11 Finance Commissions have been appointed which are as follows:

All the above 11 Commissions have submitted their reports in the years mentioned above. The recommendations of the various commission can be divided under three heads:

(*a*) Division and distribution of Income Tax and other taxes (Table 1 & 2)

(*b*) Grants-in-aid (Table 3)

(*c*) Loans to the states by the Centre (Table 3)

Table 1 : Recommendations of Finance Commissions of Income Tax

Finance Commission	*States Share of Income Tax (per cent)*	*Distribution of Income Tax to the States on the Basis of*	
		Population & other Criteria	*Tax Contribution*
I	55	80	20
II	60	90	10
III	65	80	20
IV	75	80	20
V	75	90	10
VI	80	90	10
VII	85	90	10
VIII	85	90	10
IX	85	90	10
X	77.5	20	80 (Other basis)

Table 2 : Recommendations of Finance Commission on Excise Duty

Finance Commission	*States Share of Excise Duty*	*Distribution of Excise Duty (%)*	
		On the Basis of Population	*On the Basis of Backwardness of States the percentage of the Poor in the States etc.*
I	40% of 3 duties	40	60
II	25% of 8 duties	40	60
III	20% 35 duties	40	60
IV	20% of 45 duties	80	20
V	20% of 45 duties	80	20
VI	20% of 45 duties	75	25
VII	40% of all duties	-	25% of Four Factors
VIII	45% of all duties	-	New Formula
IX	45% of all duties	-	New Formula
X	47.5% of all duties	-	New Formula

Eleventh Finance Commission

The Eleventh Finance Commissions was set up under the Chairmanship of Prof. A.M. Khusro with Mr. N.G. Jain, Mr. J.C. Jetty and Dr. Amaresh Bagchi as members of the commission. Mr. T.N. Srivastava was the member secretary. The commission was asked to submit its report by December 31, 1999 but the commission submitted its Interim Report on January 15, 2000 and final report on July 7, 2000. The commission also submitted its supplementary report on August 31, 2000.

The Eleventh Finance Commission was mandated for the first time in terms of 73rd and 74th amendments of the constitution to recommend the measures needed to augment the consolidated fund of a state to supplement the resources of the panchayats and municipalities.

Table 3 : Grants-in-aid given to States

Finance Commission		*Grants-in-aid*
I		For 7 states to cover their deficits during the period 1951-56.
II		Larger grants-in-aid for meeting developmental need of the States.
III	(*i*)	Rs. 550 crore to all States except Maharashtra to cover part of their revenue expenditure.
	(*ii*)	Rs. 45 crore for improvement of communications.
IV		Rs. 610 crore to cover deficits during the period 1966-71.
V		Rs. 638 crore to cover deficits during the period 1969-74.
VI		Rs. 2,510 crore for 14 out of 21 States to cover their non-plan deficits during the period 1974-79.
VII		Rs. 1,600 to cover deficits of a few poorer States during the period 1980-85 and also to up-grade the standard of administration.
VIII	(*i*)	A small grant of Rs. 1,556 crore for the period 1985-90 to cover deficits.
	(*ii*)	A grant of 915 crore to certain States to upgrade the standard of administration.
IX	(*i*)	Grant of Rs. 15,017 crore to cover deficits on plan and non-plan revenue account during 1990-95.
	(*ii*)	A special annual grant of Rs. 603 crore towards the Centre's contribution to the Calamity Relief Fund-totalling Rs. 3,015 crore for the 5 year period 1990-95.
	(*iii*)	A grant of Rs. 122 crore to Madhya Pradesh towards the expenditure on rehabilitation and relief of victims of Bhopal Gas leak.
X	(*i*)	Grant-in-aid of about Rs. 7,580 crore to cover deficits on revenue account during 1995-2000.
	(*ii*)	Upgradation grants of about Rs. 1,360 crore for such items as police, fire service, jails, girls education, primary education, drinking water etc.
	(*iii*)	Grants of about Rs. 1,250 crore to solve special problems of States.
	(*iv*)	Calamity Relief of Rs. 4,730 crore.
	(*v*)	Grants of Rs. 5,380 crore to local bodies *viz.* Panchayats and municipalities.

Table 4 : Components of Statutory Devolution Under the Last Five Finance Commissions

(Rs. Crores)

Finance Commission	*Tax Devolution*	*Deficit Grants*	*Other Grants*	*Total Grants*	*Total Devolution*
V	3,590 (88)	490	–	490 (12)	4,080 (100)
VI	6,940 (80)	820	930	1,750 (20)	8,690 (100)
VII	18,810 (97)	140	390	530 (3)	19,340 (100)
VIII	33,130 (93)	970	1,420	2,390(7)	35,520 (100)
IX	87,880 (83)	15,010	3,140	1,81,60 (17)	1,60,040 (100)
X	2,06,340 (99)	7,580	12,720	20,300 (1)	2,26,640 (100)

Note: Figures in brackets represent the percentage.

Salient Recommendations of 11th Finance Commission

- 28% of Net revenue proceeds of shareable Central taxes and duties should be transferred to the State. In addition to this 1.5% of this net revenue should also be transferred to the states for not imposing sales tax on sugar, tobacco and textiles. Thus, the 11th Finance Commission recommended to transfer 29.5% of net revenue proceeds of shareable central taxes and duties.

- The upper limit of transferable revenue proceeds from the centre to the states will be 37.5% of total revenue (tax and non-tax revenue).
- The formula for determining interstate division of revenue includes various factors. The weightages of these factors are : Population (10%), Deviation from the average per capita income (62.5%), Area (7.5%), Infrastructure Index (7.5%), Tax Efforts (5%) and Fiscal Discipline (7.5%). [On the basis of this formula Uttar Pradesh will get the maximum share of 19.798% in tax revenue transferable to the states, Sikkim will get the minimum share of 0.184% of this transferable tax proceeds.]

Share of Various States in Transferable Tax Revenue of the Centre
(Recommendations of 11th Finance Commission)

States	*Share (%)*
Uttar Pradesh	19.798
Bihar	14.597
Madhya Pradesh	8.838
West Bengal	8.116
Andhra Pradesh	7.701
Rajasthan	5.473
Tamil Nadu	5.385
Orissa	5.056
Karnataka	4.930
Maharashtra	4.632
Assam	3.285
Kerala	3.057
Gujarat	2.821
Jammu & Kashmir	1.290
Punjab	1.147
Haryana	0.944
Himachal Pradesh	0.683
Tripura	0.487
Manipur	0.366
Meghalaya	0.342
Arunachal Pradesh	0.244
Nagaland	0.220
Goa	0.206
Mizoram	0.198
Sikkim	0.189

- Total grant-in-aid of Rs. 35359 crore to states for solving the problem of revenue deficits. Out of this 85% amount is to be given to 15 identified States. The remaining amount of Rs. 5303.86 crore should be transferred to Promotional fund along with the same additional amount added by the Central Govt. This amount is to be distributed among all the states on the basis of their fiscal performance during 2000-05.
- Grant-in-aid of Rs. 10000 crore for local bodies during 2000-2005 *i.e.,* Rs. 2000 crore every year (Rs. 1600 crore for rural bodies and Rs. 400 crore for urban bodies).
- Repayment of specific debts given to Punjab in 1984-94 for removing terrorist activities should remain suspended up to 2005.
- Services should be brought under taxation and service tax should be included in concurrent list.
- Suggestion to modify the limits of profession tax.
- Suggestion to determine the rates of water and electricity charges on the basis of cost itself.
- Suggestion to set up pay commission on the advice of states (not necessarily after 10 years).
- Suggestion for exploring the measures to reduce the burden of pension on the treasury.
- Suggestion for establishing NCCF (National Calamity Contingency Fund) in place of NFCR (National Found for Calamity Relief).

Twelfth Finance Commission

The Twelfth Finance Commission (TFC) under the chairmanship of Dr. C. Rangrajan was appointed on November 1, 2002 to make recommendations regarding the distribution between the Union and the States of net proceeds of shareable taxes, the principles which should govern the grants-in-aid of the revenues of States from the Consolidated Fund of India and the measures needed to augment the Consolidated Fund of a State to supplement the resources of local bodies in the State on the basis of the recommendations made by the Finance Commission of the State. The terms of reference mandated the Commission to review the state of the finances of the Union and the States and suggest a Plan by which the Governments, collectively and severally, restore budgetary balance. Achieve macroeconomic stability and debt reduction along with equitable growth. Furthermore the Commission was also asked to suggest corrective measures for debt sustainability and to review the Fiscal Reform Facility introduced by the Central Government.

The Commission submitted its report on November 30, 2004 covering the period 2005-10.

The Commission recommended debt relief to States linked to fiscal reforms, doing away with the present system of Central assistance to State plans in the form of grants and loans and transfer of external assistance to States on the same terms and conditions as attached to such assistance by external funding agencies. The TFC raised the share of States in shareable Central taxes from 29.5 per cent to 30.5 per cent. Total transfers to States recommended by the TFC amount to Rs. 7,55,752 crore over the five year period 2005-2010. Of this transfers by way of share in Central taxes and grants-in-aid amount to Rs. 6,13,112 crore and Rs. 1,42,640 crore respectively. The total tranfers recommended by the TFC are higher by 73.8 per cent over those recommended by the Eleventh Finance Commission (EFC). Within the total transfers, while the share in Central taxes is higher by 62.9 per cent, grants-in-aid recommended by the TFC are higher by 143.5 per cent over those recommended by the EFC.

Criteria & Weights for determining Shares
(12th Finance Recommendations)

Criteria	*Weight (%)*
Population	25.0
Income Distance	50.0
Area	10.0
Tax Effort	7.5
Fiscal Discipline	4.5
Total	**100.0**

Share of Various States in Transferrable Tax Revenue of the Centre
(Recommendations of 12th Finance Commission)

States	*Share(%)*
Andhra Pradesh	7.356
Arunachal Pradesh	0.288
Assam	3.235
Bihar	11.028
Chhattisgarh	2.654
Goa	0.259
Gujarat	3.569
Haryana	1.075
Himachal Pradesh	0.522
Jammu & Kashmir	1.297
Jharkhand	3.361
Karnataka	4.459
Kerala	2.665
Madhya Pradesh	6.711
Maharashtra	4.997
Manipur	0.362
Meghalaya	0.371
Mizoram	0.239
Nagaland	0.263
Orissa	5.161
Punjab	1.299
Rajasthan	5.609
Sikkim	0.227
Tamil Nadu	5.305
Tripura	0.428
Uttar Pradesh	19.264
Uttaranchal	0.939
West Bengal	7.057
All States	**100.000**

Recommendations of the Twelfth Finance Commission

Restructuring Public Finances

- *Centre and States to improve the combined tax-GDP ratio to 17.6 per cent by 2009.10.*
- *Combined debt-GDP ratio, with external debt measured at historical exchange rates, to be brought down to 75 per cent by 2009-10.*
- *Fiscal deficit to GDP targets for the Centre and States to be fixed at 3 per cent.*
- *Revenue deficit of the Centre and States to be brought down to zero by 2008-09.*
- *Interest payments relative to revenue receipts to be brought down to 28 per cent and 15 per cent in the case of the Centre and States, respectively.*
- *States to follow a recruitment policy in a manner so that the total salary bill, relative to revenue expenditure, net of interest payments, does not exceed 35 per cent.*
- *Each State to enact a fiscal responsibility legislation providing for elimination of revenue deficit by 2008-09 and reducing fiscal deficit to 3 per cent of State Domestic Product.*
- *The system of on-lending to be brought to an end over time. The long term goal should be to bring down debt-GDP ratio to 28 per cent each for the Centre and the States.*

Sharing of Union Tax Revenues

- *The share of States in the net shareable Central taxes at 30.5 per cent, treating additional excise duties in lieu of sales tax as part of the general pool of Central taxes. Share of States to come down to 29.5 per cent, when States are allowed to levy sales tax on sugar, textiles and tobacco.*
- *In case of any legislation enacted in respected of service tax, after the notification of the eighty eighth amendment to the Constitution, revenue accruing to a State should not be less than the share that should accure to it, had the entire service tax proceeds part of the shareable pool.*
- *The indicative amount of overall transfers to States to be fixed at 38 per cent of the Centre's gross revenue receipts.*

Local Bodies

- *A grant of Rs. 20,000 crore for the Panchayati Raj institutions and Rs. 5,000 crore for urban local bodies to be given to States for the period 2005-10.*
- *Priority to be given to expenditure on operation and maintenance (O & M) costs of water supply and sanitation, while utilizing the grants for the Panchayats. At least 50 per cent of the grants recommended for urban local bodies to be earmarked for the scheme for solid waste managemental through public-private partnership.*

Calamity Relief

- *The scheme of Calamity Relief Fund (CRF) to continue in its present from with contributions from the Centre and States in the ratio of 75:25. The size of the Fund worked out at Rs. 21.333 crore for the period 2005-10.*
- *The outgo from the fund to be replenished by way of collection of National Calamity Contingent duty and levy of special surcharges.*
- *The definition of natural calamity to includes avalanches, cloud burst and pest attacks.*
- *Provision for disaster preparedness and mitigation to be part of State Plans and not calamity relief.*

Grants-in-aid to States

- *The present system of Central assistance for State Plans, comprising grant and loan components, to be done away with, and the Centre should confine itself to extending plan grants and leaving it to States to decide their borrowings.*
- *Non-plan revenue deficit grant of Rs. 56,856 crore recommended to 15 States for the period 2005-10.*

Grants amounting to Rs. 10,172 crore recommended for the education sector to eight States. Grants amounting to Rs. 5,887 crore recommended for the health sector for seven States. Grants to education and health sectors are additionalities over and above the normal expenditure to be incurred by States.

- *A grant of Rs. 15,000 crore recommended for roads and bridges, which is in addition to the normal expenditure of States.*
- *Grants recommended for maintenance of public buildings, heritage conservation and specific needs of States are Rs. 500 crore, Rs. 1,000 crore, Rs. 625 crore and Rs. 7,100 crore, respectively.*

Fiscal Reform Facility

- *With the recommended scheme of debt in place, fiscal reform facility not to continue over the period 2005-10.*

Debt Relief and Corrective Measures

- *Central loans to States contracted till March, 2004 and outstanding on March 31, 2005 amounting to Rs. 1,28,795 crore to be consolidated and rescheduled for a fresh term of 20 years and an interest rate of 7.5 per cent to be changed on them. This is subject to enactment of fiscal responsibility legislation by a State.*
- *A debt write-off scheme linked to reduction of revenue deficit of State to be introduced. Under this scheme, repayments due from 2005-06 to 2009-10 on Central loans contracted up to March 31, 2004 will be eligible for write-off.*
- *Central Government not to act as an intermediary for future lending to States, except in the case of weak States, which are unable to raise funds from the market.*
- *External assistance to be transferred to States on the same terms and conditions as attached to such assistance by external funding agencies.*
- *All the States to set up sinking funds for amortization of all loans.*
- *States to set up guarantee redemption funds through earmarked guarantee fee.*

Others

- *The Centre should share 'profit petroleum' from New Exploration and Licensing Policy (NELP) areas in the ratio of 50:50 with States where mineral oil and natural gas are produced. No sharing of profits in respect of nomination fields and non-NELP blocks.*
- *Every State to set up a high level committee to monitor the utilization of grants recommended by the TFC.*
- *Centre of gradually move towards accrual basis of accounting.*

Recommendations of Chelliah Committee on Tax Reforms

The Central Government appointed a high level committee of experts in 1991 under the chairmanship of Prof. Raja J. Chelliah for investigating the tax structure in the country. The committeee submitted its interim report in Feb. 1992. The first part of the Final Report was submitted in August 1992 while the second part was submitted in January 1993. Most of the recommendations of the committee were included in the budget proposals for 1993-94.

Final Report

Part I : (Main Recommendations)

1. To curtail the rate of corporate tax from 51.75% (in context to indigenous companies upto 1993-94) to 45% *i.e.,* to abolish surcharge on corporate tax.
2. From 1994-95, the above rate of 45% should be further reduced to 40%.

3. To continue the general rate of depreciations on plant and machinery at the existing level of 25%.
4. To abolish Interest Tax.
5. To continue gift tax (exemption limit should be extended from 20,000 to 30,000.)
6. Tax on agricultural income of non-farmers if it exceeds Rs. 25,000.
7. To extend VAT tax system upto manufacturing level.
8. To convert sales tax on manufacturing sector into State VAT.
9. To determine suitable targets for stable tax administration and for tax payers.
10. To start TIN in place of PAN for identification of tax payers.

Part II : (Main Recommendations)

1. To abolish zero rate of import duty on certain commodities.
2. To curtail the existing 110% import duty ceiling to 50%.
3. Single rate of import duty for all the goods should not be adopted.
4. To minimise the tax slabs.
5. (*i*) Import duty should have six slabs-5% as the lowest and 30% as the highest (*i.e.* 5%, 10%, 15%, 20%, 25% and 30%.)
   (*ii*) 50% Import duty on non-essential consumer items.
6. To adjust tax rates upto 1997-98 in a phased manner.
7. Duty free import of Wheat and Rice, but oil seeds, pulses and other agricultural products should have 10% ad valorem import duty.
8. Duty free imported items should be charged 5% duty under 'Protection' head.
9. The import duty on almond and cashewnuts should be 50%.
10. 5% import duty on inputs of fertilizers and newsprints production.
11. 20% import duty on medical equipments.
12. To provide additional protection to new industry or to new products of essential sector for a certain period.
13. To continue the Advanced licensing system for exporters.

Recommendations of Advisory Committee on Tax Policy for the 10th Five Year Plan.

- *Corporate tax should be reduced from 38.5% to 30%.*
- *Maximum Corporate tax rate should be kept similar to maximum income tax rate.*
- *Corporate tax-GDP ratio which is at present 1.55% should be increased to 2.35% by 2006-07.*
- *Income tax-GDP ratio which is at present 1.38% should be increased to 1.88% by 2006-07.*
- *The estimated figures as presented by Advisory committee are as follows:*
  (*i*) *Excise duty receipts (as % of GDP) will become 3.51% by 2006-07 (At present it is 3.16%).*
  (*ii*) *Custom duty receipts (as % of GDP) should become 2.44% by 2006-07 (At present it is 2.47%).*
  (*iii*) *Tax revenue-GDP ratio should become 10.88% by 2006-07 (At present this ratio is 8.8%.)*
  (*iv*) *State taxes (as % of GDP) has been estimated to become 6.9% by 2006-07 (At present it is 5.29%).*
  (*v*) *Total Central & State taxes (as % of GDP) should become 17.78% by the end of X plan (at present it is 14.09%).*

Modifications in the Gadgil Formula for Plan Assistance to States–Mukherjee Formula

This formula is related to the financial assistance provided to the States by the Centre under various plans. The Union Government provides economic assistance to the States in order to remove the regional imbalance existing amount the various States. Before the Fourth Five Year Plan. There was no certain formula for determining the amount of assistance. A formula was prepared in the Fourth Five Year Plan for this purpose and it was called as the 'Gadgil Formula'. This formula provided a basis for the transfer of assistance to the states. According to this formula, the distribution of the total assistance among the States was done on the following basis:

	Bases	*Weight*
1.	Population of the State	60%
2.	Per Capita Income	10%
3.	Tax Efforts	10%
4.	On going Irrigation and Power Projects	10%
5.	Special Problems	10%
	Total	**100%**

This formula was used in the Fourth and Fifth Five Year Plan for transferring financial assistance to the States. During the discussions on the Fifth Five Year Plan, many Chief Ministers of the States demanded modifications in this formula. The first modification in the formula was made in 1990 in which current projects were associated with per capita income getting 20% weightage. This modification was applied in Sixth Plan, Seventh Plan and 1990-91 Annual Plans.

In October 1990 National Development Council remodified the formula for the second time which was as follows:

	Bases	*Weight*
1.	Population	55%
2.	Per Capita Income	
	(*a*) Deviation Method	20%
	(*b*) Distance Method	5%
3.	Financial Arrangement	5%
4.	Special Development Problems	15%
	Total	**100.0%**

The above modified formula was used only for the Annual Plan of 1991-91 for giving the assistance to the States by the Centre. The Planning Commission under the leadership of the immediate Vice-Chairman of the Commission, Mr. Pranav Mukherjee, constituted a committee for re-considering the modified Gadgil formula on September 10, 1991. The current formula which was adopted by the National Development Council after accepting the modifications suggested by the Mukherjee Committee in the meeting held on September 23-24, 1991, is as follows:

Mukherjee formula was applied in the Eighth Plan for transferring central assistance to States for non-specified heads. For specified heads, it was decided to continue Modified Gadgil Formula.

	Bases	*Weight*
1.	Population (census of 1971)	60%
2.	Per Capita Income	
	(*a*) Deviation Method	20%
	(*b*) Distance Method	5%
3.	Performance	7.5%
	(*a*) Tax Efforts	
	(*b*) Financial Management	
	(*c*) Progress in the form of National objectives	
4.	Special Development Programmes	7.5%
	Total	**100%**

KELKAR REPORT

Finance Ministry had constituted a task force on Tax Reform under the chairmanship of Mr. Vijai Kelkar who submitted his **final report** to the Ministry. This report was made public on December 27, 2002. The task force has produced a comprehensive and far reaching report on the reform of both direct and indirect taxes.

(A) Recommendations on Direct Taxes

(i) For individuals

- Generalised exemption limit is raised from Rs. 50,000 to Rs. 1 lakh for all individuals and HUF tax payers.
- Senior citizens and widows would have an exemption limit of Rs. 1,50,000.
- Present 3-tier income tax structure to be replaced with 2-tier.

Upto 1 lakh	No Tax
1-4 lakh	20%
above 4 lakh	30%

- In personal income tax, retention of the present exemption for conveyance allowance upto Rs. 9,600 has been advised.
- Standard deduction and surcharge to be eliminated.
- Tax incentives for savings should be withdrawn.
- Proposal of doubling the exemption under 80C to Rs. 20,000 from Rs. 10,000 for encouraging investment in annuity-oriented pension scheme.
- Abolition of divided tax and long-term capital gain tax.
- Tax rebate on housing interest to be reduced from Rs. 1,50,000 to Rs. 5,00,000; alternative suggestions of 2% interest subsidy on housing loans upto Rs. 5 lakh.
- Income tax on agriculture income should be left to states.
- Income based deduction for medical insurance should be converted into a tax rebate at the rate of 20% subject to a maximum of Rs. 3,000.
- Tax rebate schemes u/s 88 for saving should go, as well rebate under 88B for senior citizens and under 88C for women below 65 years of age.

- The deducations for handicapped under 80DD and 80U should continue.
- 5% surcharge on Income Tax should be withdrawn.

(ii) For Corporates

- Reduction of corporate tax to 30% as compared to current level of 36.75% for domestic companies.
- Reduction of tax on income of foreign companies to 35% as compared to the current level of 40%.
- No tax on distribution of dividents by a company.
- General rate of depreciation for plant and machinery should be reduced to 15%.
- Assorted concessions given for scientific research under section 35 and benefit u/s 33 AC should 90.
- Minimum alternative tax under section 1157B to go.
- Business loss to be allowed to be carry forward indefinitely.

(B) Recommendations of Indirect Taxes

- Customs duty to be reduced to 10% for raw materials, inputs and intermediate goods and to 20% for consumer durables by 2004-05.
- Customs duty for coal, ores and other raw material should be reduced to 5% and for capital goods, basic chemicals to 8% by 2006-07.
- Custom duty on crude oil should be reduced to 8% and 15% on petroleum products by 2003-04 and further to 5% and 10% respecting by 2004-05.
- Higher duty of 150% for specified agri-products and demerit goods.
- Complete exemption of custom duty on life saving drugs, equipments defence related goods.
- Excise duty should be 14% for most of the items, 20% for motor vehicles, air conditioners and aerated water.
- Excise duty to be reduced to 6% while exempting life saving drugs, security items & agricultural products.
- Central excise duty on kerosene should be raised by Rs. one per litre.
- Uniform duty of 16% for textile fibre and yarn which should be reduced to 14% by 2004-05, 12% duty on all fabrics till 2004-05.
- Duty exemption for SSIs with turn over upto Rs. 50 lakh.
- Nationwide Value Added Tax (VAT) and a comprehensive service tax should be introduced from April 2003.

States' Financial Position

Though the fiscal deterioration of State began much later than that of the Centre, the fiscal stress of some of the State Governments is more acute and an important constraint in their development. The revenue account of States has been continuously in deficit since 1987-88. The deterioration in the revenue account of States has been more pronounced from the late nineties. As a proportion of GDP, revenue deficit of States which increased from 0.3 per cent in 1987-88 to 1.1 per cent in 1996-97, shot up to 2.5 per cent in 1998-99, following mainly the revision of pay scales of Government employees. The deficit continued to remain high till 2001-02. In 2002-02, the deficit declined to 2.2 per cent of GDP. Fiscal deficit of States as a proportion of GDP, increased from a level of 3.3 per cent in 1990-91 to 5.1 per cent in 2003-04 (RE). In BE for 2004-05, revenue and fiscal deficits of States were placed at 1.4 per cent and 3.6 per cent of GDP, respectively.

States have done relatively better than the Centre in terms of raising own tax revenue. The own tax revenue of States increased from an average of 5.4 per cent of GDP in the period 1990-95 ti 5.8 per cent of GDP in 2002-03. In contrast, Central transfers declined from 4.9 per cent of GDP to 4.1 per cent of GDP in the same period. The performance of States in mobilizing non-tax revenue has deteriorated. Non-tax revenue to GDP ratio declined from an average of 4.1 per cent in the period 1990-95 to 3.3 per cent in 2002-03. Total expenditure of States increased from an average of 16.0 per cent of GDP in the period 1990-95 to 17.1 per cent of GDP in 2002-03.

Outstanding liabilities of States as a proportion of GDP, which was 18.2 per cent in 1994-95, after initial decline in the next two years, shot up to 29.2 per cent in 2003-04 (RE). It is budgeted at 29.4 per cent in 2004-95.

Fiscal Responsibility Legislation

The FRBM Act, 2003 (as amended), which became effective from July 5, 2004 mandates the Central Government to eliminate revenue deficit by March, 2009 and to reduce fiscal deficit to an amount equivalent to 3 per cent of GDP by March, 2008. The annual targets for fiscal correction were to be specified by rules to be framed under the Act. Accordingly, Government notified the rules under the FRBM Act on July, 2, 2004, which came into force on July 5, 2004. The rules made under FRBM Act specify the annual targets for reduction of fiscal and revenue deficits, annual targets for assuming contingent liabilities in the form of guarantees and the total liabilities as a percentage of GDP.

Task Force on FRBM Act, 2003

Following the enactment of FRMB Act. Government constituted a Task Force headed by Dr. Vijay Kelkar for drawing up the medium term frame work for fiscal policies to achieve the FRBM targets. The Task Force was also asked to formulate annual targets indicating the path of adjustment and required policy measures. The Task Force submitted its report in July, 2004. The Task Force recommended a path of fiscal adjustment that is frontloaded and mainly revenue-led, with complementary reform efforts on revenue expenditure and enhanced capital expenditure to counteract the possible contractionary effects of fiscal correction.

The Task Force estimated that under the reforms scenario recommended by it tax-GDP ratio of the Centre would improve from 9.2 per cent in 2003-04 (RE) to 13.2 per cent in 2008-09. Total expenditure is estimated to come down to 14.3 per cent of GDP by 2008-09 from 15.4 per cent of GDP in 2003-04 (RE). A revenue surplus of 0.2 per cent of GDP is estimated in 2008-09. Fiscal deficit is estimated to come down from 4.8 per cent of GDP in 2003-04 to 2.8 per cent of GDP in 2008-09.

The Fiscal Responsibility and Budget Management Rules, 2004

- *Reduction of revenue deficit by an amount equivalent of 0.5 per cent or more of the GDP at the end of each financial year, beginning with 2004-05.*
- *Reduction of fiscal deficit by an amount equivalent of 0.3 per cent or more of the GDP at the end of each financial year, beginning with 2004-05.*
- *No assumption additional liabilities (including external debt at current exchange rate) in excess of 9 per cent of GDP for the financial year 2004-05 and progressive reduction of this by at least one percentage point of GDP in each subsequent year.*
- *No guarantees in excess of 0.5 per cent of GDP in any financial year, beginning with 2004-05.*
- *Specifies four fiscal indicators to be projected in the medium term fiscal policy statement. These are,*

revenue deficit as a percentage of GDP, fiscal deficit as a percentage of GDP, tax revenue as percentage of GDP and total outstanding liabilities as percentage of GDP.

- *For greater transparency in the budgetary rules mandate the Central Government to disclose change, if any, in accounting standards, policies and practices that have a bearing on the fiscal indicators. The Government is also mandated to submit statements of receivables and guarantees and a statement of assests, at the time of presenting the annual financial statement, latest by Budget 2006-07.*
- *The rules prescribe the from for the quarterly review of the trends of receipts and expenditures. The rules mandate the Central Government to take appropriate corrective action in case of revenue and fiscal deficits exceeding 45 per cent of the budget estimates, or total non-debt receipts falling short of 40 per cent of the budget estimates at the end of first half of the financial year.*

ECONOMIC SURVEY 2004-05

(*Presented in Parliament on Feb. 25, 2005 by Finance Minister Mr. P. Chidambaram*)

(A) Macroeconomic Overview

- The performance of the Indian Economy in 2004-05 so far has exceeded expectations formed at the beginning of the year. Supported by a rebound in the agriculture and allied sector and strongly helped by improved performance in industry and services, The economy has registered a growth rate of 8.5% in 2003-04, the highest ever except in 1975-76 and 1988-89.
- According to the advanced estimates of CSO released on Feb. 7, 2005, the economy in likely to grow by 6.9% in 2004-05.

Sectoral Real Growth Rates in GDP (At Factor Cost)

Item	*% change over the previous year*							
	'97-98	*'98-99*	*'99-00*	*'00-01*	*'01-02*	*'02-03 (P)*	*'03-04 (Q)*	*'04-05 (A)*
Agri. & Allied	-2.4	6.2	0.3	-0.1	6.3	-7	9.6	1.1
Industry	4.3	3.7	4.8	6.5	3.6	6.6	6.6	7.8
(*a*) Mining & quarrying	9.8	2.8	3.3	2.4	2.5	9.0	6.4	5.3
(*b*) Manufacturing	1.5	2.7	4	7.4	3.6	6.5	6.9	8.9
(*c*) Electricity, gas and water supply	7.9	7	5.2	4.3	3.7	3.1	3.7	6.3
(*d*) Construction	10.2	6.2	8	6.7	4	7.3	7.0	5.7
Services	**9.8**	**8.4**	**10.1**	**5.5**	**6.8**	**7.9**	**9.1**	**8.9**
(*a*) Trade, hotels, transport and comm.	7.8	7.7	8.5	6.8	9.0	9.8	11.8	11.3
(*b*) Financial services	11.6	7.4	10.6	3.5	4.5	8.7	7.1	7.1
(*c*) Community, social & personal services	11.7	10.4	12.2	5.2	5.1	3.9	5.8	6
Total GDP at factor cost	**4.8**	**6.5**	**6.1**	**4.4**	**5.8**	**4**	**8.5**	**6.9**

A = Advance estimates, Q = Quick estimates, P = Provisional estimates.

Source : Central Statistical Organization

- After a drought induced decline of 7% in 2002-03 the growth rate in the agriculture and allied sector bounced back to 9.6% in 2003-04 while industry maintained the higher growth of 6.6% observed in 2002-03, the services sector improved its performance significantly form 7.9% in 2002-03 to 9.1% in 2003-04. Growth in the industry and services sectors in 2003-04 was broad based with manufacturing, public utilities, the trade, hotels, transport and communication group and community, social and personal services recording higher growth than that in the previous year.

- GDP grew by 7.4% in the first quarter and 6.6% in the second quarter of the current year, compared with 5.3% and 8.6% in the corresponding quarters of the previous year. The deceleration of growth in the second quarters is on account of a negative growth of 0.8% in agriculture and allied sector, a lower growth of 8.2% in the services sector compared with 9.5% in the first quarter, and a fall in the growth of community, social and personal services. The growth in industry accelerated from 6.9% in the first quarter to 8.1% in the second quarter. Within industry, growth in manufacturing accelerated from 8% in the first quarter to 9.3% in the second quarter (The highest in any quarter since 1997-98). Despite a lower growth in the second quarter the overall growth in the first half of current year are 7% is marginally higher than the growth of 6.9% achieved in the same period last year.

(B) Fiscal Health

- Fiscal consolidation, after a promising beginning in the early, 1990s started faltering from 1997-98. Fiscal deficit of the Central Government as a proportion of GDP, after its decline from 6.6 per cent in 1996-97, rose every year to reach 6.2 per cent in 2001-02. Progress in fiscal consolidation resumed in 1990-91 to 4.0 percent 2002-03. According to provisional data in 2003-04, the ratio at 4.6 per cent was lower than the budget estimate of 5.6 per cent. A further improvement in this ratio to 4.4 per cent was budgeted for 2004-05.

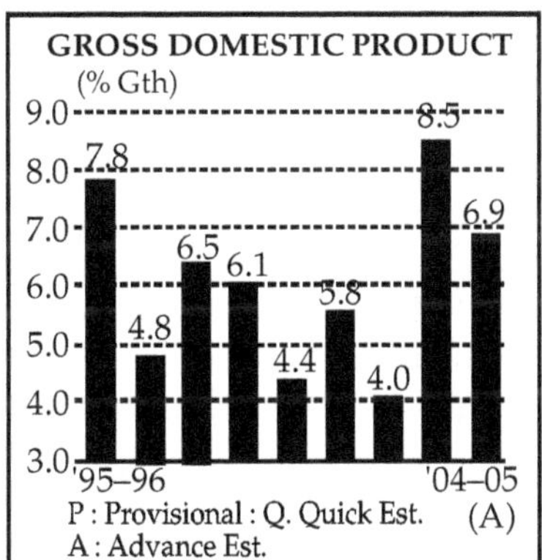

- Revenue deficit declined to 3.6 per cent of GDP in 2003-04 (Prov.), but even at this level, it was higher than the level of 3.3 per cent of GDP observed in the pre-reform year of 1990-91. The increasing share of revenue deficit in fiscal deficit distinctly reveals the deterioration in the composition of the fiscal deficit and in the quality of expenditure. The share of revenue deficit in fiscal deficit had risen from 49.4 per cent in 1990-91 to 78.0 per cent in 2003-04, which was sought to be reversed in 2004-05 by targeting a lower revenue deficit of 2.5 per cent of GDP in the budget estimates (BE).

Chapter 6

SMALL SCALE FISHERIES AND MARKET FACILITIES*

1. INTRODUCTION

This chapter is addressed, primarily, to those concerned with small-scale fisheries development in order to facilitate the planning of small-scale fish collection and marketing facilities. It is, specifically, concerned with the identification, planning and basic designs and services needed for fish collection, handling and marketing infrastructure.

This chapter is oriented to the problems as they appear in developing countries and is focused on the functional planning for small-scale fish collection and marketing facilities, the requirements necessary for the introduction of improved and efficient fish marketing operations and the reduction of post-harvest losses and marketing costs.

1.1 Characteristics of Small-Scale Fisheries

For the purposes of this study, the definition of small-scale fisheries is that of the FAO Expert Consultation on Small-Scale Fisheries Development (Rome, 1975): "Small-scale fisheries are labour intensive and are conducted by artisanal craftsmen whose level of income, mechanical sophistication, quantity of production, fishing range, political influence, market outlets, employment and social mobility and financial dependence, keep them subservient to the economic decisions and operating constraints placed upon them by those who buy their production."

Presently, it is widely accepted that, in most developing countries, small-scale fisheries play an important role as a major source of animal protein and in the provision of employment. In these countries there are estimated to be around 20 to 30 million small-scale fishermen, to which must be added their families, fish traders and a large number of people employed in related activities, all of whom depend on small-scale fisheries for a livelihood.

Despite the enormous socio-economic importance of small-scale fisheries in these countries (Emerson, 1980), and the fact that they contribute almost half of the world's fish catch for direct human consumption, it is only in the last few years that national and international organizations have begun to show more concern for this sector (Diegues, 1983).

Most small-scale fisheries are located in rural areas, on lakes, estuaries, lagoons, and coastal areas. They are characterized by being labour-intensive and conducted by artisanal fishermen whose level of income is low, but which may be supplemented by other nonfishing activities conducted by family members. Mechanical sophistication is rather poor; catch per fishing craft and productivity per fisherman is low; the fishing range is short; political influence is slight; the

* The matter has been taken from www.fao.org website title – Small Scale Fish Landing and Marketing.

catch is sold at scattered landing points and often displays with seasonal fluctuations; the small-scale fishermen often lack market power, and their alternative employment opportunities are restricted.

The importance of the family-oriented form of organization, as well as the participation of the family members in fisheries related activities should be mentioned as important characteristics of these traditional fishing communities, with most women engaged actively in sorting, processing and marketing the catch, and the children engaged in mending nets, fish processing, helping during fish landing activities and fish retailing.

Living conditions in traditional fishing communities are generally poor. Often the community lives in overcrowded conditions, in houses of poor standard (e.g., huts with thatched roofs), frequently exposed to extreme climatic conditions (cyclones, floods). Health services, drinking water supplies, roads, and transport facilities to nearby market towns are often inadequate or totally lacking.

The income level of most fishing communities is lower than many other groups engaged in rural sector activities, and, in many cases, is below the poverty line. Within the community the income structure varies according to the patterns of ownership and the family participation in fisheries related activities.

Due to the seasonality of fishing the fisherman's earnings display an uneven pattern, which often inhibits savings and leads to indebtedness.

1.2 The Traditional Fish Marketing System

In most small-scale fishing communities in developing countries, the traditional fish marketing system is characterized by fishermen landing their catches on scattered beaches, normally in small quantities. The fishermen have little bargain power in the markets, with most marketing activities being financed by fish traders who also function as a source of informal credit, providing necessary cash for the fishermen's family needs, especially during the extended seasonal periods of limited catch and income. This situation can create a strong inter-dependence between traders and fishermen which influences market decisions over the latter. Marketing relationships between fish traders and fishermen tend to be long-lasting, providing an assured market outlet to the small-scale fishermen and a source of steady supply to the trader. Fishermen/traders ties should be carefully evaluated when looking for new marketing strategies, taking into consideration the socio-economic conditions in which the traditional market system operates.

1.3 Constraints of Small-Scale Fisheries Development

Small-scale fishermen in developing countries are faced with several limitations with regard to availability of resources, fishing craft, gear and techniques, landing and marketing infrastructure, access to credit and extension services, as well as organizational constraints. This study, however, will deal mainly with the landing and marketing infrastructure needs of small-scale fisheries. There are several constraints in the traditional fish marketing system:

- fish being a highly perishable commodity, high post-harvest losses are common due to lack of proper handling on board, suitable shore-based fish handling, collection, marketing, storage and distribution facilities and deficiencies in marketing practices;
- fishing communities are sometimes located in remote areas at considerable distances from the markets. Therefore, their catches have to be collected from village to village to obtain quantities large enough to justify costs of transportation to the markets;
- poor communications make it difficult to operate efficient fish collection and distribution systems due to the problems of establishing adequate fish marketing information systems;

- inadequate or misdirected government intervention and assistance schemes, e.g., due to lack of understanding of the prevalent socio-economic conditions in small-scale fisheries; establishment of costly organizations or large installations which are little used; inadequate price policies.

2. PLANNING

The basic step in the planning of small-scale landing and marketing facilities is the identification of the fishing communities that would likely be involved. For such identification the following range of data should be collected on the spot, in the area under consideration: information on the available fish resources, socio-economic characteristics of the fishing population, existing fish marketing systems, existing basic infrastructure and facilities for fish landing and marketing activities, shortcomings of existing facilities and any existing proposals and/or plans by the local authorities for the improvement of the system. The type of information required is outlined in Appendix I.

The collection of basic socio-economic data would also be necessary as part of the total identification process, in order to assess the degree to which the small-scale fishing communities can be served and participate in the implementation of any planned marketing development programme.

This stage of the identification process should identify:

- the socio-economic profile of the small-scale fishermen and their households;
- the existing conditions, production costs, returns, credit sources and problems;
- information on fisheries resources, fishing activities, landings, number and type of fishing vessels; fish utilization patterns;
- current fish marketing practices, costs, margins and problems; market channels used and proportion of fish passing through each channel; the geographic flow of the commodities, fish collection systems;
- existing fish landing and marketing infrastructure and facilities, access roads or other communications infrastructure; ice, water and fuel supplies and their costs at landing places; suitable sites for small-scale fish landing/marketing development.

2.1 A Pragmatic Approach for the Identification of Small-Scale Fisheries Marketing Shortcomings

A practical methodology has been developed for the identification of major short-comings in the traditional fish market system, which allows a quick appraisal of the prevailing situation in such a way that specific alternative technical options can be formulated (Ruckes, 1980). The following identification catalogue, including the major problems points of producers, traders, administration and policy-makers, has been designed for the diagnosis of fish marketing systems in small-scale fisheries (Ruckes, 1980) (see Figure 1).

After the identification of major shortcomings of the existing, traditional, fish marketing practices has been conducted, a more detailed examination of each problem point should be carried out. A check list of the topics to be further studied is given as Appendix 1.

It is of particular importance to estimate the major cost of the elements involved from the point of arrival of the fish at the landing beaches, through collection points, until it is delivered to retail outlets.

During the evaluation of existing fish landing infrastructure, it is also important to ascertain to what extent small modifications of the existing system could help to improve the situation and be easily accepted by the fishing community.

2.2 Elements of Strategies

In general, development strategies for small-scale fishing communities should be oriented toward an integrated approach by linking the overall community development programme with improved fish handling, collection, marketing and distribution.

Efforts should be made to quantify the adequate, minimum, package of basic infrastructure and services needed to remedy the identified shortcomings, trying to make the best use of local, available, resources under the prevailing conditions, with fullest possible participation of local leaders and authorities, and assuring that appropriate extension, training and credit services will be available.

Small-scale fish marketing development should be aimed at increasing both fish consumption and fishermen's income, as well as higher foreign exchange earnings, if potential export markets exist. Fish consumption could be increased through a better distribution chain, provided that fish can reach the consumers at price levels within the purchasing power of the local population. Fish prices should try to reflect real marketing costs and the degree of risk involved in encouraging fishermen to produce more, as well as stimulating consumers interest in fish.

3. ELEMENTS FOR PLANNING

Careful examination of the basic data collected will provide the background material for evaluating the feasibility of the technical alternatives to be formulated. A detailed analysis of the following aspects, which are briefly discussed, is mainly aimed at facilitating the evaluation process.

3.1 General Development Stages

Fish marketing development should take into consideration the present technological level of the people in the area under consideration and the economics of scale in fish production, storage, transport and distribution. In order to keep production costs at the lowest possible level, small-scale fish landing and marketing facilities should be unsophisticated and inexpensive whilst at the same time maintaining good standards of technology and hygiene. Particular emphasis should be placed on ensuring continuity in the supply of spare parts and availability of essential skills to operate and maintain the equipment. The level of technology chosen for the fish landing and marketing operations, should be able to easily fit into the present general technological level of the country and be, technically, suitable to the local marketing conditions.

3.2 Raw Material

A detailed examination of the raw material supply, taking into consideration the different species, their estimated shelf life, physical and chemical composition, available commercial sizes, the present raw material available and the expectation of decrease, stability or the increase of landings in the future. Seasonal fluctuations of the landed volumes of each species over a period of a number of years should also be examined.

3.3 Products and Fish Marketing Aspects

If the identification survey has shown that the major defects of the system could be remedied by the construction of fish marketing facilities, the next step should be to conduct a fish marketing study which should take into account the following:

- the volume of fish to be handled and the quantity of fish products demanded during a given time period which consumers would be willing to purchase at a given price;
- consumers preferences and attitudes;
- the cost of raw material (by main species), cost of collection of fish (including ice, fish boxes weight losses) operating and transportation costs from landing places to the main distribution centers and to the retail outlets, other miscellaneous costs;
- the relation of the planned fish landing and marketing facilities capacities to the yearly quantities of raw material available;
- the expected number of days per year the fish collection/distribution scheme would be able to operate at full capacity;
- the species most likely to be marketed and type of handling to be proposed (e.g., chilled fish or live fish in wells, etc.).

It is advisable at this planning stage to assess the type of products expected to be handled, their market outlets, requirement at the markets for the products, and possibilities for product development work with the available fish species.

An assessment of the available labour resources in the area, and the identification of training requirements should also be prepared.

The role of existing or proposed cooperatives or other organizations in organizing local fishermen which could facilitate training, as well as the development of management skills, should also be examined.

3.4 Location

It is anticipated that the sites provisionally selected for the location of fish landing sites, collection points and marketing outlets will be done during this planning stage and their suitability will be re-evaluated as the planning progress.

For the site, or sites, selection the following points should be taken into consideration:

- the availability of fish and the distance from the fishing grounds;
- the availability of road, rail or waterway links, or other transport facilities between the provisional sites for fish landing places and collection points and market outlets;
- the availability and cost of land and the costs of site development;
- the availability of water, sewage disposal, fuel and, to some extent, electricity will also influence site selection;
- the provisional site should include sufficient space, both for presently planned waterside and land based installations as well as for future expansion;
- preference should be given to sites located on flat ground so as to avoid additional costs for site preparation;
- the willingness of local fishermen to make use of the proposed landing and marketing facilities.

In general, the least-cost point of location, with respect to the proposed fish marketing developments, should take into consideration the importance of costs of inputs and of their transportation to the point of fish landing and marketing; the costs of marketing, allowing for economies of scale both within the fish landing marketing facility as well as those common to the fish trade; the costs of transporting final products to consumption destinations.

3.5 Cost Estimates and Sources of Financing

Cost estimates of land acquisition, site preparation, construction equipment and services required for fish landing and marketing facilities should be prepared, taking account of project property and materials, and services and labour costs so as to reflect future variations, from present costs.

A time schedule for the acquisition and implementation of the civil works will be needed in order to establish the timing of and consequent increased funding requirements.

Cost estimates should be as complete as possible and include total capital investment, schedule of expected revenue and anticipated operating and maintenance costs, so that the funding agency may be fully advised of the expected financial situation of the project during its construction and initial operating phases.

3.6 Basic Design Concepts

During the planning stage a process flow diagram showing the sequence of operations, equipment needed and service requirements should also be prepared. This will facilitate the planning of the provisional lay-out and the assessment of total service requirements. This process flow diagram could be prepared by setting out each stages concerned in the collection of fish either at sea or on the landing beaches, followed by transport, receiving, sorting, weighing, washing, chilling, storage, etc., and could end with transport to the retail outlets (Figure 1).

In addition to this process flow diagram, an estimate of materials and energy needed during these operations should also be prepared, this will be very helpful throughout the planning process.

3.7 Small-Scale Fish Landing Facilities : Basic Designs

3.7.1 Site selection

The basic considerations regarding site selection for suitable landing and marketing facilities in a fishing area should take into account the following:

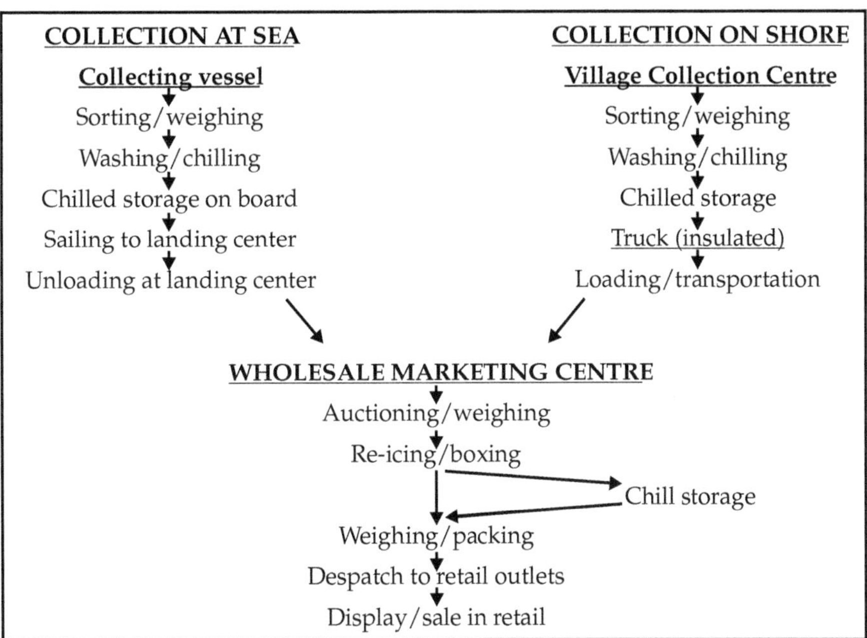

Figure 1 : Simplified flow diagram for fresh fish collection, distribution and marketing

- the proposed fish landing sites should be located at a convenient distance from the fishing grounds;
- they should be in an area that can provide safe breathing for vessels supplying fish in all weathers and at all states of tide;
- they must be located in an area providing ease of access from fishing grounds and to market outlets.

3.7.2 Services expected at landing sites

The basic service facilities required at a small-scale fisheries landing centre are:

- **Services for boats and crews :** for unloading the catch (sometimes they might be needed for 24 hours/day), for loading fishing gear, fuel, water, ice, supplies, etc.
- **Bunkerage :** a simple diesel or petrol (sometimes both) fuel station should be available at a point from which fishing vessels can take their supply directly from a pump by a hose or by hand-carried fuel containers.
- **Engine repair workshop :** facilities and qualified personnel should be available for efficient engine repairs. A properly equipped workshop should provide a basic repair service for small fishing craft and maintain a stock of spare parts most often used in the area.
- **Boat repair area :** a properly equipped boat repair area, including some boat lift out facilities, should be provided to enable fishermen to repair their vessels on shore.
- **Fishing gear repair area :** a small shed with a dry, clean floor for the repair and storage of nets and for other fishing gear should be provided.
- **Breathing services :** dock side services such as fresh water outlets should be available and properly designed to avoid wastage; dock side electricity supply outlets with power points for small tools may also be needed in more sophisticated small-scale fishing communities.
- **Food supply and washroom facilities :** it might also be desirable to provide a takeout canteen, a fishermen's food store; and sanitary facilities, including toilets and showers are necessary both for the comfort of the crews and shoreside workers and to ensure that hygiene standards are maintained.
- **Fish handling, marketing and processing facilities :** these facilities should be designed according to the type of fish and the products currently marketed as well as for any future products which may be under consideration; a more detailed description and design parameters of these facilities is listed in the following section.

3.7.3 Small-scale fish handling and marketing facilities; a design approach

The basic data needed for assessing the appropriate size and location of small-scale fish handling and marketing facilities should, partly, come from a detailed fish marketing survey. Such a survey should provide accurate data on products demand to provide for a sound design which should be adequate to take account of both present needs and expected near future developments. Design factors used should take account of available raw material, local climate, type of fish products (fresh fish, dried-salted, smoked or fermented, etc.). For example, a simple roofed platform with parking space could facilitate wholesale marketing operations in a small-scale fishing village trading in fresh fish.

3.7.4 Main construction design considerations

Technical and economic considerations are normally primary factors ruling the best form of construction of improved facilities for fish utilization, but it should be kept in mind that the selected materials and design should be adapted to suit local conditions.

Thus, for example, access roads to fish landing and marketing facilities should possibly be designed to match the quality of the trunk roads in the country by which it is intended to reach the main markets.

3.7.4.1 Layout

During the preparation of the layout of the facilities, two basic requirements should be kept in mind :

a. hygiene and functionality, and

b. economy.

The layout of a fish-marketing facility should be arranged so as to avoid the possibility of cross-contamination, have proper drainage, and allow easy access to all equipment for effective cleaning and maintenance; follow the process flow diagram in such a manner as to obtain the best quality product at the lowest possible cost; provide space for the staff managing and supervising the daily operations; provide space for the quality inspector's operations and the storage of quality control equipment.

3.7.4.2 Buildings

In designing new fish marketing premises, a smooth sequence of operations from the receipt of the fish to its loading and transportation should be achieved. All operations should be conducted off the floor, at a height convenient for workers to perform their tasks in a standing position. For example, a single storey building, at a short distance from the landing area, will enable fast and minimum handling of fish along the quay, conduct marketing operations inside the market hall, and also reduce costs for drainage and structural civil works. This type of design will also enable ready access to vehicles to the market for loading purposes.

Ample, natural air circulation should be provided, especially in the tropics, where hollow-brick walls or grills are often used and a part of the building is sometimes open, without walls. Proper long eaves for protection against sunshine and rainshowers are essential in the humid tropics. An adequate pitch of the roof and the building orientated in relation to the prevalent direction of sunshine are also important considerations.

Floors should be hard-wearing, non-porous, washable, easy to drain, non-slip and resistant to possible attack from brine, weak ammonia, fish oils and offal. The choice of flooring materials will depend on the characteristics of the materials available in the area and their cost. For this purpose, granolithic concrete, terrazo and clay tiles can be used, but clay tiles are the best. Generally speaking, the harder tiles are less absorbent but more slippery than the softer tiles which are more absorbent and less easy to keep clean. Tiles with slightly abrasive surface are quite good as they are less slippery. Light-coloured tiles are recommended because they reflect light and show up dirt. Care should be taken when laying tiles in cement that all joints between tiles are complete and permanently sealed. The junctions with the walls should be covered for ease of cleaning. A sufficient slope of 2% from the highest point to the drainage outlets will be necessary, and at least 10 cm diameter drains will be required.

Walls should be constructed of materials that have smooth, washable and impervious inside surfaces of light colour. For easy cleaning they should be rounded at the junctions with other walls and ceilings should be kept as free as possible from ledges, projections or ornamentations to avoid dust collection. Walls, whose internal surfaces are not tiled, should be finished in plaster. Walls which are intended primarily as partitions should be strong enough for fish boxes or other light equipment to be piled up against them; good partitions can be erected from prefabricated concrete blocks and finished to a hard smooth surface, especially the lower parts, which are subject to constant wear.

Doors should be of simple and functional design. The main doors should be sufficiently high and wide to permit safe circulation of internal transport vehicles, e.g., when fork lifts are used in medium size fish terminals a 2.8 m high door will be required, and the width will be between 1.5 to 2.5 m. Internal doors should be self-closing and fitted with metal kick plates at the bottom.

Lighting should provide for adequate natural light, as well as artificial overhead lighting when necessary, in order to allow personnel to perform their duties without eye strain. Maximum use of natural daylight should be always part of the design, by providing adequate windows and skylights. It is also cheaper than artificial lighting. Fluorescent lighting is particularly suitable (daylight type) for fish-market areas where a shadowless light with very little glare is required continuously for a long time; even though the initial costs are relatively higher than other lighting systems, operational costs are lower and generally such a lighting is found to be economical. A light level of 220 lux as minimum is considered adequate. All artificial lighting fixtures should be water-proofed and shielded to protect against broken glass.

Sanitary facilities should provide for both workers and production requirements. In general, at small-scale fish landing centres, sanitary accommodation should be provided for the staff and for fishermen. It is important that the rooms containing the toilets be separated from the fish working areas and be well lit. They should be equipped with wash-hand basins, but the selection and design of on-site latrine/toilets will depend on the water effluent disposal system used in the area and cultural factors, as well as costs and availability of materials. The design should provide for ample water supplies at adequate pressure for washing and disinfecting operations; a regular schedule for cleaning and standardized procedures for manual cleaning of fish landing surfaces such as floors, walls, pillars, boxes, miscellaneous handling equipment should be established.

3.7.4.3 Fish stores and ice plants

To estimate space requirements, it is necessary to have reliable information on the daily fish landings in the area, on the basis of the normal average daily quantities expected to be handled. However, care should be taken to ensure that, during the high fishing seasons, especially in seasonal fishing areas, there is adequate space to handle peak landings. An estimate of:

a. yearly average daily landed catches, and

b. average daily catches landed during the peak season

could show the degree of increase in the peak season. It would often be advisable to allow, for this contingency, a 25% excess of space requirements on top of the average daily landings found for the peak season estimate. In general, it is estimated that around 12 to 15 m^2/ton of fish boxed and displayed on the floor, can be a good compromise for estimating the space for the sales market hall.

In fish landing terminals, facilities are required to maintain the quality of fish and also to store it during seasonal peak landings. Several options are available such as chilled rooms with mechanical refrigeration to maintain a temperature of 0° – 1°C, insulated rooms, chilled or refrigerated sea water stores (or tanks), insulated containers or simple fish boxes. The choice will depend mainly on the degree of marketing sophistication required and local conditions.

Some design factors for calculating area and capacities for fish storage facilities are outlined below:

Type	*Characteristics*	
Refrigerated, chilled, rooms	Storage capacity:	3–4 m^3/ton of iced fish in boxes
	Area required:	2.1 m^2/ton of iced fish stored in boxes
	Type: Manual operation, held overnight and sold next day	
Chilled seawater storage facility (fixed)	Storage capacity:	1.25 to 2 m^3/ton of fish
	Area required:	1.5–2.7 m^2/ton of fish
Portable container with Chilled seawater system	Storage capacity:	1.57 m^3/ton of fish
	Area required:	1.5 m^2/ton of fish
Ice plants (ice maker and refrigeration plant only):		
(a) Block ice	Plant capacity:	10 ton/24 hours
	Area required:	100 m^2
(b) Flake ice	Plant capacity:	10 ton/24 hours
	Area required:	20 m^2
(c) Tube ice	Plant capacity:	10 ton/24 hours
	Area required:	30 m^2
(d) Plate ice	Plant capacity:	10 ton/24 hours
	Area required:	9 m^2

Plate and flake ice machines are normally located over the ice store enabling gravity feed of ice to the store with a consequent, considerable, reduction in ground area requirements. In all cases storage space for empty fish boxes should be provided as well as ample water supplies for cleaning and an area for box washing (preferably manually).

The ice plant should have sufficient capacity for, at least, the expected daily fish landings in the fish handling facility plus an additional production capacity if a fish collection system is also operated. The daily demand for ice can be estimated as follows:

- ice for chilling of fish to be stored and transported for marketing outside the area and for un-sold fish which will be stored overnight and sold next day..... A tons
- additional ice needed for a fish collection system.... B tons
- required daily capacity of the ice plant A+B tons

The ice store should, at least, have the capacity to handle quantities sufficient for two days of ice production. Further information can be obtained from FAO Fisheries Circular No. 735 (Myers, 1981). The capacity of the chilled room for storage of iced fish should be calculated based on the expected market turnover. At a minimum, a capacity to store the average daily throughput of the fish landing facility should be provided; additional capacity would be needed if it is planned to operate a collection system.

3.7.4.4 Space for fish preparation

In some cases, a line for gutting and/or filleting might be required. Simple concrete or stainless steel tables should be provided. If crabs or other shellfish are to be processed (boiling, shucking, cooling) small boiling pans would also be required. Space for these operations as well as ample water supplies for washing should be provided.

3.7.4.5 Water supplies

There is a need for clean, fresh or sea water both at landing points and within the market premises. Sometimes, well water or sea water supplied from nearby beaches cleaned up by simple filtering devices, such as, for example, sub-sand filters are used in certain Southeast Asian facilities. Piped city water is, however, the most convenient.

The estimated volume of water required for only the basic fish marketing operations should provide for at least:

i. for fish washing, 1 litre per kg of fish/day;
ii. 10 litres/m^2/day for cleaning the market premises;
iii. 10 litres/box/day for fish box washing;
iv. 100 litres of fresh water/person per day for personnel use; and,
v. an additional 15% of the total daily consumption of water calculated above for other uses (e.g., canteens and vehicles washing).

In addition, water for ice-making, to supply fishing boats and for cooling diesel engines and refrigeration equipment should also be considered where appropriate. Care should be taken to plan for the fact that fish markets are sometimes used twice a day, in which case an estimate of peak water demand on an hourly basis is necessary. In the design of small reservoirs and pipeline networks care should be taken to ensure that the system is capable of taking the peak demand at any time of the day, and an elevated reservoir should, possibly, have a capacity to store the maximum daily water consumption requirements. Hydraulic factors such as, for example, the frictional resistance of the water against the piping system (head losses) as well as type of piping materials to be used should be given detailed consideration. In the selection of materials, especially for seawater network systems, where the galvanic potentials between the various materials has to be considered in order to ensure that corrosion effects are minimal. Care should also be taken in selecting plastic materials, and the fillers used to make the plastic material must be relatively inert when in contact with the pumped water. Either manual or mechanically driven pumps can be used to lift water as shown in the Figures 6,7 and 8. For a more detailed study of these factors, reference should be made to FAO Fisheries Technical Paper No. 174 (Blackwood, 1978).

Mechanical pumps are driven either by an electric motor, small petrol engine or by windmills. Small, mechanical, centrifugal, pumps are suitable for lifting against not more than 7 m of suction lift. Hand operated reciprocating pumps (low-lift type), with a capacity of 35 1/ minute, and maximum lift of about 6 m height, are commonly used either for seawater or freshwater.

Installation of elevated water storage tanks (header) from which water is supplied to the usage point by gravity provide a good technical alternative for fish landing centres, where large volumes of water must be stored. They consist of mechanically or windmill driven pumps supplying water to elevated storage tanks from which water is carried by gravity through a pipeline network to users. The usual locations for elevated tanks are either on top of a tower near the facilities, on the top of the building where the facilities are housed, or on a nearby hill. The size and type of water pump to use will depend upon such factors as :

1. volume of water to be pumped;
2. whether it is seawater of freshwater;
3. vertical distance from the level the pump is installed to the level of the source of water;
4. distance from the storage tank;
5. total height against which the water pump has to work; and,
6. kind of driving power available.

The pumps should be of a self-priming type and properly mounted and protected against floods, rain and vandalism. Windmill power should be considered only where the prevailing winds blow strongly enough (wind velocities not less than 2.2 to 3 m/sec, preferably above 4 m/sec) for the required number of hours a day and dependable enough to drive the pump so as to supply the required amount of water to a storage tank almost every day, or where no other form of power supply is available or appropriate. On-site chlorination systems for rural areas, which should be simple to operate, together with a sand filter (sub-sand biological types are commonly used) are essential for providing bacteriologically safe water to fish landing centres.

3.7.4.6 Waste disposal

Waste material is one of the outputs from the operations of fish handling and marketing facilities. It is, by its nature, highly prone to spoilage and can easily contaminate the fish and fish products being handled. Therefore, it should be immediately removed from the vicinity of areas where it can easily come into immediate contact with fish and fish products and, as soon as possible thereafter, removed altogether from the fish landing site and the market/handling premises themselves and their surroundings.

This waste material can be divided into solid and liquid waste. Waste of a mainly solid type should be stored in waterlight bins with covers. Such bins should be stored in an area with a concrete base sloping to a drain. These bins and the waste they contain should be removed at frequent intervals and disposed of in such a way as to avoid pollution. Such waste might be used for animal feed or fish silages or could be buried in pits.

Waste of a prevalently liquid type (effluents) should be removed from the landing site and the handling/marketing premises by draining it into the public sewage system where such exists. The drains should be designed to remove waste water without risk of flooding. At least 15 cm diameter drain pipes are needed in fish terminals. However, local health authorities should be consulted in order to comply with possibly more strict national regulations.

In rural areas, two options are available for water effluent disposal:

a. water effluents from the landing/marketing facility could be drained into a septic tank and percolated into a field trench or soakaway;

b. water effluents could be discharged into the main sewage system (if available) after course screening to retain solids.

The first option is, presently, most widely used in rural fishing centres. Septic tanks are rectangular chambers (two or three compartments), usually sited just below ground level, receiving both water effluents from the fish handling area (after coarse screening) and those from sanitary facilities. This material is stored in the tanks for a retention time of 1 to 3 days. During this period, the solid fraction settles to the bottom of the tank where it is digested anaerobically (without oxygen), which results in appreciable reduction in the volume of the sludge and releases gases (carbon dioxide, methane, hydrogen sulphide). The effluent, although clarified to some

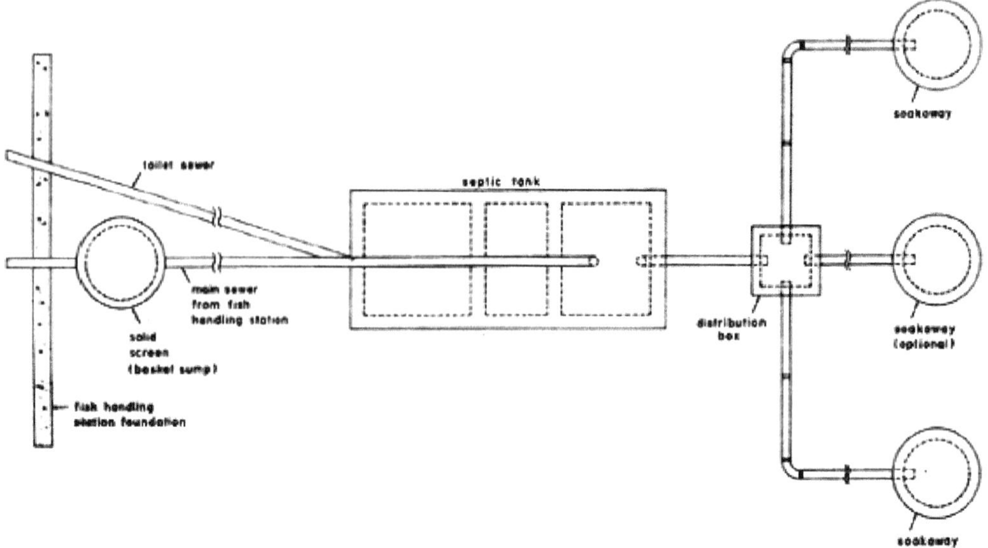

Figure 2 : Sewage disposal system (not to scale)

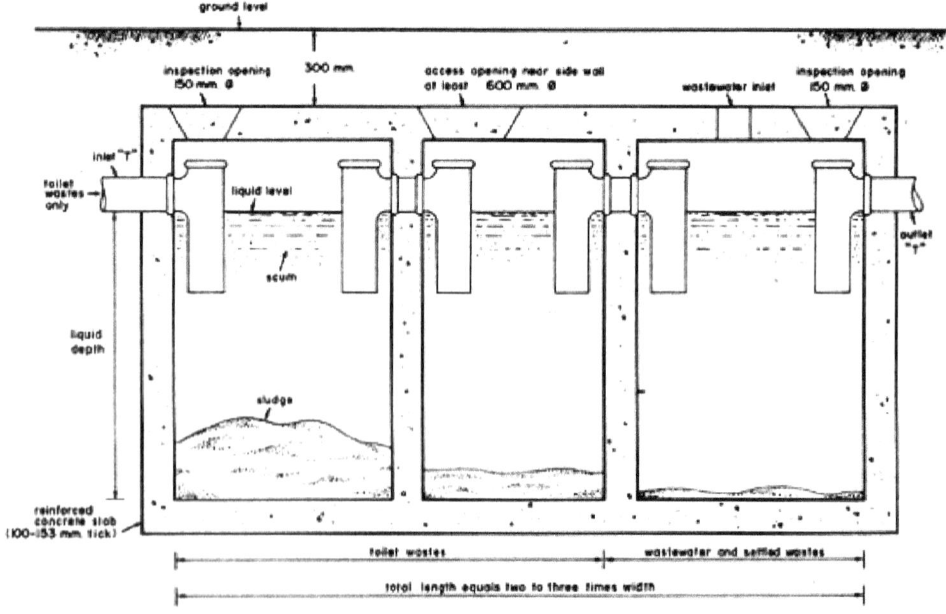

Figure 3 : A three compartment septic tank

extent, will still contain some dissolved organic solids and viable pathogens which will require further treatment in the soakaway. (A septic tank and schematic view of a sewage disposal system for fish receiving stations are provided in Figures 2,3,4). Septic tanks may be constructed of brickwork, stone masonry or concrete and, for design purposes, the overall length to breadth ration is 2 or 3 to 1. The location of the septic tank and soakaways should be at as far as possible a distance from buildings and sources of water, or from trees whose roots may damage them. Care should be taken to avoid contamination of the ground water. The minimum distance requirements for septic tanks and saokaways will depend on the type of soil and its percolative capacity. For example, in sand and gravel type of soils (very often found at landing beaches) the septic tank should be located up to 30 m away from wells and water streams. The design of septic tanks and soakaways as well as the piping system should be based on the maximum daily flow of sewage to the system, the detention period required and the frequency of cleaning the sludge in the bottom. It should be noted that soil permeability is a basic factor determining the use of on-site septic tank/soakaways/trenches sanitation systems; with impermeable soils these are not feasible. For the design of soak soakaways the rate of infiltration should be investigated in the area; however, for purposes of design, a 30 $1/m^2$ of side wall area per day infiltration rate can be assumed in landing beaches.

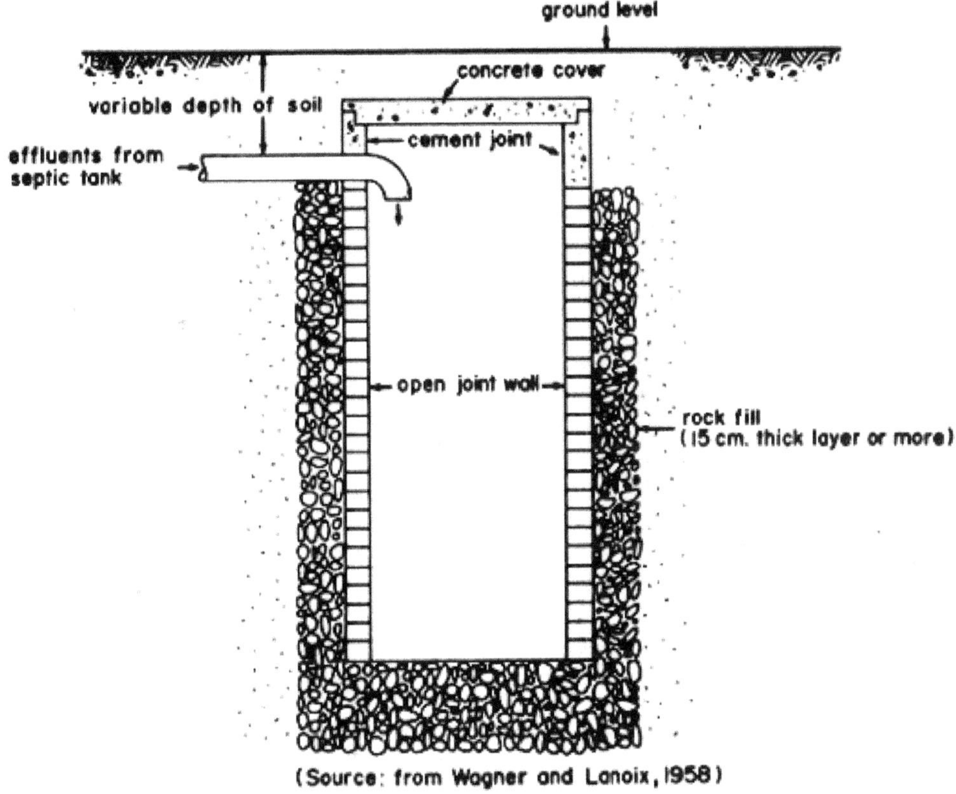

Figure 4 : Soakaway for secondary treatment of sewage

3.7.4.7 Power supply

The electricity supply should satisfy the present demand and the likely future requirements. The wiring should be of ample capacity and properly insulated. The choice of power source will depend on the local conditions. A dependable public supply power is the most convenient, however, it is advisable to have, in addition, an emergency diesel electric generator to be able to supply enough power to run the most important refrigeration equipment (chilled rooms) and water pump and lighting. When it is necessary to operate a diesel electric generator an overall energy assessment will be required to determine the capacity of the generator to be selected. Diesel electric generators have good efficiency, quick starting, relatively low capital cost especially for small ratings, and simple maintenance. The 4-stroke engine is more common for the higher speed engines (750 and 1 500 rev/min) which are desirable to drive electric generators. Fuel consumption ranges between 0.2–0.3 kg/kwh. Maintenance requirements consist of routine inspection and servicing of fuel-injection nozzles and an overhaul after about 5 000 hours of running.

Non-conventional sources of energy such as mini-hydroelectric generators, wind, biogas, photovoltaic cells and flat-plate solar collectors may be alternative options. However, careful consideration should be given to their potential application in the area under study and factors to be considered include: costs and performance in the actual village situations, availability of skilled personnel for operation and maintenance, as well as spare parts.

3.7.4.8 Parking areas

A detailed traffic study will help to quantify the area needed for parking. In small-scale fish landing facilities a convenient way of estimating is that access roads and loading yards should be large enough to absorb the anticipated peak traffic and that the loading area should be wide enough to allow for the necessary movement for the handling equipment (hand pallet trucks, trolleys, etc.) in order to move the fish quickly from the market hall to the vehicles.

3.7.4.9 Equipment

Some of the most important equipment used for small-scale fish handling marketing and distribution facilities is described in this section. It is needed to preserve the quality of the raw material and to facilitate efficient and economic handling and marketing.

(i) Weighing scales : Weighing scales should be of a heavy duty type and precise. They are needed both at landing places and also at sea for the fish collecting vessels when these are operated. Weighing scales used at landing places should be of the platform type with a counter-balanced arm and a tare device which can be set to compensate for the weight of the fish boxes. As an example, suitable specifications for a platform weighing scale would be:

- graduated in increments of 100 g;
- capable of weighing loads of up to 200–300 kg;
- incorporate a rapid tare device with a range of 50 kg;
- be constructed of corrosion resistant material;
- portable, that can be rolled on four wheels for easy movement (see Figure 12).

For weighing scales used on collecting vessels, heavy duty, spring dial, type hanging balances are quite suitable, provided that they are made of corrosion-resistant material; a tare device that allows deduction of tare weights is also useful.

(ii) Washing and sorting tables : At landing places, fish should, normally, be unloaded from the vessels, tipped into market containers, if not already boxed at sea, washed and sorted according to species and sizes, weighed into containers and chilled for either local marketing or further storage and transportation.

In order to facilitate the washing and sorting of the catch, simple tables constructed of aluminium, stainless steel, galvanized iron or concrete can be used. The primary aim of using washing and sorting tables is to speed up the operations of sorting the catch into species and to effectively remove blood and dirt by thoroughly washing the catch in clean seawater or freshwater and to reduce, to a certain extent, the number of bacteria present on the surface of the fish. After washing and sorting, the catch should then be handled according to the different needs of the market for which it is destined, e.g., removal of gills, gutting, filleting, packaging or boxing before chill storage and delivery to that market. Tables for washing and sorting should be constructed of non-corrosive metal or other impervious material, of rigid construction, with no open seams, and should also be constructed in such a way that they and the areas beneath them can be easily cleaned. In designing such tables, care should be taken to ensure that the dimensions are suited to the physical characteristics of the people who will be working at them.

(iii) Fish boxes : The design of a good fish box for efficient fresh fish handling, under tropical conditions, should take account of the varying needs it will have to satisfy on board fishing vessels, in warehousing, transport and at the landing and marketing places. It should have a constant tare weight, be strong, easy to handle, with adequate drainage, be easy to repair, have stack/nesting capabilities and be economical.

There must be an adequate stock of boxes at the landing sites, sufficient to allow the quick turn round of fishing vessels with clean boxes, thus allowing the boxes in which the fish was landed to be used in the market or by traders. An area for washing fish boxes (preferably manually) should also be provided for.

The dimensions and capacity of fish boxes should take into consideration such factors as fish size, packing density of fish and ice, drainage, suitable shape to ensure stable storage in vessels and maximum utilization of space in both at sea and during land transportation. It is desirable that boxes have sufficient length to accommodate the majority of fish caught (about 82 cm); and that the overall weight of the box and contents should be suitable for two men to lift comfortably, which is estimated to be a maximum of 67.5 kg. Detailed information about fish boxes is given in Planning and Engineering Data, Containers for Fish Handling, FAO Fisheries Circular No. 773 (Brox, 1984).

(iv) Internal transport equipment : The first step to be taken in designing the most appropriate internal transportation equipment, is to study the internal operating practices and select the equipment that will permit flexibility of movement with emphasis on the use of manual handling aids such as small four wheeled push carts, platform trolleys, hand pallet trucks, manually operated two-wheel case trucks, etc. The use of a particular handling aid will depend upon the total load to be carried, type of floor surface, cost of the equipment and maintenance considerations. More detailed information on internal transportation equipment is given in Planning and Engineering Data, Fish Handling, FAO Fisheries Circular No. 735 (Myers, 1981).

(v) Ice plants : Ice is of major importance in small-scale fish landing and marketing facilities. It is needed to maintain the quality of fresh fish during handling on board the vessel, collection, storage, transportation, marketing and distribution. Ice has good cooling capacity and its use will maintain low, chilling, temperatures, which is the most important factor in influencing the quality and storage life of fresh fish. At present, there are several types of ice plants for the manufacture of either block, plate, tube, or flake ice.

Block ice plants have the advantage that ice blocks can be transported relatively easily over long distances with low meltage losses; however, they lack flexibility in production, require large floor space and lifting devices to lift the blocks; they also have higher maintenance costs.

Flake, plate or tube ice plants (small ice) have the advantage of a more flexible production pattern, as the output of ice will start almost immediately they are switched on; handling is simplified if the flake or plate ice plant can be located on top of an insulated ice store which also reduce grand area requirements.

Small ice causes less damage to fish than crushed, block, ice but tends to melt faster and requires more storage space than block ice. In addition manpower requirements for small ice plants should be less than for block ice plants, as the labour required for the storage of ice is minimized. For example, a 10 t/24 h capacity small-ice plant requires about 29% less labour (man hours/24 h) than a block plant of the same output.

With any ice plant it is necessary to install an adequate ice store and its size will depend on the type of ice being manufactured and the local demand for ice. It can take the form of a silo or a rectangular bin but must be well insulated. According to the regularity of supply of fish and type of operations envisaged, the capacity of the ice store can be easily estimated.

(vi) Chill stores : Refrigerated chill rooms are widely used in small and medium scale fish handling, distribution and marketing operations. A suitable choice is the pre-fabricated, walk-in, chillroom (well insulated for tropical conditions) with independent refrigeration equipment and, whenever possible, having a simple but reliable system of control, which can be easily understood and operated.

Chill rooms should be designed taking into consideration the expected total weight of the product to be stored, handling methods to be employed (preferably manual), ambient temperature of the product entering the chill room, availability and cost of electricity, water and labour, and spare parts, and service/repair personnel. Whenever possible, standby refrigeration equipment should be provided for, in the event of any serious breakdown of the main refrigeration system.

In small-scale fish landing and marketing facilities, other types of chill stores are also used; these include insulated rooms, chilled or refrigerated seawater containers, cargo containers, refrigerated cabinets and small insulated containers.

Chilled seawater (CSW) containers are sometimes used for storage of higher market valve fish species and offer several advantages over traditional, boxed and iced handling techniques. They are also well suited for small pelagic species, which are caught in large quantities, which are not, normally, gutted or chilled at sea, but which require rapid handling and chilling with minimum effort, in a manner which ensures that the fish do not get crushed. CSW containers can be portable or fixed, and, in addition to their use on board fishing vessels, they find useful application in landing and marketing sites. Portable, insulated, CSW containers constructed of fibreglass or aluminium have been used for chilling small pelagic fish both on board vessels and at landing sites. HD polyethylene insulated containers with capacities of 70 to 1 000 litres can also be used in small-scale fish landing/marketing operations for the storage of chilled or frozen fish. Insulation materials often used include fibreglass or high density polyurethane, with outer and inner skins made of either aluminium or metals resistant to seawater; glass reinforced plastic (GRP); HD-polyethylene; concrete (outer skin) and marine plywood with GRP or epoxy resin coating. In order to minimize temperature stratification, it is desirable either to blow air into the containers or to recirculate the chilled water; in small-scale operations the use of a simple hand operated pump can solve this problem.

Cargo containers can also be used as chill stores for fish or ice storage and when so, refrigeration is provided either by the use of an attached individual cooling unit or by a central refrigeration plant. Cargo containers are normally available in standard sizes and are made of pre-fabricated insulated laminates fixed to an aluminium profile bearing system for the walls and roof, with the bottom frame made of steel profiles, and generally using polyurethane foam as insulation. They have several advantages in that they can easily be transferred to another location as they are designed for fast, loading/removal from a lorry chassis and only need to be sited on a leveled concrete floor and a shaded area for protection against sunshine.

Small, insulated, containers can be either locally made or factory constructed in different shapes and volumes and may be fixed or removable. There are several versions of small, insulated, locally-made, containers. These include, palm baskets lined with polyethylene film and insulated with dried coconut core or dried grass placed inside heatsealed polyethylene bags; in another version the lids are made of fiber mat and the same type of insulation materials as above are used; the "Bagamoyo" version of these insulated containers was rated only 10% lower from the thermal point of view in efficiency when compared with a factory made container (Lupin, 1986). Locally-made containers for small-scale operations at landing sites, markets, transportation, distribution and retailing outlets should:

- be light in weight, with high net volume to area ratio
- be easy to clean, handle and stack
- be cheap, easy to repair and have a reasonable life-span
- have good insulation properties
- have adequate drainage.

Several types of locally made, returnable insulated containers are used in small-scale fish landing and marketing operations. These include the use of rectangular wooden or plywood boxes, concrete tanks or oil drums insulated with expanded polystyrene or polyurethane and lined them to avoid moisture absorption; also used are expanded polystyrene boxes lined with either plywood, fiberglass, plastic or galvanized steel sheets and reinforced with an outer wooden frame.

Refrigerated seawater stores have wide application on board industrial fishing vessels, fish processing plants (canneries) and can also be used at landing sites. The basic principle of this method is to chill seawater in a heat exchanger from which it is circulated to the holding tanks. Normally a fish:water ratio of 4:1 is used and a holding temperature of -1.1°C is maintained. However, this method of chilling is not, generally, recommended for use in small-scale operations. The method is mainly used for the short term storage of particular species such as small pelagics which are too small and numerous for easy handling, sorting and boxing; moreover the equipment is expensive and requires careful maintenance and supervision when in operation.

(vii) Gutting and filleting tables : There is a need for simple and easy to clean gutting and filleting tables in small landing and marketing facilities. These should be designed so as to provide ample working space, have offal chutes or bins located as close as possible to the working stations, have impervious and corrosion-resistant surfaces, and be provided with a good water supply and drainage facilities. Tables can be made of smooth concrete or stainless metal. Wooden working surfaces should be avoided since they tend to become water-logged and are very difficult to clean effectively. As an example, the specifications for a portable metal type, gutting, filleting, table could be:

Dimensions: 3 m long by 1.2 m wide, 85 m high

Materials: Stainless steel 18/8 with fitted plastic cutting boards on each side of a central trough.

Tables should be designed to take into consideration the physical characteristics of the people who will work at them, providing a comfortable working space, and be of suitable height, to suit their body and reach characteristics.

3.7.5 Fish collecting systems

In order to meet the special needs of small-scale fisheries and to utilize more efficiently the fresh fish which they land at numerous and often scattered landing points, several fish collection methods have been attempted. These include, for example, the use of fish collection vessels, the use of road transport employing insulated trucks, or the use of modified bicycles.

The main requirements for fish collecting vessels are to have adequately insulated or refrigerated fish holds, for delivery of ice to the fish collecting beaches and the transport of the iced catch back to the main central fish depot. At the central fish depot or landing centre such vessels require adequate jetties for easy and fast unloading of fish and a supply of fuel, water, ice and fish boxes and food, as well as chill storage and transportation facilities for onward distribution of the fish to the markets. It is also necessary to have a reliable communication system in order to route collecting vessels, decide on quantity of ice and fuel to be supplied, and to obtain timely marketing information. As in all other commercial operations, fish collection schemes must be properly planned, the technical and economic feasibility evaluated and the schemes efficiently managed and operated.

A detailed fish marketing plan, including a realistic estimate of the sales and daily demand, should provide basic information for the planning stages. Fish marketing trials on a modest scale will also be required in order to assess in practice all aspects of the fish collection and distribution schemes being considered. In order to keep operational costs realistic, attempts should be made to provide for return cargo (for trucks or carrier vessels), and to work out an optimum size of the carrier vessels or trucks, keeping in mind the quantity of cargo available, the duration of the trips, fuel consumption, and maintenance.

3.7.6 Management

Basic decision making skills and supervisory capacity are vital for good management. Training in technical and organizational methods will often be required to upgrade staff skills to enable them to successfully overcome daily fish marketing problems.

A detailed plan of operations should be prepared during the planning stages and used to establish clearly, each staff member's functions. This plan should also establish detailed procedures for the systematic collection, handling, storage and distribution of landings, method of payment, issue of receipts, weighing system, daily sales and marketing expenses recording system, and the preparation of monthly financial reports. In the case of fish landing centres with fuel and oil service stations, gear and engine workshops, ice plants, supplies stores, etc., a detailed plan of operations and procedures for each component should also be prepared.

3.7.7 Live fish marketing

In Southeast Asian countries such as China, Hong Kong, Indonesia, Philippines and also the Bahamas, in the Caribbean, the live marine and freshwater fish trade, originating, predominantly, from coral reefs areas and fish ponds is an economically viable practice.

These live fish are stored on board vessels in live-fish holding tanks which are constructed so as to provide for a constant exchange of water in which the vessel sails. On arrival at the landing centres the live fish are sold to wholesalers or retailed directly to the consumers (Bahamas). In these landing centres handling space is required for the storage and marketing of the live fish. For this purpose, watertight concrete tanks with a water circulation system, which also provides adequate aeration by the spraying action of the incoming water, should be considered. For example, a live-fish collecting station and market (330 × 40 m) designed to handle 2 500 t/year was proposed for Java (Magnusson).

It included 24 concrete tanks (2 × 3 × 0.6 m each) for live-fish storage, an elevated water tank (45 m^3 capacity) and a water pump.

However, mortalities due to post-harvest handling of live marine fish have been reported to be around 50%, due mainly to injuries incurred during capture and transportation (inflicted by the fishing gear or excessive agitation of the water). An improved water circulation system on board has been proposed to lower this mortality rate (Leung, 1978).

Water used for the transportation and holding of live fish should be clean, well oxygenated, uncontaminated, and unchlorinated (small quantities of chlorine can kill fish). To hold live fish for sale (market) storage rates of 5–8 kg of live fish/100 litres of water is a practical approach for design purposes and provides a few days holding. For live fish retail shops, simple tanks provided with biological filters (gravel/sand) and aerated by small electrically driven air pumps can be a good alternative.

FAO GUIDELINES FOR RESPONSIBLE FISH TRADE

These Guidelines for Responsible Fish Trade have no legal status and are intended to provide general advice in support of the implementation of Article 11.2: Responsible international trade and article 11.3: Laws and regulations relating to fish trade, of the Code of Conduct for Responsible Fisheries (CCRF) and to assistant in further dissemination, understanding and voluntary implementation of the CCRF worldwide.

Fish and fishery products are among the most trade agricultural and food commodities with more than one third of production entering international trade. A specific feature of fish trade is the wide range in product types and markets. Significantly, one half of international fis trade originates from developing countries for which fish is an important earner of foreign exchange. Developed countries accounted for about 80 percent of the total value of imports of fish products.

Trade in fish and fishery products, is dynamic. Capture fisheries are levelling off while aquaculture continues to rise, thus affecting the nature of the sector's supply. The distribution chain, including the location and nature of processing activities, is constantly adjusting itself to changes in technology, communication and transportation. Freer trade and liberalized markets also increase the global nature of the sector. Trade is therefore more responsive to global, regional and national changes in supply and demand characteristics. The demand for fish and fishery products reflects changing consumer preferences and purchasing power, as well as demographic changes.

Article 6.14 in the General Principles section of the Code of Conduct for Responsible Fisheries in particular recognizes that: *International trade in fish and fishery should be conducted in accordance with the principles, rights and obligations established in the World Trade Organization (WTO) Agreement and other relevant international agreements. States should ensure that their policies, programmes and practices related to trade in fish and fishery products do not result in obstacles to this trade, environmental degradation or negative social. Including nutritional, impacts.*

A noted above, fish and fishery products are widely traded. Trade links production to consumption, making it necessary to ensure that sustainable management practices underpin production. The Benefits from international trade include better incomes, employment and foreign exchange. Currently the main barriers to trade are tariffs and non-tariff barriers, including technical issues related to safety and quality, certification and traceability. Other issues that continue to be of concern and have an impact on trade are subsidies that are prejudicial for trade and the environment. The improper use of ant-dumping, countervailing and safeguard measures is also a concern. In addition, producers and traders in developing countries are often in a disadvantaged position because of difficulties in obtaining market information.

The FAO Sub-Committee on Fish Trade of the Committee on Fisheries provides a forum for States to consult on technical, economic and environmental aspects of international trade in fish and fishery products, including production and consumption aspects. The FAO Sub-Committee on Fish Trade of the Committee on Fisheries also deals with issues related to technical cooperation. The FAO Sub-Committee on Fish Trade is an important forum where States can share their views on these issues, consider new developments, and recommend areas for further work.

On a global level, the WTO and organization of the United Nations (UN) system, in particular the FAO, are the main actors shaping the global trade regime for fishery products. UN organization address issues related to sustainable development, environment conservation, food safety and quality and food security. The rules governing international trade, embodied in the WTO agreements, are negotiated in the WTO. Together, the WTO, FAO and other organizations provide a frame of reference for States to cooperate in the formulation of appropriate rules and standards for international trade, including trade in fish and fishery products.

The WTO system is based on a series of agreements whose aim is to establish a rules-based framework for trade-and the liberalization of international markets for goods, services and investment. The General Agreement on Tariffs and Trade (GATT) provides for the liberalization of trade in goods through gradual reduction of tariffs, conversion of non-tariff import restrictions into tariffs (tariffication) and elimination of trade distorting domestic support. Developing States are given special consideration under GATT. They are given more time to reduce their tariffs and other obstacles to trade, and there are other special provisions designed to help them adapt to the liberalization of trade.

The Codex Alimentarius Commission was created in 1963 by FAO and the World Health Organization (WHO) to develop food standards, guidelines and related texts such as codes of practice under the Joint FAO/WHO Food Standards Programme. The main purposes of this programme are protecting the health of the consumers, ensuring fair practices in the food trade, and promoting coordination of all food standards work undertaken by international government and non-government organizations.

The World Organization for Animal Health (OIE) was created in 1924 to ensure global transparency in relation to animal diseases. The OIE collects, analyzes and disseminatges veterinary scientific information and provides expertise in the control of animals disease. The OIE develops rules and standards that can be used for protection against the introduction of diseases and pathogens. OIE standards are recognized by the World Trade Organization as the reference for international sanitary rules.

The Convention on International Trade in Endangered Species of Wild Fauna and flora (CTES) regulates international trade in species that are threatened wit extinction at the species level or that may be threatened as a result of international trade in specimens of the species. Several fish and shellfish species are listed under CITES Appendices.

11-POST-HARVEST PRACTICES AND TRADE

Article 11.2 – RESPONSIBLE INTERNATIONAL TRADE

11.2.1 The provisions of this Code should be interpreted and applied in accordance with the principles, rights and obligation established in the World Trade Organization (WTO) Agreement.

1. International trade in fish and products is covered under the international trading rules of the WTO. WTO agreement address such issues as tariffs and non-tariff measure, technical standards, including food safety and quality, rules of origin anti-dumping measures, subsidies and safeguards, trade in services, intellectual property and dispute settlement.

2. The WTO agreements are based on two fundamental principles: Most Favoured Nation (MFN) treatment and national treatment. The MFN principle requires States to accord the same treatment at the border to all similar products sourced from other WTO member States. National treatment requires that once a product entries the customs territory of another WTO member, that will treat the product no less favourable than like products produced by that importing Member State.
3. Many of the agreements are detailed and technical, but there are some dominant principles. For example, trade should be carried out without discrimination and there should be steady movement towards freer trade based on negotiations between and among members. Decision making within the WTO is based consensus among members. The WTO has established a Dispute Settlement Understanding (DSU) that enables members to sort out disagreements and solve trade disputes. States should take note of the decisions made by the Dispute Settlement Body (DSB). Taking into account the decisions made by the DSB, States should also consider whether their measures and practices relevant to trade in fish and fish products continue to be compatible with the principles, rights and obligations establishment by the WTO Agreements.
4. International trade in dynamic and States should continually assess their trade rules and international legal requirements in this light consistently with the WTO framework.

11.2.2 International trade in fish and fishery products should not compromise the sustainable development of fisheries and responsible utilization of living aquatic resources.

5. To provide a foundation for sustainable fish trade, States should adopt conservation and management measures to achieve long term conservation and sustainable use of aquatic resources. Conservation and management measures should be based on best scientific evidence available and be designed to ensure long-term sustainability of fisheries resources at levels which promote the objectives of their optimum utilization. States should also acknowledge the need to implement the precautionary approach and the application of the ecosystem approach, with the development of indicators that describe the level of biological, economic and social sustainability.
6. As responsible fisheries management measures are a necessary pre-condition for sustainable trade, states consider that in the absence of adequate conservation and management measures, increasing demand for to supply international markets can result in excessive fishing pressure, leading to overexploitation and wasteful use. This can have substantial impacts on food security and poverty, especially where there is a high level of dependence on fish in the diet. All persons and entities engaged in the international trade of fish products should ensure that their trade activities are consistant with the sustainable development of capture fisheries and aquaculture and the responsible utilization of living aquatic resource and do not undermine the effectiveness of fisheries conservation measures.
7. States that are WTO members taking measures to conserve living aquatic resources in relation to trade in fish and fish products should ensure they are compatible with WTO provisions. These provisions provide for exceptions to be made in particular circumstances to the general requirement for open trade between WTO members. Article XX states that, *"Subject to the requirement that such measures are not applied in a manner which would constitute a means of arbitrary or unjustifiable discrimination between countries where the same conditions prevail, or a disguised restriction on international trade, nothing in this Agreement shall be construed to prevent the adoption or enforcement by any contracting party of measures (g) relating to the conservation of exhaustible natural resources if such measures are made effective in conjunction with restrictions on domestic production or consumption".*

8. States should take into account the increasing demand for verification that fish products in international trade are originating from legal fishing operations and sustainable fisheries and aquaculture.
9. These trends include catch documentation and trade certification schemes developed by regional fisheries management organizations (RFMOs) and voluntary ecolabelling schemes. States should actively cooperate in developing and implementing catch documentation and trade certificate schemes, such as those developed by RFMOs, by adopting appropriate regulatory provisions and encouraging private sector collaboration.
10. Ecolabelling can provide producers of fish and fish products with an opportunity to differentiate their products. Ecolables can also have a beneficial effect on sustainable fisheries if properly designed and implemented. However, they also have the potential to create barriers to trade and may unfairly discriminate against non-ecolabelled products that are harvested sustainably. The lack of an ecolable does not imply that the fish has not been sustainably harvested. States and proponents of ecolabeling schemes should refer to the FAO Guidelines for the Ecolabelling of Fish and Fishery Products from Marine Capture Fisheries. These guidelines are applicable to ecolabelling schemes that are designed to certify and promote lables for products from well-managed marine capture fisheries and focus on issues related to the sustainable use of fisheries resources.
11. Recognizing that all States should have the same opportunities to benefit from sustainable trade, and in view of the special conditions applying to developing States and States in transition and their important contribution to international fish trade, developed States and relevant organizations, in particular the FAO, should provide developing States and States in transition with financial and technical assistance. Such financial and technical assistance should aim at capacity building in areas such as improving fishery management, and developing and implementing catch documentation, trade certification and ecolabelling schemes.
12. States should encourage ecolabeling schemes regarding sustainable fisheries within their territories to be consistant with the FAO Guidelines for the Ecolabelling of Fish and Fishery products from Marine Capture Fisheries.
13. To avoid trade-related measures compromising the sustainable development of fisheries and the responsible use of resources, States should cooperate, including through relevant RFMOs, to ensure that trade-related measures are compatible with the sustainable development of fisheries and responsible use of re sources, and consistant with WTO Agreements.
14. Trade-related measures to promote the sustainability of fisheries should be adopted and implemented in accordance with international law, including the principles, right and obligations established in the WTO Agreements, Such measures should be used only after prior consultation with interested States Unilateral trade-related measures should be avoided.

11.2.3 States should ensure that measures affecting international trade in fish and fishery products are transparent, based, when applicable, on scientific evidence, and are in accordance with international agreed rules.

15. States should promptly notify other States of measures affecting international trade in fish and fish products, including technical regulations, standards and procedures. They also should, when applicable, establish an inquiry point, and provide sufficient lead time for comments from interested States, consistant with WTO agreements.

11.2.4. Fish trade measures adopted by States to protect human or animal life or health, the interests of consumers or the environment, should not be discriminatory and should be in accordance with internationally agreed trade rules in particular the principles, rights and obligation established in the Agreement on the Application of Sanitary and Phytosanitary Measures and the Agreement on Technical barriers to Trade of the WTO.

16. States may adopt trade related measures for fish and fish products, under WTO agreements, if these measures are necessary to protect human and animal life or health, the environment, or if the measures relate to the conservation or fishery resources, States, however, must demonstrate that their measures do not result in *"arbitrary or unjusifiable discrimination between countries where the same conditions prevail, and do constitute a disguised restriction on international trade".*
17. The recognized international standard setting body for food safety and quality is the FAO/World Health Organization (WHO) Codex Alimentarius Commission (CAC) and the recognized body for animal health is the World Organization for Animal Health (OIE). Consequently, States should adopt, at a minimum, agreed Codex Alimentarius standards for fish safety and quality and OIE standards fro trade of liye fish. States also should participate actively and facilitate participation, as appropriate, by developing States in the work of the CAC's committees relevant to international fish trade, such as the Codex Committee on Fish and Fishery Products and other committees working on food additives, vertinery drugs, labelling, food hygienes, contaminants, sampling and analysis in order to ensure that the standards developed remain relevant to their objectives and CAC members.
18. Both the Agreement on the Application of Sanitary and Phytosanitary Measures (SPS) and the Agreement on Technical barriers to Trade (TBT) require Members to ensure that their technical requirements give national treatment to products imported from other Members. However, the Agreements recognize that under exceptional circumstances different treatment may be justified on the basis of objectives criteria, including scientific criteria, and reasons.
19. SPS measures should be based on an objective assessment, as appropriate to the circumstances, of the risks to human, animal and plant life and health, and take into account relevant economic factors. Members should, when determining appropriate levels of protection, take into account the objective of minimizing negative trade effects.
20. States should harmonize these measures wherever possible. States should also recognize measures that are different from theirs, where those measures produce an outcome that can be objectively assessed to be equivalent. Members are also encouraged to consult to achieve bilateral or multilateral agreements of mutual recognition.
21. All technical standards and regulations should have a legitimate purpose and States should ensure that the impact or cost of implementing these standards and regulations are proportional to their purpose. If there are two or more waves of achieving the same objective, the least trade restrictive alternative should be pursued.

11.2.5 States should further liberalize trade in fish and fishery products and eliminate barriers and distortions to trade such as duties, quotas and non-tariff barriers in accordance with the principles, rights and obligation of the WTO Agreement.

22. Barriers to trade in fish products (originating from both wild fisheries and aquaculture), including tariffs and non-tariffs measures, reduce the opportunities for States to maximize their welfare and benefits from their comparative advantage and will increase the cost of fish and fish products for consumers.

23. As fish is limited renewable resource with externalities in the production process, benefits from further market liberalization will only be produced if fisheries management that allows for the sustainable use of resources is in place. To ensure that societies maximize and sustain the benefits derived from the fisheries sector, States should seek market liberalization and improvements in fisheries management concurrently.

24. As per of market liberalization States should seek to eliminate trade and production distorting subsidies, especially those that are inconsistent with the sustainable development of fisheries and the responsible use of fish products, in particulars subsidies that contribute to over-capacity, to over-fishing, and to illegal, purported and unregulated fishing (IUU).

11.2.6 States should not directly or indirectly create unnecessary or hidden barriers to trade which limit the consumer's freedom of choice of supplier or that restrict market access.

25. Impact on trade from non-tariff measures can arise from any sources including technical measures such as requirements for product standards and conformity assessment, packaging and labelling. Sanitary and technical measures should not be more trade restrictive than necessary to fulfil legitimate objectives. Legitimate objectives are, *inter alia;* protection of human health and safety, animal or plant life or health, the environment or the consumer from deceptive practices.

26. Other market access conditions can include requirements for traceability documentation, banking and finance. Market access conditions can also reflect concerns about national and international security especially in light of the risks of terrorism. States need to be conscious of the costs and impacts on trade that may arise from conditioning market access.

27. The WTO agreements provide Members the right to implement exceptional measures under certain circumstances. States should ensure that measure that affect trade do not undermine the fundamental WTO principle, particularly those of national treatment and MFN.

28. States should avoid concealed or unannounced restrictions to international trade of fish and products and the misuse of exceptions to the fundamental principles of the WTO Agreements.

11.2.7 States should not condition access to market to access to resources. This principle does not preclude the possibility of fishing agreement between States which include provisions referring to access to resources, trade and access to markets transfer of technology, scientific research, training and other relevant elements.

29. As globalization evolves, the nature of access to markets and access to resources is expanding to include trade in services (for example, chartering or other services) and investment (including sale of fishing quota and joint ventures) and intellectual property with links to investment. States should apply the principles of the code of conduct for Responsible Fisheries (CCRF) to such trade.

30. Access to markets and access to fisheries should be negotiated on their separate merits, in a transparent manner, and in conformity with WTO principle and the relevant in UNCLOS, respectively. Distant-water fishing States seeking access to a coastal States' exclusive economic zone (EEZ) resources should not withhold market access if unsuccessful in grain access to fisheries. Coastal States have exclusive responsibility for determining how the living resources in EEZs are to be utilized. This includes determining the coastal State' sown catching capacity and, in that context, permitting other States to

access any surplus of the allowable catch in conformity with UNCLOS provisions. IN the EEZ, the coastal State has sovereign rights for the purpose of exploring exploiting, conversing and managing the natural resources, whether living or non-living.

31. Coastal States can require the payment of feeds and other forms of remuneration as part of licensing fishers, fishings vessels and equipment.
32. States should ensure that fisheries access agreements and associated services are negotiated in conformity with contractual practices, including transparency in negotiations and levels of access payments.

11.2.8 States should not link access to markets to the purchase of specific technology or sale of other products.

33. Trade in the fisheries sector includes a range of goods and services, as well as trade in fishing quotas, fishing licences, permits, joint ventures and related items. Trade takes place both as traditional cross-border trade between two companies or within the same company.
34. States should not condition access to markets to the purchase of specific technology, the provision of specific services or sale of other products. This equally applies to State-owned companies. Negotiations should be conducted in accordance with WTO member's commitments to MFn and national treatment principles. The same principles should be applied to capture fisheries and aquaculture related development assistance.

11.2.9 States should cooperate in complying with relevant international agreement regulating trade in endangered species.

35. States should fully participate and cooperate in the development, implementation and enforcement of measures to regulate trade in endangered species, in particular those adopted by the Convention on International Trade in Endangered Species of Wild Fauna and Flora (CITES).
36. CITES regulates international trade in species that are threatened with extinction at the species level or that may be so threatened as a result of international trade in specimens of the species. Several fish and shellfish species are listed under CITES Appendices.
37. States and RFMOs should cooperate with FAO in the provision of advice to CITES in the context of the 2006 Memorandum of Understanding (MOU) between the two organizations.
38. In addition to promoting good sustainable management practices, States should also facilitate effective participation by developing Stage in the development, implementation and enforcement of measures to regulate trade in endangered species, in particulars those adopted by CITES and other such measures as may be developed by relevant organization, including RFMOs, within competence, by providing assistance and through capacity building. States should adopt in their internal markets trade measures that are consistant with CITES.

11.2.10 States should develop international agreements for trade in live specimens where there is a risk of environmental damage in importing or exporting States.

39. States should note the risk posed by trade in live aquatic organisms destined for human consumption for human consumption and trade in live specimens for aquarium use and as breedings stock for aquaculture.
40. This trade may pose an environmental risk either through accidental introduction of non-indigenous species into the environment, or from the introduction of other organisms

or diseases that specimens may carry. States should assess the risk posed by such trade in a fair, transparent and non-discriminatory manner, consistent with WTO agreements and other applicable law.

41. Care should be taken to ensure that the transportation and storage of live specimens, where applicable, takes place under acceptable conditions with due concern to animal health.

42. The WTO Agreements clarify that States have the right to take appropriate risk-based measures to protect human, plant and animal life, and health and the environment. The OIE standards provide a framework for the prevention of the spread of animal diseases. States should use OIE standards, guidelines and recommendations for fish-health in live fish trade.

43. In negotiating international agreements, exporting and importing States should collaborate to minimize the environmental damage associated with trade in live species. States should encourage important and exporters to collaborate to avoid destructive practices and minimize losses.

11.2.11 Stated should cooperate to promote adherence to, and effective implementation of relevant international standards for trade in fish and fishery products and living resource conservation.

44. Within the existing frameworks of international organization, States should active participate and encourage other States to promote responsible and sustainable trade in fish products.

45. In this respect Status should promote adherence to international standards for trade in fish products. States should adopt, use or give effect to international standards of relevance to trade. Trade regulations should be consistent with relevant WTO agreement and provisions.

46. States should seek to ensure full cooperation on trade measures taken for resource conservation purposes. Such measures should be consistent with WTO rights and obligations.

11.2.12 States should not undermine conservation measures for living aquatic resources in order to gain trade or investment benefits.

47. A precondition for responsible, sustainable trade in fish and fish products is appropriate and effective fisheries management systems and a sustainable resource base, contribution to long-term food security.

48. The many activities of States, including the adoption of rules and policies regarding trade, services and investment, can undermine the conservation measures adopted by States and relevant RFMOs. States should ensure coherence between actions and arrangements to promote fisheries trade, services and investment on the one hand, and conservation objectives and action being promoted at the domestic and international levels on the other hand. These rules and policies should be consistent with the States' international obligations as established by relevant international organizations.

49. States should cooperate in the conservation and management and management of living aquatic resources in accordance with international law.

50. All States (including coastal, port, flag and importing States) should cooperate and make best efforts to deter, prevent and eliminate rate in fish and fish products originating from illegal fishing and illegal fishery activities as this trade undermines responsible trade, the sustainable use of resources and the activities of responsible operators. Importing States and port States should avoid using unilateral measures.

51. States should ensure that activities, including activities to promote trade, investment, services and the use of subsidies, do not lead to illegal fishing activities. These include illegal fishing activities arising from overcapacity. States should also ensure that the import, export or charter of vessels does not contribute to overcapacity or illegal fishing. Flag States, port States and coastal States should cooperate, including through RFMOs, as appropriate, to consider the non-discriminatory use of trade measures consistent with WTO agreements to remove incentives to illegal fishing.
52. States should support measures to deter, prevent and eliminate IUU fishing on the high seas by *inter* by *alia* exercising adequate control of vessels flying their flag and through relevant international fisheries management bodies, including through the use of measures that affect international trade, consistant with international law and WTO agreements, with a view to ensure sustainable and responsible fisheries.

11.2.13 States should cooperate to develop internationally acceptable rules or standards for trade in fish and fishery products in accordance with the principles, rights and obligation establishment in the WTO Agreement.

53. In order to facilitate responsible and non-discriminatory trade, States should participate and cooperate in the formulation of appropriate rules and standards for fish trade under the WTO frameworks, and also under other relevant framworks such as agreements regarding environmental protection and the sustainable use of fishery resources.
54. National measures should be consistent with international rules and standards, guidelines and recommendations adopted under the WTO frameworks. Of particular importance to fish trade are the standards, guidelines and recommendations of the Codex Alimentarius for human health and foot safety and those of the OIE for animal health. If States maintain measures that aim at higher levels of protection than those of the Codex Alimentarius and OIE, such measures should be based on scientific evidence and appropriate risk assessment.

11.2.14 States should cooperate with each other and actively participate in relevant regional and multilateral fora, such as the WTO, in order to ensure equitable, non-discriminatory trade in fish and fishery products as well as wide adherence to multilaterally agreed fishery conservation measures.

55. States that are members of international organization, including the WTO and regional fisheries management bodies, or have ratified or accepted binding international conventions have obligation to comply with their rules and requirements. States should actively participate in decision making processes so the agreements remain relevant to their objectives and to their members.
56. Recognizing that all States should have the same opportunities, States, relevant intergovernmental and non-governmental organizations and financial institutions should provide developing States and States in transition with financial and technical assistance on mutually agreed terms and conditions to actively participate in all aspects of the organization, particularly to develop and maintain appropriate measures and standards.
57. States should endeavour at all times to act in conformity with the provisions of international and regional organization and agreements that they are party to and avoid acting unilaterally.

11.2.15 States, aid agencies, multilateral development banks and other relevant international organizations should ensure that their polices and practices related to the promotion of international fish trade and export production do not result in environmental degradation or adversely impact the nutritional rights and needs of people for whom

fish is critical to their health and well being and for whom other comparable sources of food are not readily available or affordable.

58. Fish and fish products constitute a major source of animal protein in some States and regions. Furthermore, fish and fish products can provide an important basis for maintaining the social fabric and employment. This is the case for both developed and developing States, but may be particularly important in some developing States.
59. There are multiple objectives for the fisheries sector. States that provided assistance and State that are recipients of assistance should ensure coherence between fisheries and development policies with a view to increase the effectiveness of both policy areas.
60. Attention should be paid to the challenges facing States where market liberalization and globalization put pressure on the exploitation of aquatic resources. Both the recipient and donor States need to bring the precautionary approach and ecosystem considerations to bear when considering development assistance to specific projects.
61. States and other organization supporting initiatives related to international trade in fish products should adopt policies and procedures, including environmental and social assessments, to ensure that adverse impacts on the environment, livelihood, and food security needs are equitably addressed. Consultation with concerned stakeholders should be part of these policies and procedures.
62. States and relevant organization should cooperate with each other in the development and implementation of best practices, standards and guidelines for these activities. Changing conditions related to access to markets pose specific challenges to small-scale producers. States may give specific attention through capacity-building to these small-scale producers to organize their production and market access.

Article 11.3 Laws and regulations relating to fish trade

11.3.1 Laws, regulation and administrative procedures applicable to international trade in fish fishery products should be transparent, as simple as possible comprehensible and, when appropriate, based on scientific evidence.

63. Transparency requires that laws, regulations, administrative and operating procedures are publicly available and that decisions taken in consequence are readily understood. Transparency contributes to predictability and discourages corrupt practices.
64. Laws and regulations should avoid unnecessary requirement and duplication. States should provide plain language explanations and illustrative examples. FAO, other international and non-govenmental organization can facilitate transparency by providing information on the regulatory framework governing international trade in fish products.
65. Where laws, regulations and administrative procedures arise for technical reasons, States should ensure they are based on scientific evidence and referenced to internationally-agreed norms.

11.3.2 States, in accordance with their national laws, should facilitate appropriate consultation with and participation of industry as well as environment and consumer groups in the development and implementation of laws and regulation related to trade fish and fishery products.

66. Laws and regulations should be developed and implemented in consultation with stakeholders. Stakeholder include all those with a legitimate interest in the subject matter. The objectives of consultation should be to enable regulators to understand and accommodate the concerns of all stakeholders who will be affected. Involving all

stakeholders in the development of las and regulations fosters greater knowledge, understanding and acceptance of regulations and encourages voluntary compliance.

11.3.3 **States should simplify** their laws, regulations and administrative procedure applicable to trade in fish and fishery products without jeopardizing their effectiveness.

67. When appropriate, States should simplify regulation to make them easier to understand, apply and enforce. Complicated regulations are costly to comply with, may discourage trade and compromise legal trade and compliance.

11.3.4. **When a State introduces changes to its legal requirements affecting trade in fish and fishery products with other States, sufficient information and time should be given to allow the States and producers affected to introduce, as appropriate, the changes needed to their processes and procedures. In this connection, consultation with affected States on the time frame for implementation of the changes would be desirable. Due consideration should be given to request from developing countries for temporary derogations from obligations.**

68. Procedures for States to notify changes, to legal requirements that are technical or related to food safety issues should be complied with in a timely fashion. These procedure may required notification to other States through established procedures such as those described in the SPS and TBT agreements.

69. Where the changes affect trade in fish products of relevance to developing countries, appropriate consideration should be given to the capacity of those countries to comply. Appropriate flexibility may be needed vis-a-vis the ability of developing countries to implement necessary changes. Where needed, capacity-building should be promoted to meet and speed up the implementation of required changes.

11.3.5 **States should periodically review laws and regulations applicable to international trade in fish and fishery products in order to determine whether the conditions which gave rise to their introduction continues to exist.**

70. States should periodically review laws and regulations and the ways in which they are administrated. These is a continuing need to ensure that legal measures and regulation applying to trade in fish and fish products are effective and necessary. States should also ensure that laws and regulations are implemented in an efficient and cost effective manner.

11.3.6 **States should harmonize as far as possible the standards applicable to international trade in fish and fishery products in accordance with relevant inernationally recognized provision.**

71. States should harmonize technical and safety standards where possible and participate actively in the development of standards by the Codex Alimentarius and OIE. In the absence of harmonization, States should make every effort to recognize different regulatory processes as equivalent when they can be shown to achieve the same outcome. The same approach should be encouraged for other standards affecting international trade in fish and fish products.

11.3.7 **States should collect, disseminate and exchange timely, accurate and pertinent statistical on international trade in fish and fishery products through relevant national institutions and international organizations.**

72. States should collect and disseminate accurate and timely information, including statistical information, on international trade. This is a key element for the understanding of domestic and international markets and the impact of trade and fisheries management

policies. To improve its usefulness, the information collected by States should distinguish between aquaculture and capture fisheries. International organization, non-government organization and regional fisheries management bodies and national or regional institutional play an important role in providing the public with statistical information. Information disseminated by these organization can contribute significantly to improving cooperation with the sector.

73. Developed countries are encouraged to assist developing countries to build their capacity to collect and disseminate fishery and trade related information, including statistical information.

74. FAO and FISH INFO Network provide extensive information on fish trade. States should ensure that information to fish trade is readily available to interested parties. Information services should meet the need of stakeholder, including harvester. Processors retailers non-govenmental organization and consumers.

11.3.8 States should promptly notify interested States, WTO and other appropriate international organization on the development of and changes to laws, regulations and administrative procedures applicable to international trade in fish and fishery products.

75. States should regularly revise laws, regulations and procedures to take into consideration new evidence and technical and scientific development. Prompt, transparent and widespread notification of development and changes to laws, regulations, administrative and operating procedures is fundamental if delays and unnecessary costs and inefficient are to be avoided in international fish trade.

76. States should promptly notify and exchange information to facilitate the functioning of the trade system and to encourage compliance by States and trading enterprises. A number of WTO Agreements contain notification obligations. These obligations aid transparency and compliance. Where no such obligation apply, States should nonetheless inform trading partners directly of changes and developments relevant to the intranational trade in fish and fish products.

References

FAO. 1995 Code of Conduct for Responsible Fisheries. FAO, Rome. 41p.

FAO. **Fisheries Department.** 1998. Responsible Fish Utilization. *FAO. Techinical Guidelines for Respondents Fisheries*. No. 7, FAO. 33p.

Chapter 7

NEED FOR CONSERVATION OF MARINE RESOURCES

Introduction

Fishing has been a major source of food for millions and the fishing community has been benefited with employment and economic benefits. Only recently, when faced with evidence of overexploitation that the finite nature of fisheries was understood. In recent years, global marine fish production reached a plateau of 84 million metric tonnes per year and it became clear that fisheries resources couldn't sustain for long under often uncontrolled increase of exploitation.

The chief threats to oceans and their resources include pollution, invasive species, climate change, habitat destruction, by catch and overfishing. Overexploitation of fish stocks, changes in ecosystems, significant economic losses and international conflicts on management and trade have threatened the long-term sustainability of fisheries. Therefore, the Nineteenth Session of the FAO Committee on Fisheries (COFI), held in March 1991, recommended that new approaches to fisheries management embracing conservation and environmental, as well as social and economic, considerations were urgently. FAO was asked to develop the concept of responsible fisheries and a code of Conduct to foster its application.

Despite serious threats to the marine environment, efforts have been initiated to reverse the trends and ensure sustainability in marine ecosystems for future generations.

Sustainable Development of Fisheries

Sustainable development is defined as "the management and conservation of the natural resource base, and the orientation of technological and institutional changes in such a manner as to ensure the attainment of continued satisfaction of human needs for present and future generations. Such sustainable development conserves land, water, plants and animal genetic resources, is environmentally non-degrading, technologically appropriate, economically viable and socially acceptable" (FAO Council, 1988).

Sustainable use of marine resources means that the resource can be used indefinitely. Sustainable fishing implies a need to rebuild populations of exploited species and to promote recovery of ecosystems from effects of overexploitation. Overfishing, destructive fishing practices, bycatch of non-target species and environmental pollution have been responsible for fall in fish production.

Bycatch-Problems and Solutions

The incidental capture of non-target species known as 'bycatch' is a major fisheries management problem. It is also one of the greatest threats to the marine environment, wasting valuable natural resource causing dramatic decline of very valuable species. Different parts of the bycatch are sometimes called trash, discard and incidental catch. Juvenile fishes, turtles, sea-lions, starfishes and fish with little commercial value are caught every year by fishing boats and discarded into sea either dead or half-dead or found thrown outside the water.

Trawlers bring large volume of bycatch on board and is an expensive problem. Fishermen lose substantial revenue because of the loss of juvenile fish and non-target fish to bycatch and on top of this contributes to overfishing and declining ecosystem health endangering food security in poorer countries. The slipper lobster *Thenus orientalis,* which appears as bycatch in shrimp trawlers off Mumbai, completely disappeared from the fishery by 1995 and the fishery in yet to recover. Several other species of fish along the Indian coast has either depleted beyond recovery or continue to exist at an undesirable level.

There is growing acceptance by the fishing industry of the need to reduce bycatch. Modifying the fishing gear so that fewer non-target species are caught or it can be completely eliminated. These modifications are simple, cost-effective and simple and many times such innovations are coming from fishermen themselves. By modifying the cod-end to incorporate trash exclusion devices, bycatch could be substantially reduced. Another successful modification has been square-mesh panels in cod-ends, which exploit the behavioural differences between prawns and fish. Unlike the fishes, which herd when a trawl net is approaching and escape through the square mesh, prawn have a limited reaction to trawl. The flow of water generated by the trawl quickly forces them against the meshes and they eventually tumble along the bottom of the net into the cod-end. Fishers that seek to remove large organisms like turtles from their trawly may use a form of separator panel, which may be either flexible netting or a solid metal grid. Importers from USA have been demanding from seafood exporters certification that the shrimp exported is from trawl nets fitted with TEDs.

Overfishing

Overfishing, formerly defined as 'situations where one or more fish stocks are reduced below predefined levels or by fishing activities' means that fish stock are depleted to the point where they may not be able to recover. Biological overfishing occurs when fishing mortality has reached a level where the stock biomass has negative marginal growth. Economic or bioeconomic overfishing in addition to the biological dynamics takes into consideration the coast of fishing and defines overfishing as a situation of negative marginal growth of resource rent. Ultimately overfishing may lead to depletion in cases of subsidised fishing, low biological growth rates and critical low biomass levels. In some cases, depleted fish stocks may be restored and this is possible only them the ecosystem is intact. If the species depletion causes an imbalance in the ecosystem, it is difficult for the depleted stocks to restore to sustainable levels and other species dependent upon the depleted stocks may become imbalanced causing further problems. The non-penaeid prawn, which is the major food item for nearly 50 species of fish is harvested by trawlers and dol nets along the Maharashtra and Gujarat coast in large quantities and depriving the fish with their food may affect the population of these fishes.

Excess fishing capacity and overcapitalization reduce the economic efficiency of fisheries and usually are associated with overfishing. Overcapacity is difficult to manage directly, and usually involves in management regimes that encourage unrestricted competition for limited fishery resources. Overfishing and poor management can lead to collapse of fishery, fishing community and the industry.

Strict control over issuing fishing license and monitoring of actual fishing days that a boat is speeding in the sea (which can be monitored as actual absence of fishing boat from port) are some of the measures to control overfishing. In order to overcome the overcapacity problem, fishery managers should focus on developing or encouraging socioeconomic and other management incentives that discourage overcapacity and that reward conservative and efficient use of marine resources and their ecosystems.

Rebuilding of depleted fish stocks within a specified time with the exception of slow growing species may help in restoration of depleted species. Apart from overfishing, other factors such as bycatch, destruction of coastal breeding habits, pollution and climate change (global warming) may also be responsible for reduced fish in the sea. Deep sea resources are much more responsible for overexploitation as they are highly vulnerable to fishing due to slow growth, low fecundity and lower survival. This is true in the case of deep sea prawns exploited from fishing grounds off Quilon along the Kerala coast. Fishermen should be made aware of dangers of overfishing and should advise them to comply with government regulations.

Destructive Fishing Practices

Habitat loss as a result of undesirable fishing practices is a major factor responsible for decreased fish populations. Fishermen use certain destructive gears and fish in breeding and nursery areas very close to the shore or in backwaters where the shrimps and certain fishes spend their juvenile phase. Reducing or eliminating destructive fishing practices is essential for sustainable fishing and conservation of the resource. Minitrawling along the southern Kerala coast, Thalluvalai and Thallumadi along the southeast Tamilnadu coast, trammel net fishing of lobsters in Tamilnadu, stake nets in Kerala backwaters and estuarine systems, operation of dol nets along the Maharashtra and Gujarat coasts are typical examples of destructive fishing. Shrimps especially the green tiger shrimp *Penaeus semisulcatus* and *Parapenaeopsis stylifera* spend their juvenile phase in the nearshore areas and major portion of the catch by these gears consists of juveniles. Cyanide fishing and dynamite fishing are other destructive fishing practices adopted illegally by fishermen, all over.

Ecolabelling

By promoting sustainably caught seafood products consumers can be encouraged to demand sustainable fisheries. Ecolabelling is a major conservation alternative to promote best fishing practices by harnessing consumer purchasing power. Ecolabelling is increasingly perceived as a way to simultaneously maintain the productivity and economic value of fisheries while providing incentives for improved fisheries management and the conservation of biodiversity. The Marine Stewardship Council (MSC) founded originally as a partnership between the corporate Uniliver and WWF for Nature is leading global effort to encourage fishing establishments to apply for certification. MSC works with fishers, retailers and other stakeholders to identify, certify and promote responsible, environmentally appropriate, socially beneficial, and economically viable fishing practices around the world.

Degradation of Environment

Coastal industries release effluents in the sea or in backwaters and cause widespread pollution and damage to coastal habitats where the fish breed and spend the early phase. The Polluter Pays Principle and the Precautionary Principles shall be made applicable to industries, which are violating the standards for release of effluents prescribed by the Pollution Control Board of each State. Monitoring of effluent release by independent agencies and compliance by industries to stick to standards for effluent release fixed by Government are other measures to the implemented for protection of fish resources. Loss of nursery areas due to human encroachment of backwaters and inadequate flow of water through and dams has also affected the juvenile migration of fishes to the nursery areas.

Sustainable fishing can only be achieved by integrating conservation into fisheries management. Marine protected areas, searanching and co-management of resources by participating fishermen in decision making are some of the activities by which conservation of biodiversity and fish stocks could be achieved.

Marine Protected Areas (MPAs)

A marine protected area is an area of sea especially dedicated to the protection and maintenance of biodiversity. They have often effective in protecting and rebuilding ecosystems and populations of many marine species. MPAs include Marine parks, nature reserves and locally managed marine areas that protect reefs, tidal lagoons, mudflats mungroves as well as open water.

Protected areas will make the most effective contribution to the management of species and ecosystems when they are into management plans that cover the full life cycles and geographical range of the species involved. The design and implementation of marine protected areas should involve systems will protect their long-term interests and to improve operational integrity.

Benefits to Fisheries

- Conservation of biodiversity and ecosystems
- Probably reversing the declining fish populations by protecting critical breeding, nursery and feeding habitats
- Providing opportunities for education, training, heritage and culture; and
- Providing broad benefits as sites for reference in long-term research

Sea Ranching

Many countries are practicing stock enhancement of natural resource through release of hatchery produced, disease-free postlarvae/juveniles to nursery areas. Many countries such as Japan, Thailand and UK have reported potential benefits through searanching. Searanching of green tiger shrimp *Penaeus semisulcatus, Pinctada fucata* and seacucumber have been carried out by CMFRI. Released shrimps on recovery showed quantum jump in growth, restricted movement and recruitment to the fishery and survival. Release of 100 million postlarvae of *P. semisulcatus* has been estimated to increase shrimp production by 100 tonnes valued at Rs. 250 lakhs in an year.

Co-management

It is being increasingly recognized that fisheries can be better managed when fishermen are involved in management. The other partners who can be associated with the initiative are traders, exporters, NGOs and the Government officials. Such co-management has gained momentum since last decade, which recognizes the need for management decision to be made in collaboration with fishermen who depend upon the resource for their livelihood.

Co-management does not mean entrusting total control of fisheries management to fishermen. The responsibility of the Government and fishermen will have to be clearly defined. Many rules and regulations enacted by the Government are either not strictly enforced or not followed by fishermen. The Minimum Legal Size prescribed by Ministry of Commerce and Industry, Government of India for export of lobsters is to be strictly adhered to save illegal shipment of juvenile lobsters. However, maritime state governments are yet to bring in legislation for a minimum size for fishing. The fishing community is aware of decreasing catches but need guidance how they should manage their resource. Participation of the fishermen and other stakeholders is essential to establish a co-management regime. The fishermen will feel a greater sense of ownership of the resource, which can provide a powerful incentive to view the resource as a long term asset rather than to overfish.

CMFRI and CIFT with the funding support of MPEDA has taken up a participatory management initiative to implement co-management of the resource through several outreach programmes such as distribution of posters, pamphlets and stickers in vernacular languages with message of lobster conservation in fishing villages, village-level participatory meetings and workshops, rallies in fishing villages involving students from fisheries schools to convey the message of conservation V-notching and releasing of egg bearing lobsters back into the sea to create awareness among the fishermen the need for protection of breeding population and distribution of CIFT designed lobstor traps to wean them away from using destructive fishing gears such as trammel nets. The Mangrol Bandar Samj, Gujarat during this year declared fishing of juvenile lobsters along the Saurashtra coast as illegal and this a positive respose from fishermen who has understood the negative impact of fishing juvenile lobsters.

Conclusion

Many populations and some species of marine organisms have been severely overfished. There are widespread problems of overcapacity. There is much more fishing power than needed to fish sustainability. Fishing has tremendous impact on the ecosystem and these effects are beginning to be understood. Recent understanding of ecosystem effects of fishing has resulted in part from research on ecosystem approaches and has called for ecosystem approaches to fishery management to achieve sustainability.

The proper management of fishery resources requires a multidisciplinary approach calling on a wide range of scientific and technical knowledge. More research is needed into the biological characteristics of fishery resources, into the economic, social and cultural characteristics of the users of fishery technologies and resources and the legal tools required for implementing effective management regimes. Proper management includes measures such as licensing, closed seasons and areas, standards of fishing gear and nets etc. Good management seeks to control the fishing activity and needs political will to implement it. Conservation of biodiversity and stock enhancement programmes shall be given utmost priority and massive funding will be required for implementing these programmes.

The trader/importer/processor/restaurants make more money in B grade fish than A grade. This is because they get B grade products at 10 to 12 cents lower prices than A grade. The final fish dish served to the consumer makes no difference between A grade and B grade. For example let us take the case of Ribbonfish, the largest item of finfish exported to China in terms of quantity. When the fishes are caught in the nets and hauled into the vessel, they struggle for survival and in the process hurt each other with their powerful snouts. So there will be scratches on the body and even burst bellies through which the intestine protrudes. The real attraction for the ribbonfish is its shining silver colour, tasty meat and very cheap price. To meet the very high demand of fish at affordable for the people of China, the country has to import huge quantities of Ribbonfish from various sources. And India is the largest supplier of this much needed fish. That being the situation, the Chinese traders know that Ribbonfish will not be available with 100% silver intact unless it is caught by hook. But the fish caught by hook constitutes only a very negligible portion of the total exports from India. But it is business. They take advantage of the grading system of the fish. If India can produce to eat Chinese special dishes in pouches with Ribbonfish, the packers will get higher prices and our foreign exchange earnings will go fantastically up.

Many reputed Chinese companies are interested in joint venture projects with Indian packers for making such traditional Chinese special dishes from cheap fishes like Ribbonfish/Croaker etc. Interested packers/exporters are advised to explore and exploit the new emerging business-opportunities in China.

To get more and more production we have to protect, preserve and conserve our natural fisheries resources. Culture based capture technology must be adopted for ecological balance and to earn more money (revenue).

Chapter 8

CO-OPERATIVES FOR RESOURCE AND FISHING

FISHING COOPERATIVES

The PCC was modeled after the *Pacific Whiting Conservation Cooperative* (PWCC); a cooperative formed in 1997 by at-sea processing companies participating in the Pacific whiting fishery off the West Coast.

The 1999 Bering Sea pollock fishing season marked the first year of the Pollock Conservation Cooperative (PCC). Under the PCC agreement, pollock catcher/processor companies agree to limit their individual catches to a specific percentage of that sector's overall quota allowance, thereby allowing the fishery to be conducted at a more rational, deliberate pace. Since its inception, the PCC has helped reduce over capacity in the catcher/processor fleet and enabled participants to produce 50 percent more fish products on a per pound basis than the fleet produced in 1998 under an "Olympic-style," race for fish.

THE POLLOCK CONSERVATION COOPERATIVE

The Pollock Conservation Cooperative (PCC) is a private sector initiative by the Bering Sea pollock catcher/processor fleet to further improve conservation and utilization of marine resources. Catcher/processor companies, which operate vessels in the Bering Sea pollock fishery, formed the cooperative beginning with the 1999 fishing year. Member companies of the PCC allocate among themselves the overall quota of pollock available to the catcher/processor sector. The PCC is a harvest-only agreement among fishers. Cooperative members cannot coordinate processing, marketing or sales activities. They simply divide up a harvest quota that the government has set aside for their sector of the pollock industry.

By apportioning the allowable harvest among eligible fishery participants, the cooperative eliminates a race among fishers who would otherwise be seeking to maximize their individual share of the harvest. The race for fish can result in wasteful fishing practices. An orderly harvesting arrangement provides opportunities for fishers to maximize the amount of food produced per pound of fish harvested and to better avoid the incidental harvest of non-target species, sometimes called bycatch.

Ending the race for fish also helps resolve the problem of overcapitalization. Under the race for fish format, harvesting and processing capacity is continually added, enabling fishers to take the available harvest faster. The result is excess capacity beyond what is needed to catch and process fish in a rational manner over the course of a fishing season. There is no economic return for investing capital to catch fish faster since the amount of fish available to be harvested remains relatively constant. Overcapitalization, which is a root problem in many open access fisheries, is wasteful, inefficient and economically destabilizing.

The PCC represents a commitment among participants in the Bering Sea pollock fishery to improve upon one of the best managed, and largest, fisheries in the world. By weight, pollock landings account for approximately one-third of *all* U.S. fish harvests. Catcher/processors harvesting pollock produce fillet products used for fish and chips, frozen fish dinners, and fish sandwiches. Also, exported to Japan is pollock roe, which is available during the winter fishery. Surimi, a minced pollock used to make imitation crab products, is another primary product made from harvested pollock. Fish meal and fish oil are also produced from inedible portions of the fish. Fish meal is sold as a feed to aquaculture operations, and the fish oil is burned in boilers onboard ships to provide a clean alternative energy source.

Legal Framework for the PCC

The PCC was not formed under any specific legislative authority. The cooperative is viable because the federal government limits participation in the fishery and regulates output by setting an allowable catch level for the fishery. The PCC is simply a private contractual agreement among eligible companies apportioning individual shares of the catcher/processor sector's allocation. The Department of Justice's Antitrust Division issued a "business review letter" regarding formation of the PCC that stated in part, *"To the extent that the proposed agreement allows for more efficient processing that increases the usable yield (output) of the processed Alaskan Pollock and/or reduces the inadvertent catching of other fish species whose preservation is also a matter of regulatory concern, it could have pro-competitive effects."*

The PCC operates within the existing federal fishery management structure. Federal fishery managers and scientists establish annually a safe harvest level for Bering Sea pollock. Federal regulations divide the pollock quota among three sectors—inshore, catcher/processor, and mothership sectors. After setting aside a portion of the pollock total allowable catch (TAC) for the Community Development Quota (CDQ) program and for bycatch needs in the non-pollock groundfish fisheries, 40 percent of the remaining Bering Sea pollock TAC is allocated to the catcher/processor sector.

Improving Utilization of Fishery Resources

Since the pollock co-op formed in 1999, catcher/processors are producing about 50% more products per pound of fish harvested than in the last year of operations under the race for fish. When racing to catch fish the incentive was to catch as large a share as possible of the available common resource. At times, fishing operations sacrificed deriving the maximum yield from the catch in exchange for catching and processing as much fish as possible as quickly as possible. By apportioning the harvest among the participants, and stopping the race for fish, the cooperative, appropriately, creates the incentive to utilize fully and maximize the value of their individual allotments.

Improvements in yield have resulted, in part, from optimizing fishing operations, including slowing down harvesting to be more size selective. Larger-sized fish provide higher yields. Under the race for fish, individuals who slowed the pace of their own fishing operations forfeited harvest opportunities to those who continued racing full speed for the common quota. Under the co-op, fishers are no longer penalized for taking time to locate larger-sized fish.

Important improvements in utilization are also realized by rationalizing processing operations and placing the processing focus on enhanced recovery and not speed. The cooperative allows for more labor intensive activities that increase yields and matches up harvesting and processing activities so that processing operations, including producing fish meal from inedible portion of the fish, can keep pace with catch amounts. And the co-op creates an incentive to invest in processing equipment specifically designed to yield more food products from each pound of fish harvested.

Avoiding Incidental Catch of Non-Pollock Species

Pollock aggregate in enormous schools and are harvested using "midwater" trawl nets that are not dragged along the ocean floor. As a result, the pollock fishery is a very "clean" fishery, that is, non-pollock species account for about 1% of the catch. Since the inception of the PCC, the fleet has retained and used about 99.5% of fish harvested.

Even with this enviable conservation record, the PCC continues to work on reducing the incidental catch of non-pollock species. The PCC contracts with a private sector firm, Sea State, Inc., to monitor incidental catch. PCC member companies have waived confidentiality privileges, enabling Sea State to download proprietary catch data submitted to NOAA Fisheries on a real time basis. Sea State reviews this data and advises vessel operators of bycatch "hotspots" to avoid. Harvest cooperative members cease fishing in an area if bycatch is encountered and move to other fishing grounds. Under the race for fish, the responsible fisher often forfeited fishing opportunities if he took the time to act responsibly by relocating fishing activities.

Monitoring and Enforcement

There is extensive monitoring and strict enforcement of the North Pacific groundfish fisheries off Alaska. The U.S. Coast Guard and NOAA Fisheries enforcement agents patrol the fishing grounds and regularly board vessels for inspection. In addition, all pollock catcher/processors carry two federal fishery observers onboard at all times. Virtually all hauls by pollock catcher/processors are observed. Observers determine amounts of catch and bycatch and engage in fishery dependent research activities. To determine catch amounts, all PCC vessels are equipped with "flow scales" that weigh all fish brought onboard. The catcher/processor fleet also employs Vessel Monitoring System (VMS) technology. VMS units transmit the vessels' location to satellites, and NOAA Fisheries receives that data.

NOAA Fisheries monitors the catcher/processor sector to ensure that the fleet does not exceed its allotted quota. Using observer catch reports, Sea State monitors pollock harvests on a company-by-company basis to ensure that individual PCC members do not exceed their quota. The PCC fleet has never exceeded its sectoral allocation.

Addressing Overcapitalization and Restoring Economic Stability

The PCC has rationalized the catcher/processor sector and helped restore economic stability to this important component of the fishing industry. During the 1990s, the Bering Sea pollock fishery was overcapitalized significantly as fishers and processors continued to increase capacity in an effort to obtain the largest possible share of the available annual quota. Fishery managers failed to stop the race for fish and address the problem of overcapitalization. The result was chronic economic inefficiencies, numerous bankruptcies, layoffs and severe economic dislocation for fishers and processors.

The harvest cooperative allows fishers and processors to match harvesting and processing capacity to the percentage of the harvest available to each of the participants. In 2000, Alaska Trawl Fisheries, which operated the *F/T Endurance*, one of the 20 vessels qualified under the American Fisheries Act (AFA) to participate in the catcher/processor sector, sold its share of the pollock quota to the other PCC members and forfeited the vessel's right to fish in the U.S. Of the remaining 19 eligible pollock catcher/processors, only 15 or16 vessels participate in the pollock fishery in a given year. To avoid a "spillover" of effort to other fisheries, the AFA curtails fishing opportunities for the pollock fleet. By and large, the less efficient vessels remain at the dock.

The Rational Pace of Fishing Has Ecosystem-Wide Benefits

Throughout the 1990s, the race to catch the available pollock quota resulted in shorter, more frenetic fishing seasons. The cooperative reversed that decline. In 1998, the pollock fishing season lasted less than 100 days. Since that time, the fleet has taken twice as long to catch its quota. Fishing effort is spread out over time as vessels make fewer tows per day and limit the amount of pollock harvested per tow. The rational pace of fishing complements management actions to spread fishing out temporally and spatially to ensure that fishing activities do not compete with Steller sea lions that might be foraging for pollock.

Self-Assessments to Fund Marine Research Programs

Cooperative fishing has also led to cooperative funding of marine research. The PCC is contributing more than $1.5 million annually to fund marine research at universities, including the PCC Research Center at the University of Alaska/Fairbanks School of Fisheries and Ocean Sciences. Current research priorities include increasing understanding of natural changes in the Bering Sea ecosystem, examining factors that could explain declines of Steller sea lion populations and improving fish stock assessment models.

FISHERIES COOPERATIVES

India is endowed with 8.85 Kms of coast line and about 68.79 lakh hectates of inland water area, which provides immense scope for fish production. On account of various developmental measures so far undertaken, the fish Production increased to 55.46 lakh tonnes during 1997-98 as compared to only 11.6 lakh tonnes during 1960-61. Increased Production has enabled stepping up of export of marine products to Rx.4698.48 crores duriung 1997-98. Apart from its contribution towards foreign exchange earnings, added production has also helped in augmenting protein rich food supply, generation of employment opportunities and raising incomes of fishermen who are among the disadvantaged sections of society. Despite the considerable adbances made, the per capita consumption of fish in our contry is as low as 3.5 kgs as compared to estimated requirements of 11 kgs and there is still bast potential for fisheries development and growth.

NCDC statted promoting and developing fisheries cooperatives after its Act was amended in 1974 to cover fisheries within its purview. In order to discharge these functions effectively, the Corporation has formulated specific schemes and pattern of assistance for enabling the fishery cooperatives to take-up activities relating to production, processing, storage, marketing etc. Assistance is provided to fishermen cooperatives on liberal terms treating the activity as weaker section programme. Assistance to fishery cooperatives is provided for the following purposes.

- Purchase of operational inputs such as fishing boats, nets and engines.
- Creation of infrastructure facilities for marketing (transport vehicles, cold storages, retail outlets etc.).
- Establishment of processing units including ice plants, cold storages etc.
- Development of inland fisheries, seed farms hatcheries etc.
- Preparations of feasibility reports.
- Appointment of experts under Technical & Promotional Cell Scheme.
- Integrated Fisheries Projects (Marine, Inland & Brackish Water).

COOPERATIVE STRUCTURE

With the National Federation of Fishermen's Cooperatives (FISHCOOPFED), at top of the Cooperative structure, there are 17 Federations at the State level, 108 Central Societies at the district & regional levels and over 11,000 Primary fishermen cooperative societies.

Membership of primary societies is around 11.39 lakhs covering about 21% of active fishermen in the country.

Presently, the cooperative structure differs from state to state. While Derala has got a two tier structure (primary and apez), Maharashtra has 4 levels viz, village, district, regional and apex. Consensus is gradually emerging to develop three tire structure viz. primaries at the village level, central societies at the district level and state federations at the apex level.

NCDC'S ASSISTANCE

During 1998-99, the Corpotation sanctioned Rs. 64.28 cores and released Rs. 53.19 cores (including spillover assistance) for fisheries development through cooperatives. The State-wise details of the assistance sanctioned and released are presented in Tables -1 & 2 respectively. Cumulatively, the Corporation has sanctioned assistance of Rs.465.63 cores and released Rs.175.o7 Cores for fisheries development to cooperatives in differet States and UTS.

Impact of NCDC's Assistance

- The increase in marine fish catch of cooperatives in Gujarat, Kerala, Maharashtra, West Bengal, etc., is attributed to the support of NCDC.
- The total number of fishermen to be benefited from all the Integrated Projects is estimated to be over 3,05,200.
- The total estimated additional fish production on implementation of the projects is over 4.12 lakh tonnes per year.
- The societies are democratically managed and the members take active interest in the functioning of the societies.
- Beneficiary fishermen earn more income than non-beneficiaries. NCDC's assistance has helped fishermen by reducing drudgery of work, increase in the number of fishing days thereby increase in fish catch of fishermen. Fish marketing infrastructure helps the fishermen's cooperatives in realising better value for fish catch in the market.
- The increased fish catch by the fishermen has improved the socio-economic conditions of the fishermen perceptibly.

Procedure Involved in availing NCDC's Assistance

The societies fulfilling the conditions of eligibility should apply through the Commissioner/ Director of Fisheries/Registrar of Cooperative societies to NCDC. Advance copies of the application may be sent to NCDC's Regional Directors/Head Office. NCDC would however consider the proposals only after the proposals are duly recommended by the State Govt. Societies desiring assistance under direct financing scheme may refer to the Direct Funding Guidelines available on the Web site.

Few assisted Successful Cooperatives are as follows :

1. Kerala State Cooperative Federation for Fisheries Development Ltd. Karuvankonam Thiruvanthapurum, Kerala.
2. Karanja Macchimar Vividh Kanyakari Sahakari Sanstan Ltd. Karanja, Maharashtra.

3. Amala Machhimarvividh Karyakan Sahakari Sanstha Ltd. Amala Thane Distt. Maharastra.

The Co-operative recognises that there is an ongoing risk of IUU fish (illegal, unreported or unregulated) coming into the marketplace, otherwise known as 'blackfish'.

The Co-operative will not accept product known or suspected as coming from IUU fishing. Countries and boats or fleet of boats supplying raw material destined to products for the Co-operative must not be listed on or associated to any "black list" internationally recognized by the appropriate regional fishery management organisation (RFMO).

The Co-operative supports the use of supply chain traceability as a means of verifying the provenance and the legality of the seafood in our branded range. All Co-operative tuna is fully traceable to the catching vessel (or group of catching vessels if they are small scale artisinal boats).

Fishing Techniques

The Co-operative specifies the use of selective fishing techniques designed to minimise their effect on other species, marine animals and birds, as well as the ecosystem. All Co-operative canned Tuna is caught by methods that conform to Earth Island Institute Dolphin Safe standards.

The Co-operative only sources from tuna suppliers able to demonstrate the respect of these environmental values and legality:

- Destructive catching methods, specifically bottom trawling and drift nets are not acceptable
- Conscious effort to reduce bycatch of juveniles and unwanted species by using non intrusive repulsion measures and larger mesh net
- Commitment never to capture mammals, birds or turtles under the EII guidelines
- Commitment to eliminate and release from the catch any unwanted species alive and unharmed where possible
- No fishing in areas recognized as giving high levels of bycatch
- Reject of any shark fining practices by members of fishing crew

To confirm our specification for Tuna we are party to an agreement whereby the international conservation body Earth Island Institute monitor that claim.

The Co-operative will always give first consideration when sourcing tuna, to suppliers using selective fishing practices, such as hand line or pole and line. Yellowfin tuna (Thunnus albacares) and Albacore tuna (Thunnus alalunga) must be sourced using these catch methods only.

Recognized standards accepted by the Co-op include the Marine Stewardship Council, Earth Island Institute Dolphin Safe label and the Marine Conservation Society.

Reviewing the Policy

The Co-operative Group has created an internal management structure to drive this agenda. The Co-operative Group Fish Sustainability Teams are multi-disciplinary in constitution, drawing from a broad base of internal and external stakeholders and represent a holistic business approach to this important issue. These two teams fulfil the roles of analysing and interpreting scientific data as well as having policy and decision making powers based on the outputs of these analyses.

We regularly review our fish sourcing policy, assessing it against a wide variety of information sources. Our approach is to closely monitor the latest scientific advice, taking into account the views expressed by non-government organisations and other interested parties such as marine conservation groups.

FISH SOURCING POLICY - WILD CAUGHT

The Co-operative shares the objectives of all concerned in the fishing industry that supplies of fish need to be protected for the long term, by means of sustainable fisheries management practices.

The Co-operative is mindful of the diverse environmental and ecological issues associated with the capture of wild fish for the table.

Issues which effect both the biological sustainability of the target species and which have wider eco-system impacts such as; over-fishing, illegal, unregulated and unreported (IUU) fishing and fishing methods which are destructive to the sea bed and marine habitats are all of concern to us and our customers.

In view of these complex challenges to the responsible sourcing of wild captured fish, the Co-operative operates a strict policy to monitor and control our fish supplies.

Sustainability Checklist

The Co-operative has devised a unique method of assessing fish supplies that takes all of these issues into account and which, wherever possible, is based on scientific information and advice. This bespoke system examines in detail each product, by species, by fishery and by catching method so that we can clearly identify those species, fisheries and fishing methods which must be avoided and those which should be selected.

We recognise and support the concept of 'reject the worst, select the best and improve the rest' and will be working with other concerned stakeholders to ensure that our policies are effective, not just in identifying the problem areas, but in working through managed processes to improve fisheries wherever possible.

The Co-operative takes fish sustainability very seriously and is committed to the maintenance of an own brand range of fishery products which gives the consumer a wide choice supported by informed, accurate and highly visible information to enable them to make responsible sourcing a factor in their purchasing decisions should they wish to do so.

The Co-operative holds environmental responsibility the consumer trust, which is of paramount importance to us.

Sourcing

Our goal is to operate our fish sourcing in line with the aims and objectives of the Marine Stewardship Council, of which the Co-operative has been a member and key supporter since its launch in 1997.

The Co-operative therefore pursues a rigorous policy of selecting and monitoring its seafood suppliers. We work only with suppliers who are able to demonstrate the highest levels of good practice, sourcing from well-managed fisheries and actively avoiding vulnerable species. We do not purchase fish where the origin or method of catch is unknown.

All Co-operative suppliers are subject to audit and inspection on a regular basis to ensure that the required standards are met.

Blackfish

The Co-operative recognises that there is an ongoing crisis with IUU fish (illegal, unreported & unregulated) coming into the marketplace, otherwise known as 'blackfish'.

We aim to eradicate any possibility that illegal material should enter our supply chain and will never knowingly purchase such material or deal with suppliers implicated in practices such as:

* Exceeding quota limits
* Fishing outside prescribed areas
* Using banned fishing methods
* Capture and sale of non-target species.

In addition, the Co-operative requires it's suppliers to purchases all fish within a size specification, to discourage the use of undersized fish that are an element of those caught over quota.

Regulatory bodies are allowed free access to any information maintained by the Co-operative, which may be of help towards monitoring fish sources. We suppliers to do the same.

Fishing Techniques

The Co-operative specifies the use of selective fishing techniques designed to minimise their effect on other species, marine animals and birds, as well as the ecosystem, for example all Co-operative Tuna is caught by methods which conform with Earth Island Institute Dolphin Safe standards. We specify that drift nets are not used to catch Co-operative Tuna, nor are our fish caught using the method known as 'setting on dolphins'.

To confirm our specification for Tuna we are party to an agreement whereby the international conservation body Earth Island Institute monitor that claim.

If UK fishermen are catching the fish, we require the trawler to be registered and certified to the Seafish Responsible Fishing Scheme. Certification to this scheme demonstrates that a vessel operates to industry good practice guidelines. The scheme gives an assurance to the supply chain and consumers that fish from the vessel has been caught responsibly.

Bycatch

Most fisheries are unselective to some degree, in that they incidentally catch other species along with their target catch during the process of fishing. This non-target catch is known as "bycatch". A significant proportion of this bycatch is discarded back into the ocean. Focusing fishing efforts on a narrow range of species is not the most efficient way to harvest marine resources. Learning to process and market a diverse range of species may be a more ecologically sustainable approach in the long term.

The Co-operative will continue to work with our fish suppliers to find ways of reducing the level of bycatch discarded by, amongst other means, identifying and developing products to best utilise a wider range of the fish species commonly caught.

Reviewing the Policy

The Co-operative has created an internal management structure to drive this agenda. The Co-operative Fish Sustainability Teams are multi-disciplinary in constitution, drawing from a broad base of internal and external stakeholders and represent a holistic business approach to this important issue. These two teams fulfil the roles of analysing and interpreting scientific data as well as having policy and decision making powers based on the outputs of these analyses.

We regularly review our fish sourcing policy, assessing it against a wide variety of information sources. Our approach is to closely monitor the latest scientific advice, taking into account the views expressed by non-government organisations and other interested parties such as marine conservation groups. We also maintain an ongoing review of the IUCN Red List of Threatened Species.

Handling Your Catch

- **Land them fast.** It may be fun to play a fish on the line for minutes or even hours, but that fish will be far poorer table fare than one landed quickly. Reel the fish in quickly, and use the proper line for the fish you expect to catch: If you expect to catch 10-pound fish, you should use 10-pound line.
- **Knock them on the head.** You should stun your fish when you get them aboard. Use a miniature baseball bat or a stout wooden dowel to whack them on the head. This prevents the fish from beating themselves up in your cooler, which can damage the meat.
- **Bleed them.** This is controversial, because you are essentially bleeding out a live fish, which is no fun for the fish. I rarely do it, unless I am catching sharks, bluefish or tuna—all of which are either very large and need cooling ASAP, or which run hot internally. And you need to cool fish down quickly for the best eating quality. If you choose to do it, cut about an inch back from the tail up until you feel bone.
- **Get them on ice right away.** Even on cold days, it is important to ice your catch. Fish can spoil even at temperatures that we humans find cold. Bring a large cooler with lots of ice, and place your fish in right-side up (as if they were swimming) as soon as they are stunned or bled. Make sure your cooler has a spigot so you can drain the meltwater—you do not want the fish to rest in water.
- If you cannot bring ice, keep the fish in the shade, on a stringer in the water, bury it in the sand (mark the spot!), or try this: Bleed and gut the fish right away, then salt it all over. Wrap the fish in fresh green leaves or seaweed and put it in the coolest place you can find.

SCHEMES OF NCDC IN THE DEVELOPMENT OF FISHERIES

NCDC has been promoting and developing fisheries cooperatives after its Act was amended in 1974 to cover fisheries within its purview. The Corporation has formulated specific schemes and pattern of assistance for enabling the fishery cooperatives to take up activities relating to production, processing, storage, marketing, etc. Assistance is provided to fisheries cooperatives on liberal terms treating the activity as weaker section programme. Assistance to fishery cooperatives is provided for the following purposes:

- Purchase of operational inputs such as fishing boats, nets, and engines
- Creation of infrastructure facilities for marketing, transport vehicles, ice plants, cold storages, retail outlets, processing units, etc.
- Development of inland fisheries, seed farms, hatcheries, etc.
- Preparation of feasibility reports.
- Integrated Fisheries Projects (Marine, Inland and Brackish Water)

Future Thrust

Future thrust of the corporation is on creating infrastructure for marketing, processing and by products to reduce post harvest losses and increase in income of the fishermen.

Conditions of Eligibility for NCDC's Assistance

The society seeking financial assistance should be functional, working efficiently.

- A newly registered society can also be considered, provided it has a well designed programme of pisciculture, marketing etc.
- Primary societies should be affiliated to the regional/state level fisheries cooperative federations and the district/state cooperative union, if existing in the state.
- The society which has been classified below "C" category during the audit for the last three years will normally not be considered. However rehabilitation projects of such societies, those would be considered on merits.
- Societies with negative net worth (cumulative losses exceeding own funds) would not be eligible for assistance.
- The inland fishery societies operating in the water area leased out to them should invariably have a lease for more than three years. For availing assistance for excavation, the period of the lease should be for a minimum of 10 years.

Chapter 9

BIOSAFETY ISSUES TO BIOTECHNOLOGY

Biosafety is considered in relation to the use of genetically modified organisms (GMOs) in food and agriculture. In agriculture, (including animal husbandry, fishery and forestry), the concept of biosafety involves assessing and monitoring the effects of possible gene flow, competitiveness and the effects on other organisms, as well as possible deleterious effects of the products on health of animals and humans. Policy decisions taken in regard to biosafety may have long-term implications for the sustainability of agriculture and food security according to FAO.

Within the WTO, biosafety in relation to GMOs appears to fall chiefly under the Agreement on the Application of Sanitary and Phytosanitary Measures (SPS Agreement). There are also implications for GMOs in standard-setting under the Agreement on Technical Barriers to Trade (TBT Agreement). This Agreement covers a large number of technical measures that seek to protect consumers from economic fraud and deception and measures concerning human, animal and plant life and health not covered by the SPS Agreement, and the environment.

INTERNATIONAL STANDARD SETTING WITHIN THE FRAMEWORK OF FAO

FOOD STANDARDS: *THE CODEX ALIMENTARIUS COMMISSION*

The purpose of the Codex Alimentarius Commission is to protect the health of consumers, to ensure fair practices in food trade, and to promote coordination of all food standards work undertaken by international governmental and non-governmental organizations. The Commission's Medium-term Objectives include *inter alia* "consideration of standards, guidelines or other recommendations as appropriate for foods derived from biotechnology or traits introduced into foods by biotechnology on the basis of scientific evidence and risk-analysis and having regard, where appropriate, to other legitimate factors relevant for the health protection of consumers and promotion of fair practices in food trade".

The 23rd Session of the Codex Alimentarius Commission (June/July 1999) established an *Ad Hoc* Intergovernmental Task Force on Foods derived from Biotechnology to fulfil these objectives, which will take full account of the work of national authorities, FAO, WHO, other international organizations and other relevant international fora. The Task Force will submit a preliminary report to the Codex Alimentarius Commission in 2001, and a full report in 2003. The Codex Committee on Food Labelling is also working on the development of recommendations for the labelling of foods obtained through biotechnology.

It should be noted that Codex standards apply to all types of foods, and, for this reason, Codex will need to deal with foods of plant, animal and fish origin. The impact of feeding GMO plants to animals, and the nature of the resulting foods from these animals will also need to be addressed.

PHYTOSANITARY STANDARDS : *THE INTERNATIONAL PLANT PROTECTION CONVENTION (IPPC)*

The IPPC's purpose is common and effective action to prevent the introduction and spread of pests of plants and plant products, and the promotion of appropriate control measures. It covers both cultivated and wild plants; the direct and indirect effects of pests; and the prevention of the introduction and spread of weeds, and their control. The IPPC also covers the movement of biological control agents, and other organisms of phytosanitary concern claimed to be beneficial. The IPPC provides the global standard setting mechanism for phytosanitary measures. It may be concerned with evaluating the potential "pest" characteristics (including weediness) of GMOs, that is, whether a GMO may be detrimental to plant life or health.

At the second meeting of the Interim Commission on Phytosanitary Measures (ICPM) in October 1999, a number of members gave high priority to standard setting in relation to GMOs in particular to risk assessment and testing and release of GMOs. They indicated that this could be addressed within the framework of the IPPC. Others advocated a more cautious approach while some indicated the need to give sufficient priority to development of standards for plant quarantine. The ICPM decided that an exploratory working group would address the issues of biosafety in relation to GMOs and of invasive species and report back to the 3rd meeting of the ICPM in April 2001.

THE COMMISSION ON GENETIC RESOURCES FOR FOOD AND AGRICULTURE (CGRFA)

Since 1989, the CGRFA has regularly considered reports on technical and policy issues regarding biosafety, within the context of biotechnology as it relates to genetic resources for food and agriculture. In 1991, it requested the preparation of a draft *Code of Conduct on Biotechnology*, with the aim of maximizing the positive effects, and minimizing the possible negative effects of biotechnology. The draft *Code of Conduct*, drawn up following a survey of over 400 international experts, contained four modules, one of which was on biosafety. In 1993, noting that the CBD was considering the development of a biosafety protocol, the CGRFA recommended that FAO participate in this work, in order to ensure that the aspects of biosafety relevant to genetic resources for food and agriculture were appropriately covered. The biosafety component of the draft Code of Conduct was forwarded to the Executive Secretary of the CBD, at the request of the CGRFA, as an input to the biosafety protocol.

The CGRFA has suspended work on the draft *Code of Conduct* pending the completion of the negotiations for the revision of the International Undertaking on Plant Genetic Resources. The Eighth Session of the CGRFA, in April 1999, noting that the negotiations for the revision of the International Undertaking were expected to be completed during 2000, requested a report on the status of the draft *Code of Conduct*, at its Ninth Session, in 2001.

GMOS IN FISHERIES

The fishery sector has recognized that GMOs are a diverse class of organisms that share many common features with introduced or alien species. FAO's Regional Fisheries bodies have adopted, in principle, codes of practice on the use of introduced species and GMOs, produced by FAO's European Inland Fishery Advisory Commission (EIFAC) and the International Council for the Exploration of the Sea (ICES). The general principles in such codes of practice, which include general principles for environmental assessment, contained use, advanced notification and the application of the Precautionary Approach, have been incorporated into the FAO Code of Conduct for Responsible Fisheries. FAO continues to work with regional bodies, professional fishery associations and national governments in the harmonization and refinement of these codes, and in methods for appropriate risk assessment. A recent international meeting 1 convened by FAO and the International Centre for Living Aquatic Resources Management (ICLARM) on developing policies for aquatic genetic resources recognized "that in the formulation of biosafety policy and regulations for living modified organisms, the characteristics of the organisms and of potentially accessible environments are more important considerations than the processes used to produce those organisms". In terms of aquatic animal health and quarantine, FAO and OIE are working together. According to the SPS agreement, the accepted international standards governing the movement of aquatic animals are those of OIE. FAO closely collaborates with OIE in providing assistance to developing countries to improve their capacities in the effective application of these standards.

FOR THE NEAR FUTURE

In view of the importance of harmonizing regulations at the regional and sub-regional levels related to the testing and release of GMOs, FAO will continue to strengthen its normative and advisory work, in coordination and cooperation with other relevant organizations.

Recent advances make it likely that a diverse set of GMO-based technologies, and transgenic animals, will be brought into agricultural production environments (including for modified milk) in the near future. This will require more systematic consideration of the biosafety questions involved. At the international level, there are as yet no immediately relevant instruments. There is an evident need for harmonization over a wider range of biosafety issues involved within animal agriculture, beginning with systematic consultation among the relevant international organizations. In terms of technical advice and capacity-building, FAO will: advise member governments on regulatory issues (including in the context of implementing the Biosafety Protocol, if adopted); advise on harmonization at regional and international levels; offer legal advice for the establishment of any regulatory bodies required; assist in establishment of the capacity for risk assessment; and seek to mobilize extra-budgetary funds and to cooperate with other relevant organizations. Member countries have also requested FAO's assistance in establishing and implementing regulations regarding the quality and safety of foods derived from biotechnology; in this context, existing regulatory instruments elaborated by national food safety authorities should also be considered.

Biosafety

International efforts are underway to develop regulations governing the use and transboundary movement of genetically modified organisms (GMOs) and living modified organisms (LMOs)[1]. Foremost in this regard are the negotiations of the International Convention on Biological Diversity's (CBD) Ad hoc Working Group on Biosafety, which are designed to create internationally binding protocols on biosafety. FAO has worked and continues to work closely with the CBD and others on issues relevant to the sustainable use and conservation of aquatic genetic resources. National legislation and guidelines are also being developed, such as the Performance Standards for Safely Conducting Research with Genetically Modified Fish and Shellfish that were created by a working group organized by the US Department of Agriculture (*http://www.nbiap.vt.edu/perfstands/psmain.html*).

Biosafety, as currently discussed in the CBD, refers to environmental and human health safeguards concerning living modified organisms produced by modern biotechnology, especially biotechnology related to gene-transfer or transgenics. In the aquaculture sector, gene-transfer is still not commercially viable, but there are several pilot projects and research programmes in many parts of the world that are developing commercially important transgenic fish, such as salmon, catfish, carps, and tilapia. However, other biotechnologies, such as chromosome manipulation, sex-reversal, hybridization, and conventional selective breeding are becoming more widespread in aquaculture. In light of the fact that most aquatic biological diversity still resides in natural populations, all biotechnologies have the *potential*, both to improve greatly production, because of the "unimproved" state of wild aquatic species, and to impact adversely those wild resources. Thus a narrow scope of the current biosafety protocols. i.e. focusing primarily on transgenics, should be carefully considered. All of the wild relatives of domesticated aquatic species are still found in nature; biosafety protocols, or similar regulations, should eventually strive to protect these resources while allowing for the development of aquaculture and international trade.

The use of introduced species and genotypes is also a practice that can greatly increase production, but also has the potential to damage natural genetic diversity and ecosystems. In the CBD, "introduced species" are referred to as "alien species" and the term refers to species that are introduced into an area where they do not naturally occur, e.g. the movement of rainbow trout, *Oncorhynchus mykiss,* from North America to Europe. "Alien genotypes" is another phrase that is being used in international fora and would include in addition to alien species, those genetically differentiated populations that are transferred from one part of their natural distribution to another, e.g. the movement of rainbow trout from California to British Colombia. Hybridization of two local species that do not naturally hybridize would also create an alien genotype, e.g. the hybridization of *Colossoma*and *Piaractus* in Venezuela.

Biosafety in the following section refers only to environmental safeguards. Safeguards concerning human health fall under the domain of the FAO/WHO Codex Alimentarius Commission, the Agreement on the Application of Sanitary and Phytosanitary Measures, and the Agreement on Technical Barriers to Trade of the WTO.

1. The definition of LMOs and GMOs is somewhat controversial. The Convention on Biological Diversity uses an internationally accepted definition of LMO as an organism that has been modified by modern biotechnology; the International Council for the Exploration of the Sea uses the European Union's definition of GMOs to include organisms in which the genetic material has been altered anthropogenically by means of gene or cell technologies.

Chapter 10

FIVE YEAR PLAN BY GOVERNMENT OF INDIA

The plan of GOI is given here with intention of development of rural area students, teachers, researchers should know and work, accordingly.

Since the bulk of the oil is imported, crude oil should be priced at import parity price. And, since we have large refining capacities and export a large number of products, petroleum products should be priced at trade parity price, that is, products which are significantly exported should be priced at Free on Board (FOB) prices and products which are significantly imported, at CIF prices. This will be best accomplished by competitive pricing of oil products at the refinery gate and at the retail level. These prices, in turn, must be passed on to the consumer though there is some scope for cross subsidy.

The pricing policy for petroleum products will pose a major challenge in the Eleventh Plan, given the sharp increase in international oil prices which is yet to be passed on to the consumers. With the present domestic petroleum prices, the extent of under recovery in the petroleum sector is estimated at Rs 100000 crore. This situation is simply not sustainable. Consumer prices of petroleum products involve a significant burden of taxes. Either the taxes have to be reduced or consumer prices have to increase. Also, the subsidies on kerosene and LPG, which lead to substantial diversion, need to be reduced and also rationalized. The scope for disbursing subsidy through a smart card system while shifting, to market prices, needs to be explored.

Gas pricing is more complex as gas is not easily tradable. While the cost of imported Liquefied Natural Gas provides a ceiling on the domestic price of gas, the FOB price of gas less the cost of liquefaction and less the risk premium associated with exports constitutes longterm opportunity cost for producers. The price itself has to reflect its opportunity cost in its marginal use. Once we have adequate gas to meet the requirement of existing stranded assets, it will be possible to move to competitively determined gas pricing. In order that there is competition in the oil and gas sector, the distribution infrastructure of pipeline networks and associated facilities would have to be under open access regulated by regulators. While the oil and gas regulator has been put in place, there is no such regulator for coal. There is need to consider whether a coal regulator should also be established.

Renewables

The importance of renewable energy in the country arises from a number of factors—it increases energy security, it provides energy at local levels, improving energy security at these levels and it involves little or no Green House Gas emissions. Appropriate policies will be pursued to encourage renewables by linking subsidies, wherever required, to outcomes rather than to outlays. The Eleventh Plan will follow an integrated energy policy to incentivize appropriate choice of fuels and technologies.

Most of the programmes for renewable energy development would continue to be promoted but with a maxim of providing subsidies or incentives which are linked to outcomes rather than capital expenditure. This is important to preserve incentives for not just setting up capacities, but also operating them and encouraging cost reduction and technology development.

Some of areas which would be pushed strongly are wind power, solar applications, biomass gasification, bio-fuels development and other clean technologies. The distributed generation based on wood gasification in rural areas, coupled with biogas plants, could provide village energy security, particularly in remote areas. Such plants can provide clean cooking energy and also electricity before the grid power reaches them and later can also feed into the grid. The programme for biodiesel and ethanol from feed-stocks that do not compete with food production will be encouraged by a well-designed policy to take care of the interests of all the stakeholders in the value chain.

Energy Efficiency and R&D

There is large scope for improving energy efficiency in the country and this must be pushed vigorously through a variety of measures. These include energy auditing of large energy consumers, benchmarking with more efficient units, labelling and rating energy consuming equipment, forcing higher efficiency standards in major energy consuming sectors such as automobiles, and promoting energy efficient buildings. While labelling can help consumers buy energy efficient products, changes will be made in procurement policies so that government departments buy equipment based on life cycle costs.

We need R&D in a number of areas to augment our energy resources and provide cleaner energy. Considering the threat of climate change and the need Inclusive Growth 17 to find clean sources of energy, missions in the following areas should be mounted:

- Clean coal technologies of carbon capture and sequestration
- In-situ coal gasification
- Solar photo-voltaics and solar thermal electricity
- Cellulosic extraction of ethanol and butanol from agricultural waste and crop residues
- Improvement in the yield of Jatropha and other oilseeds for biodiesel.

Apart from these, a rigorous R&D programme will encourage development of new sources, more efficient utilization, and improvement in efficiency applications.

Education and Skill Development

Education and skill development will receive high priority in the Eleventh Plan, both to meet the needs of a growing economy and to promote social equality by empowering those currently excluded because of unequal access to education and skills to participate fully in the growth process. Public expenditure (Centre and States) on education is only around 3.6% of GDP. The National Common Minimum Programme (NCMP) had set a target of raising it to 6%.

Several steps were taken in the Tenth Plan to expand access to primary education, especially the expansion in the Sarva Shiksha Abhiyan (SSA) and the Mid-Day Meal Scheme. As a result, the number of out-of school children declined from 32 to 7 million, indicating that SSA brought an additional 25 million children into the education system during the Tenth Plan period. The Gross Enrolment Ratio (GER) for elementary schools (Classes I–VIII) increased from 81.6% in 2001–02 to 94.9% in 2004–05. However, the drop out rate has remained high. It was as high as 48.71% at the elementary level at the end of the Tenth Plan, a decline of only 5.94 percentage points from 2001–02.

The quality of teaching in our elementary schools is also not what it should be. Teacher absenteeism is widespread, teachers are not adequately trained and the quality of pedagogy is poor. The Eleventh Plan aims to correct these deficiencies and focuses on improving the quality of education at the elementary level, especially in rural areas. It also begins the process of universalizing secondary education. The massive expansion required in secondary education calls for an expansion in both public schools as well as private aided and unaided schools. While private schools must be allowed to expand and even encouraged, it should be noted that a much larger proportion of the expansion in enrolment would come from the public schools.

The action proposed in the Eleventh Plan for secondary education includes the following:

- Rapid upgradation of 15000 Upper Primary Schools to Secondary Schools, and expansion of intake capacity in 44000 existing Secondary Schools;
- Establishment of 6000 high quality model schools at the block level to serve as benchmarks for excellence in secondary schooling. About 3500 of these will be public-funded schools while 2500 would be through PPP;
- Provision for laboratories/libraries and also strengthening of the existing facilities available;
- Continuous teacher training;
- Provision for hostels and residential schools for girls; and
- A more liberal approach on the part of State Governments on allowing private schools to be set up to meet the large unmet demand for quality education.

The Eleventh Plan must also focus on the pressing need to expand capacity in our institutions of higher education and technical and professional education (engineering, medicine, law, etc.). The GER for higher education (percentage of the 18–23 age group enrolled in a higher education institution) currently is around 11% whereas it is 25% for many other developing countries. China has increased its GER in higher education from 10% in 1998 to 21% in 2005. We must aim at increasing the GER to 15% by the end of the Plan and reach 21% by the end of the Twelfth Plan. This is necessary not only to meet the needs of a growing economy, but also to meet the aspirations of younger people who see education as an essential requirement for advancement. Along with expansion, it is also necessary to aim at improvement in quality. While the best of our institutions of higher education compare well internationally, a large number suffer from serious quality problems. A general improvement in the quality of existing universities is18 Eleventh Five Year Plan necessary, including upgrading of facilities and improved methods of teaching. There is also need for a special effort to set up world class higher educational institutions. The expansion should aim at a much larger provision for science teaching.

The following initiatives will be taken in the Eleventh Plan to attain these objectives in higher education.

- Establishment of 30 new Central universities, one in each of the 16 States which do not have a Central University at present, and 14 other Central universities in different parts of the country. Some of these universities will be targeted ab initio to achieve world class standards, which will involve coverage of a wider range of subjects, including, especially engineering and medicine.
- Establishment of eight IITs, seven IIMs and five Indian Institutes of Science Education and Research.
- Since the establishment of world class institutions involves considerable expenditure on creating facilities, the scope for public–private participation in setting up these universities

will be carefully explored. The location of these institutions should take advantage of the colocation of other scientific and research institutions in certain places.

- The scope for setting up institutions of higher education in the private sector must also be explored. State governments would be well advised to adopt a supportive stance on this issue, including flexibility in charging higher fees.
- At present, fees vary across universities, but generally these have been kept very low, in many cases not even covering 5% of the operating cost. The Centre and State Governments must either be able to subsidize university education massively or try to mobilize a reasonable amount from those who can afford it by way of fees that cover a reasonable part of the running cost. Since most university students come from the top 10% of the population by income levels, they would be able to pay fees amounting to 20% of the operating cost of general university education. The fees for professional courses could be much higher. The fee levels should, therefore, be increased gradually in existing institutions but the new norms could be implemented in new institutions from the start.

Skill Development

In an economy growing at the rate of 9% plus, skill development poses major challenges and also opens up unprecedented doors of opportunity. The magnitude of the skill development challenge can be estimated by the fact that the NSS 61st Round results show that among persons of age 15–29 years, only about 2% are reported to have received formal vocational training and another 8% reported to have received non-formal vocational training, indicating that very few young persons actually enter the world of work with any kind of formal vocational training. This proportion of trained youth is one of the lowest in the world. Our Vocational Education and Training (VET) system needs to cover more trades. Qualitatively it suffers from disabilities such as poor infrastructure, ill-equipped classrooms/laboratories/workshops, below par faculty, absence of measurement of performance and outcomes, etc. Placements are not tracked, training institutions are not rated, and accreditation systems are archaic. End-of-the-training examinations and certification systems are either nonexistent or flawed.

In addition to the existing basic problem with the skill development system in the country, the urgency of skill development is underscored by the demographic changes taking place. It is estimated that the ageing phenomenon globally will create a skilled manpower shortage of approximately 46 million by 2020 and if we can take effective action on skill development, we could have a skilled manpower surplus of approximately 47 million. In an increasingly connected world, where national frontiers are yielding to cross-border outsourcing, it is not inconceivable that within a decade we can become a global reservoir of skilled persons.

Taking cognizance of this challenge, the consequential endeavour to launch effective measures will require a paradigm change in our VET system and in other forms of skill development. It is, therefore, proposed to launch coordinated action for skill development which will be a major initiative for inclusive growth and development and will consist of an agglomeration of programmes and appropriate structures. The coordinated action would aim at creating a pool skilled personnel in appropriate numbers with adequate skills in line with the employment requirements across the entire economy with particular emphasis on the twenty-one high growth high employment sectors (10 in manufacturing and 11 in services).

The coordinated action will aim to initiate and guide policy dialogue to energize, re-orient, and sustain Inclusive Growth 19 the development of skills through private and he public initiatives towards both self employment and wage employment at various levels. As a part of coordinated action, a National Skill Development Coordination Board will be set up which will act as an instrument of implementation and make appropriate and practical solutions and strategies to address regional imbalances in skill development infrastructure, the socio-economic (SC/ST/OBC, Minorities, BPL, etc.), rural–urban, gender divides and ensure that each sectoral plan have built-in long terms measures with a self-corrective mechanism.

Furthermore, it is proposed to create a 'Virtual Skill Development Resource Network', which can be accessed by trainees at 50000 Skill Development Centres, to provide web-based learning. It is also proposed to create a 'National Skills Inventory' and another database for 'Skills Deficiency Mapping' for facilitating tracking of careers and placement and for exchange of information between employers and employment seekers. The Eleventh Plan will also see repositioning of Employment Exchanges as outreach points for storing and providing information on employment and skill development, and to encourage them to function as career counselling centres.

Health and Nutrition

Good health is both an end in itself and also contributes to economic growth. Meeting the health needs of the population requires a comprehensive and sustained approach. Our health services should be affordable and of reasonable quality. The Eleventh Plan will try to strengthen all aspects of the health care system—preventive, promotive, curative, palliative and rehabilitative. This will be accompanied by emphasis on access to clean drinking water, sanitation, diet, hygiene and feeding practices, which will significantly affect the health status of the people. Public health spending will be raised to at least 2% of GDP during the Eleventh Plan period.

Both the Central and State Governments will have to augment resources devoted to health. This will be accompanied by building absorptive capacity for enhanced allocations and innovative health financing mechanisms, including health insurance for the poor, in which the premium for basic coverage will be borne by the Centre and the States. There is a strong case for experimenting with different systems of PPP and risk pooling. The Eleventh Plan aims to establish 60 medical colleges and 225 new nursing and other colleges in deficit States through PPP. Incentives linking payment to performance will also be introduced in the public health system.

The following targets have been set during the Eleventh Plan to ensure an efficient public health delivery system under the National Rural Health Mission (NRHM), which was launched in 2005.

- Over 5 lakh Accredited Social Health Activists (ASHAs), one for every 1000 population in 18 Special Focus States and in tribal pockets of all States by 2008.
- All sub-centres (nearly 1.75 lakh) functional with two Auxiliary Nurse Midwives by 2010.
- Primary Health Centres (PHCs) (nearly 30000) with three staff nurses to provide round the clock services by 2010.
- 6500 Community Health Centres (CHCs) strengthened or established with seven specialists and nine staff nurses by 2012.
- 1800 Taluka or Sub-Divisional Hospitals (SDHs) and 600 District Hospitals (DHs) to be strengthened to provide quality health services by 2012.
- Mobile Medical Units for each District by 2009.
- Functional Hospital Development Committees in all CHCs, SDHs, and District Hospitals by 2009.

- Untied grants and annual maintenance grants to every Sub-centre, PHC and CHC released regularly and utilized for local health action by 2008.
- All District Health Action Plans completed by 2008.

During the Eleventh Plan, special attention will be paid to various aspects of women's health, including maternal morbidity and mortality, and child sex ratio. Besides encouraging institutional deliveries under NRHM, Traditional Birth Attendants (TBAs) will be trained to upgrade them as Skilled Birth Attendants. Reducing travel time to two hours for emergency obstetric care will be a key social intervention.

For reducing infant mortality, focus will be on Home Based Newborn Care (HBNC) complemented by Integrated Management of Neonatal and Childhood Illnesses. HBNC will be provided by a trained Community Health Worker (such as ASHA) who will guide and support the mother, family and TBA in the care of the newborn, and attend to the newborn at home if she is sick. The strategy during the Plan is to introduce and make available high quality HBNC in all districts/areas with an infant mortality rate of more than 45 per 1000 live births.

Improving all district hospitals will be a key intermediate step in the health strategy, till health care through PHCs and CHCs is fully realized. During the Plan period, six AIIMS-like institutions will be set up and 13 medical institutes upgraded to that level.

For meeting the health needs of the urban poor, particularly slum dwellers, a health insurance based National Urban Health Mission will be launched. 1.123. The Eleventh Plan also focuses on developing human resources to not just meet the needs of the health care system, but also to increase employment opportunities and make India a hub for health tourism. This will involve reintroducing licentiate courses in medicine, and establishing medical, nursing, dental and paramedical colleges in the under-served areas.

Good governance, transparency and accountability in the delivery of health nutrition and related services will be ensured through involvement of local self governments, community and civil society groups. The Eleventh Plan aims to establish 60 medical colleges.

High levels of malnutrition continue to influence morbidity and mortality rates in the country. According to the NFHS-3 (2005–06), 38.4% of children under 3 years are stunted, 19.1% wasted and 45.9% underweight. These figures have not improved much since 1998–99 (NFHS-2); in fact the proportion of children has increased. The Body Mass Index (BMI) of 33% of women and 28.1% of men is below normal. Prevalence of anaemia is very high among young children (6–35 months), ever married women (15–49 years) and pregnant women and has increased since 1998–99 in all the three groups.

A variety of interventions consisting of dietary diversification, nutrient supplementation and public health measures involving better hygiene, sanitation and deworming will be undertaken to tackle the problem of malnutrition. The Integrated Child Development Services (ICDS) scheme, the government's main programme for addressing the problem of malnutrition, will be universalized on a fully decentralized on-demand basis and restructured in the Eleventh Plan as described below. The implementation of ICDS will be improved by giving responsibility of execution to the community and PRIs. Besides, there will be wider coverage of hot cooked meals and extensive promotion of infant and young child feeding practices. This is for all including agriculture women, children and labours.

Women's Agency and Child Rights

Inclusive growth in the Eleventh Plan envisages respecting the differential needs of all women and children and providing them with equal access to opportunities. This can only happen when women are recognized as agents of socio-economic growth with autonomy of decision-making and the rights of children are respected.

The Eleventh Plan proposes a five-fold agenda for gender equity. This includes economic empowerment; social empowerment; political empowerment; strengthening mechanisms for effective implementation of women-related legislations; and augmenting delivery mechanisms for mainstreaming gender. For children, it adopts a rights framework based on the principles of protection, well-being, development and participation. Recognizing that women and children are not homogenous categories, the Eleventh Plan aims to have not just general programmes, but also special targeted interventions, catering to the differential needs of different groups. Thus during the plan period, specific pilots for girl children and Muslim women, will be taken up.

To ensure a gender responsive health care system, maternal health services will be improved and emergency and compulsory obstetrics care will be made available within a travel time of two hours. Targets for halving prevalence of anaemia and malnutrition among women and children have been set and the ICDS scheme is being restructured and reoriented towards this end. A new component of conditional maternity benefits is proposed to target foetal malnutrition, ensure better care for pregnant women and encourage breastfeeding. Universalization of Anganwadis will be carried out, and Panchayats and mothers' groups will be given resources to ensure effective implementation of the scheme. Need based, community-run crèches will also be introduced to tackle child malnutrition.

Departing from previous Plans, the Eleventh Plan views violence as a public health issue and calls for training of medical personnel at all levels of the health care system to recognize and report violence against women and children. The Eleventh Plan also introduces the Integrated Child Protection scheme and the Scheme for Relief and Rehabilitation of Victims of Sexual Assault. New and empowering bills like the Sexual Harassment at Workplace Bill and the Compulsory Registration of Marriages Bill are expected to become law during the Plan period.

Powerty, Livelihood Security and Rural Development

Accelerating agriculture development with emphasis on watershed development in dryland areas and a special focus on small farmers will increase employment and help reduce poverty in rural areas. The Eleventh Plan also places emphasis on several other initiatives which will increase economic well-being and opportunities in rural areas. The NREGP provides income support for those in need of employment while also helping to create assets that will increase land productivity. The Bharat Nirman programme will provide electricity and drinking water to all habitations before the end of the Plan period. PMGSY will provide all-weather connectivity to all villages with a population of over 1000 or with a population of over 500 in hilly areas. SSA will provide eight years of education to all children. The other measures taken for improving education will help give opportunities to all children to fully develop their capabilities. The Rural Health Mission is directed to reach health services to all. The total sanitation campaign (TSC) should provide clean environment in villages.

When these programmes are effectively implemented, the benefits of growth would spread widely, especially in rural areas and backward regions. An important element of the Eleventh Plan strategy is to empower PRIs through special training as well as funds, supplemented by the Backward Regions Grants Fund (BRGF), to implement and oversee these programmes.

Urban Infrastructure and Urban Poverty Alleviation

India is relatively less urbanized (29.2% as on 1 October 2007 as per the projections made by the office of Registrar General of India) than other countries at the same level of development. However, since the scope for employment opportunities in rural areas is somewhat limited, accelerated growth as envisaged in the Eleventh Plan is expected to result in more rapid migration of rural populations to urban centres. To deal with the situation, a two-pronged action plan is necessary. First, the quality of infrastructure in existing cities will have to be upgraded toprovide improved municipal services to larger numbers of people and, second, new suburban townships will have tobe developed in the vicinity of existing cities as satellites/ counter magnets to reduce/redistribute the influx of population.

Not only is India less urbanized, the state of urban infrastructure, especially the availability of water and sewage treatment facilities, is much lower than what it should be. Urban transport infrastructure also leaves much to be desired. The Jawaharlal Nehru National Urban Renewal Mission (JNNURM), which commenced in the Tenth Plan, will continue to be the main vehicle for raising the level of infrastructure and utilities in the existing cities. The aim of the Mission is to create economically productive, efficient, equitable and responsive cities and the focus is on: (i) improving and augmenting the economic and social infrastructure of cities; (ii) ensuring basic services to the urban poor, including security of tenure at affordable prices; (iii) initiating wide-ranging urban sector reforms whose primary aim will be to eliminate legal, institutional and financial constraints that have impeded investment in urban infrastructure and services; (iv) strengthening municipal governments and their functioning in accordance with the provisions of the Constitutional (Seventy-Fourth) Amendment Act, 1992.

The development of satellite townships can be left largely to the private sector. However, in order to facilitate and induce such development the State Governments will need to undertake provision of trunk level infrastructure. The National Capital Region Planning Board is tasked with planning the process of infrastructure development for the areas around Delhi. Similar boards can be set up to undertake coordinated development of townships around other large cities, learning from the Delhi experience to make the planning process more effective.

Urban poverty alleviation and slum development will continue to be an important component of the Eleventh Plan. Tenth Plan schemes under JNNURM for providing affordable shelter and decent living and working conditions to the poor and for helping them to develop self-employment enterprises will be continued. The Swarna Jayanti Shahri Rozgar Yojana (SJSRY), a CSS to provide gainful employment to the urban unemployed (below the poverty line) will be implemented in a revamped form during the Eleventh Plan period.

Science and Technology as a Driver of Growth

Science and Technology are important drivers of economic growth and development in the contemporary world. The present juncture is critical for Indian science and major positive steps in this area will help the country to achieve sustained and rapid growth in the future.

In order to enlarge the pool of scientific manpower in the country and to attract and foster talent in scientific research a programme for 'Innovations in Science Pursuit for Inspired Research' (INSPIRE) will be launched. This will involve innovation funding in schools, summer camp with Science Icons (for high performers), assured career opportunity scheme for proven talent, and retention of talent in publicly-funded research. In addition, the Oversight Committee recommendation regarding Scholarships for Higher Education (SHE) providing 10000 scholarships per year for attracting talented science students of BSc and MSc courses will also be effectively implemented. Further, discipline-specific education programmes would be launched particularly in strategic sectors like, Space Sciences and Nuclear Sciences to capture talent at 'plus two' level with a view to developing quality human resources in the country for these sectors.

Universities are the cradle for basic research. To promote basic research in the country, a two pronged strategy will be adopted, aiming at: (i) Expansion and strengthening of the S&T base in the universities, and (ii) providing support to established centres of advanced research on the basis of competitive research funding for undertaking major and internationally competitive research programmes.

An effort would also be made to ensure that the wide pool of excellence available in the publically funded research institutions also participate actively as adjunct faculty in the universities and newly created academic and research institutions.

Effective mechanisms for promoting scientific research in the country need to be established in the Eleventh Five Year Plan. The peer review system will be upgraded to make it more stringent. All matters relating toScience and Technology including scientific audit as well as performance measurement of scientists and scientific institutions would need to be dealt with holistically.

Environment and Sustainability

Translating the vision of environmental sustainability will require that environment concerns are given a very high priority in development planning at all levels in the Eleventh Plan. The responsibility of enforcing environmental laws will be defined and shared with the States to broad base the enforcement effort. Enforcement mechanisms for dealing with industrial and vehicular pollution will be strengthened on the principle of polluter pays concept.

Environmental management (including municipal solid waste and sewage management) is a crucial component of urban planning and has been sadly neglected. Sewage treatment must be given focused attention. Recycling of the treated water for appropriate purposes must be incentivized. Efforts will be made to integrate sewage treatment with water conservation. Local bodies will be provided with adequate financial support for the purpose.

The River Conservation Programme will be strengthened to ensure that river flows are adequate to provide water of at least bathing quality standards. A Comprehensive Action Plan for Yamuna and a manageable number of important rivers will be implemented urgently with a stiff penalty system for violators.

Participating systems will be strengthened for sustainable use of forests in keeping with the global consensus on the need for community involvement for management of natural resources. The livelihood potential of forests will be improved for enhancing the stake of the community in conservation. The scope of wildlife management principles will be extended beyond sanctuaries and national parks. Conservation efforts will include compatible human interface with forests. Scientific aspects of ecology and biodiversity will be the basis of conservation planning.

The prospect of climate change presents a serious threat to our development over the longer term horizon. Available scientific evidence suggests that India will be one of the countries that will be severely affected by climate change. The Himalayan glaciers are already receding and the trend could intensify. There is likely to be an increase in the mean temperatures which would have adverse effects on foodgrain production with the present level of technology. The monsoon would be affected with a greater frequency of extreme events. We have to evolve a positive stance toward adaptation and also engage constructively with the international community to reach a consensus on mitigation based on fair principles of burden sharing. Development and promotion of low carbon and high energy-efficient technologies with reasonable costs will be a priority. The Clean Development Mechanism (CDM) will be used as an incentive. A National Action plan on Climate Change is being prepared. The government's effort will be towards creating an enabling environment for encouraging energy and carbon efficiency Inclusive Growth 23 in public and private sectors and towards internalizing climate change concerns in development planning.

Macroeconomic Framework

This chapter presents the basic macroeconomic projections underlying the Eleventh Plan consistent with achieving the target GDP growth rate of 9% on average over the Eleventh Plan period. These projections are not based on a single model of the economy, but on the results of several different models which have been used to explore feasible levels of growth and to derive a broadly consistent macroeconomic picture covering the broad sectoral composition of growth, savings and investment and projections of the balance of payments which are consistent with external viability of the strategy being proposed.

Growth Potential

The growth potential of the economy can be judged to some extent by the performance of the economy in the past, and especially the performance in recent years. Table 1 presents the growth rates achieved in each Plan period since the First Five Year Plan.

It shows that the economy grew at 6.5% per year in the Eighth Plan period (1992–1996) and then decelerated to 5.5% in the Ninth Plan period (1997–2001), but recovered sharply to achieve a growth rate of 7.7% during the Tenth Plan. The last four years of the Tenth Plan recorded an average of about 8.7% and this momentum has continued into 2007–08 which is the first year of the Eleventh Plan, and is likely to record a growth of around 8.5% or, perhaps, even a little higher.

Table 1 : Growth Performance in the Five Year Plans

(% per annum)

Plan Period	*Target*	*Realization*	*Plan Period*	*Target*	*Realization*
1. First Plan (1951–55)	2.1	3.5	8. Sixth Plan (1980–84)	5.2	5.5
2. Second Plan (1956–60)	4.5	4.2	9. Seventh Plan (1985–89)	5.0	5.6
3. Third Plan (1960–65)	5.6	2.8	10. Annual Plan (1990–91)	–	3.4
4. Annual Plans (1966–68)	–	3.9	11. Eighth Plan (1992–96)	5.6	6.5
5. Fourth Plan (1969–73)	5.7	3.2	12. Ninth Plan (1997–2001)	6.5	5.5
6. Fifth Plan (1974–78)	4.4	4.7	13. Tenth Plan (2002–2006)	7.9	7.7
7. Annual Plan (1979–80)	–	–5.2			

Note : The growth targets for the first three Plans were set with respect to National Income. In the Fourth Plan it was Net Domestic Product. The actual growth rates are in terms of GDP at factor cost. Average growth rates over a short period can be misleading because of fluctuations in agricultural output due to variable monsoon.

In this background, the 9% growth target for the Eleventh Plan period (2007–2011) approved by the NDC, which involved accelerating gradually from around 8.5% in the first year of the Plan to 10% by the end of the Plan period, appears entirely feasible, provided supportive policies are put in place. China has achieved growth rates exceeding 9% for two to three decades and while circumstances in India are not identical, the Indian economy has many strengths and now looks well positioned to achieve this goal. The economy has been accelerating gradually for the past fifteen years and the further acceleration needed to achieve the Eleventh Plan target is only modest. If this target is achieved, it would mean that per capita GDP would grow at around 7.6% per annum during the Eleventh Plan period. Per capita consumption growing by about 7% a year would double per capita consumption in about 10 years. Further, if growth is sufficiently inclusive, it will certainly provide an environment conducive to bringing about a broad-based improvement in living standards. Accelerated growth will also boost tax collections in both the Centre and the States, allowing the public sector to finance the special programmes needed to ensure inclusiveness.

The major macroeconomic parameters associated with achieving the average GDP growth of 9% during the Eleventh Plan are given in Table 2 with a comparative picture of the same parameters for the Tenth Plan. It may be noted that the higher growth target in the Eleventh Plan is premised upon a significant increase in the investment rate from an average of 32% of GDP in the Tenth Plan to an average of 36.7% of GDP in the Eleventh Plan. The Incremental Capital Output Ratio (ICOR) is expected to improve marginally. With the competition stimulated by openness and the expected improvement in infrastructure, ICOR can be expected to come down. Considering the acceleration in savings and investment rates in recent years, the projected growth may seem conservative. However, since private investment may fluctuate over a business cycle, a deliberately conservative target is set.

Sectoral Composition of Growth

Earlier, planning models placed heavy emphasis on estimating the sectoral composition of growth based on input-output balances established on the basis of detailed multi-sector models. Such detailed sectoral targets have little relevance in an open economy where necessary balances for all the tradable goods can be achieved through exports and imports. The only sectors where it is necessary to plan for a balance between domestic production and likely demands are the non-tradable sectors which are mainly in infrastructure, e.g. electric power, road transport, ports, airports, telecommunications, etc.

Table 2 : Macroeconomic Parameters

(at 2006–07 prices)

	Tenth Plan	*Eleventh Plan*
1. Investment Rate (% of GDPmp)	32.4	36.7
2. Domestic Savings Rate (% of GDPmp)	30.9	34.8
3. Current Account Deficit (% of GDPmp)	1.5	1.9
4. ICOR	4.3	4.1
5. GDP Growth Rate (% per annum)	7.5	9.0

Note : GDPmp = GDP at market prices.

Table 3 : Sectoral Growth in Recent Plans

(% per annum)

Sector	*Eighth Plan (1992–96)*	*Ninth Plan (1997–2001)*	*Tenth Plan (2002–06)*	*Eleventh Plan (2007–11)*
1. Agriculture	4.72	2.44	2.30	4.0
2. Industry	7.29	4.29	9.17	10–11
3. Services	7.28	7.87	9.30	9–11
4. Total	**6.54**	**5.52**	**7.74**	**9.0**

Although detailed sectoral targets are no longer relevant, it is possible to present a broad picture of sectoral growth prospects consistent with the target of 9% growth of the economy as a whole. Table 3 presents the sectoral composition of growth in the past three Plans together with indicative projections for the Eleventh Plan. It may be seen that both agriculture and industry show a marked deceleration in the Ninth Plan compared to the Eighth Plan. The deceleration in industry was reversed in the Tenth Plan but the growth in agriculture continued to be slow. The services sector accelerated sharply in the Tenth Plan.

The deceleration in agriculture, which began in the Ninth Plan period and continued in the Tenth Plan period, has been a major area of concern from the point of view of inclusiveness. With half our population deriving the greater part of their income from agriculture, faster growth in agriculture is necessary to provide a boost to their incomes. Rising incomes in agriculture will also boost non-agricultural income in rural areas, thus helping redress the rural–urban imbalance. The Eleventh Plan has therefore set a sectoral target of doubling agricultural growth to 4% per year. In this context it may be noted that agricultural growth increased from less than 1% in the first three years of the Tenth Plan to average more than 4% in the last two years and, from early indications, this will be maintained in the first year of the Eleventh Plan also. Although no firm judgement is yet possible, this growth upturn in agriculture is a promising development and suggests that the target of 4% growth in agriculture for the Eleventh Plan is eminently achievable.

Industrial performance in the Tenth Plan period improved to a respectable 9.2% from the very low growth rate of 4.3% in the Ninth Plan. This revival of industrial growth is a major achievement of the policy in recent years. However, industrial performance needs to be improved further if high quality employment in the nonagricultural sector is to be generated. Within industry, the manufacturing sector, accounting for 77% of industrial output has shown significant growth acceleration in the last two years. This revival of dynamism in industry has to be sustained to reverse the unacceptable decline in the share of manufacturing in GDP that has happened since 1991. This will also help generate more employment opportunities for the burgeoning workforce. The Eleventh Plan aims at double digit growth both in manufacturing and in industry. At the same time, it will be critical to improve the performance of the core sector (steel, coal, cement, oil, fertilizers and refined petroleum) to ensure that their supply response is adequate to sustain double digit manufacturing and industrial growth. Accelerated growth in industry will help to provide faster growth in organized sector employment, which is typically of a higher quality.

The services sector has grown impressively in successive Plans with a sharp acceleration in the Tenth Plan period, reflecting the rapid growth in high-end services spurred by the IT revolution. This has opened up new and attractive employment opportunities for our educated youth. We can expect further improvement in performance in this sector with an acceleration of services growth to 10% in the Eleventh Plan. Rapid growth in IT related services and in tourism will contribute to this outcome as will the expected expansion in health and education which should provide additional jobs for teachers, doctors and medical attendants.

Aggregate Investment

High levels of investment are critical for rapid growth and an important strength of the economy is that we are entering the Eleventh Plan after experiencing several years of a rising level of investment as a percentage of GDP. As shown in Table 4, the investment rate averaged 24.4% in the Eighth Plan period and was almost unchanged in the Ninth Plan period. Thereafter it rose to an average of 32.1% in the Tenth Plan with the last year of the Plan showing an acceleration to 35.9%. This rateof investment should normally be adequate to attain 9% growth, but in our circumstances, because of prolonged underinvestment in infrastructure and the consequent need to enhance investment in this area, the Eleventh Plan aims at a further increase in the rate of investment in succeeding years to reach 38% at the end of the Plan period, yielding an average of around 37% in the entire Plan period.

Table 4 : Investment and Savings Rates

(% of GDP)

Year	*Investment Rate*	*Savings Rate*
Eighth Plan (1992–93 to 1996–97)	24.4	23.1
Ninth Plan (1997–98 to 2001–02)	24.3	23.6
2002–03	25.2	26.4
2003–04	28.2	29.8
2004–05	32.2	31.8
2005–06	35.5	34.3
2006–07	35.9	34.8
Tenth Plan (2002–03 to 2006–07)	32.1	31.9
Eleventh Plan Targets	36.7	34.8

Note : Ratios for Eleventh Plan are at constant 2006–07 price; Ratios for earlier years' Plans are at current price; 2005–06: Provisional Estimates; 2006–07: Quick Estimates.

An important structural change in the investment behaviour of the economy in the recent past is the change in relative shares of public and private investment. As shown in Table 5, although the rate of total investment as a percentage of GDP was almost the same in the Eighth Plan as in the Ninth Plan, the composition shifted in favour of private investment. Public sector investment as a percentage of GDP declined from 8.5% in the Eighth Plan to 7% in the Ninth Plan. This meant that the share of public sector investment in total investment declined from 34.7% in the Eighth Plan to 29% in the Ninth Plan. This trend of a declining rate of public investment as a percentage of GDP continued in the first two years of the Tenth Plan but then began to be corrected in the rest of the Plan period. Private sector investment continued to be buoyant throughout the Tenth Plan period so the share of public investment in total investment in the Tenth Plan fell to 22%.

The rapid increase in private sector investment in the aggregate investment is in large part a reflection of the impact of the reforms initiated in the 1990s, which reduced restrictions on private investment and created a more favourable investment climate. It reflects the fact that the private sector has responded positively with an improvement in the investment climate. This is clearly a welcome development and should be encouraged in the Eleventh Plan. However, the declining trend in public investment as a percentage of GDP is a matter of concern. Itreflects inadequate public investment in the Ninth Plan period in many critical areas such as agriculture and infrastructure. The reversal of the declining trend of public investment as a percentage of GDP in the Tenth Plan is a welcome development which needs to be continued in the Eleventh Plan. It must be emphasized that a higher share of public investment can be justified only if there is a demonstrable improvement in the efficiency of public investment. However, it can be argued that the infrastructure gaps are so large that an increase in public investment in infrastructure will actually crowd in private investment. These objectives are reflected in the projections for the Eleventh Plan, as shown in Table 5.

Table 5 : Total and Public Investment in the Ninth and Tenth Plans

	Total Investment (% of GDPmp)	*Private Investment (% of GDPmp)*	*Public Investment (% of GDPmp)*	*Public Investment (% of Total Investment)*
Eighth Plan (1992–93 to 1996–97)	24.4	15.9	8.5	34.7
Ninth Plan (1997–98 to 2001–02)	24.3	17.3	7.0	29.0
2002–03	25.2	19.2	6.1	24.1
2003–04	28.2	21.8	6.3	22.5
2004–05	32.2	25.3	6.9	21.4
2005–06	35.5	27.9	7.6	21.4
2006–07	35.9	28.1	7.8	21.6
Tenth Plan (2002–03 to 2006–07)	32.1	25.1	7.1	22.0
Eleventh Plan	36.7	28.7	8.0	21.9

Note : Ratios for Eleventh Plan are at constant 2006–07 price; Ratios for earlier years' Plans are at current price; 2005–06: Provisional Estimates; 2006–07: Quick Estimates.

The private sector, which includes farming, micro, small and medium enterprises (MSMEs), and the larger private corporate sector, accounts for over 77% of the total investment in the economy and an even larger share in employment and output. Expansion in this sector will have to play a critical role in achieving the objective of faster and more inclusive growth. The MSMEs, in particular, have a vital role to play in making the growth regionally balanced, and in generating widely dispersed off-farm employment, some of it in rural areas. The Eleventh Plan, therefore, aims at an increase in investment in the private sector to 28.7% from an average of 25.1% in the Tenth Plan. This is best done by creating an environment in which entrepreneurship can flourish at all levels. New entrepreneurs should be able to enter and expand. In order to compete with incumbents, there must be strong competitive pressures which promote efficiency, government policies must be investor friendly, transaction costs should be low and infrastructure services of good quality should be readily available, especially to small and medium enterprises. Many elements of this environment are already in place. Licensing controls and discretionary approvals have already been greatly reduced, but remnants of the control regime remain that still need drastic overhaul, if not elimination, especially at the State Government level. Quantitative controls, where they exist, should give way to fiscal measures and increased reliance on competitive markets, subject to effective and transparent regulation. The burden of multiple inspections by government agencies must be removed and tax regimes rationalized. Since MSMEs are expected to create the bulk of new employment, it is particularly important to make the financial system more efficient in meeting the needs of our expanding private sector, and to ensure financial inclusiveness.

While encouraging private investment, the Eleventh Plan also aims at a substantial increase in public investment in critical areas such as agriculture, irrigation and water management, and in social and economic infrastructure. Public investment in infrastructure is projected to increase from 4.1% in 2006–07 to 6.8% in 2011–12. Since this increase must be accommodated with the overall increase in public investment from 7.8% in 2006–07 to 8.6% in 2011–12, it implies that public investment will shift from the non-infrastructure to infrastructure sector.

Aggregate Savings

The rise in investment projected in the Eleventh Plan is expected to be supported by a substantial increase in domestic savings. Table 6 presents the structure of savings since the

Eighth Plan period. Private savings consist of household savings, including direct investment by households, and corporate sector savings. Public sector savings consist of the savings of the government departments (Centre and States) and public sector corporations. Both components of private savings (households and corporate sector) have risen as a per cent of GDP in the Ninth Plan and also in the Tenth Plan. Corporate savings have been especially buoyant in the Tenth Plan, reflecting the very strong output and financial performance of the private sector in recent years.

The behaviour of public savings displays a more varied pattern. There was a marked deterioration in public savings in the Ninth Plan period, reflecting the sharp deterioration in the performance of Departmental Savings mainly on account of the impact of the Fifth Pay Commission which led to an increase in government salaries. Departmental Savings deteriorated from –0.9% of GDP in the Eighth Plan to –4.1% of GDP in the Ninth Plan. There was a marginal improvement in the savings of public sector undertakings from 3% of GDP in the Eighth Plan to 3.3% in the Ninth Plan, but the deterioration in Departmental Savings in the Ninth Plan meant that total public sector savings deteriorated from 2.1% of GDP in the Eighth Plan to –0.8% of GDP, a turnaround of 3 percentage points.

As shown in Table 6, both components of public sector savings showed an improvement in the Tenth Plan. Reduction in the dis-savings of government administration has been due to three main reasons: first, the impact of the Fifth Pay Commission's recommendations worked itself out in the system; second the implementation of the Fiscal Responsibility and Budget Management (FRBM) Act, and the fiscal and revenue deficit targets for 2008–09 established thereby helped introduce an element of discipline; and third, the buoyancy in tax revenues arising out of the high growth rate recorded in the Tenth Plan combined with improvements in tax administration contributed to improved savings.

Table 6 : Composition of Saving

(% of GDP)

Period/Years	*Private Sector*			*Public Sector*			
	Household Sector Sector	*Private Corporate Sector*	*Total Private Sector*	*Government Administration*	*Public Sector Undertaking*	*Total Public Sector*	*Gross Domestic Savings*
Eighth Plan							
(1992–93 to 1996–97)	17.0	4.0	21.0	(–) 0.9	3.0	2.1	23.1
Ninth Plan							
(1997–98 to 2001–02)	20.5	4.0	24.5	(–) 4.1	3.3	(–) 0.8	23.6
2002–03	23.2	3.9	27.0	(–) 4.7	4.0	(–) 0.6	26.4
2003–04	24.4	4.4	28.7	(–) 3.1	4.2	1.1	29.8
2004–05	23.0	6.6	29.6	(–) 2.0	4.2	2.2	31.8
2005–06	24.2	7.5	31.7	(–) 1.4	4.0	2.6	34.3
2006–07	23.8	7.8	31.6	(–) 0.8	4.0	3.2	34.8
Tenth Plan							
(2002–03 to 2006–07)	23.7	6.3	30.0	(–) 2.2	4.1	1.9	31.9
Eleventh Plan	23.0	7.3	30.3	0.5	4.0	4.5	34.8

Note : Ratios for Eleventh Plan are at constant 2006–07 price; Ratios for earlier years' Plans are at current price; 2005–06: Provisional Estimates; 2006–07: Quick Estimates.

The projections for the Eleventh Plan assume that the private sector savings effort will continue to be strong with some further improvement and that this will be combined with a very strong improvement in public sector savings. The scale of the public sector effort can30 Eleventh Five Year Plan be seen from the fact that the Eleventh Plan projects a combined public sector savings rate of 4.5% of GDP compared with an achieved level of only 1.9% in the Tenth Plan. The projected turnaround of 2.6 percentage points of GDP in public sector savings requires a massive effort on the part of both the PSEs and the departments, mainly, the latter. It is important to note that this favourable outcome assumes that the award of the Sixth Pay Commission will not present severe difficulties. Hopefully, since the Fifth Pay Commission had indexed salaries fully to inflation, the pay increases should be only moderate. Pay revision should also increase private savings. If some of the subsidies financed by the government can be moderated, the savings target of public administration is realizable. It is also assumed that PSEs will maintain their savings performance of the Tenth Plan. One could expect it to increase. However, a slightly lower savings rate of 4% of GDP is conservatively projected to account for government policy on oil product price subsidies and its adverse impact on oil PSUs' balance sheets.

The Balance of Payments

A key aspect of macroeconomic consistency is the viability of the Plan targets in terms of their implications for managing the balance of payments.

During the Tenth Plan period, merchandise exports moved on to a new trajectory with an annual average growth rate of 23.2%. Similarly, with a buoyant economy, merchandise imports increased by 27.8% during this period. There has been a continuous shift of exports towards technology intensive high-value manufactures, including machinery and instruments and also petroleum products, gems and jewellery. Despite substantial growth in the export of petroleum products by an annual average of 54.3% in the Tenth Plan, net oil imports increased by 26.5% due to the steep rise in oil prices of the Indian basket. Services exports increased substantially in the Tenth Plan contributing to increase in net invisibles. This provided a cushion for financing a large part of the trade deficit on the merchandise account. The current account deficit in 2006–07 was US$ 9.6 billion, i.e. 1.1% of GDP.

Based on the Tenth Plan trends, projections for the Eleventh Plan have been made and are summarized in Table 7. These projections reflect a picture of a dynamic Indian economy taking full advantage of the opportunities provided by integration with the world economy.

Exports in the Eleventh Plan are projected to grow at about 20% per year in US$ terms, which is lower than the 23% rate at which they have been growing over the Tenth Plan reflecting somewhat the lower projections of growth in world trade. Even with this moderation in export growth, the projection implies that the share of exports to GDP in the Eleventh Plan period will rise from 13.9% in 2006–07 to 22.5% by the terminal year of the Eleventh Plan, reflecting the continued integration of the economy with the rest of the world. This also implies an increase in India's share of world exports since world trade is unlikely to grow by more than 6%.

Imports on the other hand are likely to continue increasing at an average rate of 23% over the Eleventh Plan. This will help address critical capacity constraints in the plant and equipment sectors, the demand for which is expected to remain strong due to rising investment. Petroleum, oil and lubricants (POL) imports, which currently accounts for nearly 30% of total imports, will be boosted by the higher oil prices and rising energy consumption levels.

As Table 7 shows, the trade deficit is expected to go up significantly to 16% of GDP by 2011–12. This appears high but is expected to be largely compensated by rising invisible account surplus on the back of accelerating services sector exports, tourist earnings and remittances.

This surplus is projected to be as much as 13.5% of GDP and the current account deficit projected for the Eleventh Plan period therefore increases from 1.1% in 2006–07 to 2.4% of GDP in 2011–12. These projections are calibrated around an international crude oil price of US$ 80 per barrel. Financing a deficit of this order should not present a problem given the foreign capital inflows that are taking place and which can be expected to continue, barring an unexpectedly severe and prolonged downside shock to the world economy. In fact, the capital flows we have been experiencing in recent years have proved difficult to manage, leading to a sharp buildup in foreign exchange reserves and upward pressure on the exchange rate which has adverse consequences.

Table 7 : Balance of Payments

(at 2006–07 price)

Items	*2006–07*		*2011–12*		*(2007–12)*	
	US$ billion	*% to GDP*	*US$ billion*	*% to GDP*	*US$ billion*	*CAGR*
Current Account Balance						
Exports	127.1	13.9	316.2	22.5	1135	20.0
Imports	192.0	20.9	540.5	38.5	1864	23.0
Trade Balance	–64.9	–7.1	–224.3	–16.0	–729	–
Invisibles (net) of which:	55.3	6.0	190.0	13.5	616	28.0
Exports of Services	81.3	8.9	238.4	17.0	812	24.0
Imports of Services	48.6	5.3	217.9	15.5	653	35.0
Current Account Balance	–9.6	–1.1	–34.3	–2.4	–113.2	–1.9@
Capital Account						
Foreign Direct Investment (net)	8.4*	0.9	14.0	1.0	48.3	10.7
NRI Deposits	3.9	0.4	9.8	0.7	38.6	20.3
Portfolio (net)	7.1	0.8	11.2	0.8	43.3	9.7
Net Capital Flows	44.9	4.9	51.9	3.7	206.8	2.9

Note : *Foreign Direct Investment (net) includes US$ 20.41 billion inflows and US$ 11.97 billion outflows during 2006–07. @ indicates percentage to GDP.

Uncertainties and Risks

The external environment facing the economy in the Eleventh Plan period presents some uncertainties and risks which could have an impact on performance. These include the following:

- A downturn in the global economy, given the current macro imbalances and uncertainties in the financial markets.
- Uncertainty about the price of oil in world markets.
- Capital flows and exchange rate risks. This risk is more likely to be an upside risk.
- Impact of increase in food prices in the world due to increasing use of foodgrains for ethanol production.
- Continued governance problems that prevent necessary efficiency improvements in public expenditure and public investments.

Slowdown in World Economy

The growth of the world economy affects our economy directly through exports and also through the impact of global growth prospects on Foreign Direct Investment (FDI). It is therefore relevant to evaluate the medium term prospects in this regard. Although unsettled financial conditions prevailing at the end of 2007, combined with the persistence of global imbalances, have raised the prospect of a slowdown in some of the major economies over the next 12 months; this is likely to be temporary and not a medium-term phenomenon. In any case, Indian exports have shown considerable resilience and have also diversified considerably so that the impact of a temporary slowdown in some of the major industrial economies may not be significant. India's share in global trade is still small and an aggressive policy drive for expanding exports may enable us to weather the impact of a temporary slowing down of the global economy. In effect, the constraints on our growth are primarily internal and it is reasonable to assume that uncertainties affecting the world economy are not such as to have an impact on India's growth potential over the Eleventh Plan period.

High Oil Prices

The impact of higher oil prices on our growth can be significant and will depend on the strategy we adopt to deal with the situation. A 25% increase in oil price from US$ 80 per barrel can be expected to increase our import bill by 5% and the current account deficit to GDP ratio by a little over one percentage point of GDP. Base year deficit at around 1.1% of GDP is projected to increase to 2.4% of GDP by the terminal year. With higher oil prices, the deficit is likely to increase further, but would still be financiable. In fact, it would enable us to deal with foreign inflows which, at present, appear excessive. The additional foreign exchange cost of oil imports therefore should not present a serious problem. The more important issue is how we deal with the need to pass on high oil prices to the consumer. The present policy of insulating consumers from the full impact of the rise in the global price of oil is not sustainable as the public sector oil majors bear a very large burden in terms of under-recoveries and there are also fiscal costs which run into constraints posed by the FRBM Act. It is always possible to undertake some subsidization for those who are really needy, but the bulk of the under-recoveries at present are not on account of such targeted subsidization. As and when the global price is passed on fully to the consumers, it would reduce the demand for oil, mitigating the impact on the balance of payments, and also raise the general level of prices somewhat. Given the low rate of inflation at present, this one-time adjustment could be absorbed in the system without too much disruption. A price adjustment would also moderate growth to some extent in the short run but it would not significantly affect growth in the medium term.

Volatility of Capital Flows

The volatility of capital flows in a globalized world is a major problem which affects many countries. Traditionally, we have been concerned with volatility in terms of possible outflows triggering a balance of payments crisis. However, recent experience has exposed us to a new phenomenon of managing a surge of inflows. From a low of US$1 billion in foreign exchange reserves in the middle of 1991, when the economy was plunged into a balance of payments crisis, our foreign exchange reserves have reached over US$ 290 billion which is generally regarded as more than necessary for managing any conceivable downside shock to external payments. The same phenomenon is evident in many Asian countries which have seen a build-up in foreign exchange reserves.

This build-up in foreign exchange reserves was facilitated by foreign capital inflows to emerging market countries which have been very buoyant. Some part of the build-up was justified as a measure of insurance against a possible adverse turn in the external climate in future. However it is clear that the accumulation in reserves in recent years is in excess of requirements and reflects the desire to manage capital inflows without allowing them to impact on the exchange rate.

The problem posed by having to manage large inflows has been discussed in Chapter 13. It arises because of the well known 'trilemma' that it is not possible to achieve three objectives simultaneously, i.e. free capital mobility, an independent monetary policy and a stable exchange rate. Attention should be given to measures to restrain these capital flows and enhance the absorptiveMacroeconomic Framework 33 capacity in the economy to avoid running into the classical 'Dutch Disease' situation where non-tradables become over-priced and erode the competitiveness of the economy in the tradable sector.

Faced with a surge of inflows as in the past year, wehave to either protect the exchange rate by intervening in the forex market or allow the rupee to appreciate. Rupee appreciation has a negative effect on export, especially in the short run, and is therefore best avoided. However, the option of intervening to buy up foreign exchange and add to reserves also has adverse consequences. If the intervention is not sterilized it leads to a build-up of liquidity, which could lead to inflation. If it is sterilized, as has been done extensively in 2007–08, it involves a significant fiscal cost since the interest paid by the government on market stabilization bond is significantly higher than the interest earned on reserves. It may be noted that if the spread between the two rates is 5 percentage points then the cost of sterilization of US$100 billion of excess reserves is US$5 billion per year or Rs 20000 crore.

In practice, governments have rarely chosen pure strategies. Instead, efforts are made to respond in terms of a mixed strategy of allowing some appreciation in the currency, some unsterilized intervention and some effort to limit inflows at the margin. This is essentially the policy we have followed thus far, and it is on the whole well conceived.

The cautious approach to opening the capital account followed thus far as a conscious act of policy has given the government some leeway in limiting inflows of certain categories. In determining the categories of capital flows on which restrictions should be imposed, we have rightly followed the approach of exempting FDI flows completely from any restrictions associated with shortterm compulsions. We have also kept the door open for genuine Foreign Institutional Investment (FII) flows, while introducing some regulation of flows through nontransparent mechanisms such as participatory notes. This leaves external commercial borrowing (ECB) (including non resident Indian [NRI] deposits) which is subject to control and should bear the brunt of short-term restrictions, if these become necessary.

Rising Food Prices in the World

Foodgrain prices in the world markets have increased significantly in the last year due to a combination of poor harvest and a shift towards bio-fuels. The latter tendency is bound to accelerate, which suggest that food prices may remain high in the near future. This will undoubtedly put pressure on Indian agricultural prices. It should be emphasized that farmers are likely to gain from higher prices and to the extent to which low profitability of agriculture have been a cause of low farm income and adequate investment, the favourable shift in prices could actually boost domestic production and rural incomes. However, it must also be noted that consumers are hurt by higher prices and since most of the poor in India are net purchasers of food, it would be necessary to protect them from any undue increases in food prices. How to balance these conflicting objectives will present a major problem especially in the short term.

Conclusion

On balance, the macroeconomic parameters facing the economy at the start of the Eleventh Plan are very positive reflecting strong fundamentals and the economy is well set to benefit from them. There are risks and uncertainties but these can be managed, especially over the medium term. On the whole, the growth target of 9% proposed for the Eleventh Plan is feasible provided necessary supportive policies are put in place to address domestic constraints.

FINANCING THE PLAN
Financial Resources : Centre and States

Introduction

This write up presents projections of the likely availability of total savings in the economy and also of public sector resources in the Eleventh Plan period given the target GDP growth rate of 9%. The overall picture in terms of resources for the economy as a whole suggests that given recent trends, it will be possible to mobilize the savings needed to finance the gross domestic investment needed for 9% growth. As far as public sector resources is concerned, the estimates show resource availability for the Eleventh Plan of Rs 3644718 crore at 2006–07 price for the Centre and States taken together. At comparable prices, this amounts to an increase of 120.5% over the Tenth Plan realization.

These projections imply that public sector resources for the Plan will increase from 9.46% of GDP in the Tenth Plan to 13.54% in the Eleventh Plan. Thus outcome depends critically on achievement of buoyancy in tax revenue, effective control over consumption expenditure and subsidies, and an improvement in the resource mobilizing capacity of Public Sector Undertakings (PSUs) both at the Central and State levels.

Public Sector Resources in the Tenth Plan

This section presents an overview of the resources of the Centre and States in the Tenth Plan period.

Centre's Resources

The GBS to the Tenth Plan was projected at Rs 706000 crore at 2001–02 price. This included Rs 300265 crore of Central assistance (CA) to the States and UTs. With the Tenth Plan resources of Central Public Sector Units (CPSUs) projected at Rs 515556 crore, resources available for the Central Plan was fixed at Rs 921291 crore.

Table 8 shows the financing pattern of the Centre's Plan during the Tenth Plan. The realized GBS available for the Plan was 84.2% of the projected amount. Realized Central assistance to States and UTs at Rs 203117 crore was 67.6% of the projected level. As a percentage of GBS, this declined from 42.5% to 34.2%. This decline in the share of CA to States and UTs is partly a reflection of the disintermediation of the loan portion of CA following the Twelfth Finance Commission's (TFC) award in the middle of the Tenth Plan period, and partly also a result of the increasing resource transfers to States through CSS specially in health, education and rural development, which expanded well beyond what was originally projected. CPSUs achieved 71.5% of resources projected in the Plan.

The total resources available for the Central Plan, consisting of GBS for the Central Plan plus PSUs' resources, worked out to 82.5% of the projected level i.e. Rs 760327 crore at 2001–02 price.

The pattern of funding GBS in the Tenth Plan as actually realized reflects a significant deterioration of nondebt contribution compared with the Plan projections. The share of Balance from Current Revenues (BCR) in GBS was projected to be (–)0.9% but deteriorated sharply to (–)17.2%. The realized share of borrowings had toFinancing the Plan 35 increase to 115% as against the projected share of 97.1% to bridge the BCR gap.

The deterioration in BCR during the Tenth Plan as compared to the projections reflects a relatively poor performance in the earlier years of the Plan. The BCR improved continuously, however, during each of the successive years of the Plan and turned positive in 2006– 07. A robust economic growth and improved performance of the manufacturing sector helped to ensure that revenue receipts, particularly tax revenues, were buoyant throughout the Tenth Five Year Plan. Gross tax revenue of the Central Government recorded an average annual growth of 20.5%. Net of the share of the States, the tax revenues of the Centre grew at 21.4%. However, growth in non-tax revenue in the same period was moderate at 4.0%. The average annual growth of revenue receipts of the Central Government during the Tenth Plan was 16.7%.

Revenue receipts of the Centre increased by 1.1 percentage points of GDP from 9.4% in 2002–03 to 10.5% in 2006–07. Between 2002–03 and 2006–07, gross tax revenue as a proportion of GDP increased by about 2.7 percentage points, of which 0.6 percentage points was the increase in the share of the States. The gross tax GDP ratio increased to 10.3% in 2005–06 and is expected to increase further to 11.5% in 2006–07. Tax revenues (net of States' shares) increased by about 2.1 percentage points from 6.4% in 2002–03 to 8.5% in 2006–07. However, non-tax revenue fell by about 0.9 percentage points from 2.9% of GDP in 2002–03 to 2.0% of GDP in 2006– 07. The decline in non-tax revenue has been largely due to a steep decline in interest receipts by about one percentage point owing to debt swap, and debt consolidation and resetting of interest rates, and disintermediation in borrowings arising from the award of the TFC.

Table 8 : Projected vis-à-vis Realized Financing Pattern of the Plan Outlay of the Centre (including UTs)

(Rs crore at 2001–02 price)

Sources of Funding	*Projection*	*Tenth Plan Realization*	*% Realization*
1. BCR	–6385 (–0.9)	–102280 (–17.2)	–1601.9
2. Borrowings including net MCR	685185 (97.1)	683962 (115.0)	99.8
3. Net Flow from Abroad	27200 (3.9)	12966 (2.2)	47.7
4. Gross Budgetary Support for the Plan (1+2+3)	7060003 (100.0)	594649 (100.0)	84.2
5. Central assistance to States and UTs	300265 (42.5)	203117 (34.2)	67.6
6. GBS for Central Plan (4–5)	405735 (57.5)	391532 (65.8)	96.5
7. Resources of PSEs	515556	368796	71.5
8. Resources for Central Plan (6+7)	921291	760327	82.5

Note : Figures in parentheses are percentages of Gross Budgetary Support to Plan (S.No.4).

Source : Planning Commission.

The Non-Plan Revenue Expenditure (NPRE) declined by about 1.86 percentage points from 10.87% of GDP in 2002–03 to 9.01% of GDP in 2006–07 (refer to Table 9). This was mainly because of a sharp decline in interest payments of about 1.17% of GDP. Subsidies actually reflected in the Central budget declined by about 0.49% of GDP between 2002–03 and 2006–07. However, this ignores the fact that there was substantial underprovisioning in the budget for fertilizer subsidy. It also does not account for the under-recovery by petroleum companies because of the inadequate adjustment in prices to compensate for oil price increases.

During the Tenth Plan, expenditure on subsidies increased by 22.8% from Rs 43533 crore in 2002–03 to Rs 52935 crore in 2006–07. The main factor behind the increase was fertilizer subsidy, which more than doubled. While food subsidy remained stable and other subsidies increased by 29%, petroleum subsidies reflected in the budget declined in nominal terms. The apparent decline in petroleum subsidies, however, does not reflect the quasi-fiscal subsidization of oil prices through the issue of oil bonds. Subsidy rationalization, including targeting of the poor, has been an announced policy objective of the government for over a decade. However, success in this area has been rather limited.

Because of buoyant revenue receipts and some control over expenditure following the enactment of the fiscal responsibility legislation (see Box 3.1), the borrowings of the Central Government have been contained within the projected level in absolute terms. The percentage of interest payments to revenue receipts declined from 51% in 2002–03 to 34.5% in 2006–07 implying improved sustainability of the Centre's debt burden. The debt burden of the Centre has declined by almost 3.1 percentage points from 63.4% in 2002–03 to 60.3% of GDP as per 2006–07 BE.

The gross fiscal deficit of the Centre, as a percent of GDP, declined from 5.9% in 2002–03 to 4.1% in 2005- 06 and further to 3.5% in 2006–07. The gross fiscal deficit of the States, as a percent of GDP, also declined from 4.2% in 2002–03 to 2.8% in 2006–07 revised estimates (RE). As a result, the combined fiscal deficit of the Centre and States came down from 9.6% in 2002–03 to 6.4% in 2006–07 (RE). The average fiscal deficit for the Tenth Plan, as a percent of GDP, was 4.4% for the Centre, 3.5% for the States and 7.7% for the Centre and States combined. The year-wise figures of fiscal deficit are provided in Table 10.

The net inflow from abroad on government account, which is deployed for funding externally aided projects, was projected to contribute 3.9% of GBS in the Tenth Plan. However, actual realization was less than half of the projected level, thereby reducing its realized share in Plan resources to 2.2% of GBS. The fall in net inflow from abroad is due to the repayment of costlier debt in the initial years of the Tenth Plan.

The IEBR of the CPSUs was projected to provide Rs 515556 crore but the actual realization was only Rs 368796 crore or 71.5% of the projected amount. As a result the realized share of IEBR in the Central Plan resources was only 48.5%, substantially lower than the projected share of 56%.

The investment by CPSUs is financed through budgetary support provided by the Central Government, which is a part of total plan outlay and GBS, and IEBR raised by CPSUs on their own. IEBR comprises of Internal Resources, and Extra-Budgetary Resources (EBR). Internal Resources comprise retained profits—net of dividend paid to government, depreciation provision, and carried forward reserves and surpluses. EBR consist of receipts from the issue of bonds, debentures, ECB, suppliers' credit, deposit receipts, and term loans from financial institutions.

Table 9 : Non-Plan Revenue Expenditure (NPRE) and its Components

(Rs crore)

		2002–03 Actual	*2006–07 Provisional*
1.	Interest	117804 (4.79)	149553 (3.62)
2.	Pension	14496 (0.59)	21984 (0.53)
3.	Salary*	18253 (0.74)	23232 (0.56)
4.	Subsidies	43533 (1.77)	52935 (1.28)
5.	Other NPRE	73058 (2.97)	123940 (3.00)
6.	(Total) NPRE	267144 (10.87)	371644 (9.01)

Note : *2006–07 is BE. Figures in parentheses are percentages of GDP.
Source : Planning Commission.

Table 10 : Gross Fiscal Deficit

(as % of GDP)

Year	*Centre*	*States*	*Combined*
2002–03	5.9	4.2	9.6
2003–04	4.5	4.5	8.5
2004–05	4.0	3.5	7.5
2005–06	4.1	2.5	6.7
2006–07 (RE)	3.7*	2.8	6.4
2007–08 (BE)	3.3	2.4	5.6
Tenth Plan (2002–07)	4.4	3.5	7.7

Note : *3.5 (2006–07 Provisional). BE stands for Budget Estimates.
Source : Macroeconomic and Monetary Developments Mid-Term Review *2007–08,* 30 October 2007, RBI.

IEBR contributed 83% of the Plan outlay of CPSUs during the Tenth Plan. Of this, Internal Resources contributed 47% and Extra-Budgetary Resources, 36%. Inthe original projections, Internal Resources (IR) were to contribute 79% and EBR were to contribute only about 21%. However, over the Tenth Plan, realization of IR has been about half the projection. The shortfall in IR has led to increased borrowings by the CPSUs. Consequently, EBR have exceeded the Tenth Plan target by about 51%.

Resourcecs of States and Union

Territories (UTs)

The Tenth Plan resources of the States and UTs were projected at Rs 590948 crore at 2001–02 price. The realization at 2001–02 prices is placed at Rs 569233 crore or 96.3% of the projected level. The realized pattern of funding, however, shows a divergence from the projected levels (as shown in Table 11). BCR has deteriorated by about 67 percentage points over the projected level. However, with resources of the PSEs being 16% higher and borrowings 13% higher, the States' own resources have improved by 11 percentage points over the projected level.

FRBMA, 2003 and Fiscal Consolidation

The Fiscal Reforms and Budget Management Act (FRBMA) enacted in 2003, is an important institutional mechanism to ensure fiscal prudence and support for macro economic balance. According to the Rules framed under the Act, revenue deficit is to be eliminated by 31 March 2009, and fiscal deficit is to be reduced to no more than 3% of estimated GDP by March 2009. The process of fiscal consolidation under FRBMA has been continuous. It has yielded rich dividends in terms of creating fiscal space for increased spending on infrastructure and social sectors. The traditional annual budgeting has moved to a more meaningful medium-term fiscal planning framework. FRBMA provides the basic structure around which many fiscal measures have been implemented. Some of the important among these include: reducing peak rates of customs duties, rectifying anomalies like inverted duty structure, rationalizing excise duties, relying on voluntary tax compliance, introduction of State-level VAT, increasing productivity of expenditure through an outcome budget framework, and innovative financing mechanisms like creation of special purpose vehicles for infrastructure projects. The success in fiscal consolidation in the Tenth Plan has provided a good foundation to build the Eleventh Plan.

Table 11 : Core Tenth Plan Resources of States and UTs

(Rs crore at 2001–02 price)

Sources of Funding	*Projection*	*Realization*	*% realization*
1. Balance from Current Revenues	–15295	–25514	–166.8
	(–2.6)	(–4.5)	
2. Resources of PSEs	82684	95714	115.8
	(14.0)	(16.8)	
(i) Internal resources	–7760	9653	224.4
	(–1.3)	(1.7)	
(ii) Extra-Budgetary resources	90444	86061	95.2
3. Borrowings	264802	299022	112.9
	(44.8)	(52.5)	
4. State's Own Resources (1+2+3)	332191	369222	111.1
	(56.2)	(64.9)	
5. Central assistance	258757	200011	77.3
	(43.8)	(35.1)	
6. Aggregate Plan Resources (4+5)	590948	569233	96.3
7. GBS to Plan (6–2)	508264	473519	93.2
8. GBS as percentage of GDP	3.6	3.4	

Note : Figures in parentheses are percentages of Aggregate Plan Resources.
Source: Planning Commission.

Performance of the States can be analysed, broadly, in terms of three components, *viz.*, the BCR reflecting non-debt resources, States' borrowings reflecting debt based funding, and Central assistance, which is now all grant.

The BCR of the States was expected to be (–)Rs 15295 crore but the actual situation has been much worse, with a negative BCR of Rs 25514 crore. The States' own tax revenues have increased due to improvements made possible through the introduction of value-added tax (VAT). The share of Central taxes devolved to the States has also improved owing to buoyant resources of the Centre. However, compression of non-Plan expenditure has not been as expected.

Against a projected contribution of 44.8% of the Plan resources, borrowings in the Tenth Plan reached 52.5%. The higher share of borrowings is partly due to the deterioration in BCR. Central assistance to States and UTs in the Tenth Plan was 77.3% of the projected level, and its contribution to Plan resources however, has been only 35% as against the projection of about 44%. This has been a consequence of disintermediation of Plan loans to States and UTs in the wake of recommendations of the TFC and increased resource transfers through CSS.

Public Sector Resources In the eleventh Plan

Centre's Resources

There have been several important developments during the Tenth Plan that have implications for financing of the Eleventh Plan. FRBMA, 2003 is in force and the TFC award for 2005–10 is being implemented. Service tax has emerged as a very promising source of revenue. An announcement has also been made that efforts will be made to introduce an unified goods and service tax (GST) by 1 April 2010. The Sixth Central Pay Commission was constituted in October 2006 and is scheduled to submit its recommendations in 2008–09.

Effect of FRBMA

FRBMA, 2003 and the associated rules notified on 5 July 2004, enjoined the Central Government to reduce the fiscal deficit by no less than 0.3% of GDP every year and to bring it down to no more than 3% of GDP by 2008–09. The imposition of a ceiling on the fiscal deficit constrains the scope for enhancing GBS by resorting to more borrowings. While in the short to medium run this appears to constrain the government from making productive investments, it is necessary to take a more long-term view of the implications of FRBMA. Borrowings increase resource availability in the short run but they also increase the outstanding debt, and hence the interest burden of the Centre. This increases NPRE and hence reduces GBS in the future. High fiscal deficits also generate other undesirable consequences such as uncertainty about macro fundamentals which can affect investor confidence and make the climate unsuitable for private investment with adverse effects upon growth.

The projections assume that FRBMA will effectively constrain the fiscal deficit to the levels indicated, leading toa reduction in debt financing for funding of GBS for the Eleventh Plan. Accordingly, the Centre's net borrowings, which stood at 3.47% of GDP in 2006–07, are projected to decline to 3% in 2008–09 and remain at this level during the Eleventh Plan. Eleventh Plan projections on borrowings do not distinguish between external and domestic debt.

FRBMA not only prescribes the required reduction in fiscal deficit, but also a reduction in revenue deficit by no less than 0.5% of GDP every year and the elimination of such deficit by 2008–09. The imposition of a zero revenue deficit condition has an impact on total revenue expenditure given revenue receipts. This, in turn, has implications for the composition of Plan expenditure in terms of the revenue component of the Plan. The Centre's GBS deployment now involves a substantial component of revenue expenditure exceeding 80%, reflecting the large grants to States under the CSS which show up as revenue expenditure in the Centre's budget. The imposition of a revenue deficit ceiling, in addition to a fiscal deficit ceiling, makes it vitally important to control non-Plan expenditure to ensure that the total of Plan and NPRE does not exceed the permitted ceiling.

Effect of TFC

The Twelth Finance Commission's recommendations have essentially two critical implications for Plan financing. First, the TFC award increased the devolution to the States and thus improved their financial position and hence their capacity to finance the Plan. This increased capacity must be kept in mind when determining the necessary Plan transfers from the Centre to the States in the form of Normal Central Assistance (NCA). Second, implementation of the TFC's award has also led to a problem of straightforward comparison of GBS in the pre- and post-TFC period. For example, in 2004–05, GBS for the fourth year of the Tenth Plan of Rs 132292 crore included Rs 24806 crore of State Plan loans intermediated by the Centre. A straightforward comparison with GBS of Rs 140638 crore in 2005–06 gives an increase of only 6.3%. However, a proper comparison, after deducting what would have been dis-intermediated in 2004–05 under the TFC award, shows an increase of 30.8% in 2005–06 over the previous year.

The TFC's recommendations cover the period up to 2009–10, which includes the first three years of the Eleventh Plan. The projections of resources for the Eleventh Plan have been made assuming that the same proportions of resource transfers as under the TFC award will continue in 2010–11 and 2011–12. The assumption may not prove valid, as it is not possible to anticipate the nature of the Thirteenth Finance Commission's recommendations. The only possible basis on which projection can be made is to assume continuation of the existing proportions.

Effect of Service Tax

The introduction of service tax has provided a promising source of revenue, but there are some caveats which have to be kept in mind before making projections for the Eleventh Plan. First, the scope for expanding the service tax net to more and more services gets narrower as the net is widened. The contribution of the expanding net will, therefore, reduce over time. Second, the preponderance of small service providers below the taxable limit of turnover constrains the scope of revenue mobilization beyond a certain level. Third, service tax was introduced under the residuary entry No. 97 in List I in the Seventh Schedule of the Constitution and, as such, is not subject to sharing with the States. There are already strong demands for a sharing of the tax base between the Centre and the States, particularly in the context of phasing out of Central Sales Tax. Any sharing of the tax base with the States will diminish the Centre's available resources to finance the Plan. Fourth, with the declared goal of introducing a unified GST by 1 April 2010, there is considerable uncertainty about the rates, base, and mechanism for setting off (i.e. input tax credit), all of which have implications for future revenue collections.

Keeping in mind the implication of the FRBM and also the prospects for service tax, an assessment has been made of the likely GBS of the Centre, assuming that the growth rate of GDP will average 9% per annum in the Eleventh Plan, reaching 10% growth in the final year. The resource projection made by the Working Group on the Centre's resources yields a projection of GBS of the Centre which indicates that it will grow from 2.99% of GDP in 2006–07 to 5.38% of GDP in 2011–12. The average GBS for the Central Plan in the Eleventh Plan period stands at 3.97% of GDP as against 2.77% of GDP realized in the Tenth Plan.

The tax revenue (net of States' share) increases from 8.5% of GDP in 2006–07 to 10.26% of GDP in 2011–12, averaging 9.28% during the Eleventh Plan. Collection of direct tax is projected to exceed indirect tax collection, for the first time in history, from 2008–09 onwards. Corporate tax collection averages 63% of direct tax collection during the Eleventh Plan. It increases from 3.5% of GDP in 2006–07 to 5.31% of GDP in 2011–12 averaging 4.28% of GDP, that is, 1.66 percentage points increase over the average Tenth Plan realization. It may be mentioned that the projection regarding corporate taxes by the Working Group is based on the assumption of a corporate tax elasticity of 2.27. This is the weighted average of the estimated elasticity over the period 1995–96 to 2005–06 and the buoyancy estimated by the Task Force on the implementation of the FRBM Act, 2003.

Subsidies in the first year of the Eleventh Plan have been taken as per 2007–08 BE. As a proportion of GDP at current market prices, the projection assumes that the total subsidies will decline from 1.17% in 2007–08 BE to 0.93% in 2011–12. This, however, ignores the under recoveries on petroleum items because of the failure to fully pass on the effects of world oil prices and also underpayment on fertilizer subsidy. Inability to pass on Eleventh Five Year Plan increases in global oil prices to the consumers would have a substantial impact on resources for the Plan if this situation is not addressed urgently.

Table 12 presents the resources of the Centre and its funding in the Eleventh Plan. The GBS available for the Plan is estimated at Rs 1421711 crore at 2006–07 price. Central assistance to the States' and UTs' Plan works out to be Rs 324851 crore. IEBR of CPSUs is estimated at Rs 1059711 crore. The total resources available for the Central Plan are projected at Rs 2156571 crore.

Table 12 : Projection of the Eleventh Plan Resources of the Centre

(Rs crore at 2006–07 price)

	Sources of Funding	*Projection*
1.	Balance from Current Revenues	653989 (46.0)
2.	Borrowings including net MCR	767722 (54.0)
3.	Gross Budgetary Support to Plan (1 + 2)	1421711 (100)
4.	Central assistance to States & UTs	324851 (22.8)
5.	Total GBS for Central Plan (3–4)	1096860 (77.2)
6.	Resources of PSEs including borrowed resource	1059711 (74.5)
7.	Total Resources for Central Plan (5+6)	2156571

Note : Figures in parentheses are percentages of GBS to Plan (S. No. 3).
Source : Planning Commission.

Table 13 compares the funding pattern in the Eleventh Plan with the Tenth Plan realization as percentages of GDP. The imposition of the fiscal deficit ceiling ensures that borrowings, including net miscellaneous capital receipts, decline from 5.03% of GDP to 2.86% in the Eleventh Plan.

States' Resources

The FRBM legislations in the States prescribe that they should achieve a fiscal deficit of 3% of GDP by the end of 2008–09. Therefore, the gross fiscal deficit of all the States, which stood at 3.73% of GDP in 2006–07 has been projected to decline to 3% by 2008–09 and to remain at this level in the remaining years of the Eleventh Plan. This inevitably limits the scope for mobilizing borrowed resources and the States, therefore, have to look at improving revenue realization and controlling non-Plan expenditure.

Table 13 : Tenth Plan Realization and Eleventh Plan Projection of Resources of the Centre

(as % of GDP)

Sources of Funding	*Tenth Plan Realization*	*Eleventh Plan Projections*	*Increases (+)/ Decreases (–)*
1. Balance from Current Revenues	–0.84	2.31	3.15
2. Borrowings including net MCR	5.03	2.86	–2.17
3. Net Flow from Abroad	0.06	–	–
4. Gross Budgetary Support to Plan (1 to 3)	4.25	5.17	0.92
5. Central assistance to States & UTs	1.48*	1.20	–0.28
6. GBS for Central Plan (4–5)	2.77	3.97	1.20
7. Resources of PSEs	2.61	4.02	1.41
8. Resources for Central Plan (6+7)	5.38	7.99	2.61

Note : * If Plan loans intermediated to States and UTs are excluded, this reduces to 0.99% of GDP, which is the relevant figure for comparison.
Source : Planning Commission.

Table 14 : Eleventh Plan Resources of States and UTs

(Rs. crore at 2006–07 price)

Sources of Funding	*Projection*		
	State	*UTs*	*Total*
1. Balance from Current Revenues	341202	43848	385050
	(23.8)	(77.2)	(25.9)
2. Resources of PSEs	128824	–	128824
	(9.0)		(8.7)
(i) Internal resources	5692		5692
	(0.4)	(0.4)	
(ii) Extra-Budgetary resources	123132		123132
	(8.6)		(8.3)
3. Borrowings	636459	12964	649423
	(44.5)	(22.8)	(43.6)
4. State's Own Resources (1 to 3)	1106485	56811	1163296
	(77.3)	(100.0)	(78.2)
5. Central assistance	324851	–	324851
	(22.7)		(21.8)
6. Aggregate Plan Resources	1431336	56811	1488147
	(100.0)	(100.0)	(100.0)

Note : Figures in parentheses are percentages of aggregate Plan resources.
Source : Planning Commission.

The core aggregate Plan resources of the States and UTs have been projected to be Rs. 1488147 crore at 2006–07 price (see Table 14). This comprises of Rs 1163296 crore of own resources (including borrowings) and Rs 324851 crore of Central assistance. UTs account for 3.8% of the combined aggregate Plan resources of the States and UTs.

As a proportion of GDP, aggregate Plan resources of the States and UTs are projected at 5.55% of GDP, registering an increase of 1.47 percentage points over the Tenth Plan realization (refer to Table 15). The BCR, which was negative in the Tenth Plan, is projected to improve to a large positive figure of Rs 385050 crore. This represents an improvement of 1.59 percentage points of GDP perFinancing the Plan 41 year over the Tenth Plan. However, projections of resources of PSEs and Central assistance to the States show a decline of 0.19 percentage points and 0.23 percentage points, respectively as compared with the Tenth Plan.

Mobilization of resources of such a magnitude for the Eleventh Plan is contingent upon significant improvement in the States' own resources, mainly through improved BCR. The States will have to step up efforts to increase their own tax and non-tax revenue collections through better tax administration, plugging the scope for leakages and recovery of cost based user charges.

Table 15 : Core Tenth Plan Realization and Eleventh Plan Projection of Resources of States & UTs

(% of GDP)

Sources of Funding	*Tenth Plan Realization*	*Eleventh Plan Projections*	*Increases (+)/ Decreases (–)*
1. Balance from Current Revenues	–0.18	1.41	1.59
2. Resources of PSEs	0.69	0.49	–0.19
3. Borrowings	2.14	2.45	0.30
4. States' Own Resources (1 to 3)	2.65	4.35	1.70
5. Central assistance	1.43*	1.20	–0.23
6. Aggregate Plan Resources (4+5)	4.08	5.55	1.47

Note : *This is based on the figures reported by the States. Hence, it is slightly different from the figure of Central assistance to States and UTs in Table 16. If Plan loans intermediated to States and UTs are excluded, this reduces to 0.99% of GDP which is the relevant figure for comparison.

Source : Planning Commission.

As shown in Table 17, the Central assistance being transferred to the States in the Eleventh Plan amounts to 1.2% of GDP as against 1.43% in the Tenth Plan. However, as noted in the footnote of Table, if the Tenth Plan figure is adjusted to remove loans intermediated through the Centre, the Tenth Plan figure is lower at 0.99%. Besides, Central assistance is not the only means of Plan transfer. Large transfers take place through the CSS which have been greatly expanded in the Eleventh Plan. Accordingly, the States will receive larger transfer of Central resources to them.

Overall Financing Pattern

Table 18 compares the structure of financing projected in the Eleventh Plan for the Centre and States, combined with that actually realized in the Tenth Plan. The most notable feature is that the Eleventh Plan projections show relatively modest dependence on borrowings amounting to 38.9% of the total Plan resources compared with 73.9% in the Tenth Plan realization. This outcome is the consequence of tighter fiscal discipline imposed by the fiscal responsibility framework, both at the Centre and the States, and an optimistic revenue outlook driven by the buoyancies in revenue collections during the last three years of the Tenth Plan reflecting the robust performance of the economy. This is reflected in the projected massive improvement in BCR which was negative in the Tenth Plan and is projected as a large positive figure for both the Centre and the States.

Table 16 : Comparison of Tenth Plan Realization with Eleventh Plan Projection of Resources

(Rs crore at 2006–07 price)

Sources of Funding	*Tenth Plan Realization*			*Eleventh Plan Projection*		
	Centre	*States & UTs*	*Total*	*Centre*	*States & UTs*	*Total*
1. Balance from Current Revenues	–127166	–31722	–158888	653989	385050	1039039
	(–13.4)	(–4.5)	(–9.6)	(30.3)	(25.9)	(28.5)
2. Borrowings including net MCR	850382	371779	1222161	767722	649423	1417145
	(89.9)	(52.5)	(73.9)	(35.6)	(43.6)	(38.9)
3. Net Inflow from Abroad	16121		16121	–	–	–
	(1.7)	–	(1.0)			
4. Centre's GBS (1+2+3)	739337		739337	1421711	–	1421711
	(78.2)	–	(44.7)	(65.9)	–	(39.0)
5. Resources of PSEs	458530	119003	577533	1059711	128824	1188535
	(48.5)	(16.8)	(34.9)	(49.1)	(8.7)	(32.6)
6. State's Own Resources (1+2+5)		459060	459060	–	1163296	1163296
		(64.9)	(27.8)	–	(78.2)	(31.9)
7. Central assistance to States &UTs	–252539	248677	–3862	–324851	324851	–
	(–26.7)	(35.1)	(–0.2)	(–15.1)	(21.8)	–
8. Resources of the Public Sector Plan (1+2+3+5+7)	945328	707737	1653065	2156571	1488147	3644718

Note : Figures in parentheses are percentages of Resources of the Public Sector Plan.
Source : Planning Commission.

The financing plan outlined above will pose major challenges. As shown in Table 18, the total resources for the Central and State Plans taken together have toincrease from an average of 9.46% of GDP in the Tenth Plan to an average of 13.54% of GDP in the Eleventh Plan. It may be noted that while the total size of the Plan is projected at 13.54% of GDP, the total public investment in the economy is projected to be lower at 8.6% (see Chapter 1). This difference reflects the fact that a great deal of Plan expenditure finances current expenditure on various items of public service delivery which are not counted as investment. The increase of 4.08% of GDP in total resources for the Plan has to be achieved while keeping borrowing within the FRBM requirement of reducing the fiscal deficit of the Centre and States to 3% on each account. Taking account of the resources mobilized by the public sector, the combined BCR of the Centre and the States has to increase by more than the projected increase in Plan resources.

The Centre's BCR, realized in the Tenth Plan, averaged (–)0.84% of GDP. It is projected to average 2.31% of GDP over the Eleventh Plan,that is, an improvement of 3.15 percentage points of GDP. Similarly, the BCR of the States is also expected to improve substantially from (–) 0.18% of GDP as realized in the Tenth Plan to 1.41% of GDP in the Eleventh Plan. As can be seen from Table 19, the projected improvement required in the combined BCR of the Centre and States taken together is therefore 4.74 percentage points of GDP. It must be emphasized that achievement of these BCR targets is a key element in the financing of the Plan.

Table 17 : Plan Resources as Per Cent of GDP

Aggregate Plan Resources	*Tenth Plan*	*Eleventh Plan*	*Increase over Tenth Plan*
Centre	5.38	7.99	2.61
States	4.08	5.55	1.47
Centre and States	9.46	13.54	4.08
Balance from Current Revenues			
Centre	–0.84	2.31	3.15
States	–0.18	1.41	1.59
Centre and States	–1.02	3.72	4.74

Source: Planning Commission.

Underlying the projected BCR is a projection that tax revenues (net to Centre) would grow from 8.5% of GDP in 2006–07 to 10.26% of GDP in 2011–12. NPRE is expected to decline from 9.01% of GDP in 2006–07 to 6.91% in 2011–12. Thus the projected improvement of 3.15% of GDP in BCR of the Centre is expected to come slightly more from contraction in NPRE than growth in taxes.

Table 18 : Public Sector Allocation for Eleventh Plan

(Rs crore at 2006–07 prices)

Centre Sources of Funding	*Allocation*
1. Budgetary Support	1096860
2. IEBR	1059711
3. Total Centre(1+2)	2156571
States and UTs	
Sources of Funding Allocation	
4. State Own Resources	1163296
5. Central assistance to State/UT Plan	324851
6. Total States & UTs (4+5)	1488147
Total Public Sector Outlay	
7. Grand Total (3+6)	3644718

Source : Planning Commission.

The assumption of strong growth in tax revenues of the Centre and the States built into the projections is not unreasonable. Tax revenues recorded in the recent past has shown high buoyancy facilitated by tax reforms and also significant improvements in the efficiency of tax administration. These efforts will continue in the Eleventh Plan period and should contribute to achieving the targeted tax-to-GDP ratios. However, the BCR projections are equally dependent upon the ability to moderate the growth in NPRE and this aspect of the projections deserves focused attention.

There are several factors which could make it difficult to contain expenditures to the projected level. There are inevitable uncertainties associated with the impact of the Sixth Pay Commission's recommendations. Equally, if not, more important is the upward pressure of subsidies, particularly on fertilizer and petroleum, and also on food. Petroleum subsidies have not so far been factored into the plan resources calculations since they have been financed by the issue of bonds not included in fiscal deficit accounting, and some portion being borne by the oil companies themselves. However, this practice cannot be sustained indefinitely. Reform of the subsidy regime has tobe urgently taken up to keep the total subsidy, including the present off-budget subsidy, to within the ceiling of 0.93% of GDP in 2011–12 that the resources projections have built in.

Allocation of Public Sector Resources : Centre and States

The projection of the overall resources for the Eleventh Plan has been presented in the preceding section. This section focuses on the allocation of Public Sector Resources for the Eleventh Plan between the Centre and the States/UTs and the proposed sectoral distribution of the resources in keeping with the objective of achieving faster and more inclusive growth.

The projected assessment of resources of the public sector for the Eleventh Plan at Rs 3644718 crore at 2006–07 prices comprises of the Centre's share at Rs 2156571 crore and the States/UTs share at Rs 1488147 crore. The resources for the Central Plan includes the GBS component of Rs 1096860 crore and the IEBR component of Rs 1059711 crore at 2006–07 prices. Resource allocation in the Central sector according to different Heads of Development is indicated in Annexure and the ministry/department-wise details of budgetary support and IEBR are indicated in Annexure.

The Eleventh Plan resources of the States and UTs are projected at Rs 1488147 crore at 2006–07 prices, out of which States' own resources are Rs 1163296 crore and the Central assistance to States and UTs is Rs 324851 crore at 2006–07 prices. Head of Development-wise allocation for the States/UTs is indicated in Annexeure with States/UTs-wise core plan details furnished in Annexure. These allocations would be finalized in consultation with the States. Table 18 indicates the resources and allocation of public sector resources for the Eleventh Plan.

A comparison of the distribution of the total GBS in the Tenth and the Eleventh Plan has been shown in Table 19. In comparison to the Tenth Plan realization, there is an increase of 125% in the projected GBS for the Centre for the Eleventh Plan. Central assistance to State/ UT Plans for State sector programmes is about 85.6% higher than the grant component realized during the Tenth Plan. The share of the projected grant component of the Central assistance to States/UTs plan in the total GBS for Eleventh Plan has decreased slightly from what has been realized in the Tenth Plan (from 26.4% to 22.8%) primarily because a much higher allocation has been made to the CSS. The allocation to CSS has increased from 1.40% of GDP for the Tenth Plan to 2.35% of GDP in the Eleventh Plan.

Table 19 : GBS Allocation in Tenth and Eleventh Plans

(Rs crore at 2006–07 prices)

	Tenth Plan Realization		*Eleventh Plan Projections*		
	Amount	*% share in Total GBS*	*Amount*	*% share in Total GBS*	*% increase over Tenth Plan*
Central Sector	486798	73.6	1096860	77.2	125.3
Support to State Plan*	175021	26.4	324851	22.8	85.6
Total	**661819**	**100**	**1421711**	**100**	**114.8**

Note : *Grant component only.

Source : Planning Commission.

The projection of GBS allocation to different sectors, Ministries/Departments and the support to the State/UT Plan has been made in tune with the approach adopted for the Eleventh Plan for 'faster, more broad-based and inclusive growth'. The Eleventh Plan aims at putting the economy on a sustainable growth trajectory with a growth rate of 10% by the end of the Plan period by targeting robust growth in agriculture at 4% per year and by creating productive employment at a faster pace than before. The Eleventh Plan focuses on poverty reduction, ensuring access to basic physical infrastructure, health and education facilities to all while giving importance to bridging the regional/social/gender disparities and attending to the marginalized and the weaker social groups. Accordingly, a major structural shift across sectors has been proposed by allocating more resources to the priority areas identified for ensuring inclusiveness. A broad picture of the structural change in terms of sectoral allocation of resources has been shown in the Table 20.

About 74% of the total central allocation for the Eleventh Plan has been put aside for the priority sectors listed in Table 19, whereas, their share was only 55% in the Tenth Plan. It may be noted that the share of infrastructure and energy in the GBS allocation has fallen despite their being in the priority list. This is not a reflection of lack of priority but a reflection of a conscious policy to shift from public sector funding in these sectors to a strategy of increased IEBR and Public Private Partnership.

The objective of the Eleventh Plan is to increase investment in infrastructure (including irrigation) from 5% of GDP in 2006–07 to 9% by 2011–12.

To mobilize resources for allocation to the priority sectors and to realize a sustainable growth of 10% by the end of the Plan period, there is a need to substantially enhance the resources for infrastructure development, skill development and industrial growth, as well. This will contribute to raising the income levels through employment generation and will also provide impetus to the other programmes. In addition to the identified priority sectors, investment in the private sector including farming, MSMEs has a vital role in achieving regionally balanced and more inclusive growth and also the potential to generate off-farm employment. Steps are proposed to be taken to provide a comfortable and competitive environment for the MSMEs to growand some structural and regulatory changes have also been proposed to attract private and foreign investors. Keeping in mind the socio-economic diversity in the country, decentralized planning complemented with greater transparency and accountability is desirable for the overall development of the country. Also, our development strategy should be well complemented by policies for environmental protection and sustainability.

The Eleventh Plan proposes to provide Rs 324851 crore at 2006–07 prices as CA to State/UT Plans. Table 20 indicates the details of sector-wise CA component of the resources of the States/UTs. Out of the total CA to States/UTs of Rs 324851 crore at 2006–07 prices, 37% (i.e., Rs 122852 crore) has been earmarked for the Gadgil Formula drivenNCA, Special Plan Assistance (SPA) for Special Category States and Special Central Assistant (SCA) for the Border Areas Development Programme (BADP)/Hill Area Deelopment Programme (HADP)/North East Coucil (NEC), etc. The remaining 63% of Central assistant to the States is assigned to Additional Central Assistant (ACA(for various flagship programes in accordance with the priority set for the Eleventh Plan, such as the AIBP, National Social Assistance Programm (NSAP), APDRP, BRGF, and JNNURM. A new programme, the RKVY, has been introduced to incentivize the States to accord a much higher priority to the agriculture sector in their investment planning by supplementing area specific agricultural strategies and programmes.

Table 20 : Sectoral Allocation—Tenth Plan and Eleventh Plan

(Rs crore at 2006–07 prices)

S.No.	*Sectors*	*Tenth Plan*		*Eleventh Plan*	
		BE#	*% to Total*	*Projected Allocation*	*% to Total*
1.	Education	62461	7.68	274228	19.29
2.	Rural Development Land resources and Panchayati Raj	87041	10.70	190330	13.39
3.	Health Family Welfare and Ayush	45771	5.62	123900	8.71
4.	Agriculture and Irrigation	50639	6.22	121556	8.55
5.	Social Justice	36381	4.47	90273	6.35
6.	Physical Infrastructure	89021	10.94	128160	9.01
7.	Scientific Departments	29823	3.66	66580	4.68
8.	Energy	47266*	5.81	57409	4.04
	Total Priority Sector	448403	55.10	1052436	74.03
9.	Others	365375	44.90	369275	25.97
	Total	813778	100.00	1421711	100.00

Note : #Tenth Plan BE represents the actual allocation during the five years and not the original Tenth Plan projections; *Includes APDRP grant

Table 21 : Projected Central Assistance to States/UTs for Eleventh Plan

(Rs crore at 2006–07 price)

Sectors	*Programme*	*Allocation*
State Development Plan	Normal Central assistance	94720
Special Category States	Special Plan Assistance	13238
	Central Pool for North East and Sikkim	3095
Agriculture	Rashtriya Krishi Vikas Yojana	22104
	Shifting Cultivation	212
SCA	Border Area Development Programme/Hill Area Development Programme/North Eastern Council	14894
Irrigation	Accelerated Irrigation Benefit Programme	41568
Power	Accelerated Power Development & Reform Programme	12820
Urban/Local Area Development	Jawaharlal Nehru Urban Renewal Mission	41490
	Brihan Mumbai Storm Water Drain Project, Mumbai	113
	MPs' Local Area Development Programme	6985
Balanced Regional	Development Backward Region Grant Fund	25711
Elderly and Weaker Section	National Social Assistance Programme	15691
Adolescent Girls	National Programme for Adolescent Girls	886
Roads and Bridges		7972
Externally Aided Projects	Various EAPs	14273
Disaster Management	Tsunami Rehabilitation Programme	2985
Sports	Commonwealth Games, 2010	2133
E-governance	National e-Governance Action Plan	2942
Total		**324851**

Thrust Areas of the Eleventh Plan

Sectors	*Thrust Areas*
Education	Quality upgradation in Primary Education, Expansion of Secondary Education, major emphasis on upgradation of Higher Education including Technical Education, ICT throughout education system.
Health, Nutrition, Drinking	Major upgradation of rural health infrastructure, Medical education,
Water, and Sanitation	Nutritional support to children and pregnant and lactating women through ICDS, health insurance based urban health facilities, Health care for elderly, achieving sustainability, improvement in service levels and moving towards universal access to safe and clean drinking water.
Agriculture and Irrigation	Ensuring Food Security, Supporting State-specific agriculture strategy and programmes, Better seed production, Focused agricultural research, Extension, Development of modern markets.
Rural Development, Land	Universalization and improvement in programme delivery of NREGP,
Resources and Panchayati Raj	Integrated Watershed management including management of underground water level.
Social Justice and Empowerment	Special attention to the needs of SCs, STs and minorities and other excluded groups through pre and post-metric scholarship, Hostels for boys/girls, Income and employment generation opportunities, Multi-sectoral development programmes for minorities in minority concentration districts.
Physical Infrastructure	Emphasis on the public–private partnership in investment, Initiate policies to ensure timebound creation of world-class infrastructure, especially in remote and inaccessible rural areas and NE, Hinterland connectivity through improved rail and road infrastructure.
Energy	Electrification of all villages and extending free household connections to all 2.3 crore BPL households through RGGVY, Nuclear power development.
Scientific Departments	Development of satellite launch capabilities to GSLV-Mk-III, Development of new energy systems, viz. advance heavy water reactor and nanotechnology.

The overall plan outlay of all the States and UTs is projected to increase from Rs 673132 crore in the Tenth Plan to Rs 1488147 crore in the Eleventh Plan (both at the same 2006–07 price levels), an increase of 21.1% on a comparable basis. The aggregate picture indicates that the States would be allocating more than proportionate increase to social services (40.1%), transport (38.7%) and agriculture and allied activities (37.8%). The States would also be actively pursuing PPP models for infrastructure development wherever possible. The aggregate picture, it must be noted, conceals wide inter-State variations in terms of Plan sizes relative to GSDP, per capita plan expenditure and percentage sectoral outlays.

Issues in Plan Financing

Several conceptual issues arise from the present structure of the Plan financing. Important among these are classification of expenditure—which has a bearing on the overall expenditure management, the Central Plan transfers mechanism, the treatment of investment of PSUs financed by IEBR under the Plan, and the role of SPVs/PPPs and other innovative methods of raising additional resources for investment.

Classification of Expenditure

There has been much debate about the utility or otherwise of the classification of expenditure into Plan and Non-Plan. It has been argued that this distinction is illogical and, what is more, even dysfunctional. The argument against continuing with this distinction is that the focus on new schemes/new projects/new extensions to currently running schemes, etc., which alone qualify for being included in the Plan, results in neglect of maintenance of the existing capacity and service levels. Thus, while strenuous efforts are made by both the Centre and the states to achieve a large increase in Plan size, its impact is often negated by a running down of service delivery capacity already created. The problem arises from a framework that creates a distinction between Plan and non-Plan expenditure within each sector, motivated solely by the need to identify and highlight provision of incremental service.

Other perceptions that have developed around this distinction, namely, that Non-Plan expenditure is inherently wasteful and has to be minimized, that Non-Plan expenditure is different in kind from Plan expenditure, etc., are patently incorrect assessments that have nevertheless taken deep root in the process of government expenditure planning. This dichotomy also results in a fragmented view of resource allocations to various sectors. The problem has become particularly acute as government's emphasis has shifted to the social sectors where salary costs are high. Routine bans on recruitment for Non-Plan posts, ostensibly imposed to conserve expenditure, cause serious problem for service delivery in health, education, extension systems, etc. The case against the use of these categories, both on grounds of illogicality and dysfunctionality, is therefore indisputable.

At the same time, it is necessary to understand that this classification of expenditures has been used essentially as a convenient shortcut for the performance of functions that are inherent in public expenditure management. It is perhaps in the manner in which the Plan and Non-Plan distinction has been denuded of its substance over the years, rather than in any inherent conceptual inadequacy, that the causes of the present state of affairs need to be found.

The basic functions of government expenditure management, as applied to an annual budgeting process, are the following:

(i) Assessing the amount of committed expenditure, that is, expenditure whose level cannot be altered during the Budget period by any decision that might be taken (though the areas where these expenditures can be applied may still be open to alteration, and, in any case, the levels of expenditures themselves would be capable of being altered over a longer time horizon); Assistance (SCA) for the Border Areas Development Programme (BADP)/Hill Area Development Programme (HADP)/North East Council (NEC), etc. The remaining 63% of Central assistance to the States is assigned to Additional Central Assistance (ACA) for various flagship programmes in accordance with the priority set for the Eleventh Plan, such as the AIBP, National Social Assistance Programme (NSAP), APDRP, BRGF, and JNNURM. A new programme, the RKVY, has been introduced to incentivize the States to accord a much higher priority to the agriculture sector in their investment planning by supplementing area specific agricultural strategies and programmes.

(ii) The amount of headroom available within the resource envelope after the committed expenditures are provided for;

(iii) The sectors in which this 'discretionary' expenditure is to be applied; and

(iv) The process of monitoring and evaluating the efficiency and effectiveness with which the expenditures are made.

These functions would have to be performed irrespective of whether we continue to have the two categories of Plan and Non-Plan expenditures or not. These two categories, which were initially designed to enable the fulfilment of the above functions, have now acquired connotations that lead to dysfunctionality. The solution, therefore, would be to do away with these categories and, instead, devise other mechanisms that will perform the requisite functions in a rational and effective manner. We need to draw up protocols that will specify who is to perform what part of the above functions and how the entire activity will be coordinated. In this effort, there would be a need to redefine organizational mandates. This activity will be taken up during the Eleventh Plan period.

The Revenue–Capital categorization of expenditure has acquired renewed significance in the post-FRBM scenario. The FRBM law has stipulated the elimination of revenue deficit in a context where more than threefourths of Plan expenditure is revenue expenditure. Strict adherence to the FRBM stipulation would have a bearing on the ability of the Centre to formulate Plan schemes directed at national priorities and also at equalizing the availability and provision of services across States. In effect, we may be in a position where the total resources that can be mobilized within the fiscal deficit ceiling cannot be deployed into schemes which have a high revenue component.

The problem is further complicated by the fact that the categorization of expenditure into revenue and capital that has evolved over the years, does not appear to be strictly in conformity with the constitutional position on this issue. The Constitution distinguishes only between 'expenditure on revenue account' and 'other expenditure'. Over the years, 'expenditure on revenue account' has been construed to mean revenue expenditure. This could be an incorrect interpretation as expenditure met out of revenues of the Government of India (GOI) has been interpreted earlier (as is clear from reports of Finance Commissions) as expenditure that is not self financing, that is, both revenue expenditure and those capital expenditures that are not self liquidating (in the sense of not providing financial returns to the government that could be used for servicing debt). In this view, the present category of Defence Capital Expenditure, for instance, would have to be treated as 'expenditure on revenue account' only. This is in keeping with the economic classification in the national accounts where Defence capital expenditure is treated as consumption and not as investment.

It is also relevant to note that capital expenditure as currently defined is not always and invariably tantamount to investment. For example, recapitalization of PSEs, though classified as capital expenditure, is not, in most cases, investment. In fact a loss-making PSU may receive injections of equity to fund losses, in which case what is conceptually a subsidy will be classified as a capital expenditure.

Some argue that the Revenue–Capital classification is also dysfunctional from an economic management perspective as it militates against the principle of sound and efficient management of the entire expenditure in an integrated manner. Over the years, essential maintenance expenditure has become a casualty of the revenue– capital distinction.

There are other inconsistencies also with our current system of classification. For example, fertilizer and food subsidies are not counted as Plan expenditure but subsidy to socially desirable insurance schemes, and several other subsidies are also included. Exclusion of large subsidy items from Plan expenditure has the effect of avoiding any resource prioritization in the matter of determining the appropriate balance between subsidy levels and other expenditure to promote common developmental objectives.

Against the above backdrop, it would be appropriate to set up a HLC that could look into the entire gamut of issues arising from the present classification of expenditure and suggest measures for efficient management of public expenditure.

Central Plan Transfers

The last two decades have seen a massive increase in both the number of CSS as well as funds available under individual schemes. These schemes provide CA for Plan expenditure in particular sectors which are normally in the domain of the States. However, unlike normal CA which is not tied to specific programmes, funds available under CSS are tied to schemes in particular sectors and are subject to centrally prescribed guidelines. The share of normal Central Plan assistance to the States has declined as a proportion to total Plan transfers. The role of CSS has often been discussed and a view frequently expressed is that CSS should be abolished and the funds flowing through these schemes should simply be transferred to the States directly, presumably as NCA. There are two arguments against this approach. First, there is merit in using Central resources to tackle the specific obstacles that would prevent the achievement of inclusive growth and this is best done by effectively earmarking resources to support State expenditure in particular areas such as rural development, health, education, agriculture and irrigation. Unless this is done, it will be difficult to give a special impetus to these critical areas. Second, the mechanism of CSS enables the Centre to address problems as they exist in different States without being constrained by the Gadgil Formula, which would otherwise guide the transfer of untied funds.

A large proportion of the fund transfers to the States under CSS are also being routed to State and district level bodies directly by the Central Government, bypassing the State Governments. This practice is motivated by a desire toavoid delays in administrative approvals and to prevent diversion of CSS funds by the States for supporting their ways and means position. Concerns about what this does to accountability mechanisms have been expressed. In view of the growing volume of such transfers, suitable mechanisms would have to be devised to ensure there is no dilution of accountability.

Monitoring of Plan Expenditure

The existing system of accounting for Plan schemes, both for the Centre and the States, does not adequately support informed planning, budgeting, effective monitoring and decision making regarding these schemes. The current accounting system does not capture transactionoriented information. It also does not distinguish between transfers to States, final expenditure and advance payments against which accounts have to be rendered. The extant accounting framework is also not structured togenerate State-wise and scheme-wise releases of funds bythe Central Government to States and other recipients, and also the actual utilization for the intended purpose. Hence, there is a great need to design and implement a Plan scheme that would thoroughly reform this process. This Plan scheme would, inter-alia, modify the existing Code of Accounts so as to fully capture the entire range of Plan schemes operated by the GOI and the States. The Code of Accounts needs to be restructured to provide information relating to each implementing agency. While there have been stand-alone efforts by various Ministries and programmes, there is no consolidated financial information system based on the accounting structure. A comprehensive Decision Support System (DSS) and Management Information System (MIS) for effective monitoring of Plan schemes would need to be set up. This will work through a core accounting solution on a central data centre. The scheme will be implemented by the Controller General of Accounts (CGA). The details of the scheme would be finalized during the Eleventh Plan.

Scope of the Public Sector Plan

The administrative machinery through which the public sector plan is implemented has been continuously changing over the years. The setting up of PRIs, the establishment of special purpose societies and agencies and companies that have been mandated to perform special functions, are all elements of the plan implementation machinery that did not exist earlier. At the same time, some organizations that were earlier part of the public sector may have moved out of the public sector due to privatization as in the case with a few enterprises—both in the Centre and in the States. The focus on public–private partnership also creates new categories where public sector resources are made available to projects owned and managed by the private sector. In this context, the definition of the organizational boundaries of the public sector plan assumes importance, both for reasons of analytical clarity as well as inter-temporal comparability. This was a specific issue on which the Working Groups that were set up for estimation of Central and State resources were specifically required to comment.

While the Centre has consistently followed the practice of including the investment plans of a large number of PSEs in the public sector plan, the States have not adopted any uniform policy in this regard. Even in the case of the Centre, the PSEs that are included in the Plan from year to year show minor variations. Even among those PSEs that are included in the Plan, the resources estimates do not capture the entire quantum of resources available, and a portion of the available resources is treated as being outside the Plan.

As far as the States are concerned, there is no uniform practice across all States. While requesting the States to estimate resources for the Eleventh Five Year Plan, they were advised that the public sector plan of the States would have to be defined as the sum total of investments made in a State by the State Government, taken as a composite economic entity. It was explained that this economic entity would be comprised of more than one legal entity, and would include, apart from the State Government itself, other legal entities such as the State PSEs, and all urban and rural local bodies. As far as the PSEs themselves are concerned, the guidelines provided that PSEs which are separate from the government only as legal entities (through share-holding or other form of capital being wholly or substantially held by the State Government) and which make materially significant investments (i.e., other than small investments that are only incidental to or supportive of routine activities) in a direct manner (i.e., who are not financial intermediaries whose activities consist of providing either loan or equity capital to other entities who in turn make investments) should be included in the scope of the public sector plan. Unfortunately, the States have not uniformly rationalized their definitions of the public sector plan on these principles. Consequently, the resources estimates have proceeded on the same basis as in earlier plans and, to that extent, comparability across States is not what it should be.

One of the new innovations that is likely to gain importance in the Eleventh Plan period is the concept of investment in infrastructure through PPPs, which is promoted through means such as Viability Gap Funding (VGF). In such a case, the extent to which the investment was eventually made should be reflected as part of the public sector is an issue on which there is lack of clarity. If the expenditure from the public sector is viewed as a grant, then it is a revenue expenditure on the part of the government which finances investment in the private sector, the grant being a source of financing for the private investor.

Revenue Deficit Constraint

Under the FRBM Act, all the States as well as the Centre are committed to reducing the revenue deficit to zero by 2008–09. This would mean that there would be no net dis-saving on account of government administration. Infact, given the relationship between the revenue deficit and savings of the government, a zero revenue deficit would probably, in fact, result in a small positive saving from government administration. One of the problems with achieving a zero revenue deficit is that the Plan outlays of both the Centre and the States have, in recent years, had an increasing share of revenue expenditure. Inthe Central budget, in particular, the grant component of transfers to the States is revenue expenditure. As a result, the revenue expenditure component of the Centre's GBS for the Plan has risen to more than 80% in recent years. While the actuals of 2005–06 show that 79.5% of GBS was revenue expenditure, this increased to 83.7% in the revised estimates for 2006–07, and 85% in the budget estimates for 2007–08.

If the revenue deficit is to be reduced to zero from 2008–09 onwards, this would mean that the revenue expenditure component of GBS cannot exceed the BCR. Inrecent years, the Central Government has been achieving a positive BCR. However, going by the present projections, the BCR would reach only around 70% of GBS by the end of the Plan period if the gross fiscal deficit were to be taken at the ceiling of 3% of GDP. In other words, leaving the revenue expenditure share in GBS at the same levels at present would effectively mean that the total GBS would have to be reduced to a level below what would be permissible under a 3% gross fiscal deficit constraint.

The resources estimates now prepared make it very clear that the constraint of maintaining the revenue deficit at zero from 2008–09 onwards will have the inevitable consequence of very substantially limiting the overall GBS for the Plan. The overall pattern of revenue and capitalFinancing the Plan 51 expenditure in the GBS is the result of the core strategy of the Plan which seeks to substantially improve the supply of social services in the relatively backward States through increased Central funding, while at the same time providing an incentive framework for physical infrastructure to be created under PPP models. Given this strategy, and the consequent high revenue expenditure to GBS percentage, one option would be to insist on the binding nature of only the GFD to GDP constraint while accepting the inevitability of high revenue expenditure in view of the Plan strategy. The other option would be to try and identify components of what is booked as revenue expenditure in the accounts of the Central Government that lead to capital asset creation in the economy and to suitably reflect this asset creation in the accounting for the purposes of the FRBM Act. On a balance of advantage, it would appear that the former course would be preferable.

EMPLOYMENT PERSPECTIVE AND LABOUR POLICY

The generation of productive and gainful employment, with decent working conditions, on a sufficient scale to absorb our growing labour force must form a critical element in the strategy for achieving inclusive growth. Past record in this respect is definitely inadequate and the problem is heightened by the fact that the relatively higher rate of growth achieved during the last decade or so is not seen to generate a sufficient volume of good quality employment.

The Eleventh Plan provides an opportunity to focus on and diagnose the reasons for past failings observed in the employment situation and to reverse at least some of the adverse employment outcomes associated with the pattern of economic growth in the recent past.

Weaknesses in Past Performance

The basic weakness in our employment performance is the failure of the Indian economy to create a sufficient volume of additional high quality employment to absorb the new entrants into the labour force while also facilitating the absorption of surplus labour that currently exists in the agricultural sector, into higher wage, non-agricultural employment. A successful transition to inclusive growth requires migration of such surplus workers to other areas for productive and gainful employment in the organized or unorganized sector. Women agricultural workers in families where the male head has migrated, also require special attention given the need for credit and other inputs if they are self-employed in agriculture or for wage employment if they do not have land.

The approach to the Eleventh Plan had identified the following specific weaknesses on the employment front which illustrate the general failing just discussed.

- The rate of unemployment has increased from 6.1% in 1993–94 to 7.3% in 1999–2000, and further to 8.3% in 2004–05.1,2
- Unemployment among agricultural labour households has risen from 9.5% in 1993–94 to 15.3% in 2004–05.
- Under-employment appears to be on the rise, as evident from a widening of the gap between the usual status (us) and the current daily status measures of creation of incremental employment opportunities between the periods 1994 to 2000 and 2000 to 2005.
- While non-agricultural employment expanded at a robust annual rate of 4.7% during the period 1999–2000 to 2004–05, this growth was largely in the unorganized sector.
- Despite fairly healthy GDP growth, employment in the organized sector actually declined, leading to frustration among the educated youth who have rising expectations.
- Although real wages of casual labour in agriculture continue to rise during 2000–2005, growth has decelerated strongly, as compared to the previous quinquennium (1994–2000), almost certainly reflecting poor performance in agriculture. However, over the longer periods 1983 to 1993–94 (period I) and 1993– 94 to 2004–05 (Period II), the decline is moderate for rural male agricultural casual labour, from 2.75% to 2.18% per annum.
- Growth of average real wage rates in non-agriculture employment in the period 1999–2000 to 2004–05 has been negligible. Seen over the longer period of two decades (Period I and Period II), the wages have steadily increased at over 2% per annum.
- Inrespect of entire rural male casual labour, the growth in real wages accelerated from 2.55% (Period I) to 2.78% per annum (Period II) (Annexure 4.6).
- Real wages stagnated or declined even for workers in the organized industry although managerial and technical staff did secure large increase.
- Wage share in the organized industrial sector has halved after the 1980s and is now among the lowest in the world.

It is only through a massive effort at employment creation, of the right quality, and decent conditions of work for all sections of population and at all locations that a fair redistribution of benefits from growth can be achieved. This indeed is a stupendous task. Alternative policy measures focusing on different sectors and occupations, and the specific requirements of different target groups are needed to create employment on a sustainable basis.

Recent Experience Revisited

The Tenth Plan was framed against the backdrop of concerns that were posed by the employment and unemployment survey in 1999–2000 (NSS 55th Round), which showed very low growth of employment compared with 1993–94. Jobless growth therefore became a key concern and the Plan set a target of creating 50 million new employment opportunities on a current daily status basis. (For a brief description of different concepts of employment see Box.)

The results of the most recent 61st Round of NSS for 2004–05 reveal a somewhat better picture of employment growth in the Tenth Plan period than in the previous period. During 1999–2000 to 2004–05, about 47 million work opportunities were created as compared to only 24 million in the previous period 1993–94 to 1999–2000. Further, employment growth accelerated from 1.25% per annum during the period 1993–94 to 1999–2000 to 2.62% per annum during the period 1999–2000 to 2004–05. The annual increase in work opportunities increased from 4.0 million per year in the first period to 9.3 million per annum in the second period (Table 21).

The Three Kinds of Estimates of the Unemployed

Unemployment rate is defined as the number of persons unemployed per 1000 persons in the labour force. Three kinds of estimates for the unemployed are obtained following the three different approaches. These are:

(i) number of persons usually unemployed based on 'usual status' approach,

(ii) number of persons unemployed on an average in a week based on the 'weekly status' and

(iii) number of person-days unemployed on an average during the reference period of seven days preceding the survey.

The first estimate indicates the magnitude of persons unemployed for a relatively longer period during a reference period of 365 days and approximates to an estimate of chronically unemployed. Some of the unemployed on the basis of this criterion might be working in a subsidiary capacity during the reference period. The former is called as the usually unemployedaccording to the principal status and the latter, the usually unemployed excluding the subsidiary status workers (us adjusted) which admittedly will be lower than the former. The second estimate based on the weekly status gives the average weekly picture during the survey year and includes both chronic unemployment and also the intermittent unemployment, of those categorized as usually unemployed, caused by seasonal fluctuations in the labour market. The third estimate based on the daily status concept gives average level of unemployment on a day during the survey year. It is the most inclusive rate of 'unemployment' capturing the unemployed days of the chronically unemployed, the unemployed days of the usually employed who become intermittently unemployed during the reference week, and the unemployed days of those classified as employed according to the priority criterion of current weekly status.

[1]NSSO Report No. 409: Employment and Unemployment in India 1993–94: NSS 50th Round; Chapter 7. Employment Perspective and Labour Policy 65.

Despite these positive features, it must also be noted that the labour force also grew faster in the second period. However,the pace of growth in labour force in the second period at 2.84% per annum exceeded the growth in the workforce (employment) of 2.62% per annum, so that the unemployment rate increased from 7.3% in 1999–2000 to 8.3% in 2004–05.

Long-Term Trends in Employment situation : 1983 Through 2005

Some analysts have viewed the 1999–2000 survey as an 'outlier' because of the relatively depressed employment situation in that year, and have commented that 1999–2000 was a case of low statistical base, which shows up as high growth of employment in the next period 2000–05. On this view, a better assessment of trends emerges if we compare developments over two relatively longer periods, that is, 1983 to 1993–94 (period I–10.5 years) and 1993–94 to 2004–05 (period II–11 years).

Table 21 presents a comparison of the trends in employment and labour force over a longer period. It is evident that population growth decelerated in Period II as compared with Period I and this led to a deceleration in labour force growth also. However, the growth of the workforce, that is, total employment, also decelerated in Period II. Employment grew more slowly than the labour force in Period II which raised the unemployment rate from 6.1% in 1993–94 to 8.3% in 2004–05. Measured in absolute terms, the average annual increase in employment opportunities during Period II was 6.45 million, which is lower than the annual increase of 7.09 million in Period I.

The inadequate increase in aggregate employment in Period II is associated with a sharp drop in the pace of creation of work opportunities in agriculture. Agriculture should not be expected to create additional employment but, rather, to reduce the extent of underemployment and thereby increase incomes and wages of those employed in agriculture while surplus labour shifts to the non-agriculture sector. However, the increase in employment in the non-agricultural sectors was disappointing.

Table 22 : Past and Present Scenario on Employment and Unemployment

(current daily status basis)

	1983	*1993–94*	*1999–2000*	*2004–05*	*1993–94 to 1999–2000*	*1999–2000 to 2004–05*	*1983 to 1993–94*	*1993–94 to 2004–05*
	('000 person years)				*(% per annum)*			
All India								
Population	718101	893676	1005046	1092830	1.98	1.69	2.11	1.85
Labour Force	263824	334197	364878	419647	1.47	2.84	2.28	2.09
Workforce	239489	313931	338194	384909	1.25	2.62	2.61	1.87
Unemployment Rate (%)	9.22	6.06	7.31	8.28				
No. of Unemployed	24335	20266	26684	34738	4.69	5.42	–1.73	5.02
Rural								
Population	546642	658771	728069	779821	1.68	1.38	1.79	1.55
Labour Force	206152	252955	270606	303172	1.13	2.3	1.97	1.66
Workforce	187899	238752	251222	278076	0.85	2.05	2.31	1.40
Unemployment Rate (%)	8.85	5.61	7.16	8.28				
No. of Unemployed	18253	14203	19383	25097	5.32	5.3	–2.36	5.31
Urban								
Population	171459	234905	276977	313009	2.78	2.48	3.04	2.64
Labour Force	57672	81242	94272	116474	2.51	4.32	3.32	3.33
Workforce	51590	75179	86972	106833	2.46	4.2	3.65	3.25
Unemployment Rate (%)	10.55	7.46	7.74	8.28				
No. of Unemployed	6082	6063	7300	9641	3.14	5.72	–0.03	4.31

Note : Estimates both on UPSS basis and CDS basis are given in Annexure 4.1.

As shown in Table 22, the dependence of the workforce on agriculture and allied sectors declined from 61% in 1993–94 to 52% in 2004–05, that is, a decline of 9 percentage points as compared with a decline of only 4 percentage points in the period 1983 to 1993–94. Thus, work opportunities diversified away from agriculture at a faster pace during the latter period 1993–94 to 2004–05.

Table 23 shows the annual increase in the work-force by category of employment in Period I compared with Period II. There has been a sustained increase in employment opportunities since 1993–94, although at a slower ratethan in the earlier period. A notable feature is the sharp increase in the number of jobs created at regular salaried wage—from 0.98 million per year in Period I to 1.68 million per year in Period II. This is a direct consequence of the step-up in the expansion of the manufacturing and services sectors, as already discussed. However, the pace of opening up of employment opportunities for casual wage labour getting released from the agriculture sector has slowed down sharply in Period II. The annual increase in this category was 2.40 million in Period I but which declined to 0.54 million in Period II (Table 23), mainly reflecting the lower absorption in agriculture which was not offset by an expansion in other sectors. (CDS basis) (million)

Sector-Wise Employment Generation Achieved in the Tenth Plan

Though the aggregate employment generation of 47 million work opportunities in the period 1999–2000 to 2004–05 was fairly close to the target of 50 million employment opportunities for the Tenth Plan, the performance across sectors has varied.

The achievement with respect to employment creation was short of the Tenth Plan target in the agriculture sector by 0.6 million persons (8.84 million increase against a target of 9.47 million). The increase in the manufacturing sector was short by 3 million persons (8.64 million increase against a target of 11.62 million); in trade, hotel, and restaurants by 0.53 million (10.70 million against a target. Employment Perspective and Labour Policy 67 of 11.23 million); and in transport and communication by 1.47 million (4.04 million against a target of 5.51 million). In contrast, the construction, financial services, and community, social and personal services sectors exceeded the Tenth plan target of employment. In proportional terms, the largest shortfall in employment generation has been in manufacturing.

Informalization of Employment

A critical issue in assessing employment behaviour of the economy is the growth of employment in the organized sector vis-à-vis the unorganized sector. Public debate on this issue is usually conducted on the basis that unorganized sector employment is generally of low quality while organized sector employment is of high quality, and the focus of attention is on whether employment has increased in the organized sector. The inadequacy of growth in the organized sector has traditionally been illustrated using data on employment by units registered with the Directorate General of Employment and Training, which are typically large units in the organized sector. These data are presented in Table 24 below and they clearly show that whereas organized sector employment increased at the annual rate of 1.2% per year in the period 1983–1994, it actually declined at 0.3% per year in the period after 1994. This decline is shown to occur primarily on account of a decline in employment in public sector units. Employment growth in the private sector units has accelerated in the second period but the acceleration is clearly insufficient to offset the decline in public sector employment.

This issue can be explored further on the basis of data from the NSS employment surveys in 1999–2000 and 2004–05 which distinguish individuals according to type of establishment and also type of labour conditions.

To create 50 million opportunities, the Tenth Plan envisaged a contribution of 20 million from selective innovative programmes and policies leading to a changed pattern of growth in favour of the labour intensive sectors, over and above 30 million through normal buoyancy from growth.

Wage workers among the household members who answered NSSO questions in the 61st Round (2004–05) were asked as to whether the employers provide the following types of benefits to them, and also the method of receiving payment:68 Eleventh Five Year Plan and 4.5. The design of enquiry in the two rounds was not identical, and in the 55th Round it was restricted only to the receipt of PF benefits by employees.

Using this data it is possible to obtain a broad picture of employment growth in three categories of establishments, that is, private establishments hiring less than 10 workers, private establishments hiring 10 workers or more, and public sector establishments. Within each category, one can distinguish between regular employees, casual employees, and self-employed. Table 25 presents data for the set of employees other than those engaged in crop agriculture.

The picture that emerges is the following:

- Total employment in public sector establishmentshas increased by 12.2% in the period 1999–2000 to 2004–05.
- Total employment in private sector establishments hiring less than 10 workers has increased by 18.6% in the same period.Employment Perspective and Labour Policy 69
- Total employment in private sector establishmentshiring more than 10 workers has increased by as much as 45.8%!

If we treat employment in establishments hiring more than 10 workers as a measure of organized sector employment, the increase in organized sector employment in the private sector is more than the increase in the private unorganized sector.

If we limit our focus on regular employees in the larger private sector units, this category shows growth of 39.42%. However, the growth of casual employees in the larger private sector units was even faster at 58.9%.

The above conclusions emerge from unit level data of the NSS surveys focusing on the distinction between regular and casual employees. However some so-called regular employees do not have the benefits of social security. Itis also possible to define organized employment more tightly to limit it to employees who receive provident fund and social security benefits. This has been done by the National Commission for Employment in the Unorganized Sector (NCEUS). As per NCEUS estimates, 20.46 million of the 54.12 million employees working in the organized sector in 1999–2000, were unorganized workers, and the remaining 33.67 million were organized. During the next five-year period, that is, 2000–05, while the number of organized workers by this definition remained constant, the number of unorganized workers in organized enterprises increased by 8.68 million to 29.14 million (Table 26).

The NCEUS data does not correspond with those in Table 27 since they include all of agriculture workers in the unorganized sector. Of the total increase in employment of 61 million on a Usual Principal and Subsidiary Status (UPSS) basis, the increase in the unorganized sector is 52 million and the increase in the organized sector is 9 million. However, while total employment in the organized sector has increased over the period, the increase is entirely on account of what is classified as informal employment in the organized sector, that is, workers who do not have the benefit of provident fund and social security. Whereas employment of this category expanded by over 42%, employment categorized as formal was more or less stagnant.

To summarize, the recent experience with employment growth presents a mixed picture. If we focus on the most recent period, 1999–2000 to 2004–05, there appears to be an acceleration in employment growth compared with the preceding period which is consistent with expectations, given the acceleration in GDP growth. However,looking at longer term trends, this acceleration in employment growth disappears and in any case the rateof unemployment has increased throughout. Concerns about the quality of employment appear valid although different sources of data are not easily reconciled. However,it does appear that total employment provided bythe organized sector has expanded in the period 1999– 2000 to 2004–05 and this is entirely because of the growth of informal employment in the organized sector.

These trends highlight the major challenges of employment in the Eleventh Plan which can be summarized as follows:

Unemployment Among The Young and Educated

Unemployment is typically higher among the youth and the educated who look for better quality of jobs, and this phenomenon is illustrated in Tables 23 and 24. Table 23 shows that unemployment among the age group 15–29 years for both males and females and in urban and rural areas is significantly higher than the average level of unemployment of all persons.

Table 23 : Unemployment Rate among Youth (Age Group 15–29 Years)

(%)

Year	*Rural Areas*		*Urban Areas*	
	Male	*Female*	*Male*	*Female*
1993–94	9.0	7.6	13.7	21.2
1999–2000	11.1	10.6	14.7	19.1
2004–05	12.0	12.7	13.7	21.5

Source : NSSO Report No. 515(61/10/1)

Table 24 : Unemployment Rates for Persons of Age 15 Years and above by Level of Education on Current Weekly Status (CWS) Basis

(%)

General			*Level of Rural*			*male Rural female*			*Urban male*		*Urban female*	
Education	*1993-94*	*1999–2000*	*2004–05*	*1993–94*	*1999–2000*	*2004–05*	*1993–94*	*1999–2000*	*2004–05*	*1993–94*	*1999–2000*	*2004–05*
Not literate	1.8	3.0	2.7	2.2	2.7	2.5	2.2	3.1	2.8	2.2	2.0	2.5
Literate and up to primary	1.9	3.0	3.0	2.6	2.6	3.1	3.5	4.1	3.7	4.8	3.6	4.0
Secondary and above	8.3	7.3	6.5	19.8	16.9	18.2	7.0	6.9	6.2	19.6	15.8	17.9
ALL	3.1	3.9	3.8	2.9	3.7	4.2	5.2	5.6	5.2	7.9	7.3	9.0

Source : NSSO Report No. 515(61/10/1).

CDS measure has been used for analysis of trends in workforce, in keeping with the practice followed in the Tenth Five Year Plan. However this, being a person days measure (that is, a time-based measure), is not amenable, straightaway, to study of person-specific characteristics of the workers. Current weekly and the us measures study the activity status of a person over the reference period (week or the year) (CWS, UPSS, and UPS) and are therefore amenable to study of person-specific characteristics. Hence, in presenting the person-specific features of employed or unemployed persons Current Weekly Status (CWS) or the UPSS measures have been used.

The Employment Situation Among the Vulnerable Groups

It is useful to distinguish between several different groups who face special difficulties in employment.

(i) Agricultural Labour Households and Casual Labour in Rural Areas

Out of 460 million workers (UPSS), 94 million earn so little that they are below the poverty line. And if that is the lot of employed workers, the lot of the poor who are unemployed in the labour force must be worse, which is a cause of concern to the planning process.

The proportion of poor among the workers in the rural areas decreased from 25.2% in 1999–2000 to 20.3% in 2004–05. In urban areas, the incidence of poverty among the workers decreased from 22.29% to 21.22%. Though there is a net decrease by 3.5 million in the number of poor workers during 2000–05, the magnitude of poor workers at 94.3 million in 2004–05 remains very high. The incidence of poverty among the regular wage/salaried workers is much lower (around 11%) as compared to the casual labour (32%) and the selfemployed workers (17%). This suggests that all efforts should be made to increase the regular wage/salaried jobs. Employment Perspective and Labour Policy 71.

Unemployment among theruralagriculture labour households, which is the single largest segment of the poor labour households, is now 15.3% (Table 4.10). It is possible to infer that the magnitude of poverty among them may have remained the same or even increased.

Table 25 : Incidence of Unemployment among Rural Agricultural Households

(CDS basis)

Year	*Unemployment Rate (%)*
1983	7.73
1993–94	9.50
1999–2000	12.29
2004–05	15.26

There has been a slight deceleration in the growth in wages of rural male casual agricultural labour from 2.75% per annum during 1983–94 to 2.18% per annum during 1994–2005. The fall in the case of females is more steep, from 3.07% per annum to 2.10% per annum.

(ii) Weaker Social Groups : The Scheduled Castes (SCs) and Scheduled Tribes (STs)

Table 26 gives the unemployment rates among SCs vis-à-vis others (excluding SC, ST and OBC) from National Sample Surveys from 1983 (38th Round) to 2004–05 (61st Round). It is pertinent to mention here that for 1983 (38thRound) and 1993–94 (50th Round), there was no separate category of OBC and therefore the category 'others' included OBC also in these two rounds. It may be seen that between 1999–2000 and 2004–05, the unemployment rates for females, SCs as well as others, in both rural and urban areas have increased; by 0.8 percentage point in rural areas and 1.5 percentage point in urban areas. For urban males, unemployment among SCs increased by 0.4 percentage point, whereas among others it declined by 0.8 percentage point during 1999–2000 to 2004–05.

Table 26 : Unemployment Rate according to Usual Status (ps+ss)1,2,3 for Scheduled Castes during 1983 to 2004–05

(%)

Rural Areas	*1983*	*1993–94*	*1999–2000*	*2004–05*
Male	1.2 (1.4)	1.2 (1.4)	1.8 (1.8)	1.7 (1.6)
Female	0.5 (0.7)	0.4 (0.8)	0.6 (1.0)	1.4 (1.8)
Urban Areas				
Male	5.1 (0.7)	4.6 (4.0)	5.1 (4.6)	5.5 (3.8)
Female	2.9 (4.9)	4.4 (6.2)	3.1 (5.4)	4.6 (6.9)

Note : 1 CDS measure has been used for analysis of trends in the work-force, in keeping with the practice followed in the Tenth Five Year Plan. However, this, being a person days measure (that is, a time based measure), is not amenable, straightaway, to the study of person-specific characteristics of the workers. Current weekly and the us measures study the activity status of a person over the reference period (week or the year) (CWS, UPSS, and UPS) and are therefore amenable to the study of person-specific characteristics. Hence, in presenting the person-specific features of employed or unemployed persons, CWS or the UPSS measures have been used.

2 CDS estimates are not available from NSS reports. Therefore, UPSS estimates are given in this Table 27. However it should be noted that these are under-estimates of unemployment because CDS measure is the most comprehensive measure of unemployment and CDS estimates are significantly higher than UPSS estimates.

3 Figures in parentheses are the corresponding unemployment rates for others (excluding SC, ST, and OBC). Source: Report No. 516 (61/10/2), Employment and Unemployment Situation among Social Groups in India, 2004–05.

Table 28 presents a comparative picture of unemployment among STs and Others; so far as rural males are concerned, there is no change in unemployment rates between 1999–2000 and 2004–05 for SCs; but for others, the unemployment rate has declined slightly. On the other hand, during the same periods, the unemployment rate for females in rural areas has declined for SCs, and for Others it has increased by 0.8 percentage point. In urban areas, the unemployment rate for males in both categories, namely SCs and Others has declined, but for females it has increased.

Table 27 : Unemployment Rate according to Usual Status (ps+ss)1,2,3 for Scheduled Tribes during 1983 to 2004–05

(%)

	1983	*1993–94*	*1999–2000*	*2004–05*
Rural Areas				
Male 0.5	(1.4)	0.8 (1.4)	1.1 (1.8)	1.1 (1.6)
Female	0.1 (0.7)	0.3 (0.8)	0.5 (1.0)	0.4 (1.8)
Urban Areas				
Male	4.3 (0.7)	4.7 (4.0)	4.4 (4.6)	2.9 (3.8)
Female	1.5 (4.9)	1.7 (6.2)	2.8 (5.4)	3.4 (6.9)

Note : [1]CDS measure has been used for analysis of trends in the work-force, in keeping with the practice followed in the Tenth Five Year Plan. However, this, being a person days measure (that is, a time based measure), is not amenable, straightaway, to the study of person-specific characteristics of the workers. Current weekly and the us measures study the activity status of a person over the reference period (week or the year) (CWS, UPSS, and UPS) and are therefore amenable to the study of person-specific characteristics. Hence, in presenting the person-specific features of employed or unemployed persons, CWS or the UPSS measures have been used.

[2]CDS estimates are not available from NSS reports. Therefore, UPSS estimates are given in this Table. However it should be noted that these are under-estimates of unemployment because CDS measure is the most comprehensive measure of unemployment and CDS estimates are significantly higher than UPSS estimates.

[3]Figures in parentheses are the corresponding unemployment rates for others (excluding SC, ST, and OBC).

Source: Report No. 516 (61/10/2), Employment and Unemployment Situation among Social Groups in India, 2004–05.

(iii) The Children at Work

Estimates from the 61st Round reveal that 5.82 million children (age 5–14 years) work; 1.136 million in urban areas and 4.682 million in rural areas (Table 28).

Table 28 : Estimated Number of Children (5–14 Years) in the Labour Force, Workforce, and Unemployed—All India (CDS Basis)

('000)

Heads	*1993–94*		*1999–2000*		*2004–05*	
	Rural	*Urban*	*Rural*	*Urban*	*Rural*	*Urban*
Labour Force	9919	1552	7792	1447	5182	1292
Workforce	9441	1442	7203	1320	4682	1136
Unemployed	479	110	589	127	501	156
Unemployment Rate (%)	4.83	7.08	7.56	8.78	9.66	12.08

The Child Labour (Prohibition and Regulation) Act, 1986 prohibits employment of children below 14 years in hazardous occupations and processes and regulates the working conditions in other employments. Compliance with the provisions of this Act is the responsibility of Labour Sector of the Plan (Ministry of Labour and Employment). At present, the laws do not prohibit employment of children in non-hazardous occupations but children so employed must have access to education. Against this background, the Eleventh Plan Working Group on Child Labour has estimated that 3.643 million children (5–14 years) were working in the nonagricultural sector, out of which 1.219 million children were engaged in hazardous occupations. Chapter 6 (Volume II) Towards Women's Agency and Child Rights gives the comprehensive approach to deal with the problem of children at work and exposed to other risks.

The education sector has a pre-eminent role in ensuring that all children in the age group 9–14 years are at school. To the extent this goal of SSA can be ensured (now that there is a fourfold increase, at constant price, in the Eleventh Plan over the Tenth Plan, duly backed by scheme-tied revenue through a Cess), the tendency to utilize child labour at a cheap cost to increase profits from making children work, can be curbed. The Integrated Child Development Scheme (ICDS), also, has a responsibility with regard to the development of adolescent girls and thus keeping them away from wage employment.

The focus of efforts to eradicate child labour has to be location specific, confined to those pockets where employers are prone to be exploitative in accessing the cheapest cost labour. High per-capita income locations (metro towns, in particular), destinations of migrant worker families and 'industrial belts', where informal work relationships for labour-intensive occupations thrive, have therefore to be closely monitored through innovative mechanisms that provide intelligence to the enforcement agencies.

Any expansion of the Child Labour Eradication Plan has to be made only after a careful evaluation of the existing scheme with regard to:

- Its effectiveness in dovetailing SSA and ICDS;
- The ability to involve State administrations which implement the CSS pertaining to the development, education, nutrition, and protection of children;
- A purely Central Plan funded effort should be in the nature of an emergent action over a limited duration at the location, where the local administration are, by ignorance or by design, seem to be aiding the use of cheap child labour for serving the profit motive of the citizens at that location.
- A suitable form of penalization should be imposed in such local and State Governments that seem to be paying only 'lip service' to curb the problem of the use of 'cheap cost child labour'.

(iv) Women Workers

Women comprise 48.3% of the population but have only 26.1% share in the persons employed. This is presently because their share is in the labour force is only 26.4% (Table 4.14). The female labour force participation rates (LFPR) across all age groups are 25 to 30% of the male LFPR in urban areas, and 35 to 40% of male LFPR in the rural areas.

Along with lower participation rates, women face a higher incidence of unemployment than men. This is especially so for higher levels of education. While the unemployment rates between men and women do not differ much up to the primary level of schooling, unemployment among women educated up to the secondary and higher levels is much higher than among men. In the urban areas, unemployment among young women in the 15–29 years age group is much higher than for men, and is highest among young urban women in the 20–24 years age group where one among every four girls seeking work cannot find it. They are in a especially vulnerable position when they seek entry into the regular wage jobs in the unorganized or even in the private organized sector, in urban areas. This has many implications for our labour policy, particularly the gender sensitive regulations, the social policies and programmes that are designed to promote 'equality' at work.

A measure of 'underemployment' is the change in activity status of the persons employed, when the reference period for the study of time disposition is reduced from one year (the us measure) to an average day of the past week (CDS measure). Only 66% of rural women who are counted as employed on the US measure, are seen as employed on the CDS measure, whereas the corresponding proportion for men is higher at 89% (NSS Report No. 515 (61/10/1) (Part I) (September 2006) (Statement 7.2.1). The deceleration in wage rates of casual labour in agriculture between the periods 1994–2000 and 2000–05 has been higher for women (2.93% per annum to 0.93% per annum) than for men (2.79% to 1.21%). Participation in education by girls (15–19 years) in rural areas is only 33% (as compared to 47% for men), and the gender disparity increases sharply in the next age group, that is, 20–24 years. Only 1.3% of young women (15–29 years) in rural areas received formal vocational training. Such features of the labour market for women are reflected in the fact that as much as 21.7% of employed women have consumption levels below poverty line in 2004–05, that is, they are employed yet still poor. This proportion is lower among men—19.9%.

The principal reasons for low participation by women in the labour force are:

- Wage rates of women are lower than of male for comparable occupations.
- Women are denied access to certain occupations, though they may be capable of doing that work as well as the men.

- Skill development of women is not uniform across all trades; participation by them remains confined to a few labour-intensive occupations such as stitching, teachers training, etc., which forces a majority of the women to enter the labour market as unskilled labour.

Table 29 : Past and Present Macro Scenario on Employment and Unemployment—Male and Female

(CDS basis)

	1983	*1993–94*	*1999–2000*	*2004–05*	*1993–94 to 1999–2000*	*1999–2000 to 2004–05*	*1983 to 1993–94*	*1993–94 to 2004–05*
	('000 person years)				*(% per annum)*			
All India								
Population	718101	893676	1005046	1092830	1.98	1.69	2.11	1.85
Labour Force	263824	334197	364878	419647	1.47	2.84	2.28	2.09
Workforce	239489	313931	338194	384909	1.25	2.62	2.61	1.87
Unemployment Rate (%)	9.22	6.06	7.31	8.28				
No. of Unemployed	24335	20266	26684	34738	4.69	5.42	–1.73	5.02
Female								
Population	346546	430188	484837	527355	2.01	1.70	2.08	1.87
Labour Force	68011	86728	92859	110886	1.14	3.61	2.34	2.26
Workforce	61218	81151	85952	100491	0.96	3.18	2.72	1.96
Unemployment Rate (%)	9.99	6.43	7.44	9.37				
No. of Unemployed	6793	5578	6907	10395	3.63	8.52	–1.86	5.82
Male								
Population	371556	463488	520209	565475	1.94	1.68	2.13	1.82
Labour Force	195813	247468	272019	308761	1.59	2.57	2.25	2.03
Workforce	178270	232780	252242	284417	1.35	2.43	2.57	1.84
Unemployment Rate (%)	8.96	5.94	7.27	7.88				
No. of Unemployed	17542	14688	19777	24343	5.08	4.24	–1.68	4.70

Whenever equal opportunity has been given to women in recruitments, equality in wage with the men has been ensured, and an equal exposure in training has been given, the participation by women in work has improved. This is illustrated in ample measure in the IT and enabled services sectors and in various other professional services—legal, financial, commercial, education and health.

In order to promote gender equity, steps have to be taken to increase women's participation in the labour force. This has to be pursued through skill development, labour policies and also the social security framework. Significant outcomes can be expected only if the gender issue is addressed through the planning initiatives across all the 'heads of development' in the Plan, with requisite lead from the 'Women and Child Development' Head. Gender-budgeting has not, so far, received due attention.

The Eleventh Plan must seek to reduce the gender differentials by pursuing (i) target shares for women beneficiaries in the programmes for 'Skill Development initiatives', 'New initiatives at Social Security', implementation of regulations such as the Apprentices Act, 1961, the Factories Act, the Building and Construction Workers (Conditions of Service) Act, and better implementation of The Maternity Benefit Act, 1976 and The Equal Remuneration Act, 2000, and for guarding against sexual harassment at the work place. (v) Migrant Workers.

Inter-State population migration rates for the intercensus period (1991–2001) are given in Annexure4.8.The net out-migrant and in-migrant States are presented in this Annexure. Large absorbers of migrants are the States of Punjab, Haryana, Delhi, Gujarat, Maharashtra, Karnataka and Goa. The large net out-migrant States are Rajasthan, Uttar Pradesh, Bihar, Assam, West Bengal, Jharkhand, Orissa, Chhattisgarh, Andhra Pradesh, Kerala, and Tamil Nadu. Implicit in these population movements is an origin-destination migration matrix of workers. The numbers shown in the Annexure are inter-decadal and are presumably much smaller than the shorter period movements of migrant workers.

Migration itself is not an abnormal phenomenon and is common all over the world since growth centres which generate demand for labour often tend to concentrate in certain areas. However, migrant workers are the most vulnerable and exploited among the informal sector workers, and have not received any attention in the labour policy. In the States which are sources (origin) of supply of migrant workers—and most of them migrate totake up some labour-intensive, low-wage occupation—an effective and large-scale effort for vocational training in the labour intensive occupations is required. And such a programme should be amenable to the special needs of the entrants to informal labour markets. In the destination States, the focus of public policy (including Labour Policy) should be to improve the conditions under which the bulk of these in-migrants live and work. And in so far as the destination locations fail to provide certain basic minimum conditions to the new in-migrants, it would be better to restrain economic growth at such locations. In the labour and employment sector, better implementation of certain legislations pertaining to unorganized workers can protect the interests of most of the migrant workers; for example, the Building and Other Construction Workers (Regulation of Employment and Conditions of Service) Act, 1976; the Building and Other Construction Workers (Cess) Act, 1976; the Workmen's Compensation Act, 1923 and the Minimum Wages Act, 1948. An initiative has been taken recently by the government (in September 2007) with the introduction of 'The Unorganized Workers' Social Security Bill, 2007' in the Rajya Sabha.

(vi) The Self-employed and Casual Wages

Wage Employed

The self-employed and casual wage employed account for 83% of the workforce. About 20–25 million enter the labour force each year. Thus 17–21 million will enter the labour market in the non-regular wage employed category. The only strength of the self and the casual employed is their occupational skill, and the entrepreneurial skill to negotiate the price of labour putEmployment Perspective and Labour Policy 75 in by them. At present, a majority of the new entrants in this category have little or no education, not to speak of any vocational training. And many of them migrate to new locations, and to new occupations other than their traditional ones. The skill development set up of the government(s) has practically no space for them, at present. The National Skills Mission, Skill Development and Training could make a major difference by upgrading the skills of new entrants to the informal sector.

Employment Projections for the Eleventh and Twelfth Plans

The Approach Paper for the Eleventh Plan had projected an addition of 52 million to labour force in the Plan period and had called for the creation of 70 million employment opportunities. However, the projections of labour growth have been revisited in view of the latest population projections made available by the National Commission on Population and work done by the Eleventh Plan Working Group on Labour Force and Employment Projections. The projected increase in labour force during the Eleventh Plan period is now estimated at 45 million.

The employment prospects in the Eleventh Plan period have also been revised and the results are presented in Tables 30 and Table 31 with projections of labour force and employment over a longer period, 2006–07 to 2016–17, encompassing both the Eleventh and the Twelfth Plans.

As shown in Table 31, population growth is expected to decelerate through this period with a corresponding deceleration in labour force growth to 1.6% per year. However, although the labour force growth is projected to decelerate, the absolute increase in the labour force is very large. In fact, India's demographic profile is such that the expansion in the labour force in India will be larger than in the industrialized countries, and even China. The demographic dividend could be a source of global competitive advantage if it is combined with successful efforts at skill upgradation and at expansion of employment opportunities.

Table 30 : Population, Labour Force, Employment Projections

('000)

	Basis	*1993–94**	*2004–05**	*2006–07*	*2011–12*	*2016–17*
Population (age 0+)		893676	1092830	1128313	1207971	1283242
Population (age 15–59)		501760	652940	687120	760110	820570
Labour Force	UPSS	378650	471250	492660	541840	586440
Labour Force	CDS	334197	419647	438948	483659	524057
Employment Opportunities	CDS	313931	384909	402238	460310	51820
Unemployed ('000)	CDS	20266	34738	36710	23348	5853
Unemployment Rate (%)	CDS	6.06	8.28	8.36	4.83	1.12

Note : *Actual estimates derived from NSS.

Table 31 : Projected Population, Labour Force, and Employment in Different Periods

Growth rates (% per annum) and absolute increase ('000)

	Basis	*1993–94 to 2004–05**	*2004–05 to 2006–07*	*2006–07 to 2011–12*	*2011–12 to 2016–17*
Growth Rate in Population (age 0+)		1.85	1.43	1.37	1.22
Growth Rate in Population (age 15–59)	2.42	2.29	2.04	1.54	
Growth Rate in Labour Force	UPSS	2.01	1.99	1.92	1.59
Growth Rate in Labour Force	CDS	2.09	2.02	1.96	1.62
Growth Rate in Employment Opportunities	CDS	1.87	1.98	2.73	2.40
Addition to Population ('000)	UPSS	199154	35483	79658	75271
Addition to Labour Force ('000)	UPSS	92600	21410	49180	44600
Addition to Labour Force ('000)	CDS	85450	19301	44711	40398
Addition to Employment Opportunities ('000)	CDS	70978	17330	58072	57893

Note : *Actual estimates derived from NSS.

The growth of total employment overthe period has been estimated on the basis of employment projections for individual sectors which are then aggregated. These sectoral employment projections are based on sectoral GDP growth rates combined with assumptions about employment elasticity moderated by the implicit growth of productivity. The resulting projections indicate that 58 million job opportunities will be created in the Eleventh Plan period which exceeds the projected addition to the labour force, leading to a reduction in the unemployment rate to below 5%.

Over the longer period up to 2016–17, spanning the Eleventh and Twelfth Plan periods, the additional employment opportunities created are estimated at 116 million as compared to 71 million during the 11-year period from 1993–94 to 2004–05. Since the labour force will increase by 85 million in this period, a substantial part of the surplus of labour force that exists at the commencement of the Eleventh Plan could get absorbed into gainful employment by the end of the period. The unemployment rate at the end of the Twelfth Plan period is projected to fall to a little over 1%.

There are important qualifications to these projections which must be kept in mind, arising from the limitation of employment elasticity as a projection tool. The concept of employment elasticity is at best a mechanical device to project employment on the basis of projected growth of output and past relationships between employment and output. These relationships can change as a result of changing technology and change in real wages. The labour force participation rate is also subject to changes, especially because of possible changes in female participation rates in urban areas associated with advances in women's education. For all these reasons, the projected decline in the unemployment rate must be treated with caution. It could well be that the projected increase in labour demand induces greater labour supply through an increase in participation rates and also higher wages which moderate demand. However, the overall picture of an acceleration in the rate of creation of job opportunities and a reduction in unemployment rates is relatively robust, if GDP growth takes place as projected.

Sectoral employment Projection

The projected growth of employment in the Eleventh Plan and beyond is decomposed into its sectoral components.

Agriculture Employement

The agriculture sector has long been known to be characterized by underemployment, which means that with the same number of workers it is possible to generate more output. The projection for the Eleventh Plan assumes that the projected doubling of the rate of agricultural growth during the Eleventh Plan will be possible without any increase in agricultural employment. Whereas agriculture contributed 8.8 million job opportunity in the 11-year period from 1993-94 to 2004-05, it is projected to contribute no increase in the Eleventh Plan and a net decrease of 4 million agricultural workers over the Twelfth plan period (2006-07 to 2016-17). Thisf is a reasonable projection considering that the number of main workers in agriculture declined by about 1.8 crore between 1991 and 2001 and there has been a large increase in marginal workers in agriculture during 1991-2001. Rising wage differentials between the agriculture and non-agriculture sectors are also very likely to shift labour out of agriculture, and the continued growth into the Twelfh Plan period would provide sufficient pull factor from non-agriculture to encourage such a shift.

Analternative projection of agricultural employment has also been made, applying the actual employment elasticity (0.15) observed during 1993–94 to 2004–05 to the projected growth of output over the perspective period (2007–2017). This gives an estimate of employment in agriculture for 2011–12, that is, about 9 million more. In this projection there is positive growth of employment in agriculture at 0.6% per year and productivity growth is correspondingly lower at 3.4% per annum. In this scenario, employment increase would be 9 million more, with unemployment correspondingly less at only 14 million in 2011–12. However, this would be at the cost of lower productivity growth and, therefore, wages and incomes in agriculture and a larger proportion of low quality jobs.

Table 32 : Projected Increase in Number of Workers by Sector, 2007–12 and 2007–17

(CDS) ('000)

Industry	*Estimated*		*Projected*	
	1983 to 1993–94 (10½Years)	*1993–94 2004–05 (11 Years)*	*to 2006–07to 2011–12 (5 Years)*	*2006–07 to 2016–17 (10 Years)*
Agriculture	34900	8816	0	-3967
Mining and Quarrying	855	3	1	3
Manufacturing	7850	14834	11937	24516
Electricity, water, etc.	487	30	17	36
Construction	5260	10052	11922	26370
Trade, hotel, and restaurant	9190	22667	17397	34402
Transport, storage, and communication	3213	7639	9025	18764
Finance, insurance, real estate, and business services	1524	4312	3428	7472
Community, social, and personal services	11163	2624	4344	8369
Total	**74442**	**70978**	**58072**	**115965**

During the 11-year period 1994–2005, the pace of increase in per worker GDP in agriculture was only 2.24% per annum as compared to 4.35% per annum growth in aggregate GDP per worker. During the Eleventh Plan, also, the pace of productivity increase being projected is lower for agriculture than for the aggregate economy, irrespective of the scenario regarding employment growth in agriculture. Since the main employment issue in the agriculture sector is the increase in farm labour income, and not the creation of a larger number of employed workers, it would be appropriate to work towards a strategy in which there is higher growth in non-services employment opportunities in rural areas which can provide additional income for the rural workforce by providing additional non-agricultural employment.

Employment in Manufacturing, Construction, and Service

The Eleventh Plan should aim at significantly stepping up growth in employment in other sectors, countering the long-term trends observed in the past. Employment in manufacturing should grow at 4% per annum against the trend of growth in the preceding 11 years (1994–2005) of 3.3% per annum. Employment in construction should grow at 8.2% per annum against the trend of 5.9% growth, and in the transport and communication sector at 7.6% against the long-term trend of 5.3%.

These growth rates in employment in individual sectors are achievable provided they are supported by programmes for skill development, which will ensure availability of the relevant skills without which the growth of employment will probably choke. It is also necessary toensure a wider provision of social security and welfare of unorganized workers, particularly in sectors such as construction and transport.

The sectors with prospects for high growth in output, creation of new establishments and for creation of new employment opportunities (direct as also indirect) are:

Services

- IT-enabled Services

- Telecom Services
- Tourism
- Transport Services
- Health Care
- Education and Training
- Real Estate and Ownership of Dwellings
- Banking and Financial Services
- Insurance
- Retail Services
- Media and Entertainment Services

Other Sectors and Sub-Sectors

- Energy-Production, Distribution and Consumption of Horticulture
- Floriculture
- Construction of Buildings
- Infrastructure Projects Construction

Industry Groups

- Automotive
- Food Products
- Chemical Products
- Basic Metals
- Non-Metallic Mineral Products
- Plastic and Plastic Processing Industry
- Leather
- Rubber and Rubber Products
- Wood and Bamboo Products
- Gems and Jewellery
- Handicrafts
- Handlooms
- Khadi and Village Industries

The Services Sector

The services sector is currently the fastest growing sector of the economy, and employment growth in the sector has remained more than 5% per annum since the 1990s as compared with the aggregate employment growth at less than 2%. This sector has the unique opportunity to grow due to its labour cost advantage, reflecting one of the lowest salary and wage levels in the world coupled with a rising share of working age population. However, two types of initiatives are required: (i) fostering the establishment of a viable size for delivery of services based on labour intensive occupations. Only in establishments of a reasonable size (in contrast to the average enterprise size of 1.2 workers, as it exists today), with a reasonable level of occupational specialization and corresponding productivity and wage levels, is this feasible; and

(ii) a massive skill development effort, as discussed later, for vocational training of the new entrants to the labour force.

Planning initiatives in health, nutrition, care of children, care of the aged, education, skill development and expansion of social security services will create a large potential for employment for delivery of these services. Quantum jumps in the requirement of personnel, their skill and in their composition—by gender, by social group and by location in favour of the backward regions—will arise from:

- A massive increase in Central funding of education which is a four fold increase over the Tenth Plan in constant price;
- Emphasis on the next phase of SSA on improvement in the quality of education;
- Reaching these services to the districts having a concentration of SC, ST, and minorities;
- Rapid expansion in the mid-day meal (MDM) scheme to cover 60 million additional children at the upper primary level by 2008–09; and
- Enrolling one crore children in vocational education–skill development streams.

Already, a substantial increase in the number of teachers has been made, which will continue further. During the three-year period of 2002–03 to 2005–06, 0.285 million para teachers were recruited, of which 0.27 million are in the rural areas (NUEPA; Progress Towards UEE—Analytical Report 2005–06). And much of this expansion has occurred in the low per capita income States of Bihar, Madhya Pradesh, Chhattisgarh, Rajasthan, and Uttar Pradesh. Further, the expansion of mid-day meals programme, will require a substantial step up in the personnel required for delivery of such services.

The Central Government has recently announced an expansion in social security services such as: (i) Old age pension to all citizens, (ii) Life and disability cover against injury or death to either the head of, or to one earning member of each poor family; and (iii) Health insurance, so that the poor do not have to bear a high cost of medical care. These would require a commensurate expansion in the requirement of a variety of professionally trained and skilled personnel by the institutions that (i) cover risk; (ii) identify, issue and update the identity of the beneficiaries (smart cards); (iii) design specific schemes for the target groups and market the same; (iv) render medical services; or (v) reach out to the prospective beneficiaries. Most of the beneficiaries of the new Central initiatives would be the aged, the poor and the landless, and thus vulnerable to vagaries of the market. The institutional base that exists at present for delivery of the kinds of services, discussed here, is quite insensitive to the special needs of the prospective beneficiaries, and breeds 'exclusion'. While the beneficiaries could be (and should be) expected to make a contribution, howsoever small, to participate in the scheme, the personnel that areEmployment Perspective and Labour Policy 79 hired for rendering these services have to be trained to reach out to the prospective beneficiaries, in a manner that is responsible and transparent, and thus evokes her/his confidence to participate in the scheme. Moreover, some token contribution to become a member of the scheme is essential to empower these beneficiaries to lay a claim to the services that especially allocated funds by the Central Government for their benefit.

(i) Expansion of it Services in Rural Areas

Village kiosks will require expansion of IT personnel deployment across the rural areas of the country, in particular to facilitate the expansion of an IT enabled governance set-up. Such improvements are essential for keeping pace with the demand for public services that will guide the diversification of the economy away from agriculture and towards the secondary and tertiary sectors, duly supported by: (i) investments in industrial infrastructure, (ii) creation of institutional infrastructure in the rural areas, and (iii) for fostering integration of rural markets with the rest of the economy. These would require manifold expansion in a variety of matching services to be delivered by the local governments and by the village Panchayats, and that would be feasible only if the governance set-up at the local level is overhauled and handled in an IT-friendly mode.

(ii) Personal Services

The increase in the income of middle-class households in the high growth phase entails a spurt in consumption of personal services related to attire, appearance, baby care, health upkeep, personal drivers, security, care for the aged dependents, household governance and management, and so on. And such personal services have to be delivered by professionally trained, hired personnel of formal establishments, quite distinct from the earlier one-person operations. But in this area, a major effort at nurturing the right type of serving establishments at reasonable fees, training and certification of their personnel who can earn a reasonable income and thereby keep themselves above exploitation by the well off and informal employers, by way of access to the social security arrangements, is required. So far, services of a reasonable standard have by and large remained confined to the few who have a very high level of personal income, leaving the average urban consumer of such services to the vagaries of a market driven by'short-life', one or two-person establishments, thriving on profits from cheap and untrained young in-migrants to high-income locations. The local administrations, including the labour administrations, have not handled the issue in a labour–employment–income perspective, and have generally ignored the problems arising due to the law and order enforcement agencies.

Employment and Labour Policy

The employment strategy for the Eleventh Plan must ensure rapid growth of employment while also ensuring an improvement in the quality of employment. While self employment will remain an important employment category in the foreseeable future—it accounted for 58% of all employment in 2004–05—there is need to increase the share of regular employees in total employment. This category has increased from 17% of total employment in 1983 to 18% in 2004–05. It should be the focus of policy to achieve a substantial increase in the share of regular employment with a matching reduction in the share of casual employment which at present is as high as 23%.

The above analysis implies that the success of labour policy should be seen in terms of the number of regular wage employment opportunities based on some formof a written contract between the employer and the employee, that is, an increase in the number of 'formal' jobs. The potential for creation of formal employment can be fully utilized by making appropriate changes in rules and procedures. It is often said that one of the obstacles to growth of formal employment in the organized sector is the prevalence of excessively rigid labour laws which discourage such employment. Broadly, it is necessary to review existing laws and regulations with a view to making changes which would:

- encourage the corporate sector to move into more labour-intensive sectors
- facilitate the expansion of employment and output of the unorganized enterprises that operate in the labourintensive sectors.

At present, the incentives and subsidies are so designed as to strongly penalize entrepreneurs for crossing a threshold size from a micro/small to a medium/80 Eleventh Five Year Plan large unit. The excise and other taxation policies need to be reviewed in this perspective.

Changes in policies also need to be examined in regard to :

- Linking incentives with the outcomes measured in terms of employment. For example, incentives are given to a wide range of production activities primarily with the objective of promoting employment and income of workers engaged in such activities. However, such incentives are hardly ever calibrated against the benefits realized in terms of employment and wages.
- Regular wage employment, that is, formal employment, merits fiscal incentives. Such incentives already exist at a limited scale for the larger establishments, but are so designed as to make it difficult for medium and small establishments to benefit from these.

Changing labour laws is a sensitive issue and it is necessary to build a consensus. However, there are several changes short of hire and fire which should not present problems. These include:

- The locations and production activities that have a high potential for employment creation merit a differential treatment.
- Employment of women must be encouraged ensuring, inter alia, the special needs that they may have by virtue of change in working hours (night shifts, for example) or the requirements of the family, for example, child care.
- Contract labour in the domestic tariff area merits encouragement, provided commensurate steps are taken to increase social security
- Monitoring the implementation of labour laws, that is, the reporting system should be simplified and be permitted in an IT-friendly mode.

Even as steps are taken to increase the volume of formal or regular employment, it is also necessary to take steps to improve the quality of employment in the unorganized sector. NCEUS in its August 2007 Report has summarized, in the form of a 13-point Action Programme, (the main recommendations for the workers of Enterprises in the Unorganized/Informal Sector. These are presented in Box 4.2.

Unorganized sector enterprises mostly hire most workers who get released, or relocated, from crop agriculture (due to the reasons discussed earlier), and seek wage employment in the manufacturing or services sector. Any significant improvement in their income, and quality of employment, is feasible only if the institutional environment in the labour market makes it feasible for the formal sector to reach out to such workers on a decentralized basis rather than through a centralized plan programme. The large coverage (in terms of absolute numbers) through Provident Fund (43 million), Employee State Insurance (33.0 million) a variety of welfare funds (5.0 million), for beedi workers, for example) has been possible because the institutional framework created through the various Acts 5 (P.F., E.S.I., Beedi Workers Welfare Fund, etc.) recognized a relationship of those employed on regular wage, with either the employer, or the specific formal commodity market that provides work to (that is, absorbs the output of labour put in by) the unorganized enterprises' workers.

As already argued, the creation of a formal relationship between the worker and the hiring establishment, in the regular wage employment mode, is a critical factor in improving the quality of employment of the workers hired by the unorganized enterprises. In this context, the work being done by NCEUS6 on: (i) the 'employment strategy' to be pursued in respect of, and through the, unorganized enterprises, (ii) the regime of labour regulations to attract the unorganized enterprise togive a formal recognition to the multitudes of workers hired by them, and (iii) to enable them to gain access to 'social security', is of paramount importance.

5. Of course, many of these organizations have to reorient their pattern of working to the new realities of the market for wage labour in which the roleofpublicsectoris diminishing and the average number of workers hired by the private enterprises is reducing consequent upon changes in technology leading to improvement in the productivity of labour.

6. The relevant terms of reference of NCEUS are :

- Suggest elements of an employment strategy focusing on the informal sector;
- Review Indian labour laws, consistent with labour rights, and with the requirements of expanding growth of industry and services, particularly in the informal sector, and improving productivity and competitiveness; and
- Review the social security system available for labour in the informal sector, and make recommendations for expanding their coverage.Employment Perspective and Labour Policy 81

Box 4.2

A Thirteen Point Action Plan Suggested by the NCEUS for Employment in the Unorganized Sector

A. ***Protective Measures for Workers***

1. Ensuring Minimum Conditions of Work in the Non-agricultural and Agricultural Sectors : Two bills, for agricultural workers and non-agricultural workers, that specify the minimum conditions of work, including a statutory national minimum wage for all workers
2. Minimum Level of Social Security : A universal national minimum social security scheme, as part of a comprehensive legislation covering life, health and disability, maternity and old age pension to protect the workers in the unorganized sectors.

B. ***Package of Measures for the Marginal and Small Farmers***

3. Special Programme for Marginal and Small Farmer : Revival of the targeted programme focusing on small and minor farmers, with an initial thrust in the areas wherein the existing yield gap is also considered high. A special agency or a coordinating mechanism should be set up if required.
4. Emphasis on Accelerated Land and Water Management : Immediate priority to, and significant up-scaling of, programmes for land and water management. Revision of the priority sector landing policy to provide a quota for micro and small enterprises.
5. Credit for Marginal/Small Farmers : RBI to monitor, separately, credit to this segment, expansion in the outreach of credit institutions in rural areas and a credit guarantee fund to obviate the need for collateral by the marginal/small farmers in accessing the institutional credits. A 10% share for small and marginal farmers in the priority sector credit (Table below)
6. Farmers' Debt Relief Commission : The Central government to lay guidelines and provide 75:25 assistance for setting up State-level Farmers' Debt Relief Commissions, in the States experiencing agrarian distress—natural or market related.

C. ***Measures to Improve Growth of the Non-agricultural Sector***

7. Improve Credit Flow to the Non-agricultural Sector:

   Percent Sector and Sub-Sector/Purpose

18 10% for small and marginal farmers; 8% for other farmers

10.4% for micro enterprises with capital investment (other than land and building) up to Rs 0.5 million and 6% for other micro and small enterprises

12 12% on loans up to Rs 0.5 million to the socio-economically weaker sections for housing, education, professions, and so on.

40 Total priority sectors landing

8. *Encouraging SHGs and MFIs for Livelihood Promotion* : Measures to encourage growth of micro finance and SHGs in poor States and in the backward areas
9. *Creation of a National Fund (NAFUS)* : Rs 5000 crore initial corpus for an exclusive statutory agency to take care of requirements of micro and small enterprises in agriculture and non-agriculture sectors that are presently not reached by SIDBI and NABARD.
10. *Up-scaling Cluster Development through Growth Poles* : Twenty-five growth poles in the traditional industries clusters with incentives at par with SEZs

D. *Measures to Expand Employment and Improve Employability*

11. *Expand Employment through Strengthening Self-employment Programmes* : Rationalization and strengthening of the four major self-employment generation programmes with 5 million annual employment generation target.
12. Universalize and Strengthen National Rural Employment Guarantee Act (NREGA): Extension of NREGA Programmes to all districts.
13. *Increase Employability through Skill Development* : On-job-training cum employment-assurance programme to provide incentive of Rs 5000 per person to any employer willing to provide one-year training on job skill enhancement.

ANNEXURE 4.1

Population, Labour Force, Employment, and Unemployment (1993–94 to 2016–17)

(million)

Statuos	*Estimated*				*Projected*	
	1993–94	*1999–2000*	*2004–05*	*2006–07*	*2011–12*	*2016–17*
1	*2*	*3*	*4*	*5*	*6*	*7*
Population						
Age 0+	893.68	1005.05	1092.83	1128.31	1207.97	1283.24
Age 15–59	501.76	572.23	652.94	687.12	760.11	820.57
Labour Force						
UPSS						
Age 0+	378.65	408.35	471.25	492.66	541.84	586.44
Age 15–59	337.71	369.22	431.95	451.70	496.65	535.20
CDS						
Age 0+	334.20	364.88	419.65	438.95	483.66	524.06
Age 15–59	298.95	330.78	385.87	403.75	444.72	479.70

1	2	3	4	5	6	7
Employment						
UPSS						
Age 0+	371.12	398.93	459.72			
Age 15–59	330.34	360.04	420.74			
CDS						
Age 0+	313.93	338.19	384.91	402.24	460.31	518.20
Age 15–59	279.88	305.70	352.92			
Unemployed						
UPSS						
Age 0+	7.53	9.41	11.53			
Age 15–59	7.37	9.17	11.21			
CDS						
Age 0+	20.27	26.69	34.74	36.71	23.35	5.86
Age 15–59	19.07	25.08	32.95			
Unemployment rate (%)						
UPSS						
Age 0+	1.99	2.30	2.45			
Age 15–59	2.18	2.48	2.60			
CDS						
Age 0+	6.06	7.31	8.28	8.36	4.83	1.12
Age 15–59	6.38	7.58	8.54			

ANNEXURE 4.2

Annual Growth Rate of GDP per Worker

(%)

Industry	*1983 to 1993–94*	*1993–94 to 2004–05*	*2006–07 to 2016–17*
1. Agriculture	1.03	2.24	4.57
2. Minining and Quarrying	1.66	4.95	5.64
3. Manufacturing	2.29	3.31	7.27
4. Electricity, gas and water supply	3.70	5.46	7.51
5. Construction	–1.43	1.45	5.56
6. Trade, hotels and restaurants	1.06	2.69	5.68
7. Transport, storage and communication	2.06	4.94	9.77
8. Finance, insurance, real estate and business services	2.79	–0.40	4.26
9. Community, social and personal services	1.57	5.90	6.11
Total	**2.29**	**4.35**	**7.82**

ANNEXURE 4.3(A)

Percentage Distribution of Employed Persons by Category of Employment (CDS)—Rural India

('000)

	1983	*1993–94*	*1999–2000*	*2004–05*	*1983*	*1993–94*	*1999–2000*	*2004–05*	*1983–*	*1993–94*	*1999–2000*	*2004–05*
Self-Employed	64.62	62.50	59.83	62.91	61.81	60.91	59.51	66.51	63.83	62.05	59.74	63.95
Regular Employees	11.50	9.52	10.25	10.25	4.45	4.55	4.88	5.12	9.52	8.10	8.72	8.76
Casual Labour	23.89	27.92	29.92	26.84	33.74	34.55	35.61	28.37	26.65	29.85	31.53	27.29
All	100.00	100.00	100.00	100.00	100.00	100.00	100.00	100.00	100.00	100.00	100.00	100.00
Estimated Number of Workers ('000)	135203	170677	179866	197391	52695	68075	71357	80685	187898	238752	251223	278076

Note : Derived from NSS reports on employment and unemployment situation in India.

ANNEXURE 4.3(B)

Percentage Distribution of Employed Persons by Category of Employment (CDS)—Urban India

('000)

Category	*Male*				*Female*				*Persons*			
	1983	*1993–94*	*1999–2000*	*2004–05*	*1983*	*1993–94*	*1999–2000*	*2004–05*	*1983*	*1993–94*	*1999–2000*	*2004–05*
Self-Employed	40.96	42.57	42.24	45.86	39.09	40.83	41.07	43.94	40.66	42.27	42.05	45.50
Regular Employees	46.68	43.78	43.47	42.39	35.72	36.67	40.18	43.18	44.87	42.54	42.92	42.54
Casual Labour	12.35	13.65	14.29	11.75	25.19	22.50	18.75	12.88	14.47	15.19	15.03	11.96
All	100.00	100.00	100.00	100.00	100.00	100.00	100.00	100.00	100.00	100.00	100.00	100.00
Estimated Number of Workers ('000)	43067	62103	72376	87027	8523	13076	14595	19806	51590	75179	86971	106833

Note : Derived from NSS reports on employment and unemployment situation in India.

ANNEXURE 4.3(C)

Percentage Distribution of Employed Persons by Category of Employment (CDS)—All India

('000)

Category	*Male*				*Female*				*Persons*			
	1983	*1993–94*	*1999–2000*	*2004–05*	*1983*	*1993–94*	*1999–*	*2004–2000*	*1983 05*	*1993–94*	*1999–2000*	*2004–05*
Self-Employed	58.91	57.18	54.79	57.69	58.64	57.67	56.38	62.06	58.84	57.31	55.19	58.83
Regular Employees	20.00	18.66	19.78	20.08	8.81	9.72	10.87	12.62	17.14	16.35	17.52	18.13
Casual Labour	21.10	24.16	25.43	22.22	32.55	32.60	32.75	25.32	24.03	26.34	27.29	23.03
All	100.00	100.00	100.00	100.00	100.00	100.00	100.00	100.00	100.00	100.00	100.00	100.00
Estimated Number of Workers ('000)	178270	232780	252242	284418	61218	81151	85952	100491	239488	313931	338194	384909

Note : Derived from NSS reports on employment and unemployment situation in India.

ANNEXURE 4.4

Conditions of Employment of Regular Wage/Salaried Workers—2004–05

(per 1000) (UPSS basis)

Condition of Employment	*Rural*	*Urban*	*All*
No written job contract	592	592	592
Not eligible for Paid Leave	480	455	464
Neither written job contract nor eligible for paid leave	712	549	630
Not eligible for Social Security Benefit1	569	535	547
Paid a Monthly Salary	857	900	884
Non-existence of Union/Associations	513	541	531
Sample Workers	17033	26385	43418

Source : Derived from NSS 61st round (2004–05).

[1]Coverage under any of the Schemes-Provident Fund, PPF with employer contribution, Gratuity, Health care and Maternity benefits.

ANNEXURE 4.5

Distribution of Regular Wage/Salaried Workers by Type of Enterprise

Type of Enterprise	*Distribution of Workers*
Proprietory	378
Partnership	45
Employer household	49
Subtotal	472
Govt./Public sector	333
Public/Pvt.Ltd. Co.	127
Society/Trust	38
Subtotal	498
Others	18
N.R.	12
All	1000
Sample persons	43418

Source : Table 1, Appendix A, of NSS Report of 61st Round, No. 519.

ANNEXURE 4.6

Growth of Average Daily Wage Earnings in Rural India (at 1993–94 price)

(% per annum)

Category	*Rural Male*				*Rural Female*			
	1983 to 93–94	*1993–94 to 04–05*	*1993–94 to 99–2000*	*1999–2000 to 04–05*	*1983 to 93–94*	*1993–94 to 04–05*	*1993–94 to 99–2000*	*1999–2000 to 04–05*
Casual Labour in Public Works	2.28	3.81	3.83	3.15	4.10	3.83	5.03	2.01
Casual Labour in Agriculture	2.75	2.18	2.79	1.21	3.07	2.10	2.93	0.93
Casual Labour in Non Agriculture	2.38	2.34	3.69	0.62	4.08	3.47	5.06	1.32
Casual Labour in all Activities	2.55	2.78	3.59	1.51	3.13	2.40	3.19	1.21

ANNEXURE 4.7

The Working Poor in India by their Gender, Location, and Category of Employment, 1999–2000 and 2004–05

('000)

Population Segment	*1999–2000*				*2004–05*			
	Self Employed	*Regular Wage/Salaried*	*Casual Labour*	*Total*	*Self Employed*	*Regular Wage/Salaried*	*Casual Labour*	*Total*
Rural Persons	32762	2457	41466	76686	33139	2273	34125	69537
	(19.39)	(11.62)	(36.34)	(25.21)	(16.08)	(9.30)	(30.34)	(20.27)
Urban Persons	9387	4201	7531	21120	12141	5302	7321	24765
	(23.60)	(11.10)	(43.96)	(22.29)	(22.87)	(11.49)	(41.90)	(21.22)
All Males	27728	5545	31602	64875	29135	5863	27388	62386
	(19.68)	(11.18)	(36.77)	(23.47)	(17.17)	(10.24)	(31.85)	(19.94)
All Females	14421	1114	17396	32931	16145	1713	14058	31916
	(21.27)	(11.84)	(38.41)	(26.88)	(18.03)	(12.83)	(31.99)	(21.74)
All Persons	42150	6658	48998	97806	45280	7576	41446	94302
	(20.19)	(11.29)	(37.34)	(24.52)	(17.47)	(10.73)	(31.90)	(20.51)

Notes: 1. Figures in brackets are the proportion of Poor workers to total workers in that category.
2. UPSS basis.

ANNEXURE 4.8

Net Migrants Rate (1991–2001)

(%)

	States/UTs	*Male*	*Female*	*Person*
1.	Andhra Pradesh	–0.03	–0.03	–0.03
2.	Assam	–0.06	–0.09	–0.07
3.	Bihar	–0.39	–0.17	–0.28
4.	Chhattisgarh	–0.06	–0.07	–0.06
5.	Gujarat	0.22	0.09	0.16
6.	Haryana	0.40	0.35	0.37
7.	Himachal Pradesh	0.04	–0.06	–0.01
8.	Jharkhand	–0.08	–0.02	–0.05
9.	Karnataka	0.04	0.00	0.02
10.	Kerala	–0.08	–0.08	–0.08
11.	Madhya Pradesh	–0.01	0.00	–0.01
12.	Maharashtra	0.37	0.21	0.29
13.	Orissa	–0.10	–0.04	–0.07
14.	Punjab	0.20	0.07	0.14
15.	Rajasthan	–0.08	–0.05	–0.06

	States/UTs	*Male*	*Female*	*Person*
16.	Tamil Nadu	–0.08	–0.07	–0.08
17.	Uttranchal	–0.03	–0.06	–0.04
18.	Uttar Pradesh	–0.25	–0.16	–0.21
19.	West Bengal	–0.04	–0.04	–0.04
20.	Delhi	1.93	1.57	1.77
21.	Jammu and Kashmir	–0.04	–0.06	–0.05
22.	Arunachal Pradesh	0.73	0.57	0.65
23.	Manipur	–0.16	–0.13	–0.14
24.	Meghalaya	0.09	0.04	0.07
25.	Mizoram	–0.16	–0.35	–0.25
26.	Nagaland	0.05	–0.41	–0.17
27.	Sikkim	0.26	0.16	0.21
28.	Tripura	0.02	0.02	0.02
29.	Andaman and Nicobar	0.80	0.66	0.74
30.	Chandigarh	2.15	1.78	1.98
31.	Dadra and Nagar Haveli	4.29	1.90	3.12
32.	Daman and Diu	6.15	1.88	4.05
33.	Lakshadweep	1.04	0.20	0.63
34.	Pondicherry	0.74	0.94	0.84
35.	Goa	0.83	0.59	0.71
36.	NE States	0.07	–0.03	0.02

SKILL DEVELOPMENT AND TRAINING

Introduction

Skills and knowledge are the driving forces of economic growth and social development of any country. They have become even more important given the increasing pace of globalization and technological changes provide both challenges that is taking place in the world. Countries with higher and better levels of skills adjust more effectively to the challenges and opportunities of globalization.

As India moves progressively towards becoming a 'Knowledge economy' it becomes increasingly important that the Eleventh Plan should focus on advancement of skills and these skills have to be relevant to the emerging economic environment. In old economy, skill development largely meant development of shop floor or manual skills. Even in this area there are major deficiencies in our workforce which need to be rectified. In new or knowledge economy the skill sets can range from professional, conceptual, managerial, operational behavioural to interpersonal skills and inter-domain skills. In the 21st century as science progresses towards a better understanding of the miniscule, that is, genes, nano-particles, bits and bytes and neurons,knowledge domains and skill domains also multiply and become more and more complex. To cope with this level of complexity the EleventhPlan has given a very high priority to Higher Education. Initiatives such as establishing 30 new Central universities, 5 new IISERs, 8 IITs, 7 IIMs, 20 IIITs, etc. are aimed at meeting that part of the challenge of skill development. Inthis Chapter, however, it is proposed to focus on massscale skill development in different trades through specially developed training modules delivered by ITIs, Polytechnics, vocational schools, etc. The Eleventh Plan aims at launching a National Skill Development Mission which will bring about a paradigm change in handling of 'Skill Development' programmes and initiatives.

The NSS 61st Round results show that among persons of age 15–29 years, only about 2% are reported to have received formal vocational training and another 8% reported to have received non-formal vocational training indicating that very few young persons actually enter the world of work with any kind offormal vocational training. This proportion of trained youth is one of the lowest in the world. The corresponding figures for industrialized countries are much higher, varying between 60% and 96% of the youth in the age group of 20–24 years. One reason for this poor performance is the near exclusive reliance upon a few training courses with long duration (2 to 3 years) covering around 100 skills. In China, for example, there exist about 4000 short duration modular courses which provide skills more closely tailored to employment requirement.

Review of the Existing Vocational Training system

In India, skill acquisition takes place through two basic structural streams—a small formal one and a large informal one. The formal structure includes: (i) higher technical education imparted through professional colleges, (ii) vocational education in schools at the post-secondary stage, (iii) technical training in specialized institutions, and (iv) apprenticeship training. A number of agencies impart vocational education/training a various levels. Higher professional and technical education, primarily in the areas88 Eleventh Five Year Plan of agriculture, education, engineering and technology, and medicine, is imparted through various professional institutions.

There are seventeen ministries and departments of GOI which are imparting vocational training to about 3.1 million persons every year. Most of these are nationallevel efforts and individually they are able to reach a very small part of the new entrants to the labour force.

Even collectively, they provide training to about 20% of the number of annual additions to the labour force. Each ministry/department in charge of subject sets up training establishments in its field of specialization. The attempt to meet training needs through multiple authorities—labour, handlooms, handicrafts, small industry, education, health, women and child development, social welfare, tourism, etc. leads to redundancy at some locations. While each of the training initiatives has a definite area of specialization, there is need for coordination amongst these ministries/departments.

Vocational training being a concurrent subject, Central Government and the State Governments share responsibilities. At the national level, Director General of Employment & Training (DGE&T), Ministry of Labour is the nodal department for formulating policies, laying down standards, conducting trade testing and certification, etc. in the field of vocational training. At the State level, the State Government departments are responsible for vocational training programmes.

There are 1244 polytechnics under the aegis of the Ministry of Human Resource Development with a capacity of over 2.95 lakh offering three-year diploma courses in various branches of engineering with an entry qualification of 10th pass. Besides, there are 415 institutions for diploma in pharmacy, 63 for hotel management, and 25 for architecture.

There are about 5114 Industrial Training Institutes (ITIs) imparting training in 57 engineering and 50 non-engineering trades. Of these, 1896 are State Government-run ITIs while 3218 are private. The total seating capacity in these ITIs is 7.42 lakh (4 lakh seats in government ITIs and the remaining 3.42 lakh in private ITIs). These courses are open to those who have passed either Class 8 or 10 depending on the trade and are of 1 or 2 years duration, which varies from course to course. In addition to ITIs, there are six Advanced Training Institutes (ATI) run by the Central Government which provide training for instructors in ATIs for electronics and process instrumentation, offering long and short courses for training of skilled personnel at technician level in the fields of industrial, medical, and consumer electronics and process instrumentation.

In order to provide sufficient autonomy in academics, administration, finance, management, improved physical infrastructure (building, equipments) etc., the government launched a scheme for upgradation of 100 ITIs into Centres of Excellence, with effect from the year 2004–2005.

The Apprentices Act, 1961, as amended from time to time, regulates the training of apprentices. The Act serves a dual purpose—first, it regulates the programme of training apprentices in industry so as to conform to the prescribed syllabi, period of training, etc. prescribed by the Central Apprenticeship Council and second, to utilize fully the facilities available in industry for workers. As on 30.06.2006, over 20800 public/private sector establishments were covered under the Act and number of seats allocated were 2.30 lakh, out of which about 1.72 lakh seats were utilized.

Skill building activity was also initiated under the 10+2 level of school education. A scheme of prevocational education at lower secondary level was started from 1993–94 to impart training in simple marketable skills and to develop vocational interests. There are now about 9583 schools offering about 150 educational courses of two years duration in the broad areas of agriculture, business and commerce, engineering and technology, health and paramedical, home science and science and technology at +2 stage covering about one million students. Under the aegis of the Ministry of Rural Development, banks and Non-Governmental Organizations, through 2500 Rural Development and Self-Employment Training Institutes (RUDSETI), have undertaken entrepreneurship and skill building of the rural youth for self-employment in areas with a pre-existing market for the goods/services produced, with a reported success rate of 70%. The Entrepreneurship Development Initiative (EDI) and other programmes of the Ministry of Micro, Small and Medium Enterprises train about one lakh persons a year.

The unorganized sector which constitutes about 93% of the workforce is not supported by any structuralSkill Development and Training 89 system of acquiring or upgrading skills. By and large, skill formation takes place through informal channels like family occupations, on the job training under master craftsmen with no linkages to the formal education training and certification. Training needs in this sector are highly diverse and multi skill-oriented. Many efforts for imparting training through Swarnjayanti Gram Swarojgar Yojana (SGSY), PMRY, KVIC, Krishi Vigyan Kendra (KVK) and Jan Shiksha Sansthan (JSS) are in place but the outcome is not encouraging.

The quantitative dimension of the Skill Development challenge can be estimated by the following:

- 80% of new entrants to workforce have no opportunity for skill training. Against 12.8 million per annum new entrants to the workforce the existing training capacity is 3.1 million per annum.
- about 2% of existing workforce has skill training against 96% in Korea, 75% in Germany, 80% in Japan, and 68% in the United Kingdom.

The Quantitative Aspect of The Skill Shortage

The NSS 61st Round Survey on Employment and Unemployment indicates (Annexures 5.1.1. to 5.1.4) that educational institutions attendance rates (5– 14 years) drop by nearly half in the age group 15–19 years and by 86% after the age 15 years. Labour force participation rates rise sharply after the age of 14 years and reach close to 100% at the age of 25–29 years. The said results also reflect that 38.8% of the Indian labour force is illiterate, 24.9% of the labour force has had schooling up to the primary level and the balance 36.3% has had schooling up to the middle and higher level. They also reveal that about 80% of the workforce in rural and urban areas do not possess any identifiable marketable skills.

The Qualitative Aspect of Deficiencies

A basic problem with the skill development system is that the system is non-responsive to labour market, due to a demand—supply mismatch on several counts: numbers, quality and skill types. It is also seen that the inflexibilities in the course/curriculum set-up, lead to over supply in some trades and shortages in others. Of the trained candidates, the labour market outcomes as seen from placement/ absorption rates are reportedly very low. The institutional spread in the VET system shows acute regional disparity with over half of the ITIs/ITCs located in the southern States, both in terms of number of institutions as well as the number of seats. 5.16. The quality of the training system is also a matter of concern, as the infrastructural facilities, tool/kits, faculty, curriculum are reportedly substandard. The existing institutions also lack financial and administrative autonomy. The testing, certification and accreditation system is reportedly weak, and since the deliverables are not precisely defined, there is no effort at evaluating outcomes and tracking placements. The problem is further complicated with lack of industry–faculty interaction on course curricula and other factors.

The training system for capital-intensive sectors and hi-tech areas has always received a highly preferential treatment in contrast to those working in the informal sector. Further there is no certification system for a large chunk of workers, who do not have any formal education but have acquired proficiency on their own or through family tradition/long experience. In the absence of a proper certificate, these classes of workers in the informal sector are subjected to exploitation and they do not get any avenues for better employment in the market and their mobility is very restricted.

The private sector-run Industrial Training Centres (ITCs) do not seem to be any better than the ITIs, and the low-paying capacity of learners and consequently low fee structure and absence of quality consciousness are said to be major reasons for the current state of affairs.

The Planning Commission held extensive consultations with the industry, various Central ministries running training programmes, State Governments. The discussions have revealed that the present system of skill formation has certain critical gaps in that the curricula are inflexible and outmoded. There is an inadequate fitness-testing mechanism of the institutions with a mismatched fee structure and admission criteria. The capacities of the trainers are also not in consonance with the current requirement of various sectors due to various restrictions of the affiliating agencies.

The private sector does undertake in-house training programmes and to a very limited extent, trains 'outsiders'. However,such programmes are limited to catering to their own felt needs, in the nature of captive skill development. Low-paying capacity of learners and reluctance of90 Eleventh Five Year Plan industries to train workers for fear of losing them to competition has resulted in chronic deficiency in private investment in this area. All these deficiencies mentioned in the above five paras will need to be rectified during the Eleventh Plan.

Skill Development and Self-employment

NSSO 61st Round data also reveals that the proportion of persons (15–29 years) who received formal vocational training was around 3% for the employed, 11% for the unemployed and 2% for persons not in the labour force. In order to link skills developed into actual productive use thereof including self-employment, steps will be taken in the Eleventh Five Year Plan by providing adequate incentives, not necessarily monetary but in terms of skill and entrepreneurship development and forward and backward linkages to finance, marketing and human resource management, to those who are or seek to be self-employed to enhance their productivity and value addition, making it an attractive option, rather than be an option faute de mieux as at present.

Demographic Divided

The decline in the rate of growth of population in the past few decades implies that in the coming years fewer people will join the labour force than in preceding years and a working person would have fewer dependents, children or parents. Modernization and new social processes have also led to more women entering the work force further lowering the dependency ratio. This decline in the dependency ratio (ratio of dependent to working age population) from 0.8 in 1991 to 0.73 in 2001 is expected to further decline sharply to 0.59 by 2011 as per the Technical Group on Population Projections. This decline sharply contrasts with the demographic trend in the industrialized countries and also in China, where the dependency ratio is rising. Low dependency ratio gives India a comparative cost advantage and a progressively lowering dependency ratio will result in improving our competitiveness.

The unprecedented opportunity for Skill Development arises from a unique 25-year window of opportunity, called India's demographic dividend. The Demographic dividend consists of three elements of demographic trends fortuitously coinciding at a time when the economy is growing at 9% plus: (i) a declining birth rate means fewer people will be joining the workforce in coming years, than in previous years, (ii) a very slow improvement in life-expectancy around 63/64 years of age means an ageing population surviving fewer years after superannuation than in other countries, (iii) the baby-boomers generation having now crossed the age of 20, the demographic bulge is occurring at the age bracket of 15–29. All these trends combine to result in India havingworld's youngest workforce with a median age way below China and OECD countries. Figure 5.1 brings out this point very effectively. This would mean that dependency ratio, that is, the ratio of non-working population to working population will continue to be low, giving India

India Demographics Profile (2002) China Demographics Profile (2020) France Demographics Profile (2025)Skill Development and Training 91 a comparative cost advantage over others, for another 25–30 years. By that time the demographic bulge in India would be also reaching the age of superannuation, and India will also be joining the league of ageing economies.

It is expected that the ageing economy phenomenon will globally create a skilled manpower shortage of approximately 56.5 million by 2020 and if we can get our skill development act right,we could have a skilled manpower surplus of approximately 47 million (Annexure 5.2). In an increasingly connected world, where national frontiers are yielding to cross-border outsourcing, it is not inconceivable that within a decade we can become a global reservoir of skilled person power. As it is we account for 28% of graduate talent pool among 28 of the world's lowest cost economies. (Figure 5.2)

The criticality of Skill Development in our overall strategy is that if we get our skill development act right, we will be harnessing 'demographic dividend'; if we do not get there, we could be facing a 'demographic nightmare'. The Eleventh Plan takes cognizance of these and endeavours to take a slew of measures which will bring about a paradigm change in our Vocational Education System (VES).

India has the youngest population in the world; its median age in 2000 was less than 24 compared 38 for Europe and 41 for Japan. Even China had a median age of 30. It means that India has a unique opportunity to complement what an ageing rest of the world needs most. The demographic structure of India, in comparison with that of the competing nations, would work to the advantage to the extent our youth can acquire skills and seize the global employment opportunities in the future. This involves co-ordination, dialogue and discussions with the State Governments, private partners and other stakeholders, arriving at estimates of number of skilled personnel required across the sectors, aligning them with the career objectives of the youth drawing up different sector-specific modules of varying duration thereby.

Eleventh Plan Strategies

Inthe Eleventh Five Year Plan, the thrust will be on creating a pool of skilled personnel in appropriate numbers with adequate skills, in line with the requirements of the ultimate users such as the industry, trade, and service sectors. Such an effort is necessary to support the employment expansion envisaged as a result of inclusive growth, including in particular the shift of surplus labour from agriculture to non-agriculture. This can only take place if this part of the labour force is sufficiently skilled. During the Eleventh Plan it is proposed to launch a major 'Skill Development Mission' (SDM) with an outlay of Rs 22800 crores.

Skill Development Mission

In order to create a pool of skilled personnel in appropriate numbers with adequate skills in line with the employment requirements across the entire economy with particular emphasis on the twenty high growth high employment sectors, the government will set up an SDM consisting of an agglomeration of programmes and appropriate structures aimed at enhancing training opportunities of new entrants to the labour force from the existing 2.5 million in the non-agricultural sector to 10 million per year.

Mission Goal

To provide within a five- to eight-year timeframe, a pool of trained and skilled workforce, sufficient to meet the domestic requirements of a rapidly growing economy, with surpluses to cater to the skill deficits in other ageing economies, thereby effectively leveraging India's competitive advantage and harnessing India's demographic dividend.

The Skill Development Mission (SDM) will have to ensure that our supply-side responses are perpetually in sync with the demand side impulses both from domestic as well as global economies. The mission will, therefore, have to involve both public and private sectors in a symbiotic relationship, with initiatives arising from both sides with reciprocal support. Thus public sector initiatives torepurpose, reorient and expand existing infrastructure, will need involvement of private sector for management and running of Skill Development Programmes, ending with placement of candidates. Similarly Private Sector Initiatives will need to be supplemented by government by one-time capital grants to private institutions and by stipends providing fee supplementation to SC/ST/ OBC/Minorities/other BPL candidates. Thus the core strategy would consist of a two-track approach, of a public arm of amplified action through ministries and State Governments and a private arm of specific and focused actions for creating skills by the market through private sector-led action.

In case of government-led initiatives the concerned Ministries will conceptualize the initiatives for either expanding and improving existing institutions and providing them enlarged budgets and improved action plans or they will set up new generation institutions with budgetary support. For industry or service sector-specific private initiatives the entire strategic thinking and plan of action will emerge from industry associations and ministries will be involved in structuring government response and providing budgetary support. The SDM will oversee and facilitate entire process of collaborative action.

Mission Objectives and Functions

Articulate a vision and framework to meet India's VET needs :

- Assess skill deficits sector wise and region wise and meet the gaps by planned action in a finite time frame.
- Orchestrate Public Sector/Private Sector Initiatives in a framework of a collaborative action.
- Realign and reposition existing public sector infrastructure ITIs, polytechnics and VET in school toget into PPP mode and to smoothen their transition into institutions managed and run by private enterprise or industry associations. Give them functional and governance autonomy.
- Establish a 'Credible accreditation system' and a 'guidance framework' for all accrediting agencies, set up by various ministries and or by industry associations. Get them to move progressively away from regulation to performance measurement and rating/ ranking of institutions. Rate institutions on standardized outcomes, for example, percentage graduates placed, pre and post course wage differentials, dropout rates, etc.
- Encourage and support industry associations and other specialized bodies/councils and private enterprise to create their own sectoral skill development plans in 20 High Growth Sectors (Annexure 5.3.)
- Establish a 'National Skill Inventory' and another 'National Database for Skill Deficiency Mapping' on a national Web portal—for exchange of information between employers and employment seekers.

- Establish a Trainee Placement and Tracking System for effective evaluation and future policy planning.
- Reposition 'Employment Exchanges as Outreach points of the Mission' for storing and providing information on employment and skill development. Enable employment exchanges to function as career counseling centres.
- Enlarge the 50000 Skill Development Centres (SDCs) programme eventually into a 'Virtual Skill Development Resource Network' for Web-based learning.

Mission Strategies

The strategies of the Mission will be to bring about a paradigm change in the architecture of the existing VET System, by doing things differently.

- Encourage Ministries to expand existing Public Sector Skill Development infrastructure and its utilization by a factor of five. This will take the VET capacity fromSkill Development and Training 93 3.1 million to 15 million. This will be sufficient to meet the Annual workforce accretion, which is of the order of 12.8 million. In fact, the surplus capacity could be used to train those in the existing labour force as only 2% thereof are skilled. This infrastructure should be shifted toprivate management over the next 2–3 years. States must be guided as incentivizer to manage this transition.
- Enlarge the coverage of skill spectrum from the existing level. Skill Development programmes should be delivered in modules of 6 weeks to 12 weeks; with an end of module examination/certification. For calibrating manual skills a 4–6 level certification system must be established based on increasing order of dexterity of the craftsman.
- Make a distinction between structural, interventional and last mile unemployability and correspondingly set up programmes for 24 months, 12 months and 6 months duration. Encourage 'Finishing Schools' to take care of last mile unemployability.
- Establish a National Qualifications Framework, which establishes equivalence and provides for horizontal mobility between various VET, Technical and Academic streams at more than one career points. Expand VET to cover more classes and move progressively from post matric to cover 9th class dropouts and then 7th class dropouts.
- Encourage 'Accreditation Agencies' in different domains to move away from regulation to performance measurement and rating and ranking of institutions.
- Encourage institutional autonomy coupled with selfregulation and stake-holder accountability. Institutions must have freedom of action in governance, as also on the financial management.
- For standard setting and curriculum setting, establish or notify at least one 'standard setting/quality audit institution' in each vertical domain.
- Move from a system of funding training institutes to funding the candidates. Institutional funding could be limited to an upfront capital grant. Recurring funding requirement could be met by appropriate disbursement to the institute at the end of successful certification. Candidates from SC/ST/ OBC/ Minorities/ BPL, etc. could be funded in two parts—
  - (i) Stipend (monthly) to be paid to trainee
  - (ii) Fee subsidization at the end of the programme to be given to the institute after placement.

The Mission will encompass the efforts of several ministries of the Central Government, State Governments and the activity of the private arm, supported by the following institutions:

(i) Prime Minister's National Council on Skill Development, (ii) National Skill Development Coordination Board, and (iii) National Skill Development Corporation/Trust. The Central ministries which have skill development programmes will continue tobe funded as at present. However the spectrum of skill development efforts will be reviewed periodically for policy directions by the Prime Minister's Council on Skill Development. The Council will be supported by a National Skill Development Coordination Board, which will be charged with the coordination and harmonization of the governments' initiatives for skill development spread across the seventeen Central ministries and State Governments with the initiatives of the National Skill Development Corporation/Trust. State governments will be encouraged to set up State-level Skill Development Missions. A non-profit National Skill Development Corporation may be set up as a Company under Section 25 of the Companies Act, and/or a National Skill Development Trust under the Societies Act may be set up, to encourage private sector arm of the Mission.

Action Plan for Component—Government Initiative in PPP Mode

Over the years some 20 odd ministries have created an infrastructure for skill development. There are 1896 ITIs (under State Governments), 1244 Polytechnics, 669 Community Polytechnics, 9583 Secondary Schools with VET Stream and 3218 ITCs (in private sector). Besides Ministries of Rural Development (RD), MSME, Health, Tourism and several others have their own establishments. All these need to be restructured and repositioned in collaboration with private enterprises. Furthermore, new capacities are being created by the ministries. These need to be brought in PPP mode.

(i) Action Plan for ITIs

Action: Ministry of Labour and Employment

- Complete Upgradation of 500 ITIs by investing Rs 2.0–3.5 crore in each into institutions of excellence.
- Upgrade remaining 1396 ITIs in PPP mode by providing interest free loan up to Rs 2.5 crore each.
- Facilitate 1000 new ITIs in under-served regions –to be set up in PPP mode so that largely unskilled workforce of these backward areas could acquire skills and mainstream with workforce in progressive regions.94 Eleventh Five Year Plan
- Set up 500 new ITIs—in Industrial Clusters/SEZs on a demand-led basis—also in PPP mode.
- Quadruple ITI capacity by encouraging them to run two shifts or more. Introduce short term modules in 2nd Shift.
- Intensive Faculty Training Programme
- MoUs with States and ITIs defining outcomes and reforms and imposing obligation to transfer autonomy to the PPP.

(ii) Action Plan for Polytechnics

Action : Ministry of Human Resources Development (HRD)

- Upgrade 400 government polytechnics.
- Set up 125 new polytechnics in PPP mode in hitherto unserved districts.
- Run all polytechnics in two shifts to double the capacity utilization.

- Encourage much larger initiative in private sector since the demand for junior engineers is enormous and absorption and placements are nearly guaranteed.
- State governments may be encouraged to let their engineering colleges start polytechnics in evening shift to turnout junior engineers.

(iii) Action Plan for Vocational Education

Action: Ministry of HRD

- Expand VE from 9500 senior secondary schools to 20000 schools. Intake capacity to go up from 1.0 million to 2.5 million
- All VE schools must get into partnership with employers, for providing faculty/trainers, internship, advice on curriculum setting, in skill testing and certification, etc.
- Progressively move vocational education from an unviable 2–year stream, commencing after Class 10, to a stream that captures 9th Class dropouts and later on it should commence from Class 7, capturing 7th Class dropouts. Give emphasis to last mile employability related soft skills—viz., English language skills, quantitative skills, computer literacy, spreadsheet, word processing, computer graphics, presentation skills, behavioral and interpersonal skills, etc.

(IV) Action Plan for Rudestis

Action: Ministry of Rural Development

- Set up 600 RUDSETIS—one in each district
  - State governments to provide land
  - GOI to meet 75% of capital cost
  - Banks to meet 25% of capital cost plus provide the following handholding services:-
    i. Project consultancy and business counseling
    ii. Incubation Assistance
    iii. Marketing
    iv. Sourcing of Credit and Raw Material Supply.
- Focus on Entrepreneurship Development Programmes.
- Link RUDSETIS to EDI of MSME.

(v) Action Plan for setting UP A Vvirtual skill development Resource network Linking 50000 Skill Development Centres (SDCS)

Action: DIT/ MSME/ MRD/ ARI/ Ministry of Textiles etc.

- It is proposed to set up 50000 SDCs to train approximately 200 persons per centre, i.e., 10 million people per year to take skill development to the doorstep of rural populations. Location could be chosen from the following:-
  - One lakh Common Service Centres (CSCs) set up by Ministry of Telecom and IT.
  - 108000 secondary schools being given IT support.
  - Over 147000 rural post offices or Panchayat offices per 6000 Block Headquarters.
- SDC to deliver training capsules of 8–12 week duration with an end of the course certification system. Instruction material to be provided on CD ROMS with 80–120 hours of computer time, 20% of which will be online, rest offline.

- Training material will be created by participating ministries/enterprises/industry associations with help of NASSCOM.
- Mentor Groups for tutorial support in online interactive mode will be provided by service-providers engaged by industry associations.
- End of the programme toolkits will be provided by industry associations/State Governments.
- Employment Melas at the end of programmes to help unplaced trainees to get placed.

Action Plan for Component II—Private Sector Initiatives

Twenty high growth sectors of industries and services have been identified which have the ability to provide expanded employment. The Mission, in its private arm, will encompass the efforts of industry associations of these sectors to identify and quantify skill deficiencies in their respective sectors and envision the sectoral plan to meet their growing skill needs. TheSkill Development and Training 95 corporation/trust will be a PPP on skill development conceived as a non-profit entity. It will make periodic as well as an annual report of its plans and activities and put them in the public domain. The National Skill Development Corporation or the National Skill Development Trust, as the case may be, will identify areas where support and supplementation will be required from the government. In respect of each of these, the respective industry association or group of industry leaders will articulate the sectoral vision for the sectoral Skill Development Initiative. The corporation/trust will refine and validate this vision and ensure complementarity between private initiative and government action, in the light of policy directives from the Prime Minister's Council and the operational guidelines of the Coordination Board.

The Mission will engage with ten high growths sectors on manufacturing side and an equal number on services side. The Mission's dialogue with private sector industry will be focused on—(i) automobile and auto components (ii) electronics hardware (iii) textiles and garments (iv) leather and leather goods (v) chemicals and pharmaceuticals (vi) gem and jewellery (vii) building and construction (viii) food processing (ix) handlooms and handicrafts (x) building hardware and home furnishings. The Apex Industry Association in each of these sectors will evolve the Skill Development Vision & Plan for their respective sector. The following are some examples of how sectr-specific initiatives might work:

- Society of Indian Automobile Manufacturers/Automobile Component Manufacturers Association of India will engage with the Mission to enable India's Automobile and Auto-components Industry to scale up by 2015–16 from current turnover of US$ 45 billion to US$ 145 billion and current employment of 10.5 million to target employment of 25.0 million. This industry is very skill-intensive (with 90% manpower as Skilled). They will require 6.25 million technical and managerial personnel and the training requirements of these will be in Manufacturing Management, SQC, TQM, 6-Sigma, Statistical Process Control, Kaizen practices, Lean Manufacturing & Breakthrough Management. The Skill-sets required will be at strategic and conceptual skills level and will, therefore, require retrofitting of special courses in IITs/IIMs and other Engineering and Management Institutes— courses like Infotronics/Mechatronics. For their shopfloor and other skilled personnel training modules imparting Computer Aided Design (CAD)/ Computer Aided Manufacturing (CAM)/CNC Skills, skills relating to low cost automation and process improvement will be required. It may be necessary to set up one 'National Centre for Quality Management' in Automotive sector. It may be necessary to set up a network of institutes for motor mechanics. Another useful step could be two-way sabbaticals between Industry Personnel and Personnel of Training Institutes.

- Electronics Hardware Industry is growing at the rate of 25.30% and expects to scale up by 2015–16, from current turn over of US$ 30 bn to US$ 320 bn, with employment increasing from 1.5 million direct and 3.0 million indirect to 7.14 million direct and 14.00 million indirect. The industry requires 5% graduate engineers, 15% diploma-holders, 50% skilled workers and 30% semi skilled. The range of skills required will vary from chip manufacturer and VLSI design, to embedded software, to mere mechanical assembly line operations.
- Similarly Textile/Apparel and Garments Industry expects to scale up by 2015–16, from current turnover of US$ 47.0 billion to the targeted US$ 115.0 billion, raising the employment from current workforce level of 35 million to 41.5 million. Of the 6.5 million accretion is workforce only 20% will be unskilled, 50% will be semi skilled, 20% will have ITI certificates and 10% will be Management and Technical graduates. This industry alone could over next five years take over a very large number of ITIs.

The three examples described above illustrate how individual sectors may deal their specific requirements with strategies and plan of action which will be vastly different from each other. For this reason no generic template is being suggested and the vision/strategy and plan will have to be crafted within the overall policy framework.

On the services side ten High Growth Sectors have been identified separately, viz. (i) ITs or software services sector (ii) ITES—BPO services, (iii) tourism hospitality and travel trade (iv) transportation/logistics/warehousing and packaging (v) organized retail (vi) real estate services (vii) media, entertainment, broadcasting, content creation, animation (viii) healthcare services (ix) banking/insurance & finance (x) education/skill development services. Mission will engage with each of these sectors. Industry association and workout the appropriate Sectoral Skill Development Plans, strategies and deliverables.

The National Skill Development Mission's orchestration of private sector initiatives in concert with96 Eleventh Five Year Plan government action could give different results in different sectors, such as:

- Setting up of a domain-specific Apex Skill Development Institute for:
  - Domain specific Skill Development need assessment
  - Performance Rating of Institutions/Service Providers
  - Domain Standard setting and Quality benchmarking
  - Curriculum setting
  - Framework setting for end of programme testing and certification
  - Running special Skill Development Programmes in niche areas requiring superior skills
- Setting up of Regional Institutes/Workshops/Toolrooms and Online Mentoring Groups
- Retrofitting special courses in IITs/IIMs/NITs and other Engineering and Management Programmes or offering them as an elective
- Providing last mile employability training to engineering/management/ or other graduates from lesser known colleges.
- Establishing two-way Sabbatical Exchange Programme between Industry and Faculty of University/Colleges/VET schools/ITIs, etc.
- Collaborative action for faculty development
- Collaborative action for online 'Skill Developmentcontent-creation'.

- Private Management takeover of Public sector institutions *viz.* ITIs, polytechnics, vocational schools, etc.

Public-Private Partnership (PPP)

PPP Mode will be the major vehicle for absorbing public expenditure in skill development in the Eleventh Five Year Plan. Apart from the financial contribution from the government, it is necessary to create an Enabling Environment for Private Investment in Skill Training. This requires the prescription of a National framework for domain specific standards and common principles such as (i) trainer not to be the examiner/certifier, (ii) certifier not to be the accreditation agency and (iii) a strict separation of all the three as the basic feature of the mechanism. The facilities for career tracking and placement—biometric smart card based ID, and a National database for location wise availability and shortage of skilled personnel will be established. The system should provide the options of multiple entry and exit points and total mobility between vocational, general and technical streams. In order to take on board the vulnerable sections, provision for fee vouchers for BPL/SC/ST/OBC/Minority would also be made available. To overcome the regional disparities due to diverse socio-economic factors, VGF approach would be adopted to address regional imbalances through PPP.

Mission Structure

The Skill Development Mission has to be conceived in a manner which recognizes that many Ministries are involved and also many separate Industry and service sectors. The Structure consists of Prime Minister's National Council on Skill Development for apex level policy directions, a National Skill Development Coordination Board, and a National Skill Development Corporation/Trust. The Central Ministries with Skill Development programmes will operate in a Mission mode and the State governments will gear their Departments/Agencies into a State Skill Development Mission. The private sector, especially the twenty high growth sectors will actively participate as the private arm of the Mission. The composition of the Prime Minister's Council on Skill Development and National Skill Development Coordination Board are as described below.

Prime Minister's National Council on Skill Development: The Council will comprise of Prime Minister as Chairman, Minister of Finance, HRD, Industries, Rural Developoment, Labour & Employment and Housing & Urban Poverty Alleviation, Deputy Chairman, Planning Commission, Chairperson, National Manufacturing Competitive Council, Chairperson of the National Skill Development Corporation, six experts in the area of Skill Development as Members and Pr. Secretary to Prime Minister as Member Secretary.

National Skill Development Coordination Board

This Committee will comprise of Deputy Chairman, Planning Commission as Chairman, Chairperson/Chief Executive Officer of the National Skill Development Corporation, Secretaries of Ministries of Finance, Human Resource Development, Labour and Employment, Rural Development, Housing and Employment, Rural Development, Housing & Urban Poverty Alleviation. Secretaries of Four States by rotation, for a period of two years, three Distinguished Academicians/Subject Area Specialists as Members and Secretary, Planning Commission as the Member Secretary.Skill Development and Training 97 Ministries of the Central Government having Skill Development Programmes.

Line ministries/departments will continue to be responsible for the implementation of the skill development programmes, appropriately modifying them in line with the policies and strategies as decided by the Apex Committee into the Mission Mode.

Action by State Governments

- Transform Employment Exchanges to act as Career Counseling Centre
- Upgrade and strengthen State Council of Vocational Training
- Distance government and allow greater institutional autonomy. Maintain an arm's length relationship. Effectively delegate powers to local management of institutes
- Modernize the existing ITIs, etc. with better funding and enhancing the effectiveness of on-going programmes
- To cope with enhanced activities, existing vacancies in all training institutes must be filled
- Revamp the Institute Management Committee and ensure genuine PPP.
- Draw up plan for strengthening existing infrastructure (short-term, medium-term and long-term)
- Personnel Policy to ensure accountability and outcomes.

State Skill Development Missions

The State Governments may establish State-level missions to gear skill development activities in the Mission mode, with appropriate structures. Departments of the State Governments having skill development programmes will be required to reorient their skill development strategies and programmes in line with the central objectives. The Non-profit National Skill Development Corporation/Trust

The National Skill Development Corporation will be set up as a non-profit company under the Companies Act with appropriate governance structure (board of directors being drawn from the outstanding professionals/ experts). The head of the corporation will be a person of eminence/ reputed professional in the field of Skill Development. The Chairperson may also be the Chief Executive Officer of the Corporation, in which case S/he shall be known as the Chairperson-cum-Chief Executive Officer. The National Skill Development Corporation will be set up with as Government Equity with a view to obtaining about Rs 15000 crore as capital from governments, the public and private sector, and bilateral and multilateral sources for the promotion of skill development. The Corporation will be a public private partnership on skill development conceived as a nonprofit Corporation. It will make periodic as well as an annual report of its plans and activities and put them in the public domain. There may also be a National Skill Development Trust which can receive funds to be managed by the National Skill Development Corporation. The corporation/trust will be a flexible institutional arrangement to be able to deliver on jobs required by the market, related to its skill deficit, through training programmes operated or partnered by it.

Conclusion

The initiatives described above involving both the States and the Centre, often with private partnership will lead to the establishment of a credible, trustworthy and reliable training, testing and certification edifice linked to global standards and responsive to the needs of the ultimate consumers of skill. With an estimated 58.6 million new jobs in the domestic economy and about 45 million jobs in the international economy inviting skilled personnel for quality jobs beckoning the Indian youth, the government and private sector will act in a concerted manner so that these opportunities materialize and operate as an employability guarantee.

ANNEXURE 5.1.1

Current Attendance Rates in Educational Institutions per 1000 Persons of Different Age Groups during 2004–05

Category of persons	*Age groups* 5–14	15–19	20–24	25–29	5–29	0–29
(1)	(2)	(3)	(4)	(5)	(6)	(7)
Rural						
Male	835	471	114	16	532	450
Female	767	333	45	11	436	370
Person	803	407	79	13	485	412
Urban						
Male	890	593	232	40	541	484
Female	879	571	164	21	519	465
Person	885	583	200	31	530	475
Rural + Urban						
Male	847	504	151	23	534	459
Female	792	396	77	14	456	393
Person	821	454	114	18	497	427

Source : NSS Report 517, Table 6.

ANNEXURE 5.1.2

Labour Force Participation Rates by Age, Area, and Sex, during 2004–05 (CDS)

Age Group	*Rural Areas*		*Urban Areas*	
	Male	*Female*	*Male*	*Female*
(1)	(2)	(3)	(4)	(5)
5–9	2	2	3	2
10–14	58	52	50	28
15–19	486	231	370	116
20–24	853	295	755	216
25–29	950	365	944	221
30–34	959	413	979	256
35–39	963	467	975	288
40–44	954	457	973	266
45–49	949	459	962	229
50–54	928	405	926	225
55–59	890	376	802	190
60+	599	186	348	84
All age	531	237	561	150

Source : Report No. 515 (61/10/1)—NSS 61st Round (July 2004–June 2005).

The assistance, which was entirely a loan from the Centre in the beginning, was modified by inclusion of a grant component with effect from 2004-05. AIBP guidelines were further modified in December 2006 to provide enhanced assistance at 90 per cent of the project cost as grant to special category States, Drought Prone Area Programme (DPAP) States/tribal areas/ flood-prone areas and Koraput-BalangirKalahandi (KBK) districts of Orissa. Under the AIBP, Rs 34,783.7823 crore of Central Loan Assistance (CLA)/grant has been released up to March 31, 2009. An additional irrigation potential of 54.858 lakh ha has been created under the AIBP up to March 2009. As on March 31, 2009, 268 projects have been covered under the AIBP and 109 completed.

Rainfail and Reservoir Storage

Rainfall

Rainfall greatly influences crop production and productivity in a substantial way. More than 75 per cent of annual rainfall is received during the southwest monsoon season (June-September). During winter (January-February) of 2009, the country as a whole received 46 per cent less rainfall than the LPA. In the pre-monsoon period of 2009 (March-May), rainfall was 32 per cent below the LPA. During the south-west monsoon season of 2009, the country as a whole received 23 per cent less rainfall than the LPA. Central India, north-east India, north-west India and the southern peninsula experienced 20 per cent, 27 per cent, 36 per cent and 4 per cent deficient rainfall respectively. At district level, 9 per cent of districts received excess rainfall, 32 per cent normal rainfall, 51 per cent deficient rainfall and 8 per cent scanty rainfall. South-west monsoon (JuneSeptember, 2009) rainfall for the country as a whole and the four broad geographical regions.

Out of 36 subdivisions, 23 recorded deficient rainfall during the south-west monsoon in 2009. Out of the remaining 13 subdivisions, only three recorded excess rainfall and the remaining 10 normal rainfall. Out of 526 meteorological districts for which data are available, 215 (41 per cent) received excess/ normal rainfall and the remaining 311 (59 per cent) received deficient/ scanty rainfall during the season.

During the post-monsoon season (OctoberDecember) of 2009, the country as a whole has received 8 per cent above normal rainfall. Reservoir storage status

Table 33 : South-West Monson season (June to September 2009) rainfall

Region	*Actual(mm)*	*(LPA (mm)*	*Actual per cent of LPA*	*Coefficient of variation C/V per cent of LPA*
All-India	689.9	892.5	77	10
North-west India	392.1	611.7	64	19
Central India	795.4	995.1	80	14
Southern Peninsula	692.9	722.5	96	15
North-east India	1,037.7	1,427.3	73	8

Source: Indian Meteorological Department.

The total designed storage capacity at full reservoir level (FRL) of 81 major reservoirs in the country monitored by the Central Water Commission (CWC) is 151.77 billion cubic metres (BCM). At the end of monsoon 2009, the total water availability in these reservoirs was 90.48 BCM which is less than the water availability of 113.74. South-west monsoon season (June to September 2009) rainfall Region Actual (mm) LPA (mm) Actual per Coefficient of variation (CV) cent of LPA per cent of LPA the monsoon in 2008 and the 100.95 BCM which is the average of the last 10 years.

Table 34 : Monsoon performance 2001 to 2009 (June-September)

Year	*Number of meteorological subdivisions*			*Percentage of districts with normal/ excess rainfall*	*Percentage of LPA rainfall for the country as a whole*
	Normal	*Excess*	*Deficient/ scanty*		
2001	29	1	5	67	92
2002	14	1	21	39	81
2003	26	7	3	77	102
2004	23	0	13	56	86
2005	23	9	4	72	99
2006	20	6	10	60	99
2007	17	13	6	72	105
2008	30	2	4	76	98
2009	10	3	23	41	77

Source : India Meteorological Department.

Note : Excess= +20 per cent or more of LPA; Normal=+19 per cent to –19 per cent of LPA; Deficient= -20 per cent to –59 per cent of LPA; Scanty= -60 per cent to –99 per cent of LPA.

Table 35 : Reservoir storage (at the end of the monsoon season)

Item	*2009*		*2008*		*Average of last 10 years*	
	Storage in BCM	*% of FRL*	*Storage in BCM*	*% of FRL*	*Storage in BCM*	*% of FRL*
At the beginning of the monsoon season(as on June 4, 2009)	17.50	12	29.24	19	21.02	14
At the end of the monsoon season (as on October 1, 2009)	90.48	60	113.74	75	100.95 67	
Increase in Storage	72.98	48	84.50	56	79.93	53

Source : Central Water Commission

Price Policy for Agricultural produce

The Government's price policy for agricultural commodities seeks to ensure remunerative prices to the growers for their produce with a view to encouraging higher investment and production, and to safeguard the interests of consumers by making supplies available at reasonable prices. The price policy also seeks to evolve a balanced and integrated price structure in the perspective of the overall needs of the economy. Towards this end, the Government announces minimum support prices (MSPs) each season for major agricultural commodities and organizes purchase operations through public and cooperative agencies. The designated Central nodal agencies intervene in the market to undertake procurement operations with the objective of ensuring that market prices do not fall below the MSPs fixed by the Government.

The Government decides the support prices for various agricultural commodities after taking into account the recommendations of the Commission for Agricultural Costs and Prices (CACP), the views of State Governments and Central Ministries as well as such other relevant factors as considered important for fixation of support prices. The Government has fixed the

MSPs of 2009-10 kharif and rabi crops. The MSPs for paddy (common) and paddy (Grade A) have been raised by Rs 100 per quintal and fixed at Rs 950 per quintal and Rs 980 per quintal respectively. An incentive bonus of Rs 50 per quintal is also payable over and above the MSP of paddy. The MSP of arhar (tur) has been raised over the 2008-09 level by Rs 300 per quintal and fixed at Rs 2,300 per quintal while that of moong has been raised by Rs 240 per quintal and fixed at Rs 2,760 per quintal. The MSP of sesamum has been fixed at Rs 2,850 per quintal, raising it by Rs 100 per quintal. The MSPs of other kharif crops have been retained at their 2008-09 levels. The MSP of wheat has been raised to Rs 1,100 per quintal from Rs 1,080 per quintal and of barley to Rs 750 per quintal from Rs 680 per quintal. The MSPs of gram and safflower have been raised by Rs 30 per quintal each. The MSPs of masur and rapeseed/mustard have been retained at their previous year's levels of Rs 1,870 per quintal and Rs1,830 per quintal respectively.

Table 36 : Minimum support prices

(Rs. per quintal)

Commodity	*MSP 2009-10 (crop year)*	*Commodity*	*MSP 2009-10 (crop year)*
Kharif crops		*Rabi crops*	
Paddy (common)	950 + Rs. 50 per quintal bonus	Wheat	1,100
Paddy (Gr.A)	980 + Rs. 50 per quintal bonus	Gram	1,760
Jowar (Malindi)	860	Masur (lentil)	1,870
Maize	840	Rapeseed/mustard	1,830
Arhar (Tur)	2,300	Barley	750
Moong	2,760	**Other crops**	
Cotton (F-414/H-777/J34)	2,500a	Sugarcane	129.84b
Groundnut in shell	2,100		

Source : Department of Agriculture & Cooperation.

Notes : a staple length (mm) of 24.5-25.5 and Micronaire value of 4.3-5.1;
b Fair and Remunerative Price.

Price Support Scheme (PSS)

The Department of Agriculture & Cooperation is implementing the Price Support Scheme (PSS) for procurement of oilseeds and pulses through the National Agricultural Cooperative Marketing Federation of India Limited (NAFED), which is the Central nodal agency, at the MSP declared by the Government. NAFED is also the Central agency for procurement of cotton under the PSS in addition to the Cotton Corporation of India (CCI). NAFED undertakes procurement of oilseeds, pulses and cotton under the PSS as and when prices fall below the MSP. Procurement under the PSS is continued till prices stabilize at or above the MSP.

During 2009-10 (up to January 4, 2010) NAFED has procured 64,802 metric tonnes of various oilseeds costing Rs 278.07 crore under the PSS.

Market Intervention Scheme (MIS)

8.41 The Department of Agriculture & Cooperation implements the MIS on the request of State/Union

Table 37 : Procurement made by NAFED under the PSS during 2009-10

(up to January 4, 2010)

Sl. Commodity	*Crop season No.*	*MSP (Rs per quintal)*	*Quantity procured (in metric tonnes)*	*Value (in Rs lakh)*
1. Ball Copra	Season -2009	4,700	1,250	638.38
2. Milling Copra	Season -2009	4,450	47,916	23,200.92
3. AP Copra	Season -2009	3,900 510	219.30	
4. Cotton	Kharif-2009-10	2,850 & 3,000	1,408	405.37
5. Sunflower Seed	Kharif -2009-10	2,215	13,718	3,343.08
TOTAL			**64,802**	**27,807.05**

Source : Department of Agriculture & Cooperation.

Territory (UT) Governments for procurement of agricultural and horticultural commodities that are generally perishable in nature and not covered under the PSS. The MIS is implemented in order to protect the growers of these commodities from having to make distress sales. In the event of a bumper crop and glut in the market, prices tend to fall below economic levels/cost of production. Procurement under the MIS is made by NAFED as the Central agency and by State-designated agencies.

During 2009-10, the rates of most of the horticultural crops ruled to the benefit of growers. Thus only a couple of proposals were received, one from the Government of Karnataka for procurement of arecanut and another from the Government of Mizoram for procurement of passion fruit.

Progress of Agriculture-Sector Schemes/Programmes

Agriculture being a state subject, State Governments have an important role and responsibility for increasing agriculture production, enhancing productivity and exploring the vast untapped potential of the sector. Simultaneously, the Central Government must supplement the efforts of State Governments and a number of Centrally sponsored and Central-sector schemes are being implemented for the enhancement of agricultural production and productivity in the country, and to increase the income of the farming community.

(i) Macro Management

The Macro Management of Agriculture Scheme (MMA) was formulated in 2000-01, by bringing together under one umbrella 27 Centrally sponsored schemes relating to cooperatives, crop production programmes, watershed development programmes, horticulture, fertilizer, mechanization and seeds. The Scheme has been revised during 2008-09 to improve its efficacy in supplementing/ complementing the efforts of the States towards enhancement of agricultural production and productivity. The role of the Scheme has been redefined to avoid overlapping and duplication of efforts and to make it more relevant to the present agricultural scenario in the States in order to achieve the basic objective of food security and to improve the livelihood system for rural masses. The Revised MMA comprises 10 sub-schemes relating to crop production and natural resource management. Some of the salient features of the revised Scheme are:

- the practice of allocating funds to States/UTs on historical basis has been replaced by new allocation criteria based on gross cropped area and area under small and marginal holdings;
- assistance is provided to the States/UTs as 100 per cent grant;

- the subsidy structure has been rationalized to make the pattern of subsidy uniform under all the schemes implemented by the Department of Agriculture & Cooperation;
- the revised subsidy norms indicate the maximum permissible limit of assistance. States may either retain existing norms, or increase them to a reasonable level provided that the norms do not exceed the revised upper limits specified;
- two new components have been added, namely (a) pulses and oilseeds crop production programmes for areas not covered under the Integrated Scheme of Oilseeds, Pulses, Oil palm and Maize (ISOPOM) and (b) Reclamation of Acidic Soil along with the existing component of Reclamation of Alkali Soil;
- the permissible ceiling for new initiatives has been increased from the existing 10 per cent to 20 per cent of the allocation;
- at least 33 per cent of the funds is required to be earmarked for small, marginal and women farmers;
- active participation of all tiers of the Panchayati Raj institutions (PRIs) would have to be ensured in the implementation of the Revised MMA including review, monitoring and evaluation at district/sub-district level.

(ii) National Food Security Mission (NFSM)

With a view to enhancing the production of rice, wheat and pulses by 10 million tonnes, 8 million tonnes and 2 million tonnes respectively by the end of the Eleventh Plan, the Centrally sponsored NFSM has been launched from the rabi 2007-08 season. The three major components of the Mission are NFSM-rice, NFSM-wheat and NFSM-pulses. The Mission aims to increase production through area expansion and productivity enhancement; restore soil fertility and productivity; create employment opportunities; and enhance the farm-level economy to restore confidence of farmers. The NFSM is presently being implemented in 312 identified districts of 17 States of the country.

Focused and target-oriented technological intervention under the NFSM has made significant impact since its inception, reflected in the increase in production of rice and wheat in 2008-09.

(iii) Rashtriya Krishi Vikas Yojana (RKVY)

The RKVY, a flagship scheme of the Government in the agriculture and allied sectors was launched in August 2007 to reorient current agricultural development strategies to meet the needs of farmers and rejuvenate the agricultural sector so as to achieve 4 per cent annual growth during the Eleventh Five Year Plan. The scheme has an envisaged outlay of Rs 25,000 crore for the Plan period in the form of Additional Central Assistance (ACA). Funds to the tune of Rs 4,133.69 crore were released to the States/UTs during 2007-08 and 2008-09. For the current year, a sum of Rs 4,100.00 crore has been allocated of which Rs 3,243.76 crore has been released to the States by December 31, 2009. Up to 83 per cent and 85.95 per cent of the allocations for 2007-08 and 2008-09 respectively have been utilized by the end of November 2009.

During 2008-09, the areas of focus in the agriculture sector were seeds, fertilizers, IPM testing laboratories, horticulture, farm mechanization, extension, crops, marketing and cooperatives. A welcome feature observed during 2008–09 was that States have stepped up activities in the animal husbandry, dairy and fisheries sectors. Further, about 25 per cent of the approved funds was earmarked for projects related to these allied sectors. Besides these, projects related to micro irrigation, agricultural research, watershed and others were also approved.

Apart from the RKVY, there are many other programmes and policies responsible for growth of agriculture and allied sectors in the States; however, the RKVY is expected to play a major role. The RKVY also incentivizes States to allocate more for agriculture and allied sectors in their plans. The States have indeed stepped up allocation to agriculture and allied sectors. Allocation to agriculture and allied sectors was 5.11 per cent of total State Plan Expenditure in 2006-07. This has gone up to 5.84 per cent in 2008-09 (revised estimates[RE]/Approved).

(iv) Isopom

The Ministry of Agriculture has restructured oilseeds, pulses, oil palm and maize development programmes into one Centrally Sponsored Integrated Scheme of Oilseeds, Pulses, Oil Palm and Maize which is being implemented in 14 major States for oilseeds and pulses, 15 States for maize and 8 States for oil palm. About 75-80 per cent area of pulses is already in the NFSM-Pulses districts under 14 States.

The Oil Palm Development Programme under ISOPOM is being implemented in the States of Andhra Pradesh, Karnataka, Tamil Nadu, Gujarat, Goa, Orissa, Kerala, Tripura, Assam and Mizoram. The year-wise targets and achievements for the period 2007-08, 2008-09 and 2009-10 in respect of area coverage under oil palm through implementation of the Oil Palm Development Programme are given in Table 38.

The area under maize cultivation is 81.80 lakh ha with production of 192.80 lakh tonnes in 2008-09. About 90 per cent of the maize cultivated in kharif is rainfed. Maize is cultivated mainly for food, fodder, feed and industrial use. Under ISOPOM, the Maize Development Programme is being implemented in 15 States, namely Andhra Pradesh, Bihar, Chhatisgarh, Himachal Pradesh, Jammu & Kashmir, Gujarat, Karnataka, Madhya Pradesh, Maharashtra, Orissa, Punjab, Rajasthan, Tamil Nadu, Uttar Pradesh and West Bengal.

(v) National Rainfed Area Authority (NRAA)

The Government of India has also constituted the NRAA to give focused attention to the problem of rainfed areas of the country. The Authority is an advisory, policymaking and monitoring body charged with examining guidelines in various existing schemes and in the formulation of new schemes including all externally aided projects in this area. Its mandate is wider than mere water conservation and covers all aspects of sustainable and holistic development of rainfed areas, including appropriate farming and livelihood systems approaches. It would also focus on issues pertaining to landless and marginal farmers, since they constitute the large majority of inhabitants of rainfed areas. The NRAA has formulated common guidelines for the Watershed Development Project and is in consultation with all the States for its implementation as per instructions contained in the guidelines.

Table 38 : Targets and achievements in area coverage under oil palm through implementation of the Oil Palm Development Programme

Year	*Target (ha)*	*Achievement (ha)*
2007-08	29,580	21,330
2008-09	31,500	26,178
2009-10	16,711	9,594
	(up to October 2009)	

Source : Department of Agriculture & Cooperation.

(vi) Drought Management

During the year 2009-10, drought/scarcity/ drought-like situation has been declared in 334 districts by 14 State Governments. The States have ready availability of funds under the Calamity Relief Fund (CRF) for taking immediate necessary measures in the wake of natural calamities including drought. For natural calamities of severe nature, the State Governments can seek additional assistance from the National Calamity Contingency Fund (NCCF), by submitting a detailed Memorandum with relevant details. Several steps were taken to mitigate the hardship being faced by the States due to the drought situation. Some of the important measures were:

- States were requested to prepare alternate plans for unsown/germination-failed areas with shortduration/alternate crops;
- The Diesel Subsidy Scheme was launched to provide supplementary protective irrigation to save the standing crops (50 per cent of the cost of the subsidy with cap of Rs7.50/ litre given by the States was borne by the Central Government);
- Use of Truthfully Labelled (TL) seeds, relaxation of age for seed varieties and distribution of mini kits were allowed under the NFSM, RKVY; z Area-specific approach was adopted to achieve higher production through provisioning of inputs like fertilizers, credit and pest control measures in areas with higher rainfall;
- Agricultural advisories for appropriate crop programmes were telecast/broadcast through the media for the benefit of farmers. Scientists from ICAR institutions, Krishi Vigyan Kendras (KVKs) as well as experts of the National Rainfed Area Authority (NRAA) helped the States in their efforts to counter the impact of deficit rainfall/ drought on agriculture;
- Zonal conferences and a Rabi Campaign Programme with the State Governments were held to enable formulation of an appropriate action plan for the rabi season;
- Funds were made available under Centrally sponsored programmes like the RKVY, NFSM, NHM, MMA and ISOPOM to enable taking up of an agricultural reconstruction programme.

Allied Sectors

Horticulture

Under the Technology Mission for Integrated Development of Horticulture in the North Eastern Region during 2008-09, an additional area of 1,48,071 lakh ha has been brought under different horticulturalAgriculture and Food Management 193 crops. Further, infrastructure facilities for improving production and productivity of crops such as model nurseries, community tanks, tube wells, greenhouses, model floriculture centres, mushroom units, vermi-compost units, training of farmers/ trainers, training of women and market infrastructure and processing units, which are project based, have also been created. Apart from introduction of improved production technology in traditional crops, a significant contribution of the Mission has been in the promotion of commercial cultivation of potential crops, namely citrus, fruits, banana, pineapple, strawberry, kiwi, apple, passion fruits; anthuriums, roses, liliums, orchids and other cut flowers; and high value vegetable crops. The most remarkable development under the scheme has been the expansion of area under specific crops in the States and in clusters which will facilitate easy marketing access in the future.

A proposal for implementation of a pilot project for Replanting and Rejuvenation of Coconut Gardens in Thiruvananthapuram, Kollam and Thrissur districts of Kerala and the Union Territory of Andaman & Nicobar Islands has been approved.

Micro Irrigation

A centrally sponsored scheme on micro irrigation (MI) was launched in January 2006 for promoting water-use efficiency by adopting drip and sprinkler irrigation. All States and Union Territories and all horticultural as well as agricultural crops are covered under the scheme. The National Committee on Plasticulture Applications in Horticulture (NCPAH) provides the required technical guidance in association with Precision Farming Development Centres (PFDCs) at 22 locations. The PRIs are involved in selecting the beneficiaries. Since its inception, about 10 lakh ha has been covered under drip and sprinkler irrigation and a sum of Rs 1425.23 crore has been released as Government of India share (40 per cent of the total cost) in the scheme.

National Bamboo Mission (NBM)

The Mission intends to establish 195 bamboo bazaars and 10 retail outlets (showrooms) in different metropolitan cities by the end of 2010-11, to promote marketing of bamboo and its products.

Rubber

India is the fourth largest producer of natural rubber (NR) with an 8.9 per cent share in world production in 2008. Productivity is further being improved through the Rubber Plantation Development Schemes in the Eleventh Five year Plan. The Schemes provide subsidy on planting, supply of critical inputs with price concession, assistance for soil and water conservation and generation and distribution of quality planting materials.

In 2008-09, the estimated export of NR was 46,926 tonnes against an import of 77,616 tonnes. The export of NR is promoted through Export Promotion Schemes, which include participation in international trade fairs, assistance to exporters to participate in trade fairs and, organizing buyer-seller meets.

Coffee

Among plantation crops, coffee has made significant contribution to the Indian economy during the last 50 years. Indian coffee has created a niche for itself in the international market, particularly Indian Robusta, which is highly sought after for its blending quality. Arabica coffee from India is also well received in the international market.

In India, coffee is cultivated in an area of around 3.94 lakh ha. The post-monsoon crop estimate for the 2009-10 season is estimated at 2.90 lakh tonnes comprising 0.95 lakh tonnes of Arabica and 1.95 lakh tonnes of Robusta. The current year's production is about 10.6 per cent more than the previous year's.194 Economic Survey 2009-10.

Animal Husbandry, Dairying and Fisheries

The livestock and fisheries sector contributed over 4.07 per cent of the total GDP during 2008-09 and about 26.84 per cent value of output from total agriculture and allied activities. The Eleventh Five Year Plan envisages an overall growth of 6-7 per cent per annum for the sector. In 2008-09, this sector contributed 108.5 million tonnes of milk, 55.6 billion eggs, 42.7 million kg wool and 3.8 million tonnes of meat. The 17th Livestock Census (2003) has placed the total livestock population at 485 million and total of poultry birds at 489 million. The 18th Livestock Census has been conducted throughout the country with the reference date of October 15, 2007, results of which are awaited.

India ranks first in world milk production, its production having increased from 17 million tonnes in 1950-51 to 108.5 million tonnes by 2008-09. The per capita availability of milk has increased from 112 grams per day in 1968-69 to 258 grams per day in 2008-09, but is still low compared to the world average of 265 grams per day (Table 39). About 80 per cent of milk produced in the country is handled in the unorganized sector and the remaining 20 per cent is equally shared by cooperatives and private dairies. Over 1.33 lakh village-level dairy cooperative societies, spread over 265 districts in the country, collect about 25.1 million litres of milk per day and market about 20 million litres. The efforts of the Government in the dairy sector are concentrated in promotion of dairy activities in non-Operation Flood areas with emphasis on building cooperative infrastructure, revitalization of sick dairy cooperatives and federations and creation of infrastructure in the States.

A major programme for genetic improvement of cattle and buffaloes named the National Project for Cattle and Buffalo Breeding (NPCBB) was launched in October 2000 to be implemented over a period of 10 years in two phases of five years each with an allocation of Rs 402 crore and Rs 775.9 crore respectively. The NPCBB envisages genetic upgradation and development of indigenous breeds on priority basis. At present, 28 states and one UT are participating in the project. Financial assistance to the tune of Rs 485.73 crore was released to these states up to 2008-09. During the current financial year, Rs 93.31 crore has been released under the scheme to the implementing agencies till December 2009.

Livestock insurance

A Centrally sponsored scheme for livestock insurance is being implemented in all the States with the twin objectives of providing a protection mechanism to farmers and cattle rearers against loss of their animals due to death and to demonstrate the benefit of livestock insurance to the people. The scheme benefits farmers (large, small and marginal) and cattle rearers having indigenous/crossbred milch cattle and buffaloes. In 2009-10, Rs 23.28 crore has been released up to December 2009 and 13.16 lakh animals have been insured up to 2008-2009. The scheme has been extended from 100 districts to 300 districts from December 2009, covering all States.

Table 39 : Production and per capita availability of milk

Year	*Per capita (grams/day)*	*Milk production (MT)*
1990-91	176	53.9
2000-01	220	80.6
2005-06	241	97.1
2006-07	246	100.9
2007-08	252	104.8
2008-09	258	108.5

Source : Department of Animal Husbandry and Dairying.

Poultry

Poultry continues to play an important role in providing livelihood support and food security, especially to the rural population. India produces more than 55.6 billion eggs per year, with per capita availability of 47 eggs per annum. As per the estimate provided by the Food and Agriculture Organization (FAO) for 2008, the annual chicken meat production in India is around 2.49 million tonnes. The value of exports was around Rs 422 crore during 2008-09. Eggs and poultry are among the cheaper source of animal protein. During 2009-10, a new Centrally sponsored Poultry Development Scheme with an outlay of Rs 150 crore was launched. The scheme, through its Rural Backyard Poultry Development component is expected to cover below poverty line (BPL) sections of the society to help them gain supplementary income and nutritional support. In order to encourage entrepreneurship skills of individuals, a Poultry Venture Capital Fund is also being implemented covering various poultry activities.

Animal wealth in India has increased manifold and animal husbandry practices have also changed to a great extent. With increased trade activity, the chances of ingress of exotic diseases into the country have also increased. With improvement in the quality of livestock through launching of extensive cross-breeding programmes, the susceptibility of this livestock to various diseases, including exotic diseases, has increased. To ensure maintenance of disease-free status and compliance with the standards laid down by the World Animal Health Organization, major animal health schemes and programmes have been initiated. Further, for control of major livestock and poultry diseases, the Government of India provides financial assistance to States/UTs in their efforts to prevent, control and contain animal diseases and also to strengthen veterinary services including reporting of animal diseases. All avian influenza outbreaks reported were effectively controlled and the country was free from avian influenza in October 2009. Control and containment operations for the recent outbreak reported on January 14, 2010 in Khargram block of West Bengal are in full swing.

Fisheries

Fish production increased from 7.1 million tonnes in 2007-08 to 7.6 million tonnes in 2008-09. Fishing, aquaculture and allied activities are reported to have provided livelihood to over 14 million persons in 2006-07 apart from being a major foreign exchange earner.

Table 40 : Production and export of fish

	Fish production (million tonnes)			*Export of marine products*	
Year	*Marine*	*Inland*	*Total*	*Qty ('000 tonnes)*	*Value (Rs crore)*
1990-91	2.3	1.5	3.8	140	893
2000-01	2.8	2.8	5.6	503	6,288
2003-04	3.0	3.4	6.4	412	6,087
2004-05	2.8	3.5	6.3	482	6,460
2005-06	2.8	3.8	6.6	551	7,019
2006-07	3.0	3.8	6.8	612	8,363
2007-08	2.9	4.2	7.1	541	7,620
2008-09	2.9	4.7	7.6	603	8,608

Source : Department of Animal Husbandry & Dairying, 2009-10

Feed and Fodder

Adequate availability of feed and fodder for livestock is very vital for increasing milk production and sustaining the ongoing genetic improvement programme. It is estimated that there is green fodder shortage of about 34 per cent in the country. To increase the availability of fodder, the Department of Animal Husbandry & Dairying is implementing a Centrally sponsored Fodder Development Scheme throughout the country to supplement the efforts of the

States. Financial assistance to the tune of Rs 719.76 lakh (up to December 2009) has been provided to the States during 2009-10. A Central Minikit Testing Programme is also being implemented under which minikits of latest high-yielding fodder varieties are distributed free of cost to farmers for their popularization. During the current year (2009-10) 9.23 lakh minikits have been allotted to the States for distribution to farmers.

Credit and Insurance

Agricultural Credit

In order to provide adequate and timely credit support from the banking system to farmers for their cultivation needs, including purchase of all inputs, in a flexible and cost-effective manner, the Kisan Credit Card Scheme (KCC) was introduced in August 1998. About 878.30 lakh KCCs have been issued up to November 2009. The Scheme includes a reasonable component of consumption credit and investment credit within the overall credit limit sanctioned.

From kharif 2006-07, farmers have been receiving crop loans up to a principal amount of Rs 3 lakh, at 7 per cent rate of interest. Additional subvention of 1 per cent will be paid from the current year, as incentive to those farmers who repay shortterm crop loans on schedule resulting in bringing down the rate of interest to 6 per cent per annum.

In January 2006, the Government announced a package for revival of short-term Rural Cooperative Credit involving financial assistance of Rs 13,596 crore. The National Agriculture and Rural Development Bank (NABARD) has been designated as the implementing agency for the purpose. States are required to sign memorandums of understanding (MoUs) with the Government of India and NABARD, committing to implementing the legal, institutional and other reforms as envisaged in the revival package. So far twenty-five States have executed MoUs with the Government of India and NABARD. This covers 96 per cent of the primary agricultural credit societies (PACS) and 96 per cent of the Central cooperative banks (CCBs) in the country. As on November 2009, Rs 7,051.75 crore has been released by NABARD as the Government of India share for recapitalization of 37,303 PACS.

Government is implementing a rehabilitation package for 31 suicide-prone districts in the States of Andhra Pradesh, Karnataka, Kerala and Maharashtra involving financial outlay of Rs 16978.69 crore. An amount of Rs16,953.04 crore has been released under this package till September 2009. For the state of Kerala, the Government is implementing separate packages for the development of the Kuttanad Wetland Eco-System and mitigation of agrarian distress in Idukki district with an outlay of Rs1,840.75 crore and Rs.764.45 crore respectively.

A debt waiver and debt relief scheme for farmers announced by the Government in the Union Budget 2008-09 is under implementation. Direct agricultural loans disbursed by scheduled commercial banks, regional rural banks and cooperative credit institutions up to March 31, 2007, overdue as on December 31, 2007 and which remained unpaid until February 29, 2008, are eligible for debt waiver or debt relief as the case may be. About 3.68 crore farmers have benefited from the scheme involving debt waiver and debt relief of Rs 65,318.33 crore.

Agricultural Insurance

The frequency and severity of droughts, floods, cyclones and erratic climatic changes accentuate uncertainty and risk in agricultural production and livestock breeding in India. The National Agricultural Insurance Scheme (NAIS) is being implemented since rabi 1999-2000, as part of the strategy for risk management in agriculture with the intention of providing financial support to farmers in the event of crop failure as a result of natural calamities, pests and diseases. The scheme is open to all the farmersloanee and non-loanee-irrespective of their size of holding. Loanee farmers are covered on compulsory basis in a notified area for notified crops. For nonloanee farmers, participation in the scheme is on voluntary basis. The scheme envisages coverage of all food crops, oilseeds and annual commercial/ horticultural crops, in respect of which past yield data are available for adequate number of years. The scheme is being implemented by 25 States and two Union Territories. During the period from rabi 1999-2000 to rabi 2008-09, 1,347 lakh farmers over an area of 2,109 lakh ha have been covered, insuring a sum of Rs 1,48,250 crore.

The pilot Weather Based Crop Insurance Scheme (WBCIS) is being implemented in 13 States to provide insurance protection to farmers against adverse weather incidences which are deemed to adversely impact crop production. During five crop seasons (from kharif 2007 to kharif 2009), about 21.77 lakh farmers have been covered under the pilot scheme and claims to the tune of about Rs 388 crore have been paid against a premium of about Rs 444 crore. 8.79 The Coconut Palm Insurance Scheme (CPIS) has been launched on pilot basis during 2009-10 in selected areas of Andhra Pradesh, Goa, Karnataka, Kerala, Maharashtra, Orissa and Tamil Nadu. The pilot scheme will continue during 2010-11. To benefit from the scheme, a farmer should have at least 10 healthy nut-bearing palms in the age group 4 to 60 years in contiguous area/plots and to have been enrolled by the State Agriculture/Horticulture Department or Coconut Development Board (CDB) or any other such agency under a rehabilitation/ development/expansion scheme. The Agriculture Insurance Company of India (AIC) which is implementing the scheme is responsible for making payment of all claims within a specified period. The CDB administers the scheme.

Marketing and Extension

Agricultural Marketing

Organized marketing of agricultural commodities has been promoted in the country through a network of regulated markets. Most of the State and Union Territory Governments have enacted legislations (Agriculture Produce Marketing Committee Act) to provide for regulation of agricultural produce markets. There are 7,139 regulated markets in the country as on March 31, 2009. The countryAgriculture and Food Management 197 has 20,868 rural periodical markets, about 15 per cent of which function under the ambit of regulation. The advent of regulated markets has helped mitigate the market handicaps of producers/sellers at the wholesale assembling level. But rural periodic markets in general and tribal markets in particular have remained outside the developmental ambit of the APMC Act.

The Ministry of Agriculture has formulated a Model Law on agricultural marketing for guidance of and adoption by State Governments. The legislation provides for establishment of private markets/yards, direct purchase centres, consumers'/farmers' markets for direct sale and promotion of public-private partnership in the management and development of agricultural markets in the country. Provision has also been made in the law for constitution of State Agricultural Produce Marketing Standards Bureaus for promotion of grading, standardization and quality certification of agricultural produce. This would facilitate pledge financing, direct purchasing, forward/ futures trading and exports. Sixteen States/UTs have amended their APMC Acts and the remaining States are in the process of doing so (Table 41). APMC Model Rules based on the Model Law are under formulation in consultation with States. Extension reforms

The Government supports transfer of agricultural technologies and information to the farming community through various initiatives. The Support to State Extension Programmes for the Extension Reforms scheme launched in 2005-06, aims to make the extension system farmer driven and farmer accountable by way of new institutional arrangements for technology dissemination in the form of an Agricultural Technology Management Agency (ATMA) at district level. The ATMA has active participation of farmers/farmer groups, nongovernmental organizations (NGOs), KVKs, PRIs and other stakeholders operating at district level and below. Up to January 2010, 595districts-level ATMAs have been established. Gender concerns are being mainstreamed by mandating that 30 per cent of resources on programmes and activities are allocated for women farmers and extension functionaries. Since inception, out of a total of 10.19 crore farmer beneficiaries, 25.80 lakh women farmers (25.34 per cent) have participated in various extension activities under the scheme.Further, the Mass Media Support to Agriculture scheme is focusing on the use of Doordarshan infrastructure for providing agriculture-related information and knowledge to the farming community. The other component of the mass media initiative is use of 96 FM transmitters of All India Radio (AIR) to broadcast area-specific agricultural programmes with 30-minute radio transmission in the evening, six days a week. The Kisan Call Centres scheme provides the farming community agricultural information through toll-free telephone lines. A country-wide common eleven digit number "1800-180-1551" has been allocated for the Kisan Call Centres . The Agri-clinic and Agri-business Centres Scheme launched in 2002 provides extension services to farmers through agriculture graduates on payment basis by setting up of economically viable self-employment ventures. NABARD monitors the credit support to Agri-clinics through commercial banks. Provision of a creditlinked back-ended subsidy at 25 per cent of the capital cost of the project funded through bank loan as well as full interest subsidy on the bank credit for the first two years has recently been approved under the scheme. The subsidy would be 33.33 per cent in respect of candidates belonging to Scheduled Castes (SC), Scheduled Tribes (ST), women and other disadvantaged sections and those from the north-eastern and hill States. Under the scheme, 19,854 unemployed agriculture graduates have been trained up to December 2009.

Table 41 : Progress of reforms in agricultural markets (APMC Act) as on 31.12.2009

Sl. No.	*Stage of Reforms*	*Name of State/ Union Territory*
1.	States/ UTs where reforms to the APMC Act have been undertaken as suggested.	Andhra Pradesh, Arunachal Pradesh, Assam, Chattisgarh, Goa, Gujarat, Himachal Pradesh, Jharkhand, Karnataka, Madhya Pradesh, Maharashtra, Nagaland, Orissa, Rajasthan, Sikkim, Tripura
2.	States/ UTs where APMC Act has been partially reformed) by amending the APMC Act/ resolution	(a) Direct Marketing: NCT of Delhi (b) Contract Farming: Haryana, Punjab and Chandigarh (c) Private Markets: Punjab and Chandigarh
3.	States/ UTs where there is no APMC Act and hence not requiring reforms	Kerala, Manipur, Bihar*, Andaman & Nicobar Islands, Dadra & Nagar Haveli, Daman & Diu and Lakshadweep
4.	States/ UTs where APMC Act already provides for the reforms Tamil Nadu	
5.	States/ UTs where administrative action has been initiated for reforms	Mizoram, Meghalaya, Haryana, Jammu & Kashmir, Uttarkhand, West Bengal, NCT of Delhi and Pondicherry

Note : *APMC Act has been repealed with effect from September 1, 2006.

Food Management

The main objectives of food management are procurement of foodgrains from farmers at remunerative prices, distribution of foodgrains to consumers, particularly the vulnerable sections of society, at affordable prices and maintenance of food buffers for food security and price stability. The instruments used are the MSP and central issue price (CIP). The nodal agency which undertakes procurement, distribution and storage of foodgrains is the Food Cororation of India (FCI). Procurement at MSP is open-ended, while distribution is governed by the scale of allocation and its offtake by the beneficiaries. The offtake of foodgrains is primarily under the targeted public distribution system (TPDS) and for other welfare schemes of the Government of India. Offtake of foodgrains under the TPDS has been increasing in the last five years and has gone up from 29.7 million tonnes in 2004-05 to 34.8 million tonnes in 2008-09.

Procurement of Foodgrains

Overall procurement of rice and wheat which was 35.8 million tonnes in 2006-07, increased marginally to 37.6 million tonnes in 2007-08. However, increased MSP along with various other steps taken by the Government has resulted in record wheat procurement of 22.69 million tonnes in 2008-09 25.38 million tonnes in 2009-10 (April to December). As regards rice, the procurement in 2008-09 was 32.8 million tonnes and 22.9 million tonnes in 2009-10 (April–December). The record procurement of rice and wheat during 2007-08, 2008-09 and 2009-10 (April December) has resulted in comfortable foodstock availability to meet the TPDS needs and buffer stocks norms.

Table 42 : Procurement and offtake of wheat and rice (million tonnes)

	2004-05	*2005-06*	*2006-07*	*2007-08*	*2008-09*	*April-Dec.*	
						2008-09	*2009-10*
Procurement of Wheat and Rice under the Central Pool							
Rice	24.0	26.7	26.3	26.3	32.8	22.1	22.9
Wheat	16.8	14.8	9.2	11.1	22.7	22.7	25.4
Total	41.6	42.4	35.8	37.6	55.5	44.8	48.3
Offtake of Wheat and Rice for the TPDS							
Rice	16.6	19.2	21.2	22.6	22.2	14.9	18.1
Wheat	13.1	12.2	10.4	10.9	12.6	8.1	14.4
Total (A)	29.7	31.4	31.6	33.5	34.8	23.0	32.4
BPL (Rice+Wheat)	17.5	15.6	14.2	15.1	15.7	10.5	12.4
APL (Rice+Wheat)	6.7	8.3	8.7	9.0	9.6	6.1	12.5
AAY (Rice+Wheat)	5.5	7.4	8.7	9.4	9.5	6.4	7.4
Offtake of Wheat and Rice for Other Schemes							
Welfare Scheme (B) 10.6	9.7	5.1	3.9	3.4	2.0	2.9	
Open sales/ Exports (C)	1.2	1.1	0.0	0.02	1.2	0.1	0.5
Total (A+B+C)	**41.5**	**42.1**	**36.7**	**37.4**	**39.5**	**25.1**	**35.8**

Source : Department of Food and Public Distribution.Agriculture and Food Management 1990.

As in earlier years, procurement of foodgrains by the FCI continues to be higher in the States of Punjab, Haryana, Uttar Pradesh and Andhra Pradesh. These four States accounted for nearly 69.7 per cent of the rice procured for the Central Pool in 2006-07, 69.46 per cent in 2007-08 and 67.47 per cent in 2008-09 (Table 42).

Punjab and Haryana which accounted for 91.1 per cent of procurement of wheat for the Central Pool in 2007-08, accounted for 66.88 per cent in 2008-09 and 69.53 per cent in 2009-10, indicating an increased share in procurement by other states (Table 8.20).

The overall procurement of coarse grains in the kharif marketing season (KMS) 2008-09 has increased to 13.75 lakh tonnes due to a substantial increase in MSPs of coarse grains in KMS 2008-09 (Table 42).

Decentralized Procurement Scheme (DCP)

A number of states have opted for implementation of the (DCP) introduced in 1997, under which foodgrains are procured and distributed by the State Governments themselves. Under this scheme, the designated States procure, store and issue foodgrains under the TPDS and welfare schemes of the Government of India. The difference between the economic cost fixed for the State and the CIPis passed on to the State Government as subsidy. The decentralized system of procurement has the objectives of covering more farmers under MSP operations, improving efficiency of the PDS, providing foodgrains varieties more suited to local tastes and reducing transportation costs.

Courtsey: Economic Survey 2009-10 Planning Commission GOI.

Chapter 11

VISION : 2015 (GOI)

Summary of Ministry of Processing Industry MFPI has been supporting a range of initiatives for growth of processing industry. The vision has been given here.

THE FOOD PROCESSING SECTOR – THE VISION : 2015 (GOI)

The contribution of agriculture to India's GDP at the time of Independence was 70% and it accounted for 85% of total employment. The share of agriculture in the country's GDP has been gradually declining since then. At present, the contribution of agriculture to GDP is about 25%, but it still engages about 70% of the population. The annual average rate of growth of agricultural GDP has also declined from around 3.5% during mideighties to mere 1.5 % during 2006-07.

It is estimated that if the country has to maintain a GDP growth rate of over 8%, the agricultural sector has to grow at the rate of at least 4%. The country has a huge potential for growth in agriculture with about 160 million hectares of arable land and diverse agro climatic conditions, suitable for cultivation of a wide variety of crops.

While the productivity needs a definite improvement it is increasingly becoming evident that only a vibrant food processing sector can lead to increasing farm gate prices and thus increasing income levels, reduction in wastages and increasing employment opportunities.

India currently produces about 50 million tonnes of fruits, which is about 9% of the world's production of fruits and 90 million tonnes of vegetables, which accounts for 11% of the world's vegetable production.

Though India has a strong raw material base, it has been unable to tap the potential for processing and value addition in perishables like fruits and vegetables. Only about 2 percent of the fruits and vegetables in India are processed, which is much lower when compared to countries like USA (65%), Philippines (78%) and China (23%).

Even, within the country, share of fruits and vegetables processed is much less when compared to other agricultural products such as milk (35%) and Marine Products (26%).

More importantly the lack of processing and storage of fruits and vegetables results in huge wastages estimated at about 35%, the value of which is approximately Rs. 33,000 crore annually.

A developed food processing industry would not only reduce the wastages, but would also increasingly fetch remunerative income to farmers which is another problem before the agriculture sector at present.

At present the food processing sector employs about 13 million people directly and about 35 million people indirectly. In 2004-05, food processing sector contributed about 14% of manufacturing GDP with a share of Rs 2,80,000 Crores. Of this, the unorganized sector accounted for more than 70% of production in terms of volume and 50 % in terms of value.

On the export front, India has 1.5% (INR 360 Bn. in 2003-04) share of global agricultural exports (approximately USD 522 Bn. or INR 24,000 Bn.), despite its leadership in agricultural production.

The meat poultry, dairy, fruits, vegetables and also taken into consideration because they may be integrated with fisheries and an important future plan for rural development.

Opportunities for Food Processing

Food products today are the single largest component of household consumption expenditure. The current food consumption in India is estimated at Rs 8,60,000 Crore. Processed food account for Rs.4,60,000 Crore and share of primary processed food (includes packed fruits and vegetables, packed milk etc) is at Rs. 2,80,000 Crore.

Changing age profile, Increase in income, Social changes (Increasing number of working women), Life style factors, Organized Retail outlets are all factors favouring the growth of the food processing sector.

Food processing sector generates significant employment. The multiplier effect of investment in food processing industry on employment generation is 2.5 times than in other industrial sectors, higher than any other sector.

Constraints

While the food-processing sector offers several opportunities, it faces constraints as well, such as:

- Low income and the high share of basic food in the household consumption.
- Socio cultural factors such as preference in India for freshly plucked/cooked food, variation in food habits across the country, easy availability of raw materials for cooking, preference for consumption of food at home etc.
- Low productivity, high wastage.
- Inadequate infrastructure for sorting, grading, packing etc. in addition to the high cost of raw material (at processor's level).
- APMC Act which restricts sourcing materials from farmers.
- Lack of a common policy on Contract farming.
- Lack of trained man power for various stages of processing, storage, marketing and branding.
- Lack of access to modern technology.
- Low share of sale of food products through organized retail, which are the usual drivers of quality, scale and integration.
- Access to Credit for farmers as well as small and medium food processors is a key issue. Over 75% rely on informal credit at very high interest rates leading to increase in cost of production affecting competitiveness.
- Inability to attract investment by large corporate houses who complain of unreliable sources of supply of raw material.
- Inability to induce investor confidence.

- Low inflow of Foreign Direct Investment, in spite of the permission for 100% FDI in the food processing sector (except in food retailing, alcoholic beverages and plantations)
- Inability of Government Schemes to have the desired impact on productivity, technology and market arrivals

Thrust Areas

The vision 2015 of the Government of India for the food processing sector aims at:

- Enhancing and stabilizing the income level of the farmers
- Providing choice to consumers in terms of greater variety and taste including traditional ethnic food
- Providing greater assurance about safety and quality of food to consumers
- Promoting a dynamic food processing industry
- Enhancing the competitiveness of food processing industry in both domestic as well as international markets
- Making the sector attractive for both domestic and foreign investors
- Achieving integration of the food processing infrastructure from farm to market
- Having a transparent and industry friendly regulatory regime
- Putting in place a transparent system of standards based on science

The specific targets would be to increase:

- The level of processing of perishables from 6% to 20%
- Value addition from 20% to 35%
- Share in global food trade from 1.5% to 3%, in the next 15 years

An estimated investment of Rs. 100,000 Crores is required to achieve the above Vision of which Rs. 45,000 Crores is expected to come from the private sector, Rs. 45,000 Crores from Financial Institutions and Rs. 10,000 Crore from Government.

Review of Tenth Plan Schemes

The MFPI has been implementing several schemes for the development of food processing in the country which are as follows:

- Scheme for Infrastructure Development
- Scheme for Technology Upgradation /Establishment /Modernization of Food Processing Industries
- Scheme for Quality Assurance, Codex Standards and Research & Development
- Scheme for Human Resource Development
- Scheme for Strengthening of Nodal Agencies
- Scheme for Backward and Forward Integration and other Promotional Activities

REFERENCE

Eleventh Five Year Plan, Report of Working Group on Food Processing Sector, MFPI, GOI.

Chapter 12

TWELFTH PLAN : 2012-17 - FEW IMPORTANT ISSUES

The plan of the Government of India has been given so readers will work accordingly to achieve the target and will make it success.

TWELFTH PLAN 2012-17 BY PLANNING COMMISSION, GOVERNMENT OF INDIA

The variation in performance across States suggests that State-level responses and implementation play a very significant role in determining agricultural performance. However, to the extent that available technology limits potential growth, it will be difficult to maintain high growth rates where productivity has increased close to potential levels. This is relevant because the Eleventh Plan strategy gave much greater flexibility to States and focused more on yield gaps within existing technology, rather than emphasizing new technologies and supporting these.

The growth acceleration since 2005 has therefore been much stronger in states with lower productivity and less irrigation. This suggests that the strategy may be correcting the past relative neglect which caused rain-fed farming, covering over 60 per cent of arable land, to perform well below potential. It is a matter of concern that the recent growth revival has been weak in areas with high land productivity, not only in relatively more irrigated states such as Punjab, Haryana, Uttar Pradesh and West Bengal that had green revolution success, but also in less irrigated states such as Kerala, Himachal Pradesh and Jammu & Kashmir where high productivity reflects a high-value cropping pattern based on horticulture. These States together contribute about 35 per cent of national agricultural output from 20 per cent of arable land, but none of them have been able to surpass growth rates achieved in the past. Even Gujarat, a low productivity state that sustained near 10 per cent growth for almost a decade through better water use and rapid adoption of Bt cotton hybrids, slowed down perceptibly in the Eleventh Plan as Bt adoption saturated and yields reached a plateau. Clearly, growth is more difficult to accelerate at higher productivity levels without new technology, particularly if past patterns of growth have taken a toll on natural resources.

OUTPUTS, INPUTS AND PRODUCTIVITY

The Eleventh Plan had made four conscious choices. First, with technology fatigue evident, it funded research better but emphasised on getting more from existing technology. Second, since one size does not fit all, it decentralised plan funds to encourage initiatives at State and lower levels. Third, aware of low public investment and food security needs, it increased Centre's spending on these, particularly in disadvantaged regions. Fourth, noting farmer distress, it tried to focus not just on production but also on farm incomes, stressing service delivery and suggesting encouragement of group activity with land and tenancy reforms put back on the agenda.

Compared to the original green revolution that built on the best, this strategy sought to deliver faster growth, that is, more inclusive, more stable and less concentrated spatially. Nonetheless, there is a wide demand for a 'second green revolution' with more irrigation and better crop-specific technologies, with some even claiming that Bt cotton has been the only recent success.

The Twelfth Plan accepts the proposition that a greater technical thrust is needed, and the strategy for agriculture should take this into account In order to provide a snapshot of the Eleventh Plan performance and give indication of what the Twelfth Plan should do differently, long-run data on growth of output by sub-sector and also rates of growth of input use and productivity. Since performance is almost invariably discussed in the context of well-defined policy periods, those chosen for this table are same as in the Eleventh Plan document:

(*i*) Pre-Green Revolution (1951–52 to 1967–68);

(*ii*) Green Revolution proper (1968–69 to 1980–81);

(*iii*) Wider technology coverage (1981–82 to 1990–91) when focus shifted from intensification of Green Revolution in best areas to its spread to new areas;

(*iv*) Early liberalization period (1991–92 to 1996–97) when relative prices became an additional focus, both because agriculture was expected to gain from reduced trade protection to industry and also with Minimum Support Prices (MSP) used for active growth promotion rather than just passive price support. The other three periods in the table are subsequent Plan periods;

(*v*) Ninth Plan (1997–98 to 2001–02);

(*vi*) Tenth Plan (2002–03 to 2006–07) and (*vii*) Eleventh Plan (2007–08 to 2011–12).

For each of these periods, the average of annual growth rates is presented for each variable chosen. As noted above, growth of agricultural GDP at 3.3 per cent was short of the 4 per cent target for agricultural GDP but was faster than that in the Tenth or the Ninth Plan, though lower than the period from 1981–82 to 1996–97. The growth rates for individual crops shown in Table are for gross value of output and not value added, but they present a valid basis for inter-period comparisons.

1. Growth of total value of output in agriculture proper (crops and livestock) during the Eleventh Plan averaged 3.8 per cent per year which was the highest among all seven periods considered.
2. Total non-horticulture crop output grew marginally faster than target (2.8 per cent against 2.7 per cent target) mainly because of foodgrains (3.1 per cent actual against 2.3 per cent target), oilseeds (4.5 per cent against 4 per cent) and fibres (10.7 per cent against 5 per cent).
3. Horticulture at 4.7 per cent was only marginally short of the 5 per cent target.
4. Growth of output from livestock (4.8 per cent) was again highest amongst all the periods considered but this performance, and even more, so for fishing (3.6 per cent), fell short of the ambitious 6 per cent target set for these two sub-sectors.
5. Growth of forestry was expectedly slower, pulling down the growth of total value of output in agriculture and allied to 3.6 per cent, but this too was the highest among all the seven periods considered.

Growth in intermediate inputs has accelerated steadily reaching 4.3 per cent per annum during the Eleventh Plan, which was much higher than growth of output and over twice the growth rate of intermediate input use during 1981–97. The more rapid growth in input use explains why despite the faster growth of the gross value of output during the Eleventh Plan at 3.6 per cent than in the period 1981–82 to 1996–97 (about 3.0 per cent), GDP in agriculture (which is a value added concept) grew more slowly. In other words, agricultural growth became more input intensive in the Eleventh Plan. This suggests the need to re-look policies relating to inputs, especially fertiliser and power. Policies towards input use need to distinguish between traditional inputs such as seed, feed and organic manure and modern inputs such as chemical fertiliser, pesticides and farm power. With low seed replacement, underfed farm animals and soils short of organic carbon, projections by working groups for the Twelfth Plan suggest that past growth of these traditional inputs should be improved upon. However, these working groups also project lower growth of 'modern' inputs than observed during the Eleventh Plan. For example, 2016–17 requirements of chemical fertiliser and farm power are placed at levels that imply annual growth for both fertilisers and 'modern' energy at about 4.5 per cent. These exceed corresponding the Eleventh Plan projections but are much less than the Eleventh Plan actual. Reduced fertilizer and fuel subsidies would be consistent with the desired moderation in trend of these inputs. Restraint is also needed on pesticides use which rose sharply in the Eleventh Plan after years of being subdued. In parallel with high growth of intermediate inputs, there was acceleration in growth of the net capital stock in agriculture and allied sectors during the Eleventh Plan. As shown in Table 12.4 (item IV), Net Fixed Capital Stock in agriculture expanded at 6.0 per cent per year, much faster than in the previous two Plans.

The public component of capital stock increased by 3.6 per cent while the private component increased at 7.5 per cent per year, both showing acceleration compared to the previous two Plans. However, public investment in agriculture, which was stepped up very substantially in the last three years of the Tenth Plan, stagnated in the Eleventh Plan. This was mainly because of a large shortfall in planned investment in irrigation. As a result a key part of the Eleventh Plan strategy to achieve 4 per cent agricultural growth which was to increase public investment in agriculture to 4 per cent of agricultural GDP and thereby achieve growth of public sector capital stock in agriculture at least equal to the required 4 per cent growth of total capital stock has not fructified. Clearly, to attain 4 per cent agricultural growth in the Twelfth Plan will require firmer commitment to ensure realisation of this unattained the Eleventh Plan objective. Private investment in agriculture has accelerated over the past three Plans. Private investment averaged 15.6 per cent of agricultural GDP in the first four years of the Eleventh Plan as against expected 12 per cent. The main driver of this was a large relative price shift in favour of agriculture, showing that farmers respond to price incentives. If calculated in current price terms rather than constant, private investment averaged 13 per cent of agricultural GDP—only slightly higher than expected. Nonetheless, total capital stock in agriculture grew more than expected.

While private investment in irrigation and water-saving devices did increase, the largest increase was in labour-saving mechanisation. This was a natural response to growing labour scarcity which is reflected in rising wages. Table also shows growth rates of the two other factors of production in agriculture: land and labour. Not unexpectedly, while capital stock has grown quite rapidly throughout, the other two factors have not. As far as labour is concerned, the measure shown is employment in agriculture by usual status estimates of the National Sample Survey (NSS), which is available almost annually since 1987–88 but requires interpolation for earlier years. Combined with Census data, these show continuous increase of agricultural employment till 1994, although at varying rates of growth and at a particularly sharp rise in early 1990s when there was slow-down in rural nonagricultural employment. Agricultural employment fluctuated in the next decade, but has clearly declined after 2004–05. NSS employment data for 2007–08 and 2009–10 show clear evidence of an accelerated shift of rural labourers to non-agricultural work, which in itself is not an undesirable development. For land, the measure shown is gross cropped area which, despite the loss of nearly 3 million hectares of arable land to non-agricultural uses since 1990–91, has increased in all periods excepting a slight dip in the Ninth Plan. This is because cropping intensity has increased almost continuously. However, cropped area growth which averaged 0.9 per cent per annum till 1990–91 has averaged only 0.2 per cent subsequently. The growth rates of partial productivity of land, labour and capital taking GDP agriculture and allied as numerator. Labour productivity growth has historically been low, averaging 2 per cent per annum or less except during 1981–90 when it reached 3 per cent. Labour productivity jumped to nearly 5 per cent during the Eleventh Plan. The accelerated shift of rural labour to non-agriculture caused real wages to rise at about 5 per cent annually between 2004–05 and 2009–10, according to the NSS, and latest reports of the Commission of Agricultural Costs and Prices (CACP) suggest even faster growth of real wages in the last three years of the Eleventh Plan at almost 8 per cent per year. The trend in real wages in 2011–12 prices, as estimated by CACP, is shown in Figure 12.2. Labour saving mechanisation, a significant contributor to the sharp increase of private investment in the Eleventh Plan period, was a natural response to tighter labour markets and rising wages. But, while mechanisation helped farmers to cope with labour scarcity, it exacerbated a decline in capital productivity. Private capital stock in agriculture has increased twice as fast as agricultural GDP since the Ninth Plan and, although mitigated by terms of trade gains and a debt write-off, continued investment with declining capital productivity may not be sustainable. While greater private investment in farming is desirable where it reflects both an ability to invest and a desire to increase farm productivity, the same phenomenon can become a source of distress if farmers keep investing to cope with shrinking natural resources, more frequent adverse weather and less assured labour supply, and do not get adequate returns for this investment.

The Eleventh Plan had tried to address this in two ways: first, increase public investment to lessen the private burden and add economies of scale; and second, rework architecture of the Plan spending on agriculture to make it more decentralised and flexible but also more coordinated locally to improve total productivity of private resources by better service delivery in all areas from extension to input supply and marketing. However, as noted earlier, public investment did not increase. And, although combined Plan expenditure of Centre and States in agriculture did increase from 1.9 percent of agricultural GDP in the Tenth Plan to 2.9 percent in the Eleventh, this was relatively small and left research, education and extension under-funded, leaving much to be desired in the quality of service delivery.

Nonetheless, growth of land productivity did increase significantly. Having climbed from about 1 per cent per annum before Green Revolution to over 3 per cent during 1991–97, land productivity growth had decelerated to below 2 per cent. This rebounded to over 3 per cent during the Eleventh Plan. Total factor productivity (TFP) improved during the Eleventh Plan. Individual factor productivity, weighted by a range of factor shares suggest that TFP growth during the Eleventh Plan was back to around 1980s level applying factor shares of 30 per cent land, 40 per cent labour and 30 per cent capital give the following averages of annual TFP growth: 0.7 per cent in pre- Green Revolution period, 0.8 per cent during Green Revolution period, 2.2 per cent during the wider coverage period, 1.8 per cent during early liberalisation, 1.4 per cent during the Ninth Plan, 0.6 per cent during the Tenth Plan and 2.0 per cent in the Eleventh Plan.

Although these estimates must be treated as tentative since data on factor shares is not robust, it does suggest that the deceleration of TFP in agriculture observed in the previous two Plans, which had caused widespread apprehension, may have been reversed in the Eleventh Plan. In other words, the Eleventh Plan architecture, with the Rashtriya Krishi Vikas Yojana (RKVY) as core, appears to have delivered despite adverse weather, a public investment shortfall and implementation gaps. The strategy of spreading known technology wider had paid.

SUB-SECTOR-WISE PERFORMANCE AND ISSUES

Crop Sector : In addition to above, two indicators worth highlighting in the crop sector are the pace and pattern of crop area diversification and trends in yields/hectare of important individual crops. There has been gradual but sustained shift in cropping pattern away from coarse cereals and pulses towards other crops over the last four decades. Area under coarse cereals had declined by 18 million hectares and that under pulses by nearly 2 million hectares from earlier peaks to end of the Tenth Plan. During the Eleventh Plan, there was further decline of 2 million hectares in area under coarse cereals but area under pulses reversed earlier decline to reach a new peak in 2010–11. Noting, that technology and price policy had neglected pulses earlier despite their importance as source of protein, special attention was given to pulses in both the National Food Security Mission (NFSM) and RKVY, the two major schemes launched during the Eleventh Plan. Cotton gained most area, followed by fruits and vegetables, with rice area steady, an increase in wheat area and decline in area under oilseeds and sugarcane.

Although area under coarse cereals and oilseeds declined during the Eleventh Plan, both these crop groups averaged over 4 per cent output growth. This was because growth of yields per hectare accelerated across almost all crop groups, especially those mainly rain-fed (Not only did coarse cereals and oilseeds yields increase faster during the Eleventh Plan than in any of the earlier periods, so did pulses yields. Apart from hybrids in case of maize, and to less extent in bajra, these yield increases came mainly from better seed quality, higher seed replacement and better practice rather than from new crop technology or more irrigation.

Yield growth of cotton, another largely rainfed crop, was also respectable although it was down sharply from a spectacular performance during the Tenth Plan following adoption of Bt hybrids. With more than 90 per cent of cotton area now under Bt hybrids, and cotton yields more than doubling over the last decade, there is no doubt either about general farmer acceptance or its being a clear case of technological transformation unlike other rainfed crops. But disagreements continue about the extent to which Bt contributed to this yield increase and on wisdom of India's total dependence on Bt hybrids rather than the Bt varieties used in the rest of the world. There are also legitimate complaints of non-availability of non-Bt seeds, for example in Vidharbha. Genetically modified organisms (GMOs) therefore remain controversial, as was evident in case of Bt Brinjal. Nonetheless, since significant breakthroughs in production technologies are required to cope with increasing stress, particularly for rainfed crops, it is necessary to remain abreast with latest advances in biotechnology. It is, therefore, time to put in place scientifically impeccable operational protocols and a regulatory mechanism to permit GMOs when they meet rigorous tests that can outweigh misgivings, while simultaneously noting that many feasible advances in biotechnology do not in fact involve GMOs.

Moreover, the Eleventh Plan experience is that continuous less-visible efforts by farmers to adapt and improve can be made effective. The NFSM, which aimed to reduce gaps between potential and actual yields, was designed to aid farmers in their own efforts by demonstrating and supporting a wide range of interventions. This seems to have worked. For example, growth in wheat yields nationally was negligible during the Ninth and the Tenth Plans but increased to 3 per cent in the Eleventh Plan. Even in Punjab, where it was believed that wheat yields had reached a plateau below 4.5 tonnes per hectare, yields increased steadily during the Eleventh Plan to reach 4.9 tonnes, accompanied by wider use of conservation practices such as laser levelling, zero tillage and raised beds. Rice yield growth was also higher in the Eleventh Plan than in any period after 1991, with Assam, Bihar, Chhattisgarh, East Uttar Pradesh and West Bengal contributing 80 per cent of this, again with growing awareness of conservation practices. For example, many States are now using RKVY to mainstream the System of Rice Intensification (SRI) that was not officially accepted till 2004 and was only small part of NFSM.

Livestock and Fishery

Livestock contributes 25 per cent of gross value added in the agriculture sector and provides self-employment to about 21 million people. Rapid growth of this sector can be even more egalitarian and inclusive than growth of the crop sector because those engaged in it are mainly small holders and the landless. Growth of livestock output averaged 4.8 per cent per annum during the Eleventh Plan recovering from an average of 3.6 per cent in the Ninth and the Tenth Plans.

Growth, of dairying, which is the main constituent of livestock sector though slightly higher than the 4 per cent averaged since 1990, was short of demand. With over 75 per cent of cattle located in rain-fed areas, the major issue is access to feed, fodder and drinking water which is becoming increasingly scarce. The problems of the sector are compounded by growing numbers of unproductive male cattle. Developing a strong fodder base needs intensive effort and innovation in institutional aspects of pasture protection and management and usufruct sharing. There is little concerted effort in this area at present as it is too fragmented across various departments to be able to provide the technical inputs, institutional designs and adequate investments to make a meaningful impact. Richer farmers with access to groundwater irrigation can grow irrigated fodder and increase herd size. Poorer livestock owners, dependent mainly on commons and agriculture residues, end up underfeeding the animals. This problem raises questions about the present breeding strategy that focuses almost exclusively on induction of breeds that are high yielding, but are much less tolerant to adverse conditions in extensive livestock systems.

These issues, which also affect owners of small ruminants, poultry and even those involved in inland fishery, came to the fore during the Eleventh Plan following the drought of 2009. The consequent high inflation in feed and fodder, that also led to high inflation in prices of livestock products, revealed a need for much greater coordination not only between agencies responsible for livestock and those responsible for crops that sustain livestock, but also with other policies, for example, trade policies that influence feed and livestock product prices. RKVY provided a window which cut across departments to allow States to focus on fodder shortages and restored growth of livestock output much quicker than in earlier droughts. Nonetheless, underlying problems remain, as does so called protein inflation. The Twelfth Plan must address these problems by involving dairy cooperatives in breed and feed issues, revisit breeding strategies and make fodder development higher priority in both animal husbandry and crop programmes.

India produces about 65 billion eggs annually and production growth has accelerated from around 4 per cent per annum during the 1990s to over 5 per cent during the Tenth and the Eleventh Plan. This acceleration has been achieved despite new challenges such as periodic outbreaks of avian influenza and the biofuels effect on international prices of maize, the main poultry feed, which has now transmit into the domestic economy. One reason for this vitality has been the growth of a large and vibrant commercial poultry sector with adequate economies of scale and fairly good backward and forward linkages. Besides eggs, this commercial poultry sector also produces over 2 million tonnes of broiler meat which is an increasing part of total meat production of about 5 million tonnes. Meat, with production growth at over 5.5 per cent per annum during the Eleventh Plan, is the fastest growing segment in the livestock sector.

The performance of the fisheries sub-sector has been impressive on the whole, with growth more than 5 per cent per annum during the 1980s and 1990s, but growth in this sub-sector has been decelerating since mid-1990s. The main reason for this has been stagnation of marine fishery, a phenomenon which is expected to continue. The major growth in fisheries in recent years has come from the inland fisheries, with particularly rapid development of brackish water aquaculture. This has been linked to prawn cultivation for export, although there is also strongly growing domestic demand for fresh water fish. Fish prices more than doubled during the Eleventh Plan, a higher inflation than either crops or any other livestock segment, despite a small acceleration in production growth compared to the Tenth Plan. A problem in this sector is that although a National Fisheries Development Board was set up, responsibilities are still not clearly defined between this and the Department of Animal Husbandry, Dairying and Fisheries. This has in particular meant an inability to realise the vast potential of inland fresh water fishery. Fish production can be enhanced 2 to 4 times in rain-fed water bodies, whether irrigation reservoirs, natural wetlands or ponds and tanks created by watershed development or Mahatma Gandhi National Rural Employment Guarantee Scheme (MGNREGS). If fully harnessed, these can secure over 6 per cent fishery growth in the Twelfth Plan.

EMERGING IMBALANCES

Although the discussion so far suggests that agricultural performance did improve during the Eleventh Plan, experience of the Eleventh Plan also points to emerging imbalances in agriculture which call for a long-term strategic reorientation.

Subsidies vs Public Investment

The Eleventh Plan document had highlighted that public investment in agriculture as per cent of agricultural GDP had halved between the 1980s and in the end of the Ninth Plan while, simultaneously, budgetary subsidies to agriculture had doubled as proportion of agricultural GDP. The tendency for subsidies to increase much faster than public investment was checked to some extent during the Tenth Plan, but it reappeared again during the Eleventh Plan. Budgetary subsidies to agriculture (excluding food subsidy, which should be treated as a consumer subsidy) increased from an average of 4.1 per cent of agricultural GDP during the Tenth Plan to average 8.2 per cent in the first four years of the Eleventh Plan. Actual subsidies to agriculture were higher in both periods since CSO books budgeted subsidy on domestic urea manufacture entirely to industry and because part of the power subsidy received by agriculture is not budgeted but borne by utilities. Compared to these numbers, public investment in agriculture averaged only about 3 per cent of agricultural GDP during both Plan periods. The imbalance between subsidy expenditure and expenditure on public investment raises the issue whether a shift away from subsidies and towards greater public investment would not be beneficial. The usual argument for reducing subsidies is that it will improve the fiscal deficit, but that is not the relevant point in this context, there is a need to shift from subsidies to public investment aimed at increasing land productivity on the grounds that this would produce better agricultural outcomes and would also be more inclusive. This is particularly important in the context of strategies for combating the effect of climate change where public investment in conservation and management of water resources will be crucial.

There are also other uses of resources in agriculture which could be promoted if agricultural subsidies are restrained. The Eleventh Plan document had pointed to trade-offs that subsidies might have with other non-Plan revenue expenditures, particularly staffing of essential farm support systems such as extension. Moreover, capacity and skill shortages have made upgrading agricultural universities an urgent need. The Eleventh Plan had aimed to increase spending on agricultural education and research from 0.6 to 1 per cent of agricultural GDP, but this remains less than 0.7 per cent—a large gap in a very important area that is miniscule in relation to subsidies.

Table 12.1 : Public Sector Capital Formation and Subsidies to Agriculture (Centre and States)

(in Rs. crore and as per cent to GDP from agriculture and allied at current prices)

	Public GCF Agriculture and Allied		*Budgetary Subsidies (CSO)*		*Food Subsidy*		*Total Fertiliser Subsidy*		*Subsidy on Indigenous Urea*		*All other Agriculture Subsidies*	
Tenth Plan												
2002–03	9,563	2.0	43,597	9.0	24,176	5.0	11,015	2.3	7,790	1.6	16,196	3.3
2003–04	12,218	2.2	43,765	8.0	25,181	4.6	11,847	2.2	8,521	1.6	15,258	2.8
2004–05	16,187	2.9	47,655	8.4	25,798	4.6	15,879	2.8	10,243	1.8	16,221	2.9
2005-06	20,739	3.3	51,065	8.0	23,077	3.6	18,460	2.9	10,653	1.7	20,181	3.2
2006–07	25,606	3.5	59,510	8.2	24,014	3.3	26,222	3.6	12,650	1.7	21,924	3.0
Eleventh Plan												
2007–08	27,638	3.3	85,698	10.2	31,328	3.7	32,490	3.9	12,950	1.5	34,830	4.2
2008–09	26,692	2.8	1,56,823	16.6	43,751	4.6	76,603	8.1	17,969	1.9	54,438	5.8
2009–10	33,237	3.1	1,39,248	12.9	58,443	5.4	61,264	5.7	17,580	1.6	37,121	3.4
2010–11	34,548	2.7	1,50,170	11.8	63,844	5.0	62,301	4.9	15,081	1.2	39,106	3.1

Note: Public sector agricultural GCF and GDP are from CSO, National Accounts Division; budgetary subsidies, are also from CSO and are based on the economic and purpose classification of Government expenditure. Food and Fertiliser subsidies are from budget documents of the Central Government. 'All other agriculture subsidies' in the table are defined as budgetary subsidies (CSO) *plus* subsidy on indigenous urea *minus* food subsidy. This is because CSO classifies food subsidy as subsidy to agriculture but classifies subsidies on indigenous urea as subsidy to industry.

Another, very important reason why subsidies should be rationalised and restrained is that some of these subsidies could actually be doing harm. A case for subsidies exists if there is clear evidence that some input is being underused. Conversely, when with there is clear evidence of overuse of a subsidised input, there is a case to reduce or even eliminate the subsidy. Today, there is clear evidence of overuse. Data from all over India, especially from the prime green revolution areas, show that high use of chemical fertilisers and power is causing excessive mining of other soil nutrients and of groundwater, and that this is also leading to loss of quality of both soil and water. There is of course about 20–25 per cent of the country's arable area, located largely in North-East, East and Central India, where use of these inputs is so low that further intensification is desirable *per se*. But with nearly 90 per cent of fertilisers and 95 per cent of farm electricity currently being used outside this area, there can be no doubt that the present subsidies are actually encouraging practices that need to be discouraged. Any proposal for reducing subsidies will be opposed by farmers on the grounds that output will fall if the subsidy cut reduces input use. This is true unless other investments are made simultaneously but such investments would indeed be facilitated by the resources released. Efforts were made in the Eleventh Plan to encourage more efficient practices without actually reducing the quantum of subsidy. For example, many States have undertaken separation of feeders so that electricity supply for agricultural use can be treated differently from that for rural non-agricultural use, and stricter scheduling imposed on the former while maintaining its lower price. Similarly, the Centre introduced a new scheme, the 'National Project on Management of Soil Health & Fertility' (NPMSH&F) to promote soil testing and issue of soil health cards to farmers, aimed particularly to spread awareness of micronutrient deficiencies resulting from excessive and unbalanced fertiliser use and to encourage balanced and judicious use of chemical fertilisers in conjunction with organic manures to maintain soil health and fertility. Moreover, in order to rationalise fertiliser subsidies, a nutrient-based subsidy (NBS) system was adopted to subsidise fertiliser products uniformly on basis of nutrient content, rather than set product-wise subsidies and separate maximum retail prices (MRPs) for each product. The objective was to reduce deadweight of the fertiliser control order, set nutrient specific subsidies that maintain desirable NPK balance, and evolve a subsidy protocol to encourage both development of new complex fertiliser products (including micronutrients) and more investment in the sector.

These initiatives have had some success in particular regions, but they do not as yet show up in national data in terms of higher additional output per unit additional use of these inputs. Moreover, NBS roll-out was seriously flawed since urea was kept out of its ambit. Urea prices remain controlled with only a 10 per cent rise at the time of adoption of the NBS in 2010. Meanwhile prices of decontrolled products doubled. The fixity of the urea price naturally worsened the NPK balance. Also, there has been very little product innovation. The subsidy bill has increased because resulting higher urea demand has been met entirely by imports at a unit subsidy twice that on domestic output, with little incentive to expand domestic capacity. The NBS as rolled out has been counterproductive because urea has not been included.

The fertiliser subsidy is now much higher than all other subsidies to agriculture put together. While this is partly because fertiliser consumption rose over 30 per cent during the Eleventh Plan, the main reason is that world prices of all fertilisers and feedstock have doubled since 2006. With world fertiliser prices very sensitive to demand from India, which is not only the world's largest importer of fertilisers but also dependent almost entirely on imports for feedstock, improving efficiency of fertiliser use must be a the Twelfth Plan focus, almost as important as the issue of water use efficiency taken up in another chapter.

A New Road Map for Fertiliser Policy

A broad idea of what is necessary is evident from a few key indicators about the price of urea, the most important and politically sensitive fertiliser in India. At the world level, urea prices had averaged about 80 per cent of world wheat price during the 25 years before 2005. Since then, they have been fluctuating wildly at much higher levels and world urea prices are now over 150 per cent of world wheat price. In comparison, the price of urea in India has been declining continuously in relation to wheat MSP—from over 150 per cent during the 1980s, to 75 per cent in 2005, to only 41 per cent currently. While MSP of wheat for 2012 was 90 per cent of April–June average of world reference price of wheat, the MRP for urea was only 21 per cent of world reference price of urea.

Similarly, achieving the recommended national 4:2:1 NPK balance has proved elusive, again partly because urea (main source of N) is priced cheap relative to other fertilisers. World prices of DAP (main source of P) and MOP (main source of K) have fluctuated around 150 per cent and 100 per cent of world urea price over the last 30 years with no obvious trend. Relative prices of P to N were similar in India as globally, and K much cheaper, till decontrol in 1992 made these more expensive. The MRP for DAP and MOP in India were 194 per cent and 92 per cent of urea MRP before NBS, after which these have risen sharply again. Voluntary MRP for these are now 380 per cent and 230 per cent of urea MRP. Unless corrected soon, this large distortion in NPK prices is bound to reduce crop productivity.

One way out of the present conundrum is to bring urea into NBS and decontrol its prices. But this has not been possible so far and fertiliser decontrol both in 1992 and again in 2010 excluded urea with counterproductive effect. The reason for this is not just opposition to rise in urea prices, but also issues related to domestic urea industry. For example, subsidy provided to

N for decontrolled fertilisers in the present NBS formula is based on the weighted average of subsidies on imported (around $320/ tonne) and indigenous (around $160/tonne) urea. Three consequences would follow if urea prices were decontrolled fully with the subsidy on both imported and domestic urea equated to this (around $200/tonne). First, the domestic urea industry as a whole would get a windfall gain, and there may be consequent audit objections, since average unit subsidy on domestic urea is presently half that on imported. Second, notwithstanding this, that part of urea industry which uses feedstock other than gas would complain that they could become unavailable since their present subsidy is more than the weighted subsidy. Third, since post-subsidy price of urea would tend to settle at import cost less the weighted subsidy; this would, with world urea prices now about $420/tonne, not only double from the present MRP of Rs. 5,310 per tonne but also be subject to the very large fluctuations in world urea prices that have been evident since 2005.

Although political opposition to decontrol is mainly on the third point above, the other points, which relate to differences in costs of production between different Indian producers and between Indian costs and world prices, have historically been at least equally important impediments to reform in this sector. This is unfortunate since India's fertiliser industry, although at disadvantage on feedstock, is largely efficient and can play a key role both in ensuring future nutrient supply and in the effort to increase fertiliser-use efficiency. However, with more than half of its revenues coming from subsidies and with Government also allocating scarce feedstock cheaply, industry effort currently is more to meet pre-set requirements and lobby, rather than to either secure long-term feedstock sources or develop new products and services for its customer base. This needs to change, and one way that this can be done is by reducing industry's dependence on Central subsidies, allowing greater space for it to set prices. The industry's present cost structure is such that no subsidy would be required on over 70 per cent of domestic urea production if urea MRP was allowed to rise to MSP for wheat or paddy. This level of urea MRP would reduce subsidy by about 15,000 crore annually and bring domestic NPK price parities in line with corresponding world parities while still leaving absolute fertiliser prices in India at about half international levels.

Of course, if this were all, urea prices would more than double with all its negative consequences. It would be politically unpopular even with the 5–10 per cent extra increase in MSP that would be required to compensate increases in cost of production. There would definitely be some loss of output as result of lower urea use and farmers unable to avail MSP increase would suffer loss of income. But these negatives can be neutralised and a win-win outcome ensured if the saving in subsidy is ploughed back to develop suitable location and crop-specific packages with adequate price incentives so that farmers do not suffer income loss and yet are encouraged to use appropriate combinations not only of NPK but also organic matter and required micronutrients.

However, for this, the architecture for public intervention will need to go well beyond NBS. Designing and contracting suitable packages will require stability in prices of basic NPK in relation to crop MSPs and also considerable location-specific input, both scientific and operational. The Centre will need to ensure some insulation of domestic prices of straight fertilisers from their large world price fluctuations and devolve many functions and most of the savings from reduced urea subsidy to States. States, in turn, will need to involve universities and local bodies to design suitable local packages of products and subsidies and then contract directly with industry.

Cereals Production and Build up of Stocks

Another major imbalance that emerged during the Eleventh Plan was between production and consumption of cereals, particularly rice and wheat on the one hand which led to rising stocks and rising consumption of edible oils and pulses which led to imports. Cereals production increased by 37 million tonnes (8 million tonnes coarse cereals, 11 million tonnes rice and 18 million tonnes wheat) between 2006–07 and 2011–12. This was the result of several factors, including the NFSM, an Eleventh Plan initiative to increase production, combined with remunerative prices and an expanding and effective procurement machinery in Madhya Pradesh for wheat and Chhattisgarh for paddy. However, although NFSM exceeded targets and per capita production has bounced back beyond earlier highs, much of the increase has been absorbed by increase in Government stocks. There are lessons that need to be learnt from this for the Twelfth Plan.

The rapid accretion of stocks between 2006–07 and 2008–09 was because cereals output responded quickly to policy, both NFSM and MSP, rising from 203 million tonnes in 2006–07 to 220 million tonnes, accompanied by even larger increase in procurement, from 36 million tonnes to 59 million tonnes, while off-take from public stocks rose only from 37 to 39 million tonnes. Consequently, market availability declined during this period, increasing grain prices, the dominant source of food inflation till 2009–10 (Table 12.8). Availability contracted further in 2009–10 because of drought which caused output to fall back to 203 million tonnes. Rice and wheat relative prices eased somewhat in the subsequent two years because output increased even more rapidly than during 2006–09 to reach 240 million tonnes in 2011–12 and because this time rise in procurement (to nearly 73 million tonnes) was less than output and off-take increase (to 56 million tonne) was relatively much more. Nonetheless, procurement exceeded off-take throughout the Eleventh Plan, even during 2009 drought, and present stocks are clearly too high. Costing about Rs. 5 per kg per year to store, these are tying up huge resources that could have been put to better use.

One important point to emerge is that although food inflation is usually ascribed to production shortfalls, policy decisions on MSP and on pricing and quantum of PDS and open market sales can be even more important. This is of course true of rice and wheat prices that are directly affected by such policies, but there are indirect effects as well. For example, milk, eggs, fish and meat had almost no effect on food inflation from 2004–05 till 2008–09, but have contributed most to food inflation subsequently (Table 12.8). As discussed earlier, much of this was due to feed and fodder shortages that the 2009 drought exacerbated. But the high build-up of rice and wheat stocks may in this context have contributed additionally. Substitution effects from lower availability of rice and wheat appear to have pushed up real prices of coarse grain to levels that compare with and most likely influenced inflation in livestock products. To maintain rapid agricultural growth, it will be necessary to continuously assess both MSP and trade policy in light of domestic production trends, paying attention to such wider linkages, so as to minimise undue production imbalance and the inflationary pressures resulting from these. Another important and related issue is the likely future demand for food. The Twelfth Plan Working Group on Crop Husbandry, Demand and Supply Projections, Agricultural Inputs and Agricultural Statistics has made projections for foodgrains and other food items by the terminal year of the Twelfth Plan, that is, 2016–17 (Table 12.9) which would suggest that present levels of cereals production already exceed likely demand at the end of the Twelfth Plan. These projections are based on actual past patterns of observed demand and the fact that cereals consumption per capita has declined since at least mid-1990s. However, it is also the case that India has very high levels of malnutrition and, although there are many reasons for this, deficiencies in calorie intake remain one of the most important. With cereals supplying over 50 per cent of total calorie intake even now, falling cereals consumption is the main reason why per capita calorie intake has not increased despite rising incomes. It is not just that the share of cereals in total food expenditure is falling; even poor people are reducing the share of income spent on all foods in order to meet other non-food needs. In such a situation, where there is a disjunction between such a basic element of human development as nutrition and other demands in an increasingly consumerist society, there is need to ensure that minimum nutrition requirements are actually met. This is the goal of the proposed National Food Security Act (NFSA) under which a majority of the population will be entitled to some very cheap cereals. This is likely to increase cereals demand from those projected in Table 12.9, but nonetheless cereals demand is unlikely to rise much faster than population.

This means that agricultural production must diversify during Twelfth Plan so as to satisfy both tastes and nutrition. In particular, MSP policy should be more restrained for rice and wheat and made more effective in case of pulses and oilseeds where India is a net importer. Although MSP for pulses and oilseeds have been increased substantially in recent years, farmers are still not encouraged enough to put in the effort and resources required to substitute for current imports of these commodities. This is primarily because procurement efforts in these commodities, which are currently not part of Public Distribution, simply do not offer farmers the certainty that they have from procurement effort in rice and wheat.

Public Distribution System

The Eleventh Plan period witnessed significant improvements in administration of the Targeted Public Distribution System (TPDS). A nine-point action plan has been useful in elimination of large number of ghost ration cards, reduction in leakages and greater transparency in the conduct of TPDS operations. While carrying forward these initiatives with greater vigour, there is a need for rejuvenated approach towards the TPDS during the Twelfth Plan period. The foremost amongst those is the move towards facilitating rights-based approach under TPDS by enacting the National Food Security Bill (NFSB). The Bill has been introduced in the Parliament and is expected to provide food and nutritional security, in human life-cycle approach, by ensuring access to adequate quantity of quality food at affordable prices to people to live a life with dignity. This would require strengthening of existing infrastructure and taking up new initiatives and schemes. Reforms in the TPDS would be crucial as it would bring about more efficiency in the system with enhanced transparency and accountability. Entitlements of foodgrains are expected to shift from per household basis to per capita basis. One of the important challenges for implementation of NFSB would be proper identification of beneficiaries which may be based on the ongoing Socio-economic and Caste Census. Another important initiative required during the Twelfth Plan is the end-to-end computerisation of the TPDS operations with the help of a comprehensive Plan scheme. This should not only address current challenges but also facilitate proper tracking foodgrains and lifting by consumers using Aadhaar numbers or adopting innovative methods like smart cards. The up-scaling of the TPDS for proper implementation of NFSA is an opportunity to expand PDS coverage to include coarse cereals, pulses and edible oils and thereby bring scale and certainty to their procurement. However, given that consumption and production patterns vary greatly from state to state, this is probably something that can be done better by the States themselves than by any Central agency. Nonetheless, as part of PDS reform, the Central Government could moot the idea not only of decentralised procurement but also the innovative methods of transferring food subsidy. One option could be that, while the Centre continues to bear responsibility for delivering adequate quantities of cereals to every State, these may be priced close to market and food subsidy transferred to the States as recommended by the High Level Committee on Long Term Grain policy in 2002. Alternatively, subsidy could be credited directly to the bank accounts of the beneficiaries or the FPS dealers using authentication mechanism of Aadhaar numbers. Other option could be to have a comprehensive electronic benefit transfer system whereby subsidy is loaded on to a smart card and consumers have a choice of commodities or fair price shops. These initiatives are expected to bring down leakages significantly as there would be little incentive left for intermediaries to divert the PDS foodgrains into the open market. While implementing these measures, it would be pertinent to address the issue of viability of FPS and improve their functioning. The Gross Budgetary Support for the Department of Food and Public Distribution is Rs. 1,523 crore for the Twelfth Five Year Plan.

Consumer Welfare and Protection

Consumer welfare has been one of the core concerns of the Government since the post-Independence period. Policies have been designed and legislations enacted to protect the interests of consumers and grant them the rights of choice, safety, information and redressal. For the Twelfth Plan period, it would be apposite to expedite formulation of a comprehensive National Consumer Policy in conformity with the UN guidelines on consumer protection. Secondly, there would be a need to revisit existing legislations administered by the Department of Consumer Affairs so as to bring the provisions in line with the changes in the economy, trade, business and consumer expectations. This, inter alia, includes amendments in Bureau of Indian Standards Act and Forward Contracts (Regulation) Act. There is also a need to conceptualise a National Policy for Quality Infrastructure covering standardisation, testing and legal metrology so as to provide the infrastructure for development of definitive standards, systems of legal metrology and conformity assessment. The commodity futures markets need to be strengthened to enable it to serve the dual purpose of price discovery and risk management. Besides, a structured system of information, counselling and mediation need to be put in place with emphasis on rural consumers. The data analysis and price monitoring also need to be more comprehensive and structured so as to make informed decisions on market intervention. The Gross Budgetary Support for the Department of Consumer Affairs is Rs. 1,260 crore for the Twelfth Five Year Plan.

MAJOR CHALLENGES AND PRIORITIES DURING THE TWELFTH PLAN

The main lesson from the performance in the Eleventh Plan is that while there has been a welcome turn-around from the deceleration that was evident in the decade to 2005, and while several indicators have shown marked improvement and potential to build upon, several policy imbalances exist that can prove to be major handicaps. There are also other formidable challenges, for example, a shrinking land base, dwindling water resources, the adverse impact of climate change, shortage of farm labour, and increasing costs and uncertainties associated with volatility in international markets. The Twelfth Plan will need to face these challenges boldly. The key drivers of growth will remain:

1. viability of farm enterprise and returns to investment that depend on scale, market access, prices and risk;
2. availability and dissemination of appropriate technologies that depend on quality of research and extent of skill development;
3. Plan expenditure on agriculture and in infrastructure which together with policy must aim to improve functioning of markets and more efficient use of natural resources; and
4. governance in terms of institutions that make possible better delivery of services like credit, animal health and of quality inputs like seeds, fertilisers, pesticides and farm machinery.

In addition, certain regional imbalances must be clearly addressed. A national priority from view of both food security and sustainability is to fully extend Green Revolution to areas of low productivity in the eastern region where there is ample ground water, and thereby help reduce water stress elsewhere. Rain-fed areas continue to be at a disadvantage, and their development still requires some mindset changes.

FARM VIABILITY : SECURING ECONOMIES OF SCALE AND BETTER MARKET ACCESS AND RETURNS

Farm profitability is central to achieving rapid and inclusive agricultural growth. Improved agricultural prices were an important driver in success of the Eleventh Plan. But slower growth of demand in some major sub-sectors, combined with higher input costs due to world price trends, could cause this driver to be more muted in Twelfth Plan unless offset by increase in productivity. The reports of the Commission on Agricultural Costs and Prices show low net farm revenue for many crops, particularly rain-fed. Diversification towards higher value crops and livestock remains the best way not only to improve farm incomes and accelerate growth, but also to reduce stress on natural resources which form farmers' production base. This needs better infrastructure and emphasis on integrated farming systems, combining crops and livestock, including small ruminants, for different location-specific endowments. This also requires innovative institutional and contractual arrangements so that smallholders have the requisite technology and market access.

(A) The Centrality of Smallholdings

Small farms typify Indian agriculture and this predominance continues to increase. Agriculture Census 2005–06 reported the average size of an operational holding at only 1.23 hectare, with farms less than 2 hectares comprising 83 per cent of all holdings and 41 per cent of area. No agricultural development Plan can be credible unless it is relevant to this vast majority of farmers. Also, 12 per cent of rural households are now female headed with even smaller holding, and the feminisation of agriculture poses special problem.

An important step that would help small and marginal farmers is to reform the tenancy laws. These were originally meant to help small and marginal farmers but now operate against them. Even limited legalisation of agricultural tenancy and freeing the land lease market with proper record of ownership and tenancy status will help such farmers. Some small farmers may lease out land to shift to other occupations, provided they were assured that they could resume the land if they wished. Some large farms may lease in land and even employ the small owner on his own farm to grow specific crops under supervision. Moreover, a stark reality of India's farm situation today is that while land hunger continues unabated amongst the poor and uneducated, especially female, educated young men in richer households are leaving agriculture. The rapid rise of wages for rural casual labour during the Eleventh Plan period has further increased the relative cost of cultivating with hired labour. Many large and absentee owners are leaving land under-cultivated which could be leased out if they were assured of retaining ownership.

The Eleventh Plan had set out in detail the key elements necessary to make land policy effective for equity and efficiency. These are:

1. Modernisation of land records must be both time-bound and comprehensive. Full digitisation of land records, including GIS maps, should be completed with required survey/settlement by end of the Twelfth Plan, during which pilots should also be initiated to enable movement towards a Torrens system in the Thirteenth Plan.
2. Although there is no strong case to change existing ceiling laws, there are several pending implementation issues that can and should be addressed as land records are

modernised.

3. Land issues in tribal areas require urgent and special attention.
4. Although no major new redistribution of agricultural land is likely, it is possible to ensure that all rural households have at least homesteadcum- garden plots.
5. Tenancy should be legalised in a 'limited' manner. Prescribed rents, if any, should allow a band wide enough for rents to be contracted mutually over contract periods long enough to encourage investment by tenants while protecting ownership rights so that landowners have incentive to lease out land rather than keep this underutilised or fallow.
6. Small and marginal farmers, particularly women, lack adequate access to credit, extension, insurance and markets. While every effort should be made to strengthen delivery of public services in their favour, the intervention likely to be most potent is support to group action by farmers themselves. It was suggested that subsidies in Government schemes give preference to group activity.

Most of these issues, as well as the associated matter of consolidating fragmented holdings in course of survey/settlement, are in the State domain and progress is uneven. Ongoing efforts of Ministry of Rural Development (particularly, Department of Land Resources) and Ministry of Tribal Affairs also address some of these issues, although not necessarily related directly to agriculture. However, there was little progress during the Eleventh Plan on the suggestion to redesign schemes so that subsidies favour group activity among small and marginal farmers. In fact, a criticism of the Eleventh Plan schemes has been that these diluted earlier specific support for such farmers.

Almost all the Twelfth Plan working groups set up by the Agriculture Division of Planning Commission have strongly recommended that the Twelfth Plan should put special focus on building capacity that encourages group formation and collective effort by small, marginal and women farmers, rather than simply provide additional subsidy to individuals in these categories. Existing group activity takes many forms depending on purpose. From lower tiers of formal cooperative structures in credit, marketing, dairy and fishery, extending to self-help groups (SHGs), farmer clubs, joint liability groups (JLGs) and, more recently, to producer companies. For simplicity, these can all be termed Farmer Producer Organisations (FPOs).

The Twelfth Plan Working Group on Disadvantaged Farmers, including women has provided evidence-based assessment of the ground situation. New insecurities of tenure from urbanisation and industrialisation are impacting small farms which are efficient but lack adequate access. Its main recommendation is that a collective approach should be promoted in agriculture for small and women farmers at all points of the value chain. It cites many successful examples that stretch from the Gambhira farmer's collective in Gujarat, initiated in 1953 and still going strong, to several initiatives of women's group farming in Andhra Pradesh such as one initiated by Deccan Development Society in 1989 and another initiated by a UNDP-GoI project in 2001 and sustained since 2005 by the Andhra Pradesh Mahila Samakhya (APMSS). The most recent success story is the collective farming initiative launched in 2007 under Kudumbashree jointly by Kerala Government and NABARD. Success of these in increasing production and empowering women point to a need for States to experiment with (*i*) channelising NGO strength in mobilising people to encourage small holders to shift from an individual to a group-oriented approach; and (*ii*) facilitating land access by groups of disadvantaged farmers with appropriate arrangement for provision of inputs, including credit. Financing such experiments should be permissible under RKVY.

Since land access was the most difficult part in all the above efforts, the Working Group has suggested that, except distribution of homesteads to the homeless which should have the highest priority, future Government land distribution should be to groups of landless and women farmers rather than to individuals. This could take the form of long-term lease which would expire if the group broke down, for which it would be necessary to legalise tenancy at least for this purpose. Moreover, an innovative suggestion of both this Working Group and the Working Group on Marketing is to set up Public Land Banks (PLB) at Panchayat level. Landowners could 'deposit' uncultivated land and receive regular payments from the PLB varying by period of deposit and rents actually obtained with the guarantee that this 'deposit' can be withdrawn with suitable notice. The PLB could then lease out to small and women farmers or their collectives. A form of 'limited' tenancy aimed at fuller agricultural use of available farm land and to slow down speculation in such land for future non-agricultural use, this idea excludes leasing to corporate entities. However, to set up PLBs will require some initial seed capital and a clear legal framework. If States provide the legal framework and the necessary guarantees, the seed capital could also be permissible under RKVY. 12.64. Access to finance, especially by small holders, is crucial for improved agricultural performance. Credit flow doubled in the Eleventh Plan but mainly by credit deepening, with little increase in farmer coverage and still leaving 60 per cent of farmers without institutional credit. There are several ways in which credit access can be widened. Primary Agricultural Co-operative Societies (PACS) still have the widest coverage and must be made more memberdriven and less dependent on higher tiers. Joint Liability Groups (JLGs) are still the most appropriate mechanisms for farmers and livestock owners who have productive assets but cannot access credit because they have no land records, are located too far from banks or have last mile problems. The SHG Bank Linkage programme is still the most appropriate financial mechanism to extend credit to marginal and dry land farmers as this allows better income smoothing since SHGs provide space for diversity in loan purposes and sizes, enabling financing of a variety of activities that such families select as part of livelihood strategies when income from agriculture is low.

Commercial banks have not supported JLGs or SHGs as much as they could have, preferring instead to comply with priority sector requirements by offering bulk finance through Non-Banking Financial Companies (NBFC) and Micro-Finance Institutions (MFI). However, NBFC–MFI lending is mainly individual and based on standard products imposing short repayment schedules which did not dovetail with cash flows from agriculture. This caused multiple borrowings, increased risk to borrowers and led to a backlash. The solution is to restore the principle of group decisions by borrowers both in the borrowing process and in use of borrowed resources. This need not exclude NBFC– MFI so long as shortcuts are avoided. For example, NABFINS, a NBFC promoted by NABARD, lends only to groups and uses a Business Correspondent (BC) Model that also provides working capital to second level institutions like cooperatives and producer companies which aggregate, add value and market commodities. The SHGs have a stake in these second level institutions which help expand their livelihood base.

Small and marginal farmers face problems not only with shrinking land assets and with credit; they have difficulty in accessing critical inputs for agriculture such as quality seeds and timely technical assistance. In this situation, FPOs offer a form of aggregation that leaves land titles with individual producers and uses the strength of collective planning for production, procurement and marketing to add value to members' produce through pooled resources of land and labour, shared storage space, transportation and marketing facilities. These also improve bargaining power of small farmers and, most importantly, reduce transactions costs of banks and buyers to deal them. Investing in such group efforts has strong externalities. The Twelfth Plan Working Group on Agricultural Marketing, Infrastructure, Secondary Agriculture and Policy for Internal and External Trade has in fact suggested that an institutional development component, along lines of NABARD's farmer club scheme, be introduced in all Centrally sponsored schemes to specifically target FPO formation among small producers, especially tribals, *dalit*s and women. It notes that a majority of FPOs that are likely to emerge as a result of such an intervention will remain focused on addressing issues of crop planning, technology infusion, input supply and primary marketing. But, with adequate support for business development, about one fourth to a third would seek to leverage presence further up the value chain, most likely at the lower end (for example, setting up pack houses, grading centres, small cold stores, drying or quick freezing plants). Larger FPOs, for example, existing cooperatives could provide this support and in fact could aim bigger, but issues may be different. For example, the National Dairy Development Board's SAFAL has had only limited success although the wide network and logistics of milk cooperatives make these obvious incubators for village-level aggregation of other perishable products. Therefore, the Twelfth Plan must try to mainstream support for FPO formation and capacity building using all credible agencies for the purpose: existing cooperatives, NABARD and the Small Farmers' Agribusiness Consortium (SFAC).

(B) Issues in Expanding Agricultural Marketing and Processing

A major problem facing cultivators is that they do not get remunerative prices because of uncertainties caused by inadequate market information, unnecessary controls, lack of physical infrastructure and price volatility—both domestic and global. In order to provide adequate incentives to farmers, the Twelfth Plan will have to focus on leveraging the required private investment and also policies that make markets more efficient and competitive.

Reforming the Agricultural Produce Marketing Committee (APMC) Acts should therefore have priority as emphasised in the Eleventh Plan and the Mid-term Appraisal. The introduction of the Model Act in 2003 was directed towards allowing private market yards, direct buying and selling, and also to promote and regulate contract farming in high-value agriculture with a view to boost private sector investment in developing new regularised markets, logistics and warehouse receipt systems, and in infrastructure (such as cold storage facilities). This is particularly relevant for the high-value segment that is currently hostage to high post-harvest losses and weak farm-firm linkages. While many States have moved towards adoption of the Model Act, actual progress has been limited. Often the permissions given are subject to unacceptable restrictions which make them ineffective. Vested interests in maintaining the existing *mandi* system intact are very strong. In view of the slow progress, the Ministry of Agriculture set up a Committee of State Ministers in-charge of agricultural marketing. The Committee submitted a 'First Report' in September 2011 which has been circulated to all States and UTs. The report calls for 'speedy reforms' of Agricultural Produce Market Committees (APMC) Act across different States along with 'time-bound development' of marketing infrastructure. Calling for a ten-year perspective plan to improve infrastructure of backward and forward linkages for agriculture production and marketing, the report has suggested that agricultural marketing be given access to priority sector lending. Thus, the process to secure necessary amendments in APMC Acts and thus create the enabling legal environment is still ongoing. The Twelfth Plan will need to fasttrack modernisation of *mandi* infrastructure, with adequate provision of communication and transportation, and also empower small producers through their organisations and marketing extension. Post-harvest losses, probably average 10 to 25 per cent, being particularly high in horticulture, livestock and fisheries. Very large investments are required in developing agricultural markets, grading and standardisation, quality certification, warehouses, cold storages and other post-harvest management of produce to address this problem. Such large investments are possible only with the participation of the private sector which, in turn, require freedom from controls on sales/purchase of agricultural produce, its movement, storage and processing. Many new initiatives were taken up during the Eleventh Plan, including both terminal markets under Public–Private Partnership (PPP) mode in the National Horticulture Mission (NHM) and a model of public sector investment combined with professional management by stakeholders as exemplified by NDDB's fruit and vegetable wholesale market at Bengaluru and APEDA's Modern Flower Auction Houses.

The Twelfth Plan Working Group on Horticulture and Plantations which studied the matter in detail has observed that participation by traders, wholesale buyers, exporters and processors has actually been very low in all these new initiatives because of reluctance to be subject to transparent operating procedures. It has come to the conclusion that the present model of Market Sector Reforms which is trying to create space for a new set of modern markets in coexistence with much less transparent procedures in APMC regulated markets is unlikely to result in any major private investment in modern marketing infrastructure. In its view, to break the barrier of reluctance to participate in business of modern markets it is necessary as part of marketing reforms to define and introduce a common Standard Operating Procedure (SOP) for all markets: both the new modern markets envisaged as well as existing regulated markets under APMC Acts. Therefore, it proposes that managements of existing regulated markets must be made to adopt the modern marketing model: that is, undertake the auction function themselves and all payments to sellers ensured by the Market Committee through a system of bank credit limits of the buyers. This would involve redefining the role of APMC management with introduction of SOP and an open policy of registering buyers; permitting setting up of private markets in APMC areas; removal of interstate barriers to allow an unified national market, either by using entry 42 of the union list or at least for sealed container cargo; and single point levy at first point of sale.

While this entire area of regulation of agricultural product markets is thus in some flux and movement is still slow, an important initiative in the Eleventh Plan involved setting up a Warehouse Regulatory and Development Authority (WRDA) to set standards and modernise warehousing. The aim is enlarged use of negotiable warehouse receipts that can be linked to e-trading, both spot and future, so that farmers have an alternative to *mandis*. However, so far less than 300 warehouses have been registered and there is yet no effective coverage of perishable products. Cold storages have recently been brought under WRDA but minimum standards are yet to be set. This may be as difficult as meeting the requirement of cold storage additional capacity estimated at around 32 million tonnes over the next decade. Present cold storages are of inadequate quality, most domestic component manufacturers do not have certified performance ratings, BIS standards do not exist for many critical components of cold chain infrastructure and critical storage conditions prescribed internationally for cold chain structures have yet to be validated for many Indian agro-climatic conditions or cultivars.

Although India ranks second in world production of fruits and vegetables, only 6–7 per cent of this is processed, compared to 65 per cent in US and 23 per cent in China. A well-developed food processing industry is expected to increase farm-gate prices, reduce wastage, ensure value addition, promote crop diversification, generate employment opportunities and boost exports. Further, issues concerning food processing industry are dealt with in Chapter 9. The private sector needs to invest much more in creation of warehousing capacity, cold storages and supply chains. In this context, the Planning Commission had also set up a Committee on Encouraging Investments in Supply Chains including provision for cold storages for more efficient distribution of farm produce, which submitted its report in May 2012. The Committee has indicated that with regard to foodgrains, the Department of Food and Public Distribution has initiated steps for creation of 17 million tonnes of additional storage capacity including 2 million tonnes in the form of silos. This additional capacity is expected to take care of public sector's warehousing requirement during the Twelfth Plan. The Committee has recommended to exempt perishables from the purview of APMC, provide freedom to farmers and make direct sales to aggregators and processors, introduce electronic auction platforms for all the *mandi*s where daily transaction is above Rs. 10 crore, and replace licensees of APMC markets with open registration backed by bank guarantees to ensure wider choice to growers and to prevent cartelisation by traders. The Committee has recommended encouraging largescale private investments in the cold chain sector using PPP Model with Viability Gap Funding besides providing budgetary support and capitalising on schemes such as Rural Infrastructure Development Fund (RIDF). An Inter-Ministerial Group on Cold Chain Infrastructure and Allied Sectors has been set up by the Government to facilitate implementation of these recommendations.

There is merit in planning part of such investment as infrastructure to reduce waste and enlarge markets rather than wait for corporate investment in processing or retail. The extent of wastage is not easily ascertainable and new research suggests that some of the older estimates were quite likely exaggerated, especially if quality loss leading to lower prices is not counted as waste. Also, the experience so far is that corporate entrants have not fared very well in the competition with incumbent traders since existing trading margins, although high, are in fact much less than, for example, in the USA. However, there is no doubt that modern storage and logistics do reduce waste. If such infrastructure also improves farm shares, social returns could exceed the private and justify subsidies. Subsidy rates, increased recently to 25–50 per cent, are now quite high and policy should be clear on whether the goal is just capacity targets or wider market access and improved marketing efficiency. If the latter, eligibility criteria need to be specified and also linked clearly with marketing reform. Social returns to subsidy will be more if access to both the infrastructure and to markets is more open. The real test is whether these can spawn and sustain enterprise in aggregation, grading and processing at the bottom, preferably by FPOs, but also by lead farmers and even by existing commission agents.

The recent decision to open up debate on FDI in retail must be seen in this context. With multibrand retail already open to the domestic corporate sector, FDI in retail should not be viewed as an entirely new disruptive factor affecting traditional retail. It will only add depth and competition to the present situation. Deeper pockets and technology, and the compulsions to invest in supply chain development which is not there for domestic modern retail may accelerate investment in logistics, quicken consolidation of retail trade and create new proprietary supply chains. It must be emphasised that FDI alone will not resolve back-end issues related to modernising agricultural markets that have so far muted the domestic corporate effort and investment. FDI has an added potential to link farmers to wider markets by expanding exports. However, the Eleventh Plan had also noted the legitimate concern that if front-end investment outpaces backward linkage, the outcome could instead be more imports and lower farm prices. The introduction of FDI will increase, not lessen, the importance of priorities identified above: marketing reforms, aggregation at the bottom and public funding of stand-alone infrastructure.

With less than 40 per cent of farm produce presently consumed in urban areas and much less processed, use of public funds to improve market efficiency will have a positive effect on farm growth. There are benefits in coordinating this effort with other steps to encourage corporate investment in this area. For example, the NHM was designed based on a concept of adequately sized area clusters so that processors could plan capacities based on anticipated future fruit production that would in turn ensure markets for farmers when trees finally bore fruit. But processors have preferred to wait and watch while farmers, not sure of adequate market for any single crop, have usually chosen to diversify their production basket. Most clusters have therefore not developed in the manner intended. A larger thrust to modernise processing and retail will require bringing more synergy between corporate actors and farmers, particularly in infusion of technology and capital at the farm end.

The Ministry of Agriculture has proposed a RKVY window for Public–Private Partnership for Integrated Agricultural Development (PPPIAD) for States to facilitate 'large scale integrated projects led by private sector players with a view to aggregating farmers and integrating agricultural supply chains.' The idea is to leverage corporate interest and marketing solutions to part-finance mobilisation of expertise to form FPOs and infuse technology and capital to enhance farm production and value addition. This is in line with views of various working groups, and needs to be piloted. But since this will in effect be public subsidy to contract farming, it is necessary to be clear on what should and should not be subsidised. First, project selection should go beyond where contract farming would normally occur; that is, give priority to proposals involving FPOs composed mainly of small and marginal farmers in less accessible and rain-fed locations. Second, tangible assets that are property of the corporate partner cannot be subsidised by RKVY. Only stand-alone assets of farmers or their FPOs should be subsidised. Third, a transparent project selection mechanism will be required to rank proposals, for example, by assigning marks based on States' priorities to deliverables offered, with outcome indicators for subsequent monitoring. If this works, it might be a game changer, not only to form FPOs and widen farm-industry linkage but also to fast-track desirable changes in cropping patterns.

(C) Credit and Cooperatives

The Twelfth Plan Working Group on Institutional Finance, Cooperatives and Risk Management has projected the demand for credit during Twelfth Plan at between Rs. 31,24,624 crore and Rs. 42,08,454 crore, depending on the methodology used. At the higher end of these estimates, that is, assuming agriculture growth at 4 per cent and ICOR at 4.5, the size of the credit requirement in the Twelfth Plan period translates into about double the flow during the Eleventh Plan, that is, Rs. 8 lakh crore per year, as against the level of Rs. 4.68 lakh crore achieved during 2010–11. This projected level of credit appears feasible in view of the Eleventh Plan achievement. As against credit flow of Rs. 2,29,401 crore in agriculture during 2006–07, the total institutional credit flow to agriculture in 2011–12 was Rs. 5,11,029 crore. But despite this very robust growth, many issues continue to confront agricultural credit, particularly in the area of financial inclusion necessary for ensuring inclusive growth. Agricultural credit continues to neglect certain sub-sectors, the flow of term lending is dwindling and there is inordinate increase in the share of indirect finance. Credit dispensation by institutions to small and marginal farmers has been disappointing, including by the Cooperative Credit Structure (CCS) which has traditionally catered to relatively smaller farmers.

On these issues, the working group has pointed to the need for more objective assessment of credit requirements for direct and indirect financing of agriculture and also to redefine the priority lending sectors. It has suggested updating of KCC databases with priority analysis of KCC percentage provided to the small and marginal farmers and more intensive use of ICT applications to track the flow of credit and transmission losses, with reference to such farmers.

Some ongoing and emerging changes appear to hold promise of triggering off better financial inclusion for banking activity:

1. The Core Banking Platform provides seamless connectivity which, with the telecom infrastructure, brings a new architecture to access financial services.
2. The BC model, together with mobile phones, can along with post offices provide significant last mile connectivity.

3. Mandating payments (for example, of wages under the National Rural Employment Guarantee Act, pension dues and so on) through formal channels, including post offices, is helping to reach financial services to those so far not reached.
4. The enormous economies of scale generated by SHG Federations (each of 150–200 SHGs) is enabling banks to give larger loans for housing and health facilities for their members. A variety of insurance services are also being made available, including life, health, livestock and weather insurance.
5. The UID project of the GoI with biometric identity may facilitate easier opening of bank accounts, although this has yet to happen. The financial health of the Long-term Cooperative Credit Structure (LTCCS) continues to deteriorate with accumulated losses of Rs. 5,275 crore by March 2010, resulting in erosion of 59 per cent in owned funds. A quick decision is warranted on the implementation of the revival package for the LTCCS too on the lines of the Short-term Cooperative Credit Structure (STCCS).

Notwithstanding, the relatively improved financial health of the STCCS following implementation of the revival package, its share in total institutional credit continues to show a declining trend. The package for STCCS was conditional to radical restructuring of coops into autonomous, democratic and self reliant institutions without intrusion of politics and bureaucracy. The States have not implemented these recommendations with full seriousness. Therefore, Cooperative Sector Reforms should continue to be insisted upon during the Twelfth Plan.

In the interest of strengthening of the ground level tier, there is also need for considering disciplined refinancing of PACS as stand-alone institutions, provided that these are member driven. PACS still have the widest coverage and the recent development of financing PACS through commercial banks needs to be widened, deepened and strengthened, especially in cases where higher tiers of the STCCS are weak and not in a position to fund them.

(D) Farm Income Variability : Managing World Price Volatility and Climate Risk

The Eleventh Plan document had noted that farmers are now subject to much greater risk than what Indian farmers have been used to in the past. The frequency and severity of risks in agriculture have increased on account of climate variability and this has been accompanied by much greater variability of world prices and their quicker transmission into the domestic economy. On price variability, it had recommended much greater co-ordination between MSP and trade policies and for putting in place a system whereby tariffs on imports and exports of farm products could be varied quickly in response to world price movements rather than having to take recourse to outright bans which hurt both farmers and trade. On climate variability, it had recommended going beyond current insurance measures and to put in place a tertiary mechanism for management and assessment through climate forecasting and mapping of agricultural losses.

World agricultural prices rose sharply during the Eleventh plan period, with inflation about 9 per cent per annum in US dollar terms and price volatility much higher than before, accompanied by even higher world inflation in fuels and fertiliser. It is now generally agreed that among the several factors that contributed to this were more frequent weather shocks, policies to promote biofuels and increased demand on commodity future markets as a result of speculation and portfolio diversification. There is also consensus that linkage between agricultural prices and price of oil is now very strong and may cause high volatility to persist. As compared to this, domestic Indian agricultural prices were much less volatile and domestic prices of fuel and fertiliser were increased much less than corresponding international prices. Indian farmers were thus relatively better protected against both higher price volatility and higher costs. However, this has involved repressing inflation in fuel and fertiliser and required bans on exports during world-price spikes. Co-ordination between MSP and tariff policy is still very weak. For example, while other aspects of a recent CACP suggestion for oil palm development can be met by ongoing schemes, the proactive tariff support required is a sticking point. These will need to be addressed during the Twelfth plan.

On the climate side, a number of initiatives taken by the Indian Space Research Organisation (ISRO) and the India Meteorological Department (IMD) during the Eleventh Plan have significantly improved the scope and quality both of climate data and of other remote sensing tools. Although IMD's long-range forecasts of the monsoon still have a very large margin of error, its shorter-range products not only have greater accuracy but cover an array of agro-meteorological variables with fairly high resolution. There is also much better co-ordination today between ISRO and IMD on one hand and the Ministry of Agriculture, corresponding State departments and NARS on the other. For example, Department of Agriculture and Cooperation (DAC) has set up a Mahalanobis National Crop Forecasting Centre with ISRO collaboration to augment present crop forecasts and assessment with regular remote sensing, GIS and Global positioning System (GPS) data.

With better satellite products, an Eleventh Plan innovation was the Integrated Agro-Meteorological Advisory Service (IAAS) which now issues regular weekly Agro-Met Advisory Bulletins up to district level on field crops, horticulture and livestock. This involves agricultural universities to collect and organise soil, crop, pest and disease information and amalgamate this with weather forecasts to assist farmers in their decisions. Though still of very variable quality from district to district, and limited since district is too big a unit for useful advisory, a 2009–10 NCAER study concluded that this brought large savings to farmers. In the Twelfth Plan, a Gramin Krishi Mausam Seva (GKMS) will be launched to extend IAAS to block level, initially on experimental basis. Also, IMD will implement the Monsoon Mission aimed at generating better seasonal monsoon rainfall forecasts in different spatial ranges.

In a parallel Eleventh Plan initiative, that took advantage of IMD experience with Automatic Weather Stations technology, Government launched a Weather Based Crop Insurance Scheme (WBCIS) through the Agricultural Insurance Corporation (AIC). Initiated as a pilot in Kharif 2007 in 70 hoblis of Karnataka for 8 rain-fed crops, by 2010–11 the Scheme was being implemented in 17 States and covered more than 67 lakh farmers growing crops on 95 lakh hectares spread over 1,010 blocks in 118 districts. At present WBCIS has about one-third the coverage of the National Agricultural Insurance Scheme (NAIS), the main crop-insurance vehicle. Based on results of crop-cutting experiments, this has been in operation since 1999–2000. Although a useful device, especially for farmers growing relatively risky crops, the main problem with NAIS is that it is not actuarial insurance. Premiums for most important crops are fixed at all-India level irrespective of risk and Central and State Governments pay for the entire excess of claims over premium received. Moreover, being compulsory for all borrowers from banks in States where it is in force, and with relatively few non-loanee farmers involved, it mainly insures banks against default following poor harvest. Further, its popularity with farmers is limited since crop-cutting experiments delay claims/ payments until well after harvest and risk covered is only of yield shortfalls at the block level. \ For these reasons AIC is also piloting a Modified National Agricultural Insurance Scheme (MNAIS) since 2010 that aims to (*i*) reduce the insurance unit from block to village panchayat with higher indemnity as proportion of threshold yield, (*ii*) move to actuarial premiums supported by upfront subsidies instead of NAIS practice of Government paying the entire excess of claims over premium, and (*iii*) extend insurance cover to situations such as failed sowing, cyclonic rains and localised calamities, such as hailstorms and landslides. The main problem is lowering insurance unit which although good for farmers increases the cost and effort on crop-cutting experiments exponentially.

As a result, the Government of India is currently implementing four schemes, that is, NAIS, MNAIS, WBCIS and another pilot Coconut Palm Insurance Scheme (CPIS). Only NAIS is being implemented as a full-fledged scheme and the other three are being implemented on pilot basis. The pilot programmes will be evaluated early in the Twelfth Plan for future revisions/ modifications to evolve a *National Agricultural Insurance Programme*. For this, the following will be necessary. First, define what should be the core programme which Government should set up and what should be left to companies to devise their own insurance products. Second, to examine the trade-off between competition and benefits of risk pooling, that is, a centralised reinsurance system. Third, arrive at an optimum mix between weather-based insurance and those dependent on yield measurements whether by crop-cutting experiments or remote sensing.

Some suggestions, based mainly on the Twelfth Plan Working Group on Institutional Finance, Cooperatives and Risk Management, are:

1. Taking as core the ongoing NAIS, modifications being made through the pilot MNAIS should be continued. The high cost of lowering the insurance unit should be dealt with progressively in Agriculture consultation with States. Centre may share part of the cost of crop-cutting experiments in the short-run but should shift to new technologies such as satellite imagery in the long run.
2. The issue of private-sector involvement in agricultural insurance can be creatively addressed, for example, through a system of co-insurance under which the AIC is lead insurer (with underwriting responsibilities and contacts with multiple agencies).
3. Weather-based insurance should continue, again focused on customisation and innovation such as double trigger (weather and yield) and indexplus products, with State Governments choosing what to subsidise. Roll-out of AWS can be demand-led and private sector also involved but with mandatory accreditation from a competent third-party designated by Government to ensure consistent and high-quality weather data. Further, Terrestrial Observation and Prediction Systems (TOPS) platforms need to be pilot tested.
4. Other innovative products such as community based mutual insurance, savings-linked insurance, a properly designed product fort contract farming arrangement and so on can help establish insurance culture, especially if linked to FPO formation.
5. Agriculture insurance, being specialty insurance with huge Governmental intervention should be seen more as a social instrument of the Government rather than a commercial instrument, hence is unlikely to be effectively administered unless backed by a statute.
6. To protect non-insured farmers from extreme financial distress, Government may consider 'Catastrophe Protection.' A blanket Life Insurance cover could be devised for at least small/marginal farmers (including tenant farmers) to meet liabilities to banks or other RFIs in the unfortunate eventuality of death and to secure some financial support to families of the deceased. Premia on such group/blanket insurance could be funded by Central/State Governments and financing banks, in full or in part.
7. Crop losses arising out of natural calamities are presently compensated by Government funding or concessions like loan/interest waivers/deferments. This practice is fraught with inefficiency, besides crippling repayment ethics. It is, therefore, necessary that dealing with loan losses should be internalised within the banking system through the constitution of Relief and Guarantee Funds and Stabilisation Funds (set up partly with Government funding, by diversion of subsidies for loan repayments and so on).

AGRICULTURE RESEARCH AND EDUCATION

Agricultural research has played a vital role in agricultural transformation and in reducing hunger and poverty and its role in the Twelfth Plan will be crucial. The Eleventh Five Year Plan had noted that research in the past had tended to focus mostly on increasing yield potential by more intensive use of water and biochemical inputs, paying less attention to either the long-term environmental impact of this approach or to methods and practices for efficient use of inputs and natural resources. But now that limitations of this approach were evident, there appeared to be lack of any clear agricultural research strategy or to assign definite responsibilities and prioritise the research agenda rationally. It had proposed that ICAR institutes undertake basic, strategic and anticipative research, focusing particularly on problems of rain-fed agriculture, while SAUs concentrate on generating required manpower and on applied and adaptive research to address local problems. It had emphasised that research should shift from a commodity based approach to a farming systems approach through convergent efforts of R&D agencies within each agro-climatic region to address local problems identified by stakeholders, including development agencies. It had also stressed the need to enhance spending on NARS and proposed to raise this to 1 per cent of agriculture GDP by end of the Plan period.

As it turns out, research spending at 2006–07 prices, although reaching nearly 0.9 per cent in 2010–11, averaged only 0.7 per cent during the Eleventh Plan. At current prices, it was even less, averaging only 0.64 per cent during the Eleventh Plan. Part of the reason was a shortfall of about 20 per cent in the Centre's Plan expenditure from that originally targeted, but the main reason was inadequate spending by States. While Centre's expenditure (non-Plan and Plan, including RKVY) increased 68 per cent in real terms between the Tenth and the Eleventh Plan periods, corresponding States expenditures increased only 22 per cent. In particular, non-Plan spending on SAUs increased less than 17 per cent, less than required to meet the pay commission awards in most States. Consequently, most SAUs are understaffed and underfinanced. This is undoubtedly the most serious problem confronting NARS.

Nonetheless, new SAUs continue to be created, especially in animal husbandry, which lack adequate staff, have little infrastructure and are grossly underfunded. Emphasis has to be laid on arresting proliferation and improvement, especially in core disciplines like modern biology, to ensure a steady supply of quality human resources. ICAR should specify minimum standards, and meeting these standards could be an eligibility condition for States to get RKVY funding.

Significant contributions of public-sector research during the last decade have included breakthroughs in basmati varieties, improved wheat varieties resistant to rust including rice, improved varieties of soybean, Bengal gram, mustard, chickpea and single cross hybrid maize; which have led to higher growth in these crops. Similarly, although most Bt cotton hybrids that are commercially successful are from private producers, these are based mostly on public material. With respect to natural resource management, public research claims significant contribution in developing resource conservation technologies like integrated farming, micro-irrigation, laser levelling, zero tillage and agricultural practices to improve efficiency of nutrients and water, including in situ rain water harvesting. In fruits and vegetables, better varieties and hybrids, disease management and multiplication of planting material and in livestock and fisheries, disease management technologies (vaccines and diagnostics), feed and fodder management, improving reproductive health and production of fisheries seed.

Broadly, although NARS has yet to respond to changes suggested in the Eleventh Plan, there are signs of some new research priorities and agendas. As example of new collaborative research, ICAR launched the 'National Initiative on Climate Resilient

Agriculture (NICRA)' in February 2011 as a network project with several collaborating institutions with a view to enhance resilience of Indian agriculture to climate vulnerability through strategic research and technology demonstration. The research on adaptation and mitigation covers crops, livestock, fisheries and natural resource management. The project aims to enhance resilience through development and application of improved production and risk management technologies. It plans to demonstrate site-specific technology packages on farmers' fields for adapting to current climate risks and to enhance the capacity of scientists and other stakeholders in climate resilient agricultural research and its application. This will be continued during the Twelfth Plan. 12.100. For the Twelfth Five Year Plan, the ICAR has proposed a number of new initiatives in its manner of functioning, such as extramural funding for research, creation of funds for agri-innovations and agri-incubation and setting up of an Agriculture Technology Forecast Centre (ATFC). To improve staff strength and quality it has proposed an Adjunct Professor Scheme, Agriculture Sciences Pursuit for Inspired Research Excellence (ASPIRE), e-courses and more post-doctoral fellowships. Modernisation of SAU farms is also contemplated. In particular, it has proposed the following new thrusts:

- *Conceived Research Platforms*: Research consortia platforms are proposed for focused, time bound multi-disciplinary research in areas of 'Agro Biodiversity Management; Genomics; Seed; Hybrids; GM Foods; Biofortification; Plant Borers; High Value Compounds/Phytochemicals; Nanotechnology; Diagnostics and Vaccines; Conservation Agriculture; Waste Management; Water Management; Natural Fibre; Health Foods; Precision Farming, Farm Mechanisation and Energy; Secondary Agriculture and Agriincubators.' These will involve partnership of ICAR with R&D organisations inside and outside NARS. Inter-departmental platforms for research in these priority areas and also capacity building in basic sciences, remote sensing and medium range agri-advisory services will be fostered involving CSIR, DBT, ICMR, DRDO, DST research institutes as well as general universities and Ministries of Environment, Space and Earth Sciences.
- *National Agricultural Education Project*: A National Agricultural Education Project for Systemic Improvement in Higher Agricultural Education and Institution Development is proposed to be undertaken as an externally-funded project to improve education quality in State Agricultural Universities.
- *National Agriculture Entrepreneurship Project*: Another externally-funded project is proposed in order to build an ecosystem for nurturing entrepreneurship development through translational research for technology commercialisation, management of technologies for commercialisation, research for breakthrough technologies for accelerated growth and higher-economic impact.
- *Farmer FIRST*: In order to make technology delivery process more effective through the existing 630 Krishi Vigyan Kendras, this new initiative will enhance farmers–scientist contact through multi-stakeholders' participation to move beyond production and productivity to privilege the complex, diverse and risk prone reality faced by most farmers.
- *Student READY*: A one-year composite programme, the Rural Entrepreneurship and Awareness Development Yojana (READY) is proposed with the objective to develop professional skills for entrepreneurship: knowledge through meaningful hands-on experience in project mode; confidence through end to end approach in product development; and enterprise management capabilities including skills for project development and execution, accountancy and national/international marketing.

- *Attracting and Retaining Youth in Agriculture (ARYA)*: This initiative will be implemented with a youth-centric approach, targeting areas of agriculture research which can be converted into viable economic enterprises and build capacities to attract rural youth to agriculture. The Twelfth Plan allocation for ICAR is of a size that will allow spending on NARS to reach 1 per cent of agriculture GDP by end of the Plan provided States fund SAUs similarly. The above ICAR proposals can have priority if defined in terms of deliverables, rather than areas. Also, NARS should address the following issues on priority basis during the Twelfth Five Year Plan:
- Strengthening soil organic carbon (SOC) research, particularly on the quality of organic matter and microbial activity, physical properties of SOC, validation and refinement of models and SOC dynamics under different land uses and management regimes.
- Developing Models and technology interventions on rational use of inputs, especially nutrients and irrigation water, under diverse agro-ecologies through interdisciplinary and farmer participatory mode in order to enhance their use efficiency, as also farm profits.
- The Expert Group on Pulses has been critical of NARS. Efforts to enhance the yield potential of pulses, by analysing physiological and biochemical limitations of the current crop and designing more efficient types, is a priority which should also involve improving the nutritional quality of pulses and reducing various anti-nutritional factors.
- Another priority continues to be the development of heat resistant varieties of wheat.
- Greater thrust needs to be given to post-harvest management, secondary agriculture and value addition, along with by-products and waste management. The agricultural technologies which have been developed and matured in the Eleventh Plan should be taken for commercialisation in the Twelfth Plan. Accordingly, the human resource development including para-technicians should be emphasised.
- Private agriculture input and seed companies use the research products of public system to generate profits. The public research system should seek a share in such profits which is possible if the public research system takes due care in protecting its intellectual property rights under the Protection of Plant Variety and Farmers' Rights Authority (PPVFRA). This requires development of an appropriate pricing mechanism and preparing a suitable licensing system.

NATIONAL MISSION ON EXTENSION AND TECHNOLOGY MANAGEMENT

The extension system of State agricultural departments is the weakest link in the chain between research and the farmer. Large number of vacancies of extension workers in the State Agriculture Department was one of the gravest concerns expressed by the Eleventh Plan document. During the Eleventh Plan, efforts were initiated to improve extension services by extending Central support to State extension reforms. This has resulted in 604 Agriculture Technology Management Agencies (ATMAs) to be established across the country with 21,000 new posts sanctioned with Central assistance at State, district and block levels. Also, since a continuous problem plaguing extension has been lack of organic link between the research system and the extension machinery, R&D linkage guidelines were jointly brought out by the DAC and ICAR and sent to all States and SAUs. The basic thrust of these guidelines were to get ATMAs and KVKs to work together at the district level and below, keeping in view the priorities reflected in Comprehensive District Plans. Although neither has delivered full results, there is now much greater acceptance that things must be done together.

Seed is also an area where NARS made much greater effort than in previous recent Plan periods.

Along with seeds, farm mechanisation was also highlighted earlier as a source of the Eleventh Plan labour productivity gains. In view of emerging labour shortages in many states, there is demand to expand custom hiring services, as well as for new implements. During the Twelfth Five Year Plan it is proposed to give a co-ordinated thrust on seeds, farm mechanisation and extension through a new *Mission on Extension and Technology Management*. This should also have a component to fund ICAR research platforms to find solutions to problems thrown up by extension and requiring expertise beyond SAU.

(A) Seeds and Planting Material

Three major yield successes during the last decade relate to cotton, maize and basmati rice. Agriculture 33 These were driven by new seeds of which cotton and maize hybrids were mainly from private sector while basmati rice varieties were almost entirely public. Increased adoption of hybrids in cross-pollinated crops like cotton, maize, pearl millet and sorghum has been led largely by the private sector, which accounts for three-fourths of hybrids developed so far in the country. But there is discernable change in role of public sector in development of hybrids after 2001–02. Till 2001–02, private sector developed 150 hybrids of cotton compared to 15 by public sector; 67 hybrids of maize compared to three in public sector. In the next seven years, public sector increased its share from 8 per cent to 19 per cent in cotton, from 4 per cent to 40 per cent in maize and from 25 per cent to 58 per cent in rice, with similar changes in other crops. In parallel, public production of quality seeds of varieties have increased rapidly in recent years, expanding the public share in total seed use. Production of quality seed doubled from 140 lakh quintals in 2004–05 to 280 lakh quintals in 2009–10, contributing significantly to the Eleventh Plan yield performance. Private sector accounted for 39 per cent of this seed production. Nonetheless, the ratio of quality seed to total seed use by farmers is still much lower than norm and there is considerable scope to raise crop productivity by raising this ratio.

There are several pending issues regarding seeds. For example, at present there is no regulatory mechanism to protect farmers against non-performance, say poor seed germination rate. The Seeds Bill, 2004, introduced in Parliament in 2004, is still under consideration of the Parliamentary Standing Committee on Agriculture. It aims to regulate the quality of seeds and planting material of all agricultural, horticultural and plantation crops to ensure availability of true to type seeds to Indian farmers; curb the sale of spurious, poor quality seeds; protect the rights of farmers; increase private participation in seed production, distribution and seed testing; liberalise import of seeds and planting materials while aligning with World Trade Organization (WTO) commitments and international standards. Comprehensive and authentic databases on seed production and trade in India by public and private sectors as required under the seed and plant variety laws need to be built up. The seed chain and the norms for quality control should be followed without any compromises or shortcuts.

At present, the public sector is responsible for most valuable germplasm while private seed agencies concentrate on more remunerative high value seed segment. Under the circumstances, clear protocols need to be developed for sharing precious germplasm with the private sector on payment of royalty, while ensuring their conservation and preventing possible erosion of the national interest in the context of international agreements on plant variety and intellectual property rights. If this can be done, there is vast scope to expand linkages between the private seed industry and public research institutions to take advantage of the positive aspects of both the segments for the benefit of farmers. ICAR needs to revisit procedures for variety identification, release and notification to cover private and farmers' varieties and also to avoid bias in favour of varieties evolved by the testing institutions. The number of seed testing centres in the country should be expanded rapidly, if necessary in PPP mode and with third party oversight, to reduce the time taken in assessment and refinement of varieties and hybrids and technologies for production and protection of crops. There is also a need for 'Phytosanitary' certification, especially for export/ import of seeds. The State Seed Corporations may establish at least one such certification centre in each major State.

The DAC made the present assessment of seed requirement during the Twelfth Plan for its proposed Seed Mission with respect to some of the major crops which brings out that even excluding requirements arising from possible shift to hybrids, seed production of varieties will need to increase by about a third to meet the projected increase in seed replacement rates. Since seed-production planning should be done with a long-term perspective (considering the viability of the seed) and also to keep buffer stock of seed to meet eventualities of natural calamities that require replanting, the actual production requirements may be higher. To meet the seed demand for 45 major crops produced within the country and required under diverse conditions, seed hubs need to be identified to produce seed and supply the same to the farmers in each area. This will save cost of transportation. Public agencies will also need to strengthen infrastructure for seed processing, storage, transportation and distribution.

Adequate availability of quality seeds is a particular challenge for farmers in rain-fed areas where rainfall risks are high and productivity depends crucially on timely sowing within a short rainfall window. The seed system must be capable of providing seeds of contingency or alternative crops during prolonged dry spells. With protection of crop diversity important in rain-fed areas, strengthening and improving local-seed systems and linking these to NARS is a necessity for productivity enhancement.

An important part of the new Mission will therefore be to better integrate farmers with production and distribution of quality seeds through, for example, seed village programmes and by encouraging NGOs to help FPOs take up seed production. Therefore, capacity building will be vital to success. Fodder seeds that are presently neglected and scarce will need to be emphasised. Equally, the Mission must be enabled to convey to NARS accurate feedback from farmers on seed suitability.

(B) Farm Machinery

Wages have increased significantly in recent years and with labour accounting for more than 40 per cent of variable cost, many farm organisations report that shortage of labour is obstructing operational efficiency. Animal power is also declining, with commercial banks reluctant to extend loans for bullocks. This has naturally led to an increase in farm mechanisation. However, farm mechanisation has so far been biased in favour of tractors and been concentrated in irrigated-command areas paying little attention to the needs of farmers in dryland areas and the scope for introducing small machines that might be useful to meet their needs. Considering the farm sizes and prevailing skills, farm mechanisation penetration would have to be enhanced through promotion of custom hiring models as well as individual ownership. While draft animal power based implements and manual tools should be owned by individual farmers (with appropriate financial incentives, for example, off season employment for animal power by integrating some services such as 'manure transport' with MGNREGS), expensive machinery should be promoted thorough custom hiring. This could be done by promoting machinery service centres involving existing FPOs or by groups of farm youth trained in machinery operation and maintenance. Greater impetus is needed to develop needbased and regionally differentiated farm machinery. Ongoing efforts by NARS need to be suitably strengthened with appropriate participation of commercial agricultural machinery manufacturers. Financial incentives could be linked to requirements thrown up by extension experience from different locations or from FPO demand. The Mission should identify and convey to NARS the critical mechanisation gaps and, in particular, specific local requirements related to machinery for soil and water conservation and gender-friendly implements.

(C) Strengthening Extension

During the Eleventh Plan, the task of strengthening and restructuring agricultural extension was approached through a wide mix of different initiatives. The context for this was that while public sector extension arrangements have weakened, the number and diversity of private extension service providers have increased in the last two decades. These include the media, NGOs, producers associations, input agencies and agri-business companies. Many provide better and improved services to farmers, but their effective reach is limited and most poor producers are served neither by public nor private sector in many distant and remote areas. Notwithstanding the important role being played by private sector extension, there are also concerns with regard to wholesomeness of information, given equity and long-term implications. Although setting up ATMAs in almost all districts was the single most important achievement, this went hand-in-hand with efforts to enhance quality through domain experts and regular capacity building. Other efforts included interactive ways of information dissemination, public–private partnerships and pervasive and innovative use of ICT/Mass Media. Efforts were also made to involve agri-entrepreneurs, agri-business companies and NGO experts to bolster public extension. Most of these efforts will have to continue in the Twelfth Plan since extension is a continuous process. But, in view of the initial broken down condition, there are considerable gaps even after the subsequent effort. For example, an evaluation of ATMAs by the Agricultural Finance Corporation in 2009–10 found that although 52 per cent of respondent farmers said that they gained knowledge of new practices and technologies from this, only 25 per cent felt that this had helped to increase production. It is perhaps time to conduct a country-wide extension census to identify extension resources (manpower, infrastructure, expertise) available in public and private sectors.

It is also necessary to continue with experimentation. There are number of models which have been successfully implemented in several States and countries which can be tried as pilots by ATMA and then expanded. Many civil society organisations have successfully experimented with community managed extension systems with members of the local community acting as agents of agricultural extension. In the Community Managed Sustainable Agriculture (CMSA) model of Andhra Pradesh, members of the village community have been trained and developed as Community Resource Persons (CRPs). CRPs adopt elements of sustainable and eco-friendly agricultural practices in their own farms and are in a better position to motivate and convince other farmers than normal extension workers. Working with agricultural scientists and extension personnel under the broad ATMA umbrella, CRPs can help technology transfer and diffusion. Agricultural extension covering crops and allied sectors is primarily the responsibility of the States and it is expected that States should drive the extension reforms process. Any national effort in this regard can only support States' efforts. Moreover, as noted by the Twelfth Plan Working Group on Agricultural Extension, while public policy in agriculture increasingly recognises importance of public– private partnership in extension, the experience so far is that PPPs have been the exception rather than the rule. States must adopt PPP, but this is not substitute for strengthening the public extension system. Future collaboration between public and private players will have to focus more on the public sector's ability to set standards and monitor progress so that these standards are enforced on all players, including public extension agents, while providing institutional training and support.

An important task of the new Mission should therefore be to consult with States so as to evolve a standards and regulatory framework for certifying and validating extension activities by all players, including public extension agents. MANAGE and SAMETIs should take the leading role in driving extension reforms at the National and State levels respectively. The corporate sector should be encouraged to involve itself in this effort and in agricultural extension in general, if only as part of their Corporate Social Responsibility (CSR). Even more important than funding under CSR, the corporate sector can support by providing adequate extension training to their extensive promotion network of distributors and dealers so as to meet required standards.

The Twelfth Plan Working Group on Agricultural Extension has noted that although ATMAs exceeded targets on training, demonstrations and exposure visits, the number of farm schools set up was well below target and that matters were lagging also on strengthening and extending Farmer Advisory Committees at every level. Since active involvement of farmers in planning and executing extension reforms was a key ATMA goal, the new Mission must concentrate on this and on feedback, particularly on technology and on agricultural plans at district and lower levels. A critical aspect of this will be ATMA–KVK coordination and more intensive ICT use.

Extension services must also be gender-sensitised, and this will require joint efforts, involving the Mahila Kisan Sashaktikaran Pariyojana component of the National Rural Livelihood Mission (NRLM) under MoRD, the Project Directorate for Women in Agriculture of ICAR and National Gender Resource Centre in Agriculture (NGRCA) of Ministry of Agriculture (MoA). Further, since the present extension system does not pay adequate attention to livestock, fishery and fodder and separate extension machinery for animal husbandry and fishery is not feasible in many states, this function will need to be integrated with ATMA with suitable KVK and NGO backstopping. Indeed, convergence should be a basic goal of the new Mission, both on the side of technology dissemination and feedback as well as for planning integrated agricultural development. The ultimate objective of the Mission should be to upgrade ATMA from a society operating as an adjunct to line agricultural departments to an independent entity with technical capability to offer local solutions and deliver feedback to NARS on location specific technology needs. The larger trends of public policy point towards decentralised governance of natural resources and the promotion of growth with increasing emphasis on district (and lower) level planning. It is necessary to see decentralised planning as an iterative planning—doing—learning—planning cycle rather than as simply a onetime activity. The challenge is to institutionalise this process and ensure that the agency facilitating planning also has accountability in the overall outcome. ATMAs are a natural choice for such an agency in the present context.

SPECIFIC PLANS AND OBJECTIVES FOR THE MAJOR SUB-SECTORS

(A) Livestock

For achieving growth rate of 5–6 per cent per annum the animal husbandry sector would need to address important challenges during the Twelfth Plan. These include delivery of services, shortage of feed and fodder and frequent occurrence of deadly diseases. Compared to its contribution in the economy livestock sector has received much less resources and institutional support. Livestock extension remains grossly neglected. The country still lacks adequate facilities and the infrastructure for disease diagnosis, reporting, epidemiology, surveillance and forecasting. Livestock markets are underdeveloped, which is a significant barrier to commercialisation of livestock production. Besides, the sector is also coming under significant pressure of increasing globalisation of agri-food markets. Although there is demand for Indian meat products in international markets, lack of international processing standards is a hindrance. Unfortunately, schemes on modernisation of slaughterhouses and by-product utilisation have not been effectively implemented. In the animal husbandry sector, the major priority areas during Twelfth Five Year Plan will be breed improvement, enhancing availability of feed and fodder and provision of better health services, including proper breeding management. Conservation and perpetuation of diverse local germplasm, which are adaptable to Indian climate conditions and resistant to various endemic diseases, will be another important area, with clearer focus on sub-sectors such as small ruminants that have so far been neglected.

An important Twelfth Plan initiative is the National Dairy Plan (NDP), which has already been launched as a central sector scheme with credit support from the International Development Association (IDA). To be implemented by the National Dairy Development Board (NDDB) through a network of End Implementing Agencies (EIAs), mainly dairy cooperatives and producer companies, this aims to (*i*) increase productivity of milch animals and thereby increase milk production and (*ii*) provide rural milk producers with greater access to the organised milk-processing sector. These objectives would be pursued through adoption of focused scientific and systematic processes in provision of technical inputs, supported by appropriate policy and regulatory measures.

An important sub-component of (*i*) above will be scientific progeny testing and pedigree selection of bulls for semen required in artificial insemination (AI) services. It is planned to make available about 900 high genetic merit bulls for replacement of bulls maintained at all 'A' and 'B' graded semen stations and thereby achieve 100 per cent high genetic merit bull replacement at these semen stations by end of the Twelfth Plan. It is estimated that this would produce some 100 million high-quality disease- free semen doses annually. Taking NDP into account and, with RKVY incentives for States to substantially enhance public sector investment in agriculture and allied sector during the Eleventh Plan, the Department of Animal Husbandry, Dairying and Fisheries (DAHDF) has also decided to redesign its schemes. It aims to provide more flexibility to States while reducing the number of Centrally Sponsored Schemes (CSS) and reorientating these to secure better programmatic focus.

On genetic improvement in bovines, the current major programme is the 'National Project for Cattle and Buffalo Breeding (NPCBB)' which is being implemented since October 2000. Unlike NDP, which aims to provide breeding services from the dairy side, NPCBB is administered as part of States' veterinary services. DAHDF proposes to continue NPCBB in this present form since the DAHDF target is to expand the artificial insemination programme from present coverage of about 25 per cent of breedable population to 50 per cent, which will require an expansion of AI services beyond the about 35 per cent coverage planned for under NDP. This is because NDP will not cover all States and there are likely to be farmers not covered by dairyled breeding services even in States covered by NDP. Moreover, States have already established Livestock Development Boards (LDBs) in the present format to implement bovine breeding programmes with a stated focus on development and conservation of important indigenous breeds. The critical requirement is that NPCBB and States' efforts through LDBs share common standards and protocols with NDP in progeny testing, pedigree selection and to improve conception rates. If so, resources are sufficient to achieve 5 per cent growth of milk production in the Twelfth Plan.

Since standards and protocols will be the key to success on the breeding side and basic commonality will have to be brought between NDP, LDBs and NPCBB, there is need for some architectural redesign during the Twelfth Plan. Therefore, although NPCBB will continue, this will be as a component of a new *National Programme for Bovine Breeding and Dairy (NPBBD)* which will subsume all DADF existing schemes on dairy development. Thus, NPBBD will have two main components, namely National Programme for Bovine Breeding (NPBB) and Dairy Development. The component for Dairy Development will mainly focus on States/areas not covered under NDP and, in addition to existing support areas, convergence will be attempted in a phased manner so that dairy cooperatives which are not part of NDP also offer breeding and extension services. It is hoped that such combined activities in respect of dairying with breeding will be more effective in extension of artificial insemination services, feed management and marketing of good quality of milk which are essential for improving productivity and income of farmers. In the meantime, NPBB will continue existing NPCBB functions through LDBs and the veterinary side with two areas of focus: first, to harmonise breeding standards and protocols; and, second, to achieve the so far unrealised stated focus on development and conservation of important indigenous breeds.

The main programme on the veterinary side will be an expanded scheme for *Livestock Health and Disease Control*. Such an expansion is necessary because occurrence of diseases like foot and mouth disease (FMD), hemorrhagic septicemia (HS), brucellosis, mastitis, blood protozoon and so on, have been accentuated with introduction of exotic breeds. Taking into account the economic losses from these diseases, and also those of small ruminants (PPR or peste-des-petits ruminants), particularly to small, marginal and landless farmers including women farmers, it is necessary to have a strong focus on national control programmes for all major animal diseases, backed by epidemiological analysis and assessment of the animal diseases in different agroclimatic regions. Unrestricted movement of livestock, as well import of germplasm, and changes in ecosystems due to climate change are adding to occurrence of diseases. The availability of improved, potent and efficacious vaccines meeting international Twelfth Five Year Plan standards against major prevalent diseases can enable better management, containment and control of the diseases. The new programme will associate all ICAR institutes specialising in animal diseases and, in consultation with the State Governments, formulate and implement more effective strategies for control of different diseases. The third major programme of DADF will be the *National Livestock Mission (NLM)*. Apart from bovine breeding, dairying and livestock health schemes, DADF runs a plethora of other schemes relating small ruminants, poultry, piggery and fodder development which although of extreme importance, especially to small, marginal, landless and women farmers, have so far not received focused attention. The multiplicity of small schemes in these livestock sectors has been a major constraint since this limits the capability of states to effectively access funding under various schemes. In order to provide greater flexibility to states in formulating and implementing various projects, it is proposed to merge these schemes with the main objective of achieving sustainable development and growth of the livestock sector. The NLM will have an important mini-mission of feed and fodder, with an objective to substantially reduce the gap between availability and demand. The deficit of dry fodder (10 per cent), concentrates (33 per cent) and green fodder (35 per cent) continues to be high, although availability of feed resources has improved somewhat. The forage and fodder seed need varietal and quality improvement alongside better availability. The NLM will encourage seed companies and SAUs to take up forage seed production on a priority basis. Developing common property resources, including grazing land and wasteland, and better utilisation and enrichment of crop residues/agricultural by-products is the other priority. Ration balancing, which is being promoted under NDP, will also be promoted under this minimission on feed and fodder. The NLM will also have an additional minimission relating particularly to development of small ruminants, but also covering poultry, piggery and other minor livestock species. While subsuming some of the existing Central Sector Schemes for poultry, small animals and fodder development, the objective will be fuller development of the animal biodiversity available in our country, which is a rich treasure of germplasm. NLM will also focus on predominantly non-descript pig populations, concentrated in NE region and eastern region there have poor productivity. Indian poultry industry is well equipped and organised to achieve target growth rate of 11 per cent for commercial broilers and 7 per cent for layers although it failed to diversify in favour of duck, quail, turkey and emu production. Need-based import of grandparent stock of reputed international brands may be continued with strict enforcement of bio-security measures. Rural poultry sector however, needs financial, infrastructure and technological support to raise the present 2 per cent growth rate to 3 per cent. All these, including the conservation of threatened breeds, will be covered by NLM in a flexible but more focused programmatic manner. Other issues that NLM will address include livestock insurance and extension and any innovative initiative proposed by states for development of the livestock sector, for example, to deal with unhygienic slaughtering and processing. If State Governments notify minor veterinary services accordingly, shortage of human resources of veterinary staff could also be supplemented by recruitment of para-vets, similar to that of ASHA, to provide minor veterinary services and supplement the livestock-extension activity in the States. In this context, it might be noted that as public-sector spending is enhanced for development of livestock, there is need for continuous assessment of the efficacy of AI and of animal health programmes in terms of success rates, lactating efficiency and of potential and actual yield per animal. (B) Fisheries Potential of fisheries sector in providing quality food and nutrition, creating rural livelihoods, advancing socio-economic development in the rural and far flung areas is widely demonstrated and globally recognised as a powerful tool for poverty reduction and fostering rural development. Annual fish production has reached to the level of 8.30 million Agriculture tonnes during 2010–11 (P). Annual export earning has also touched record US$2.9 billion mark contributing about 17 per cent to national agricultural export. About 14.5 million people are engaged in fishing, aquaculture and other allied activities of which about 75 per cent are in inland fisheries and the remaining in marine fisheries.

In marine fisheries, uncontrolled fishing capacity has led to over-exploitation of the coastal resources. The estimated potential of the offshore waters offers opportunities which calls for upgradation of the fleet as well as skills and capacities of the fishers and incentives to promote diversified fishing in the offshore waters. Implementation of Monitoring, Control and Surveillance (MCS) as a new programme in the ensuing Plan is expected to bring more discipline and regulate the activities so as to maintain the growth rate in a sustainable manner. There is a need of additional infrastructure and also upgradation of facilities infrastructure for landing and breathing facilities of marine fishing fleet and for domestic marketing that have been the main reasons for post-harvest losses. Freshwater aquaculture, which contributed to the 'Blue Revolution' in the country in late 1970s, is now almost stagnating in terms of species diversification and yield rates due to less focus on sustainable development of inland capture fisheries in past Plans; increasing pressure on the resources, including habitat degradation; and multiple use of inland water bodies with least priority to fishery requirements. Average yield rates are around 1,000 kg/ha/ yr, against potential of 3–4 thousand kg/ha/yr. The efforts to raise productivity should, however, be accompanied by formulating guidelines and regulatory measures for the judicious use of critical inputs keeping in view the principles of the FAO Code of Conduct for Responsible Fisheries. Enhance the productivity and production of fishes. But, there are no organised brood-stock production and management facilities in the country. Therefore, there is need to set up brood banks in each State with one at the Central level. There is need to promote commercial fish feed mills and indigenously formulated fish feeds with locally available ingredients by supporting the private players with enhanced capital subsidy especially in the States where there are no feed mills. Adequate infrastructure is not available for disease diagnosis and treatment for fish disease management. There is a strong need for capital investment as well as support for the State Governments in capacity building and managing the disease diagnostic laboratories. There is also a need for creating a disease surveillance and communication agency/ mechanism at National level along with its wings at suitable regional locations to build awareness and send alerts to the stakeholders. This agency shall have adequate regulatory powers to ensure the disease control. The gradual decline of Freshwater Fish Farmer's Development Agencies (FFDAs) and Brackish water Farmer's Development Agencies (BFDAs) and their resultant poor performance coupled with weak extension services has impacted the overall growth of aquaculture in the country. Rejuvenation and consolidation of the two field-level agencies (FFDA and BFDA) into a single agency— Fisheries and Aquaculture Development Agency or can undertake extension of technologies, promote networking of farmers and fishers (mainly from reservoirs) and provide effective liaison between the farmers and developmental and other extension agencies such as the Krishi Vigyan Kendras and the ATMAs as well as sourcing the public finance for fishers.

An important initiative of Government of India for development of fisheries sub-sector has been to launch 'National Fisheries Development Board' (NFDB) as a Special Purpose Vehicle (SPV) in the year 2006 for implementing fishery developmental schemes in an integrated manner. The scope of NFDB would be expanded to include management of fish diseases and creation of related infrastructure which is a gap in the present scenario. During the Twelfth Plan, the existing CSS on inland and marine fisheries (except welfare of fishers) will be merged with NFDB to facilitate expansion of fisheries through integration of a wide array of activities, but with its main focus on inland fresh water fishery. The schemes will be implemented under the aegis of NFDB removing any duplication or overlap of efforts. This clear demarcation of work, it is hoped will enable the growth rate of the sector to rise to 6 per cent during the Twelfth Plan. DADF would focus its efforts on policy, regulation and welfare of fishers, and will implement the scheme relating to welfare of inland and marine fishers. The DADF will also handle the strengthening of fisheries data base, implementation of the proposed scheme on Monitoring, Control and Surveillance (MCS), all fisheries policy and legal matters, coordination with the sister Ministries/Departments at the Centre and the States to make the sector's foundation more robust and sustainable and build stronger linkages between research and development. Future course of fisheries management will have to work at two fronts—sustainable utilisation of healthy resources and rehabilitation of threatened resources by habitat restoration and appropriate conservation measures. Climate change and its possible impact on fisheries and fishers is again an additional challenge. Thus, the future course of management will require highest level of compliance of acts and regulations, extensive adoption of BMP and implementation of CCRF (Code of Conduct for Responsible Fisheries introduced by FAO) which would be possible only through the cooperation and active participation of resource user communities as partner in the development and management process.

(C) Horticulture

With increasing per capita income, Indians are consuming more of fresh and processed horticultural products indicating growing scope of horticulture by improving crop productivity and efficiency in the value chains. The initiatives taken in the horticulture sector during the Tenth Five Year Plan have helped in achieving high growth in production. During the Eleventh Five Year Plan, the growth rate of horticulture is expected to be 4.7 per annum, slightly short of the projected 5 per cent. There has been a marked push to the expansion in area under horticulture crops since taking up of a number of initiatives for horticulture development through NHB, TMNE (NE) and then NHM in 2005–06.

However, in quest for area-expansion efforts, the states have neglected due thrust on increasing productivity of existing orchards through technology infusion or by capital investment in fertilization, input management, plant protection and farm mechanisation. The area expansion programmes have also lacked the proper backward linkage with supply of quality seed and planting material. Even where Nursery Act exists, it has not been enforced effectively. A proper system of accreditation and rating of nurseries, with clearly defined protocols, is the most important priority and will have to be put in place during the Twelfth Plan. Adequate attention to post-harvest management and market development and processing has yet to pick up and is the weakest aspect of diversification towards high-value products resulting in frequent and sharp fluctuations in prices of fruits and vegetables in domestic market. As discussed earlier, marketing sector reforms implemented by States have so far not resulted in efficient marketing of perishables, or put in place transparent system of auction and price discovery. There are huge logistic gaps between production clusters and marketing centres, often at long distance, and private sector investment in post-harvest management and in marketing infrastructure has not come forward to the desired extent. There is also lack of proactive steps to enhance export competitiveness for high-end export destinations. The availability of adequate regular, uninterrupted, affordable power supply for setting up infrastructure like tissue culture labs, seed processing plants, bio control labs and post-harvest management units like cold storages, ripening chambers and so on is a constraint which needs to be addressed at least in and around horticulture clusters. Since horticulture operations are cost intensive and hi-tech, horticulture growers need to be provided affordable credit with higher ceiling and insurance against risk.

The horticulture development missions depend on a loose set-up of Technology Support Groups for technology inputs. This has proved inadequate. Many States do not have adequate technical trained manpower to implement programmes. Unless State Governments fill up vacant posts and create additional posts to provide necessary technical input, it should be deemed that they are uninterested and the mission wound up in those States. During the National Horticulture Mission will integrate the several existing schemes in this sector and aim at holistic growth of horticulture sector, including bamboo, through area-based regionally differentiated strategies, which include research, technology promotion, extension, post-harvest management, processing and marketing, in consonance with comparative advantage of each State/region and its diverse agro-climatic features. The Mission will also facilitate marketing reforms discouraging payment of unnecessary market levies and encouraging private investment for setting up horticulture produce markets. While continuing existing efforts, and aiming at 5 per cent growth of horticulture production during the Twelfth Plan, the main objective will be to build required capacities at State level, and assess their seriousness, so that the horticulture development related activities can be transferred fully to States by end of the Twelfth Plan.

Another objective will be to improve horticulture statistics which continue to be weak, lacking both a validated methodology for data collection of horticulture crops and adequate machinery to collect such data. Generation and dissemination of quality data can also help in averting frequent situations of gluts and shortages and exploitation of such situations by the middlemen and speculators. DAC needs to take up a one-time horticulture census with the objective of generating reliable base line data. Further, as recommended by NSSO committee on improvement horticulture statistics, there is need to set up an extensive network of Horticulture Information Systems (HIS) with proper data units in all relevant districts and at State and Centre level covering all relevant aspects. To facilitate this, at least 3 per cent of Mission funds should be earmarked for this purpose.

(D) Food Grains and Oil Seeds

Since cultivated land is limited, with potential for only marginal future increase through higher cropping intensity or development of cultivable wasteland, future increase in production will have to come mainly from yield improvement. Declining average annual growth of food grains yields from 3.2 per cent in 1980s to 1.6 per cent in 1990s and further to only 0.6 per cent during the Tenth Plan, taking this well below population growth, had led to widespread concern about future food security. The issue was, therefore, analysed fully with several alternatives considered and the *National Food Security Mission (NFSM)* was formulated for the Eleventh Plan. This was based on an assessment of yield gap data then available, and was focused on increasing yields in low-yield districts using a variety of known interventions, with particular attention to availability of quality seeds. Although this has paid off, with food grains yield growth increasing to 3.3 per cent during the Eleventh Plan, a valid question regards continuation of NFSM is whether yield gaps are still large?

A committee set up under Chairmanship of Chief Minister of Haryana has recently examined the issue and suggested continuing with the strategy to bridge the gap between real and potential yields. The analysis of gap between potential and achieved yields presented to this committee suggests that there is considerable potential of increasing yields even in high productivity irrigated areas with the current technology. For these areas, the strategies will need to concentrate on propagation of balanced use of fertilisers and application of micro-nutrients, water and soil-saving technology. In case of wheat, however, there is need to step up research to develop varieties resistant to temperature. The major yield gaps are due to management practices. Other reasons for this gap need to be ascertained through specific studies and addressed through appropriate interventions. In addition to enhancing productivity of food grains in the low productivity areas, it is equally important to stabilise the productivity gains in these areas as well as in areas where productivity levels are comparatively high. With these issues in mind, the *National Food Security Mission (NFSM)* will be revamped during the Twelfth Plan. While the Eleventh Plan approach of focused attention on identified districts and crops in a location specific, target-oriented manner will continue, greater attention will be put in most areas to shift from exclusive focus on individual crops to the cropping system/farming system approach. In particular, the Mission will be extended to cover coarse cereals and fodder, in addition to wheat, rice and pulses as at present. The Mission contemplates that promotion of package of practices in compact blocks in a hand holding approach would not only help in enhancing the production and productivity of a region but also help in changing mindsets of farmers due to its positive large-scale impact. This approach will ensure inclusion of all farmers in the compact block irrespective of their size of holding or social status and will be compatible with other efforts that encourage strengthening of institutions, including building of farmers organisations and FPOs. The Mission will also build upon the Eleventh Plan experience regarding conservation agriculture. However, the main way in which NFSM will be extended during the Twelfth Plan is through greater emphasis on strategic-area development. The two programmes that were started as RKVY sub-components in the Eleventh Plan namely, the 60,000 pulses village programme and the intensive millets production programme will largely be shifted into NFSM. On another sub-component of RKVY—Bringing Green Revolution in Eastern India (BGREI)—a view will be taken by DAC in consultation with States regarding format of its continuation during the Twelfth Plan. Also, some additional districts in Himachal Pradesh, Uttarakhand and the north-eastern region will be included to provide a specific thrust on foodgrains cultivation in hill areas. Such restructuring of RKVY and NFSM will address the problem of bridging the existing large gap between potential and realised rice yields in eastern States and the challenge of increasing pulses production. Since BGREI allows components which are not part of NFSM, and since development of the eastern region requires significant investments in power and marketing infrastructure, the final design of how to proceed on the relative contributions of RKVY and NFSM will need to be decided in consultation with the States. Also, since a counterpart of expanding rice production in eastern States is to reduce rice area and resulting groundwater stress in the North-West, a decision will have to be taken on what components of the latter effort should be stressed in NFSM/RKVY. Preliminary targets under the NFSM for the Twelfth Plan are enhancing production by additional 25 million tonnes of foodgrains consisting of 10 million tonnes of rice, 10 million tonnes of wheat, 3 million tonnes of pulses and 2 million tonnes of millet. Also it aims to expand fodder production to meet the demand both of green and dry fodder. In all probability, the requirement of sufficient quantity of dual purpose feed and fodder will require raising this target to 30 million tonnes, with additional production of coarse cereals put at 7 million tonnes. All these targets are less than was actually achieved during the Eleventh Plan and are consistent with demand forecasts. This would amount to targeting 2–2.5 per cent increase in foodgrains production in the Twelfth Plan. Another consequence of the expanded scope of NFSM will be to absorb the pulses and maize components presently in the Integrated Scheme for Oilseeds, Oil palm, Pulses and Maize Development. During Twelfth Five Year Plan, it is proposed to replace this scheme with a new *Mission on Oilseeds and Oil Palm* which will be launched with a preliminary target to increase the production of oilseeds by at least 4.5 per cent per annum, that is, the same rate of growth as actually achieved during the Eleventh Plan. The core of this Mission will therefore be to continue past efforts with a clearer focus on oilseeds. However, since production of oilseeds has not been able to match the increasing demand of edible oils, resulting in persistence of a huge gap between demand and production of edible oils in the country, the Mission will also aim to expand area under oil palm to realise the latent potential of the oil palm in the country. This part of the Mission will fully consider a proposal made recently by CACP and incorporate whatever is feasible.

NATURAL RESOURCES

(A) Water

The water resource potential of India is assessed as 186.9 million hectare meter, mostly from rainfall. With annual availability still more than utilisation and with its uneven spatial and temporal distribution leading to floods/droughts in some or other parts of the country every year, there is a strong demand to fully utilise this potential as soon as possible. The total States proposals on investment in Irrigation and Flood Control for the Twelfth plan add up to about Rs. 4,00,000 crore, which alone would amount to over the 4 per cent of cumulative GDP from agriculture and allied sectors being targeted as total public investment in this sector during the plan. Recognising both the criticality of irrigation for agricultural growth and the potential available, the Centre's Twelfth plan gross budgetary support for development of water resources (including on AIBP) is being stepped up to Rs. 1,09,552 crore from the Eleventh plan actual expenditure of Rs. 41,427 crore.

However, the performance in respect of creation and utilisation of irrigation facilities during the Eleventh Five Year Plan was not satisfactory. The original Eleventh Five Year Plan target for creating irrigation potential was 16 million ha. This was subsequently revised to 9.5 million ha, which has been achieved. However, utilisation out of the created potential is expected to be only 2.7 million ha. The ever increasing gap between created potential and its utilisation is an issue that is a Twelfth Plan priority, steps to address which are discussed in another chapter.

In recent decades irrigation facilities have increasingly been created through exploitation of groundwater deployment. However, non-judicious exploitation of groundwater for irrigation purposes in India is already showing signs of crisis in many parts of country. Studies report that more than 26 cubic miles of groundwater has already disappeared from underground aquifers in large areas of Haryana, Punjab, Rajasthan and Delhi, between 2002 and 2008 (NASA 2009). Global Runoff Data Centre, University of Hampshire and International Earth Science Information Networks have projected that around 30 per cent area of India falls in the extreme water scarce zone having less than 500 m3/person/year supply of renewable fresh water. The information from the Central Ground Water Board reveals that situation has worsened in most of the states since 2004. The groundwater level has been declining annually by about 4 cm during the past decade, often resulting in drying of rivers and wetlands and contamination with arsenic, fluoride and other toxic substances. This requires effective regulatory framework and participatory watershed development, especially because groundwater extraction is often highly unfavourable to the small farmers who cannot keep investing to tap deeper aquifers. Apart from developing appropriate regulatory framework, and people's participation, the need of water saving devices and crop planning cannot be overemphasised. Micro-irrigation coverage will be given priority both in irrigated and rain-fed areas, as part of comprehensive local planning.

(B) Watershed Development

Watershed development has long been one of the major channels directing public investment to natural resource base and production systems in rain-fed agriculture. From their earlier emphasis on soil and water conservation, the focus in case of watershed projects is shifting towards livelihood security and income generation. It is also now generally accepted that to be effective, the watershed development and soil conservation investments have to be complemented with farming systems investments in a watershed-plus framework that takes into account the diversity of rain-fed agriculture. However, despite considerable emphasis on this in the Eleventh Plan design and development of common guidelines, actual performance in regard to watershed development was poor during the Eleventh Plan. Since all watershed development programmes have been transferred to the Department of Land Resources, the Ministry of Agriculture has to redefine its initiatives for rain-fed farming and sustainable agriculture.

The National Rainfed Area Authority was constituted with the specific objective of integrating schemes/programmes and activities of various Departments of the Centre and the State Governments with regard to dryland farming as well as providing technical back stopping for watershed development in a comprehensive manner. The authority was expected to play a major role in training of the officials associated with the watershed development projects and also take a lead role in social mobilisation which is critical in the success of the watershed development programmes. It was also expected to take up studies for evaluation of the implementation of projects by the States. So far Departments both at the Central and State level has not taken much interest in associating NRAA either in evaluation of the programmes or for providing technical input for these. NRAA expertise will be better utilised during the Twelfth Plan.

(C) Land and Soil Health Management

Land is the prime natural resource of which 140.02 million hectares are net sown area. Since 1990–91 there is gradual but sustained decrease in net sown area from 143 million hectare to 140 million hectares with corresponding increase in fallow land. The demand from non-agricultural uses like industrial and urban requirement as well as speculative demand on account of rising land value is putting pressure on availability of land for agricultural use. There is an urgent need for State Governments to lay out clear policies to protect productive agricultural land and provide specific guidelines on preservation of commons and their protection. There are also other important institutional and policy issues concerning land: proper recording of land titles, easing tenancy rigidities, computerisation of land records as well as addressing declining size of holdings.

An important aspect of land is its degradation in terms of mechanical, chemical and biological. Widespread and continuing erosion of country's natural resource base is threatening the sustenance of agriculture sector's growth rate. Over 120 million ha have been declared degraded or problem soils (NAAS 2010). Conservation agriculture (CA), integrated nutrient management, carbon sequestration, erosion control, saline and alkaline soils management, legislation for soil protection, development of remote sensing and GPS-based Decision Support System (DSS) and amelioration of polluted soil are required to rejuvenate deteriorated soils.

(D) Use of Fertilisers and Pesticides

Fertiliser consumption in the country has been increasing over the years and now India is the second largest consumer of fertilisers in the world, after China, consuming about 26.5 million tonnes of NPK. However, imbalanced nutrient use coupled with neglect of organic matter has resulted in multinutrient deficiencies in Indian soils. These deficiencies are becoming more critical for sulphur, zinc and boron. As nutrient additions do not keep pace with nutrient removal by crops, the fertility status of Indian soils has been declining rapidly under intensive agriculture and is now showing signs of fatigue, especially in the Indo-Gangetic plain. Potassium is the most mined nutrient. Sulphur deficiencies are also showing up in all parts of the country especially in the southern region. In a comprehensive study carried out by ICAR through their Coordinated Research Project on Micronutrients, Toxic and Heavy metals, based on an analysis of 2,51,547 soil samples from different states, it was found that 48 per cent of these samples were deficient in zinc, 33 per cent in boron, 13 per cent in molybdenum, 12 per cent in iron, 5 per cent in manganese and 3 per cent in copper. The micronutrient deficiency is a limiting factor lowering fertiliser response and crop productivity. As a result of over-emphasis on chemical fertilisers and imbalanced fertiliser use, efficiencies have become abysmally low: hardly 35 per cent for N, 15–20 per cent for P and only 3–5 per cent for micronutrients like zinc, resulting not only in high cost of production but also causing serious environmental hazards. At this rate, the National Academy of Agricultural Sciences has estimated that for meeting the food needs of the country by 2025, India may have to increase NPK supply to over 45 million tonnes from the current level of 26.5 million tonnes and of organic manures from 4 to 6 million tonnes. The Twelfth Plan envisages NPK demand at 34–36 million tonnes by 2016–17, but the more important priority should be to give much greater emphasis than hitherto on fertiliser use efficiency and soil health. Restoration of soil health requires initiatives for continuous monitoring of soil health, measures to arrest decline of soil health, creating adequate facilities for soil testing, fertilisers testing, developing and upgrading testing protocols, ensuring judicious and efficient use of fertilisers and pesticides. Judicious use of fertiliser requires adequate soil testing facilities. By 2010–11 there were 1,049 soil tests labs in the country with a soil analysis capacity of 106 lakh soil samples per annum. The State Governments have issued 40.8 million soil health cards to the farmers by October 2011. Although a massive achievement in fairly short time, this remains far below the requirement of soil testing capacity. To augment the capacity the State Governments need to utilise resources from Rashtriya Krishi Vikas Yojana and also engage State Agricultural Universities, Agricultural Produce Marketing Committee and other institutions. There is need for widespread awareness creation for soil-test–based fertiliser use by involving State Agricultural Universities and KVKs and NGO and other stakeholders.

Measures to soil health improvement need to be comprehensively centred on addition of soil organic matter in substantial quantities over time. The efforts for production and use of available biological sources of nutrients like bio-fertilisers, organic manure, bio-compost for sustained soil health and fertility and improving soil organic carbon and so on as alternative inputs have been inadequate so far. For promotion of these inputs in conjunctive use with chemical fertilisers, and to promote organic farming we need to formulate and define standards for unregulated organic and biological inputs and bring them under quality control mechanism and define/ upgrade standards and testing protocols.

Similarly, use and availability of safe and efficacious pesticides and their judicious use by the farming community is critical to a sustained increase in agricultural production and productivity. Quality of pesticides is monitored by the Central and State insecticide inspectors who draw samples of insecticides from the market for analysis in the 68 State Pesticide Testing Laboratories (SPTLs) that have a total annual capacity of 68,110 samples in 23 States and one Union Territory. However, sale of low quality/spurious pesticides by dealers is widespread and is an issue that States need to handle with seriousness. Further, since use of synthetic pesticides needs to be confined to target control in the right quantity and at the right time, presence of pesticides residue in food commodities is becoming a serious food safety matter. DAC implements a scheme for monitoring pesticide residues and sharing outcomes of the sample analysis with State Governments as well as advising States to take necessary action including promotion of the Integrated Pest Management (IPM) approach, which emphasises a safe and judicious use of pesticides. Many NGOs, however, represent that sporadic promotion of IPM is not helping in establishment of sustainable agriculture practices and that Non-Pesticidal Management (NPM) of pests is the only sustainable answer.

NATIONAL MISSION FOR SUSTAINABLE AGRICULTURE

A major new mission that will be launched during the Twelfth Plan is the National Mission for Sustainable Agriculture (NMSA). Conceived originally as part of the National Action Plan on Climate Change (NAPCC), this aims at transforming Indian Agriculture into a climate-resilient production system through adoption and mitigation of appropriate measures in the domains of both crops and animal husbandry. Since a number activities relating to sustainable agriculture are already parts of other proposed missions, NMSA as programmatic intervention, will primarily focus on synergising resource conservation, improved farm practices and integrated farming for enhancing agricultural productivity especially in rain-fed areas. Key deliverables under this mission will be developing rain-fed agriculture, natural resource management, enhancing water and nutrient use efficiency, improving soil health and promoting conservation agriculture.

Nonetheless, since sustaining agricultural productivity through climate and other challenges to the natural resources base is the focus of this mission, it will have to go beyond its programmatic interventions to bring mind-set changes required in transiting from the past focus on irrigated, chemical intensive agriculture. The recent ICAR network project on National Initiative on Climate Resilient Agriculture (NICRA) provides some insights on requirements of adaptation. NMSA can collaborate with ICAR on specific matters regarding adaptation to climate change. The key to this is a paradigm shift that moves towards a knowledge-based, farmer centric and institutionally supported system where the Government is prime mover and facilitator to demonstrate at scale the overall strength and impact of rain-fed agriculture packages that have slowly emerged through several years of grass-roots work by Government and civil society organisations and have shown the strength of combining water and other interventions at a micro-level. The starting point of NMSA must be an accurate assessment of the natural resource, comprising water, land, climate and biodiversity, which determine the opportunities for livelihoods of the people.

(E) Design of NMSA

While the decision to launch the National Mission for Sustainable Agriculture (NMSA) is quite historical, there are design issues both in view of the fact that the Ministry of Agriculture no longer has a watershed development component in its programmes and because there are strong differences on the matter of fertiliser and pesticides use. While the current National

Mission on Micro-Irrigation, the National Project on Management of Soil Health and Fertility and the Rainfed Areas Development Programme (RADP) window in RKVY can be merged with NMSA, none of these address fully the issues that have been raised by the Twelfth Plan Working Group on Natural Resources Management and Rainfed Farming. Its main recommendation is to observe the following:

1. Focus on stabilising and securing diverse cropping by bringing a focus on 'Rainfall Use Efficiency' as central to policy as against mere use efficiency of applied water. This shift calls for two major focal areas: a. Promote measures for in-situ conservation and efficient use of rainwater b. Invest in shared and protective/supportive irrigation.
2. Harness the inclusive growth potential in the so far untapped Agronomic and Management Innovations that are aligned to enhancing sustainability of natural resources, reducing costs, increasing efficiency of resource use and improving total factor productivity. System of Rice Intensification and non-pesticidal management (NPM) of pests as mentioned in the Approach Paper and options evolving in conservation agriculture are some examples.
3. Strengthen the extensive livestock systems depending wholly or partly on commons and agriculture residues through intensive efforts in improving health care, feed, fodder, drinking water, shelter, institutions and so on. The domain of public policy and intervention must shift to these from the present almost exclusive focus on high yielding breeds.
4. Invest in decentralised and local institutional capacities that enable a shift away from onetime Planning to 'iterative Planning—implementation— learning cycles' anchored by local institutions.
5. Enhance institutional capacities in local governance and resource management, particularly related to Commons and strengthen Panchayat Raj, cooperatives and other stakeholder institutions.

Such institutional base is a prerequisite for evolving location and agro-ecology specific mechanisms of programme designing, credit access, filling in infrastructure gaps, marketing and so on. The specific recommendations of this working group, including the setting up of a National programme on rain-fed farming, could be another component of NMSA, financed by resources currently expended under the scheme of Macromanagement in agriculture which housed the watershed development schemes of DAC and will now have to be wound up. This component could mainstream the learning that has emerged from the International Assessment of Agricultural Knowledge, Science and Technology for Development (IAASTD) along with ICAR's National Initiative on Climate Resilient Agriculture (NICRA).

PLAN FINANCING

Expenditure on Agriculture and Allied Sectors

During the Eleventh Five Year Plan, a combined Plan outlay of 1,36,381 crore (at 2006–07 prices) by the Centre, States and UTs was envisaged for the agriculture and allied sectors. The realisation is estimated to be 1,30,076 crore at 2006–07 prices, that is, 95 per cent of projected Plan. The priority to agriculture and allied sectors in allocation of resources in the combined Plan of Centre, States and UTs has been around 5.6 per cent in the Eleventh Plan, an improvement over 3.6 per cent during the Tenth Plan. At present about 50 per cent of the agriculture and allied sectors plan in the country is being financed by the Centre, including expenditure on Rashtriya Krishi Vikas Yojana (RKVY).

FINANCIAL PERFORMANCE OF THE MINISTRY OF AGRICULTURE

The outlay and expenditures of the MoA and its three departments, DAC, DAHDF) and Department of Agricultural Research and Education (DARE), which implement plans and programmes for development of agriculture and allied sectors. The Ministry is likely to realise 88 per cent of the outlay at current prices. A noticeable feature is that RKVY, which was initiated in 2007–08, accounted for 38 per cent of MoA's total plan expenditure in 2011–12(RE).

DAC with utilisation of around 94 per cent of projected outlay for Eleventh Plan at current prices has shown a better performance. The NHM fell short of targets mainly on account of below par performance in grounding the Terminal Market Complexes. The NFSM and horticultural programmes except NHM have achieved the envisaged financial targets and expenditure on agricultural insurance exceeded the Eleventh Plan projection because of demands arising from the drought of 2009. DAHDF incurred major shortfall in the Plan expenditure. One of the reasons for this was the attempt to introduce a large number of schemes with small outlays during Eleventh Plan which faced problems in their conceptualisation, formulation and approval at various stages. Inadequate staff in the State implementing Departments and resulting limitations on absorption capacity of the States to implement the programmes has also been responsible for the shortfall. Both DAC and DAHDF also transferred increasing amounts through State/ District level autonomous bodies, which will need to be avoided in future since this limits the capacity of States to plan comprehensively for agriculture development. Plan realisation is expected to be around 77 per cent in the case of DARE.

TABLE 12.2 : Outlays and Expenditure of MoA and Its Three Departments (DAC, DAHDF and DARE)

	DAC	*DAHDF*	*DARE*	*RKVY*	*WDPSCA*	*Total*
Eleventh Plan proposed (Current Prices)	41,337	8,174	12,588	25,000	240	87,339
2007–08 Actual	5,769	782	1,280	1,247	40	9,118
2008–09 Actual	6,545	865	1,630	2,887	39	11,966
2009–10 Actual	6,827	871	1,707	3,761	40	13,206
2010–11 Actual	10,208	1,096	2,522	6,720	40	20,585
2011–12(RE)	8,654	1,357	2,850	7,811	50	20,722
Total Eleventh Plan Actual	38,003	4,970	9,989	22,426	209	75,597
% utilisation during Eleventh Plan	92	61	79	90	87	87

RASHTRIYA KRISHI VIKAS YOJANA

The National Development Council (NDC), in its meeting held on 29 May 2007 resolved to initiate a special Additional Central Assistance Scheme viz. Rashtriya Krishi Vikas Yojana (RKVY). The purpose behind this programme was to encourage States to draw up District and State agricultural plans and also increase their own spending on the sector so as to reorient agricultural development strategies for rejuvenating Indian agriculture during the Eleventh Plan (2007–12). RKVY is preferred by States for its inbuilt flexibility in selecting interventions and setting State specific targets. One objective of RKVY during the Eleventh Five Year Plan was incentivising States to increase expenditure on agriculture and allied sectors. State plan expenditures (excluding RKVY receipts) as percentage of GDP in agricultural and allied increased from 1.0 per cent in the Tenth Plan to 1.4 per cent in the Eleventh Plan. State plan expenditures on agriculture and allied sectors (excluding RKVY) have also increased as percentage total plan spending by States, from about 5 per cent during the Tenth Plan to over 6 per cent during the

Eleventh Plan. RKVY was therefore successful in motivating States to pay greater attention to agriculture, besides providing increased Central assistance for the sector. RKVY as assistance was particularly useful for the funds-starved animal husbandry, dairying and fisheries sectors. Projects amounting to over Rs. 5,000 crore were sanctioned under RKVY for these sectors during the Eleventh Plan, about 20 per cent of the total sanctioned RKVY projects, and more than spending on DAHDF's schemes. This has provided a substantial push to these sectors which account for a significant contribution to the agricultural GDP.

However, preparation of Comprehensive District Agriculture Plans (C-DAPs) has been a weak area in many states, partly due to lack of capacity at District/State level. Although there are reservations regarding quality and effective capability of district level planning and project design, this was an original NDC intention and must be fully implemented during the Twelfth Plan. At least 25 per cent of projects sanctioned by SLSCs should originate from the district level, preferably approved by District Planning Committees. For the purpose, suitable units will have to be formed involving ATMA/KVK/ SAU and any other technical support unit that States may specify. As mentioned earlier, it is necessary to see decentralised planning as an iterative planning—doing—learning—planning cycle rather than simply a one-time activity. The challenge is to institutionalise this process and ensure that the agency facilitating planning is also accountable for the outcome.

Further, while there is very strong anecdotal evidence of the early success of RKVY, a detailed impact assessment of the scheme is needed for further experience and learning. Moreover, two modifications are desirable in the present practice. First, there should be a proper committee to examine and vet all projects proposed to the SLSC. Second, that at least this vetting committee or even the SLSC work closely with, and preferably be coterminous with, State level bodies that select MoRD projects, particularly for watershed development. This would permit better convergence and better project selection. Many States have requested changes in the allocation criteria of RKVY and some have objected to opening of new windows within the RKVY. A decision has been taken that no more than 20 per cent of RKVY funding will be in such windows of national importance. A decision has also been taken that at least 40 per cent of RKVY spending should be on hard infrastructure spending. A meeting of all States will be held to discuss proposals for changes in allocation criteria.

Finally, future RKVY design needs to be seen in the context of many pending key reforms. Despite efforts by the Central Government, progress in agricultural marketing, extension and cooperative reforms continue to be sluggish. Delivery of services has not been efficient due to lack of staff at various levels. State Agricultural Universities (SAUs) need greater funding support from the State Governments. Inadequacy of agricultural infrastructure hampers achievement of growth potential of the agriculture sector. During the Twelfth Plan RKVY will need to be reoriented to facilitate such market reforms, higher expenditure on SAUs and for infrastructure development, besides emphasising effective formulation and implementation of District Agriculture Plans. These could be incorporated by changing the current eligibility conditions and allocation formula for RKVY. The proposed meeting of all States as mentioned above will need to be held before these changes in RKVY are proposed to Cabinet.

AGRICULTURAL STATISTICS

Statistics are the hard input into planning. There are numerous gaps in agricultural statistics hampering the agricultural development planning some of which include reliable and timely availability of forecasts of agricultural crops especially foodgrains, reliable statistics for small areas like blocks and Panchayats, estimates of agricultural production losses due to pests, diseases, floods and drought, good estimates of production of minor crops including spices, condiments, medicinal plants, floriculture and so on, estimates of requirement of foodgrains for seed, feed and industrial use, harvest and post-harvest losses in agricultural production and estimates of meat production. Further, the available estimates generated through sample surveys suffer from organisational and operational problems bringing in inconsistency in these surveys. The Vaidyanathan Committee has recommended setting up a National Centre for Crop Statistics, independent of the present system, for providing reliable quick estimates at the National and State level. This should have high priority since not only are there strong doubts about quality of present data among experts, the large increase in number of crop-cutting experiments for insurance purposes may further vitiate the system. An independent source of high-quality data is vital for improving the quality of agricultural statistics in India.

The existing database relating to horticulture sector needs to be strengthened as mentioned earlier in the horticulture section. Cost of production data for animal husbandry products also needs improvement. Development of appropriate methodology for estimation of feed consumed by livestock will help in updating ratios currently used by the National Accounts Division. Similarly, the existing methodology for generation of fishery statistics needs fine tuning. For ascertaining the reliability of land use statistics in the context of diversion of agriculture land to other uses for residential, industrial, urbanisation, roads and so on, there is a need for conducting a study for checking the land records through khasra registers/other records of those villages where the area have come under diversion of agriculture land to non-agriculture uses particularly in the vicinity of the metropolitan cities.

Pilot studies need to be undertaken for perfecting remote sensing techniques and GIS/GPS tools to develop reliable estimates of area under agro-forestry area under crop production, land-use planning, land development and precision farming and so on.

All in all, the Twelfth Plan objective is to continue with the decentralisation thrust of RKVY, while reducing number of Centrally Sponsored Schemes. As discussed in relevant sections above, this vision on decentralisation could extend to fertiliser and food subsidies also. While doing this, the main Twelfth plan foci are:

- Bringing scale through development of Farmer Producer Organisations
- Emphasising technology, both on the research and development sides
- Stressing standards and protocols and standard operating procedures in every scheme
- Improving statistics and evaluation
- Initiating a shift towards sustainable and climateresilient agriculture, not only through NMSA but more generally by laying emphasis on rain-fed areas and bringing about shifts of water-intensive rice cultivation from water-stressed North-West India to Eastern India.
- Preparing for faster growth through a more diversified agriculture, with investment in the necessary modern infrastructure required for perishable products.

As shown in Table 12.13, States have indicated that they will more than double their plan expenditure on agriculture and allied sectors from Rs. 1,11,824 crore during the Eleventh plan to Rs. 2,26,500 crore during the Twelfth Plan. The Centre shall also more than double its plan expenditure. The allocation for RKVY is being raised to Rs. 63,246 crore for the Twelfth Plan from actual expenditure of Rs. 22,426 during the Eleventh Plan. The indicative Twelfth Plan Gross Budgetary Support (GBS) for all other schemes of the MoA is Rs. 1,11,232 crore. This is against corresponding the Eleventh Plan actual expenditure of Rs. 53,171 crore. Refer to Table 12.12 for department-wise break-up, excluding RKVY:

Table 12.3 : Gross Budgetary Support (Department-wise)

Department	*Gross Budgetary Support (GBS) (Rs. Crore)*
Department of Agriculture and Cooperation (DAC)	71,500
Department of Agriculture and Research Education (DARE)	25,553
Department of Animal Husbandry, Dairying and Fisheries (DAHDF)	14,179

Table 12.4 : Comparison of States Outlay and Expenditure for Eleventh and Twelfth Plan

(Rs. in crore at current prices)

Name of State	*Eleventh Plan Outlay*		*Eleventh Plan Expenditure*		*Twelfth Plan Outlay*		*Increase in Twelfth Plan over Eleventh Plan Expdr. (%)*
	Agriculture & Allied Sector	*% of Total Plan*	*Agriculture & Allied Sector*	*% of Total Plan*	*Agriculture & Allied Sector*	*% of Total Plan*	
Andhra Pradesh	3,487.44	2.4	9,510.46	6.0	17,138	5.0	80
Arunachal Pradesh	752	9.5	617.71	5.7	1,114	5.3	80
Assam	877.86	2.1	2,335.56	7.8	3,272	5.9	40
Bihar	3,672.73	4.8	4,805.33	6.3	15,613	6.0	225
Chhattisgarh	4,613	8.6	5,637	12.7	8,284	6.9	47
Goa	211.76	2.5	325.39	3.6	1,046	3.9	221
Gujarat	9,092.94	0.7	8,879.8	6.9	19,712	7.8	122
Haryana	1,638.82	4.7	2,733.02	5.7	6,288	5.4	130
Himachal Pradesh	1,470.08	10.7	1,642.82	12.1	2,174	9.7	32
Jammu & Kashmir	1,818.21	7.0	892.98	3.5	2,843	9.7	218
Jharkhand	3,130.53	7.8	2,319.85	5.9	4,157	3.8	79
Karnataka	8,426.85	8.3	10,484.4	7.7	19,824	8.9	89
Kerala	2,649.11	7.8	2,931.54	7.6	8,831	11.5	201
Madhya Pradesh	3,408.18	4.8	6,057.09	7.3	17,076	8.5	182
Maharashtra	9,507.64	5.9	10,636.4	7.3	19,325	7.03	82
Manipur	386.55	4.7	234.04	3.2	643	3.1	175
Meghalaya	735.52	8.0	845.2	9.8	2,114	10.7	150
Mizoram	536.31	9.6	387.86	7.1	346	2.8	
Orissa	1,230.29	3.8	3,580.37	8.2	8,387	7.4	134
Nagaland	434.31	8.3	725.08	11.3	1,795	13.8	148
Punjab	1,309.13	4.5	1,410.77	4.0	1,524	2.9	8
Rajasthan	2,919.07	4.1	5,990.67	6.2	7,255	5.6	21
Sikkim	260.43	6.9	228.27	6.4	469	4.1	106
Tamil Nadu	7,831.57	9.2	8,170.01	8.8	20,680	10.0	153
Tripura	798.51	9.0	858.79	11.3	980	6.8	14
Uttar Pradesh	19,146.37	10.6	14,164.8	7.8	24,354	8.5	72
Uttarakhand	2,478.5	8.4	2,079.25	10.0	2,673	5.9	29
West Bengal	1,846.50	2.9	3,339.26	5.1	8,583	5.5	157
Total	**94,670.21**	**3.6**	**1,11,824**	**7.2**	**2,26,500**	**7.1**	**103**

India has become one of the fastest growing economies in the world over the last two decades, undoubtedly aided in this performance by economic reforms. The striking aspect of India's recent growth has been the dynamism of the service sector, while, in contrast, manufacturing has been much less robust, contrary to the experience in other emerging market countries, where manufacturing has grown much faster than GDP; this has not happened in India. Consequently, manufacturing sector's contribution to the GDP has stagnated at 16 per cent, raising questions about India's development strategy, especially its implications for generating adequate employment. Additionally, employment in manufacturing declined in absolute terms from 55mn to 50mn between 2004 and 2005 and 2009–10, after having grown by 25 per cent between 1999 and 2000 (44mn) to 2004–05 (55mn).

The Eleventh Plan period was marked by unfavourable global economic conditions brought on by the financial sector crisis of 2007–09 followed by the risks of sovereign debt crisis mid-2011 onwards. While this led to slackening demand, exchange-rate volatility and economic uncertainty, domestic difficulties such as poor implementation and delayed reforms also slowed the growth of the Indian manufacturing sector. The year 2009–10 witnessed a fleeting return of manufacturing buoyancy largely on account of a few sectors such as the automotive sector along with a revival in cotton textiles, leather and food products. This brief spurt, however, has now moderated. The net result is that the share of the manufacturing sector in the country's GDP continued to be stagnant, a trend now observed for nearly three decades and remained relatively lower than other emerging and developed economies

Table 12.5 : Rate of Growth of GDP at Factor Cost at 2004-05 Prices (per cent)

	2007-08	*2008-09*	*2009-10PE*	*2010-11QE*	*2011-12AE*
Agriculture, Forestry and Fishing	5.8	0.1	1	7	2.5
Industry	9.7	4.4	8.4	7.2	3.9
Ministry and Quarrying	3.7	2.1	6.3	5	–2.2
Manufacturing	10.3	4.3	9.7	7.6	3.9
Electricity, Gas and Water Supply	8.3	4.6	6.3	3	8.3
Construction	10.8	5.3	7	8	4.8
Services	10.3	10	10.5	9.3	9.4
GDP at Factor Cost	9.3	6.7	8.4	8.4	6.9

Source : CSO.

Table 12.6 : GCF in Industry

(Rs. Crore at 2004-05 prices)

	2004-05	*2005-06*	*2006-07*	*2007-08*	*2008-09*	*2009-10*	*2010-11*	*CAGR (Eleventh Plan*)*
Mining	37,322	52,259	60,456	68,372	57,045	65,984	70,389	3.9%
Manufacturing	3,44,517	4,04,928	4,74,405	6,11,928	4,20,506	5,98,445	6,40,982	7.8%
Construction	54,445	57,531	95,799	1,15,157	88,523	86,290	98,426	0.68%
Total Industry	4,89,584	5,79,391	7,07,029	8,81,464	6,65,067	8,52,999	9,13,051	6.6%
Share of GCF in Industry as % to Total GCF	48.4	49	51.8	54.9	42.5	49.6	48.3	

Source: Economic Survey 2011-12; *CAGR has been calculated for a period of four years.

Further, India was not able to fully leverage the opportunities provided by the dynamics of globalisation that resulted in a dramatic shift of manufacturing to developing countries over the last decade. The increasing gap in both, the sectoral share of manufacturing and the competitiveness of the manufacturing sector in India, compared with countries, such as China, is testimony of that (Figure 13.2). This shift of manufacturing capacities from developed nations to rapidly developing economies (RDEs) is likely to continue. It is estimated that by 2025 RDE production will account for over 55 per cent of global production compared to 36 per cent presently. Hence, India's ability to capitalise on this by capturing a disproportionate share of such a shift in global economic setting through an accelerated growth rate will be imperative.

PERFORMANCE REVIEW OF THE MANUFACTURING SECTOR

Growth Rate

The manufacturing sector averaged a growth of 7.7 per cent (till 2009–10) during the Eleventh Plan (refer to Table 13.1). Growth peaked at 14.3 per cent in 2007–08 and then started decelerating. The decline in manufacturing growth was primarily responsible for the slowdown in GDP in 2011–12.

Industry

Initial deceleration in industrial growth was largely on account of the global economic meltdown. Fragile economic recovery in US and European countries, and subdued business sentiments affected the growth of the manufacturing sector. Rising interest rates and appreciation of the rupee during the Eleventh Plan period also contributed to this slow down. It is significant to note though, that volatility of manufacturing growth has become more pronounced over the last five years. An important implication of this is the need for greater flexibility both in policy and non-policy factors which have a bearing on the manufacturing sector.

Investment

Investment and capacity additions are critical for sustained industrial growth. National accounts data clearly indicate a moderation in the growth of gross capital formation (GCF) in industry (Table 13.2). The rate of growth of GCF in four broad sectors of industry comprising mining, manufacturing, electricity and construction averaged 10.9 per cent during 2004–11, almost the same as the rate of growth of GCF in the economy as a whole. For manufacturing to grow faster than other sectors in the economy, rate of GCF in manufacturing will have to be higher.

Employment

Employment in manufacturing increased from 44 million to nearly 56 million between 2000–01 and 2004–05. However, employment in manufacturing reduced by 5 million between 2004–05 and 2009–10. The net increase in employment over the decade 2000–01 to 2009–10 was around 6 million, that is, a 13 per cent increase over 10 years. Manufacturing in India contributes to only ~11 per cent of total employment. This compares unfavourably to other emerging economies where the share of employment in manufacturing range from 15 per cent to 30 per cent. One hundred and eighty-three million additional income seekers are expected to join the workforce over the next 15 years. Agriculture cannot be expected to provide more jobs. Manufacturing must provide a large portion of the additional employment opportunities required for India's increasing number of job seekers. Unless manufacturing becomes an engine of growth, providing at least 70 million additional jobs, it will be difficult for India's growth to be inclusive. Since the pattern of development of the manufacturing sector so far has not delivered the desired growth in output and employment, a change in strategy is required. This Plan is a description of the strategy, and the process for its implementation, without which the national objectives cannot be achieved.

Manufacturing needs to grow at higher than GDP growth to capture better share of GDP

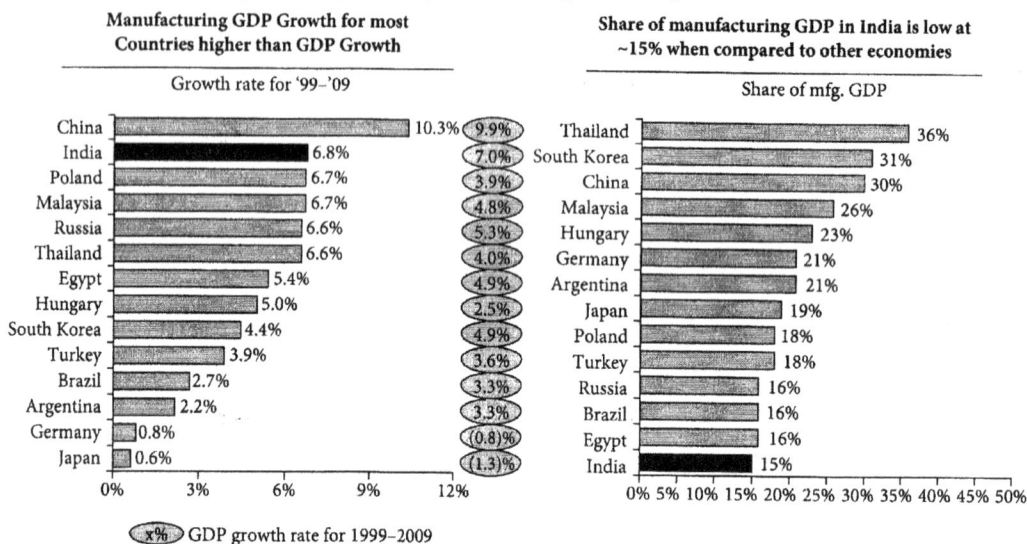

Manufacturing GDP Growth for most Countries higher than GDP Growth

Growth rate for '99–'09

Country	Manufacturing growth	GDP growth rate for 1999–2009
China	10.3%	9.9%
India	6.8%	7.0%
Poland	6.7%	3.9%
Malaysia	6.7%	4.8%
Russia	6.6%	5.3%
Thailand	6.6%	4.0%
Egypt	5.4%	4.9%
Hungary	5.0%	2.5%
South Korea	4.4%	4.9%
Turkey	3.9%	3.6%
Brazil	2.7%	3.3%
Argentina	2.2%	3.3%
Germany	0.8%	(0.8)%
Japan	0.6%	(1.3)%

0% 3% 6% 9% 12%

(x%) GDP growth rate for 1999-2009

Share of manufacturing GDP in India is low at ~15% when compared to other economies

Share of mfg. GDP

Country	Share of mfg. GDP
Thailand	36%
South Korea	31%
China	30%
Malaysia	26%
Hungary	23%
Germany	21%
Argentina	21%
Japan	19%
Poland	18%
Turkey	18%
Russia	16%
Brazil	16%
Egypt	16%
India	15%

0% 5% 10% 15% 20% 25% 30% 35% 40% 45% 50%

Source: Economic intelligence Unit, Data Monitor, Euro-monitor, World Bank Work Development Indicators, BCG analysis

Figure : Contribution of Manufacturing to GDP Very Low in India

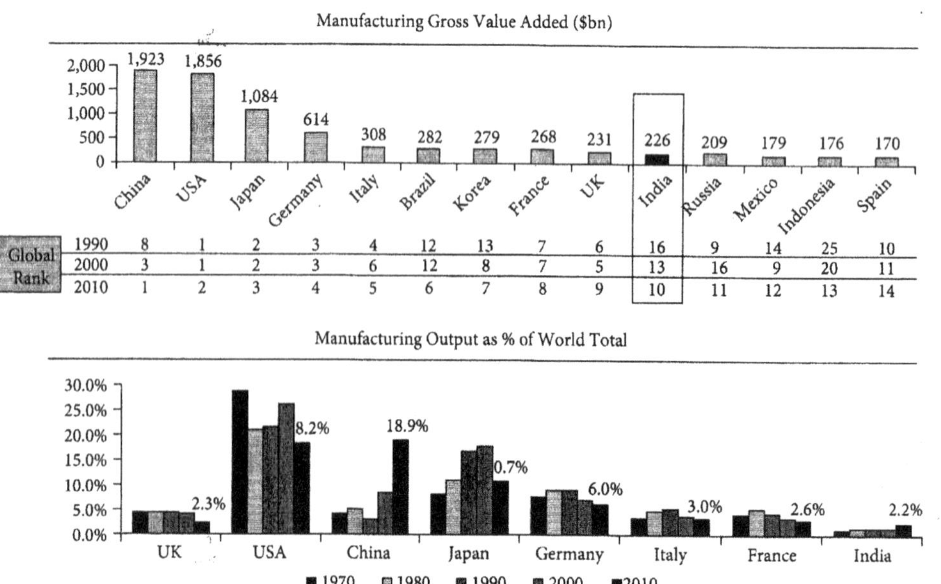

	China	USA	Japan	Germany	Italy	Brazil	Korea	France	UK	India	Russia	Mexico	Indonesia	Spain
Manufacturing Gross Value Added ($bn)	1,923	1,856	1,084	614	308	282	279	268	231	226	209	179	176	170
Global Rank 1990	8	1	2	3	4	12	13	7	6	16	9	14	25	10
Global Rank 2000	3	1	2	3	6	12	8	7	5	13	16	9	20	11
Global Rank 2010	1	2	3	4	5	6	7	8	9	10	11	12	13	14

Source: UN National Accounts Main Aggregates Database.

Figure : India and Global Manufacturing States

OBJECTIVES FOR THE TWELFTH PLAN AND BEYOND

In order to create a paradigm shift in the manufacturing sector, it is essential to consider the objectives over a longer timeframe, such as 15 years. The National Manufacturing Policy, which was introduced in 2011, states these objectives and these are the underlying objectives that the Plan aims to achieve as well. These objectives are:

1. Increase manufacturing sector growth to 12–14 per cent over the medium term to make it the engine of growth for the economy. The 2 to 4 per cent differential over the medium term growth rate of the overall economy will enable manufacturing to contribute at least 25 per cent of the national GDP by 2025.
2. Increase the rate of job creation in manufacturing to create 100 million additional jobs by 2025. Emphasis should be given to creation of appropriate skill sets among the rural migrant and urban poor to make growth inclusive.
3. Increase 'depth' in manufacturing, with focus on the level of domestic value addition, to address the national strategic requirements.
4. Enhance global competitiveness of Indian manufacturing through appropriate policy support.
5. Ensure sustainability of growth, particularly with regard to the environment.

REALISATION OF OBJECTIVES NEEDS A PARADIGM SHIFT

The Eleventh Five Year Plan as well as Plans that preceded it aimed at establishing a strong manufacturing sector but this has not happened. This suggests that a radical change in the policy approach is needed.

Comparison with the performance of other countries shows that the countries that managed to catch up with the earlier industrialised, high-income countries were the ones whose governments proactively promoted structural change. Industrial policy, and with a special focus on manufacturing, is back on the national agendas of many countries and we need to consider what lesson we can draw given our particular circumstances. In other words, the critical question now is not whether there should be an industrial policy but what should be the architecture of the industrial policy.

Industrial policies, where they have succeeded, have generally not been an outcome of Centrally planned economies but of economies that have had the active involvement of private enterprises and other non-governmental stakeholders. Successful strategies evolve from ongoing productive interactions between government and producers. Therefore, the government must improve the process of interaction, collaboration and learning amongst producers and itself. This is very different from the paradigm of Indian industrial policy prior to India's economic reforms commencing in the 1980s. In that era, industrial planning was a topdown control activity with Government determining who should produce what, where and how much and also what technology they should use. The roadmap for the Twelfth Plan and beyond can definitely not be a return to this type of planning.

Nature of Industrial Policy

The Question of 'Industrial Policy'

The Government of India needs a strategy to accelerate the growth of the country's manufacturing and industrial sectors to meet the goals and obtain the outcomes mentioned. The concept of 'industrial policy' has varied across countries and also over time. In India, industrial policy becomes assaulted under a stifling system of bureaucratic controls through licenses and quotas for industrial production. There is no doubt that these controls were highly dysfunctional and needed to be dismantled but the mere removal of these controls and reliance on markets alone was not sufficient. The collapse of the Soviet Union and the ascendancy of Western free-market approaches to economic growth which was fashionable for a time in the 1990s implied abandonment of any concept of 'industrial policy' altogether. However, this is not the recipe which delivered rapid industrial growth for many of the post-war success stories, whether we think of Japan or Korea or, more recently, China. In planning a strategy for rapid growth of industry in India we need to learn from these success stories and apply them suitably to our circumstances.

Paradigms of Industrial Policy

Countries that have succeeded in growing the competitiveness and scale of their manufacturing sectors have adopted different policy approaches. However, a common element in their approaches has been a close coordination between producers and government policymakers, with Governments playing an active role in providing incentives for domestic industrial growth and in relieving constraints on industrial competitiveness. The process by which this coordination has been achieved has differed according to the political structure of each country's economy (Figure 13.3). In Japan the coordination between Government and industry (and within Government) was very successfully orchestrated by MITI in partnership with Japanese industrial associations. In South Korea, the Chaebol and the Government collaborated to create world-class and world-scale winners. In Singapore, the Government identified industries to be developed and created ecosystems (skilled human resources, tax regime, Government incentives and so on) to support growth of competitive enterprises in the country. In China, the large State Owned Enterprise (SOE) sector has enabled the Chinese Government to adopt a very muscular 'industrial policy'. Along with preferential treatment to domestic companies, large investments in technology development/acquisition, massive investments in infrastructure and restraints on its exchange rate, China's industrial policy has been remarkably successful. Germany's manufacturing sector remains very successful in spite of high labour costs and a strong currency because collaboration between stakeholders in the German industrial system is deeply embedded in policymaking processes and also within industrial enterprises. A deeper analysis of such successes (Japan, Korea, China and Germany) of 'industrial policy' and also of its failures (India, the Soviet Union and some instances in Latin America) reveals the essence of successful industrial policy. Firstly, 'industrial policy' is a web of ongoing changes that facilitates the growth of a competitive industrial/manufacturing ecosystem in the country. Secondly, Governments have a key role in facilitating the process of learning and collaboration between producers and policymakers. Thirdly, and this is key, it is the quality of this process of collaboration and the speed of learning and execution in the system that enables the system to improve its competitiveness faster than other countries' systems. Government policymakers must have the skills and orientation to facilitate and coordinate, rather than to control. Industrial policy will not produce a competitive manufacturing ecosystem if the orientation of the Government and its functionaries is to control and micro-manage. It will also fail if Government and its functionaries do not master the skills and build institutional capabilities for better coordination within Government, smoother collaboration with industry (which must be organised in line with the industrial–political economy of the country, as mentioned before) and, above all, faster learning. The paradigm we must adopt is to build an ecosystem for rapid learning and capability building, which will encourage entrepreneurship and support innovation, and which will provide the system-wide processes to support collaboration and build stronger value chains with depth. This paradigm requires a change in the mindset of Government functionaries from being 'controllers' to 'facilitators', from 'resource allocators' to 'knowledge managers' and from 'scheme managers' to 'continuous learners'.

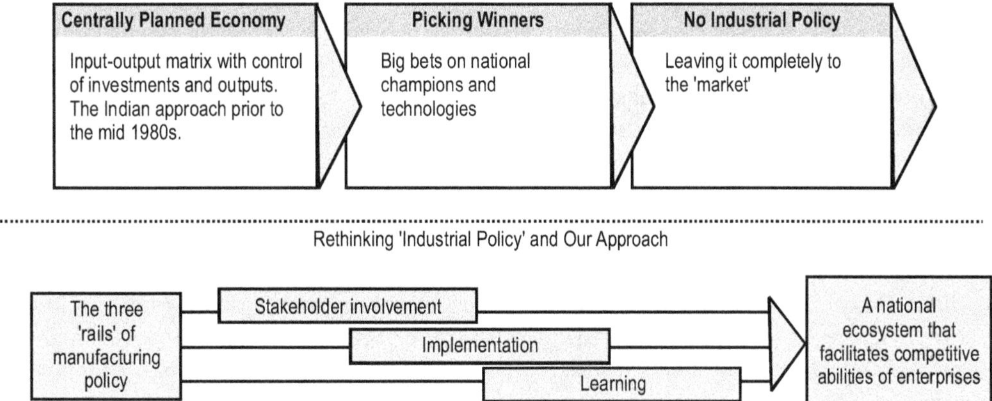

Figure : New Approach to Industrial Policy

Essential Features of a Manufacturing Ecosystem that Learns

A dynamic manufacturing ecosystem has three features that enable it to learn and grow.

1. Firstly, it must have depth (value addition) in manufacturing processes. A manufacturing sector, no matter how large, that is composed mostly of low value addition assembly industries, cannot create new technological capabilities. It may compete on low costs on account of scale and low labour costs, but it can easily lose these advantages to other countries which have even lower labour costs. Also, merely having R&D capabilities, without the wherewithal around them to convert ideas into manufactured products will not enable the growth of manufacturing industries.
2. Second, it must combine four capabilities: human skills, embodied technology in hardware, knowledge (intellectual property) and a large and demanding customer base. All four components grow together to create a productive and competitive industry.
3. Third, it must have a range of different sized firms, especially small and medium sized ones. Small firms provide the first stages for skill development. They take up larger numbers of people into the industrial workforce with less capital investment, and they provide nurseries for experimentation too. Some of these small firms can grow into specialised, internationally competitive, medium sized firms. Such firms are the backbone of the German industry, and also the strength of India's internationally recognised automotive component, pharmaceutical and IT sectors.

Firms operating in such an ecosystem would be able to flourish in an open competitive global economy. While there is a case for special support for strategically chosen industries for a limited Industry 57 period, the only way the industry can demonstrate competitiveness is to be able to export to global markets within a defined period.

In addition to the three features described above, there are five processes that enable the ecosystem to learn.

- Firstly, learning is accelerated through *the interaction of the diverse components of the system*: R&D with producers, both with customers, producers with institutes for skill development, and interactions amongst adjacent sectors and technologies that spur new combinations and innovations. Thus complexity breeds further technological development and growth. This requirement translates into the strategies for building clusters, and linking research and development institutes with *producers.*

- Second is the process of Innovation. Innovation can be spurred by several enablers that create 'safe-failing' spaces for experimentation. These enablers include early stage risk capital, incubators and quick exit/bankruptcy laws. Analysis reveals that the Indian industrial ecosystem has inadequate support systems for experimentation and innovation.
- Third is a regime of Standards. Standards are an embodied learning of the ecosystem. They enable firms, small ones in particular, with a base of knowledge, and also act as means to reduce transaction costs with their customers and suppliers, domestically and globally.
- Fourth is an IP regime. Like Standards, a good IP regime provides a base of knowledge for researchers and producers to develop upon further without having to reinvent the wheel. An IP regime also provides incentives for taking risks by assuring rewards.
- The fifth category of processes that enable systemwide learning and continuing improvement are a class of processes such as total quality management, total productive maintenance, business excellence and so on. In fact, such processes have been the foundations for the rapid, country-wide growth of productivity and competitiveness of the Japanese and Korean industry. The power of such processes has been realised by some sectors of Indian industry too, such as the auto industry, steel industry and so on.

The Architecture of a Strategy to Accelerate Growth of Manufacturing

Manufacturing enterprises, unlike IT and financial services enterprises, involve the production and movement of material goods. They, therefore, require good physical infrastructure to be competitive and this means improving transportation, uninterrupted power and adequate land to build. Moreover, the materiality of manufacturing activities also results in more regulations—of safety, pollution, factory inspections, labour conditions—and hence a more complex administration structure too. The quality and efficiency of the physical and administrative infrastructure is a basic requirement for productive manufacturing enterprises. This is a major weakness in India at present. The thrust in Government's New Manufacturing Policy (2011) to create good infrastructure for manufacturing enterprises along transportation corridors is, therefore, overdue.

Good physical infrastructure and smoothly functioning administrative infrastructure are threshold requirements for Twenty-first century manufacturing enterprises to compete in the international arena. However, these will not be sufficient. Competitive manufacturing, requires the development of complex capabilities—technologies, skills and management abilities to coordinate diverse interactions and processes of learning. Such capabilities can be learned and improved. Continuous improvement in these capabilities is the key to sustainable competitive advantage, even absent advantages from raw materials required for manufacturing, as Japan and Korea have demonstrated. Therefore, the thrust of Government strategy must be on the enrichment of the composition of these capabilities in the country's manufacturing ecosystem.

Three Components of India's Manufacturing Strategy and Plan

India's Manufacturing Plan strategy in the Twelfth Plan must be built around three components. The first are **capabilities and processes** that go across many, if not all sectors of manufacturing, and that build into the ecosystem the processes for rapid learning and building of capabilities. The second component has to be the plans to **strengthen the performance of selected sectors**. The selection of these sectors is done by a combination of top-down and bottom-up analysis. From the top, certain sectors appear more important to meet the goals of the Plan for more employment, for example, to produce goods that India needs for its strategic security. On the other hand, the capabilities created by Indian entrepreneurs in some sectors provide potential for more growth, and they should be supported. For example, the pharmaceutical and auto parts sectors. Thus the Plan, at present, has identified 18 such sectors.

India's sectoral strategy has to be broad-based, covering many sectors, to achieve the large-scale growth that India needs in manufacturing. India cannot achieve its goals by 'picking winners'. In each of these sectors, a sector strategy is required to grow capabilities and relieve constraints. Such sector strategies should be formulated jointly by the associations of producers in the sector (and other principal stakeholders too) and the relevant Government department. They should describe the opportunity for the sector and the actions required from the producers themselves, along with support from Government policies.

The third, vital, component of the Strategy is the **institutional ability for effective consultation and collaboration** between producers and public policymakers and implementers and the systemic reform of existing systems and processes within the Government. The strength of this process has been found to be the common factor in the success stories of all countries that have built large, competitive manufacturing sectors. 13.26. Lack of co ordination amongst government ministries, and the relatively poor quality of interaction between business associations and government—which is constrained by the competition amongst associations, and the orientation, by and large, towards lobbying and financial sops—prevents improvement in the process of collaborative learning and capability building that India needs to grow its manufacturing sector.

The challenges to developing and implementing a cohesive manufacturing strategy in democratic India are many. Cohesion can be brought about through more effective coordination amongst agencies, and more effective consultation amongst stakeholders. Apart from this, the Government will also require specialised skills such as consensus building and programme management to manage this process. Government should consider a 'Backbone Organisation (BBO)' to facilitate this process.

ISSUE IDENTIFICATION AND STRATEGIES TO ADDRESS THE VARIOUS CROSS-CUTTING ISSUES

The focus of this Plan has specifically been on transforming the approach to align the varied stakeholders to a common national goal, instead of having silo-limited views on individual sectors and individual goals .In order to achieve this coordination between the various sectors, and to identify the underlying causes of the slow progress of manufacturing, a set of thematic 'cross-cutting' issues were identified in addition to the major sectors of manufacturing. The 'cross-cutting' issues affect the growth of manufacturing across sectors. They fall into two categories: one category is those issues that 'industry' ministries and industrial enterprises have responsibility to address, albeit in collaboration with other stakeholders; and the other category is those broader issues that affect the economy overall in which the responsibility primarily lies with other ministries. In the first category is the weak development of human resources, of which a vast quantum is essential to achieve our goals. Another key issue, common to all sectors, is depth within the country of technology in the sector's supply chain. Yet another is a set of the infrastructural challenges, both physical and administrative, related to acquisition of land and water management, and the business regulatory framework, in which industry has a key role to play in developing and implementing solutions in consultation with other stakeholders. These cross-cutting issues have been identified in the National Manufacturing Policy recently approved by the Cabinet. This Plan describes the actions to be taken in all these areas and a process for their implementation and monitoring.

The second category, that of external inputs to industry that affect the economy as a whole too, and which are managed outside industry, includes four principal constraints on the growth of manufacturing: transport infrastructure, power, cost and availability of credit, and the exchange rate. Transport infrastructure and power have a direct bearing on the competitiveness of manufacturing. Energy and logistics are critical requirements for competitive manufacturing operations. While significant investment were made in transportation infrastructure in the Eleventh Plan, Indian industries continue to suffer from severe infrastructure handicaps compared with the infrastructure available to manufacturers in other countries. Ports are already close to fullcapacity utilisation resulting in extremely inefficient turnaround times and similarly roads suffer from congestion resulting in heightened costs. Unreliable and inadequate power supply continues to be a serious impediment in India in spite of the considerable efforts made to enhance power generation capacity in the country. Improving the supply and quality of both transport infrastructure and power are essential requirements for attaining the targeted growth rates for manufacturing in the Twelfth Plan and beyond. Adequate availability of low-cost credit is a vital requirement for sustainable manufacturing growth. Continued monetary tightening due to the recent turn of global events has resulted in a high cost of capital, adversely impacting manufacturing investment and growth in India. Cost of capital is key for ensuring competitiveness, especially of exports, of the manufacturing sector and needs to be carefully managed through a more balanced blend of fiscal and monetary measures. Specifically for MSME's, access to credit continues to remain a challenge and besides a host of measures to facilitate greater flow of credit to this segment detailed in Section 5, the overall pool of available capital needs to be enlarged to include alternate sources of capital such as private equity, venture capital and so on. Cross-cutting Groups

Finally, the exchange rate is an enormously important factor affecting the international competitiveness of a country's manufacturing sector. Large fluctuations in exchange rates can disrupt the management of supply chains. Monetary and fiscal authorities need to be cognisant of the impact that such fluctuations have on the growth of manufacturing.

TECHNOLOGY AND DEPTH

A principal objective of the Twelfth Plan must be to increase 'depth' in manufacturing, to increase domestic value addition, and meet national strategic requirements. The technological depth of the country's manufacturing sector goes up when it becomes an active player in more parts of the manufacturing value chain (research, development and production). Depth defined in these terms increases synergies across the value chain and also strengthens the overall trade position. It may be noted that depth is not necessarily required in all sectors. There is merit in being part of a global value chain but substantial part of industry must have technological depth. Depth in technology is extremely important for a country to sustain its competitive advantage in a global economy. It is not only important from the point of view of greater value addition, but it is also required to attract new industries and maintain competitive advantage of current industries.

The key requirements for improving technology and depth are to:

- Provide an enabling environment for domestic enterprises to invest in technology creation, technology absorption and achieve higher value addition.

- Ensure availability of demand for products developed and/or manufactured indigenously.
- Provide enabling environment for foreign enterprises to invest in manufacturing and research activities in the country, in the areas in which the country needs foreign technology.
- Mitigate the risks of MSMEs investing in technology development and technology upgradation.

Status and Key Challenges

Lack of depth in technology is one of the foremost issues affecting the growth of manufacturing sector in the country. India's R&D spend is 0.9 per cent of GDP, whereas China, UK and Israel spent about 1.2 per cent, 1.7 per cent and 4.3 per cent, respectively. India needs to increase its R&D expenditure to improve its depth. The private sector finances 70 per cent of the total R&D spending of China, 65 per cent in United States and 75 per cent in Korea and Japan, while Indian private sector funds only 25 per cent of the total R&D spend. As majority of private sector funding of other countries is towards industrial R&D, Indian corporate sector needs to increase its spending on industrial R&D (see chapter on Science and Technology). The key challenges faced by Indian industries are:

- The Indian Industry has not given sufficient importance to the documentation of knowledge and creation of IP. As a result, not only were opportunities lost to create IP, but we lost IPs to other countries, such as in traditional agricultural products (IPs filed by western countries on neem, turmeric and basmati rice, which India has contested). Our regulatory framework, speed of award of IPs and the enforcement of IP regulations needs improvement. India's approach on IP, hence, needs to distinguish between shaping the framework for IP creation and improving its IP management processes.
- Though there is an improvement in the industry-academia collaboration in creating patents/technologies, still there is a large scope for improvement.
- While FTAs signed with other countries are favourable for some products, they often create a distortion in the market in terms of inverted duty structure for other products.
- Many segments of the industry, especially MSMEs, have limited information and access to risk capital for sourcing/developing and internalising new technologies.
- The weak attention to standards not only invites dumping of sub-standard products by other Industry countries.

Examples of Weak Domestic Standards Leading to Influx of Sub-standard Products in the Country

(A) Absence of Standards

In the absence of technical standards, it becomes easy to import poor-quality products into the country. This hurts the domestic industry as the domestic industry is unable to match the price of these poor-quality products; it also exposes consumers to the harmful effects of spurious products. In the absence of such standards, it would not be possible to make such technical regulations which would curb import of poor-quality products. Some of the examples include mobile telephones, batteries for the mobile telephones, digital blood pressure measuring equipment, decorative lights (imported from China during Diwali festival), medical equipment and so on. Mobile telephones: Lack of manufacturing standards and testing/sampling labs are prompting dumping by foreign manufacturers. For example, till 2009, there was no standard mandating all imported mobile phones to have an IMEI number. As a result, Chinese handsets without IMEI numbers had a market share of about 13 per cent at that time.

(B) Lack of a clear framework for voluntary and mandatory compliances

In some situations, where Indian Standards exist for products or processes, the Central Government has not notified them for mandatory compliance.

Toys: Standards have been laid out for safety of toys such as quality of plastics and paints, electrical and mechanical hazards, migration of heavy elements (Lead, Cadmium) and so on. However it is not mandatory to comply with, and hence toys from other countries are being dumped in the Indian market. Structural steel: This is used in building damns, bridges and so on. Standards in the manufacturing of structural steel are voluntary and lack the need to specify end use. The lack of compliance to such voluntary guidelines and the absence of the need for requisite certification lead to dumping of poor grade structural steel competitive products for domestic as well as global customers. They compete with global manufacturers in local as well as in global markets. The Government policies for large enterprises can focus on:

- Improving IP regime
- Ensuring human resource availability by establishing institutions for technology education and research, educational institutions and so on.
- Ensuring access to critical raw materials but also makes it difficult for the industry participants to benefit from each other's learning and improve their technology depth.
- Absence of national agenda and policy framework to support innovation.

A Systems Improvement Framework

It is essential to set the context before moving to the recommendations. Government support is essential to enable a country's industrial ecosystem to gain depth because technological learning takes a long time, requires large investments and is risky. Support to the enterprises should be in such a way that it motivates and enables enterprises to learn and develop complex capabilities and not become complacent and inefficient, which was the outcome of the industrial policy adopted by India until the 1980s. capture the generic policy levers that should be moved for faster growth of manufacturing over the next five years. The specific policy interventions must be tailored to fit the requirements of sectors by a process of industry— Government consultation which, as has been emphasised before, will be the key to 'get it right'. MSMEs and large enterprises will require different kind of interventions from Government. MSMEs play a critical role in innovation, thanks to their nimbleness and their ability to experiment with new technologies on small scales. However, they often suffer from lack of funds, inability to take risks associated with technology developments and the difficulty of attracting skilled manpower. Policy interventions for MSMEs must be tailored to their conditions. Government policies for MSMEs should therefore help them improve their technological capabilities by focusing on:

- Providing access to risk capital.
- Setting up of standards for the industry.
- Improving Industry/research institute/academia interaction, mostly in clusters.
- Stimulating demand/providing scale through preferential treatment in government purchases.

Strategies for Change

Some high impact strategies for India at this time to accelerate the development of technological depth in the manufacturing sector have been analysed. These should receive special attention in policymaking and implementation. Creation of coherence amongst existing institutional agencies towards developing national priorities for indigenous technology development. Several countries like China and Singapore have followed a comprehensive approach to identify critical technologies to be developed indigenously and have formulated mechanisms to Process that Enables Learning Policy Levels:

1. Interaction between diverse components of the system—R&D, producers, customers, Government, institutes of skill development and so on.
   - Cluster development
   - FDI and JVs
   - Industry/research institute/academia partnership
   - Higher education in the country
2. Creating 'safe-failing' spaces for experimentation by firms
   - Access to risk capital, technology funds
   - Subsidy on interest costs
   - PPP model of funding
3. Creating a regime of 'Standards'
   - Setting up a system of National Standards benchmarked to International Standards
4. IP regime, which helps firms to build on each other's innovation
   - Effective 'IP' regime
   - Improving awareness of IP
5. **System-wide improvement:** Processes such as 'Total Quality Management'
   - Mainly the firm's role to adopt such tools and increase organisational learning
   - Nation-wide, and State-wide campaigns to improve 'Total Quality' in all enterprises, including MSMEs, should be sponsored by Government through institutions such as Quality Council of India.

Manufacturing Ecosystem Infrastructure

Ecosystem Infrastructure Policy levers:

1. **Physical infrastructure**
   - Cluster development
   - Special manufacturing zones (NIMZ)
2. **Improving capabilities**
   - Skill development
   - Total quality management
   - JV, Technology transfer, FDI 3. Creating the manufacturing ecosystem
   - Developing MSMEs
   - Common facilities through clusters
   - Developing Standards

- Availability of quality human resources
- Demand availability for manufactured products
- Modular industrial estates/laboratories near premier technical institutions with the required plug and play facilities. setting up of a Technology Acquisition and Support Fund.

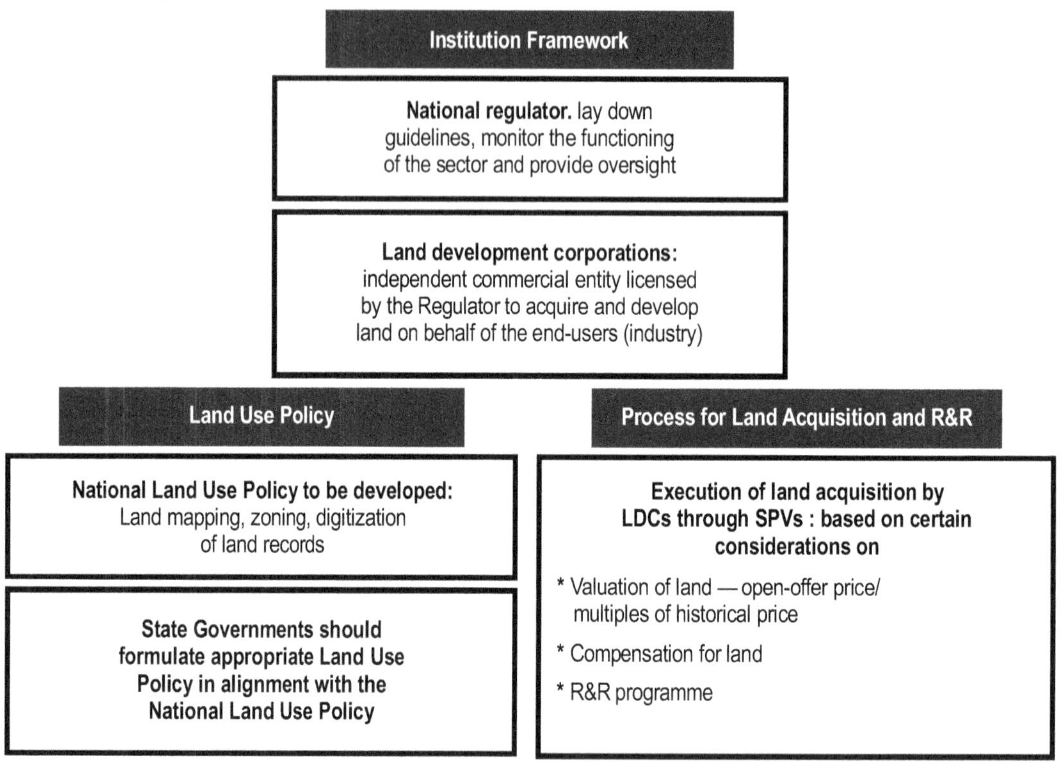

Institution Framework
National regulator. lay down guidelines, monitor the functioning of the sector and provide oversight
Land development corporations: independent commercial entity licensed by the Regulator to acquire and develop land on behalf of the end-users (industry)
Land Use Policy
National Land Use Policy to be developed: Land mapping, zoning, digitization of land records
State Governments should formulate appropriate Land Use Policy in alignment with the National Land Use Policy
Process for Land Acquisition and R&R
Execution of land acquisition by LDCs through SPVs : based on certain considerations on
* Valuation of land — open-offer price/ multiples of historical price
* Compensation for land
* R&R programme

Figure : Strategy for Land Issues

On the other hand, large enterprises handle complex technologies and manufacture globally Industry 63 ensure that these technologies were funded and incubated.

In India, we have various agencies like the Department of Science and Technology, NMCC and the Planning Commission working in this area. Connections between these agencies remain weak as they continue to function in silos, resulting in a cluttered approach to technology development. To make this process more robust and comprehensive (including funding and incubating projects), the present process/institutional arrangements should be reviewed and fine-tuned/restructured. The industry, as key stakeholders, should be involved and consulted in the design of new arrangements. Create 'Safe-failing' Spaces for Companies to Engage in Innovation Government participation in funding of research through a 'Technology Fund' or 'Technology Upgradation Fund' is an important instrument for reducing the risk for firms in investment in research. The structure of the 'Technology Fund/ Technology Upgradation Fund' has to evolve over a period of time. Traditionally such funds have been operated in the form of Government grants or schemes. However, they can be more effective in producing outcomes if they were managed by professionally managed investment entities.

The ways in which the Government could provide/redesign fiscal incentives for R&D activities are:

- *Tax credit instead of tax incentives*: With the imposition of Minimum Alternate Tax (MAT) of 20 per cent, companies are unable to avail full benefit of weighted deduction. Equivalent benefits of weighted deduction on R&D spend should be treated as tax credit and be allowed to be set off against Tax and/or MAT payable.
- *Credit on inputs/capital goods used for R&D outside the factory premises:* The Cenvat Rules provide that credit can be availed on inputs and capital goods if they are used in the factory of the manufacturer. Enterprises having R&D facility separately from manufacturing facility will not be able to claim Cenvat benefits on inputs and capital goods used for R&D. This anomaly should be removed and the Cenvat benefits to be available for inputs and capital goods used in R&D, even if the R&D is carried out in a different premises, as long as linkage between manufacturing and R&D activities can be established. Due to this lacuna, assesses with sizeable investments in R&D facilities outside their factory of manufacture will not be entitled to avail Cenvat credit on investments and certain operating expenses. Consequently, this forms a disincentive to setting up of R&D centres by increasing the costs of setting up such centres. The tax incentives should be provided in such a way that they do not penalise existing enterprises that do not operate in special economic zones or particular locations/States. To ensure a 'level playing field' to all domestic manufacturers and to provide a wider stimulus by the incentives, the tax incentives should be available for all enterprises involved in a specific activity rather than for a few enterprises operating in some specific locations. Knowledge sharing should be improved between the industrial and financial sectors. The financial sector works with many industrial sectors and thus can see patterns and, with its perspective, obtain insights that are not available to people within industrial institutions. There are several programmes like Small Industry Business Research Initiatives (SIBRI), Technology Development Board (TBD), Biotechnology Industry Partnership Programme (BIPP) and Biotechnology Industry Research Assistance Programme (BIRAP) which promote early stage innovations and PPPs. These institutes should work more often and closely with financial sector institutions to share knowledge that can improve policies for the manufacturing sector. Twelfth Five Year Plan Strengthen the IP Regime and Systems to Leverage IP A strong intellectual property regime is a prerequisite for creation of global IP from India. It has also become a requirement under WTO. While the importance of IP for creation of innovations in the industry is well understood, the question is whether developing countries will get penalised given that they are starting with a low base compared to developed countries. Various alternatives like 'utility model' of patents (as China has) to manage this need to be examined to put in place an efficient model that can help generation and protection of incremental innovations in Indian manufacturing.

Given the need for a strong IP regime from a long-term point of view, the following steps need to be taken:

- Improve IP management and protection mechanisms.
- Develop global information database on IPs accorded.
- Strengthen and modernise the process of patent examination and according patents. 13.52. Also, in order to leverage the benefits of IP.
- Build awareness about IP through education and training.
- Create national IP mission to continually evolve the IP strategy of the nation.
- Encourage joint IP filings by industry/academia/research institutes.
- Encourage the formation of companies specialising in IPs (through tax incentives).

- Exempt income tax for the income generated from domestic IPs. Strengthen Partnership between Industry and Academia/Other Research Institutes to Create IPs Domestically Industry–academia partnerships are relatively weak in India compared to many other countries.

The partnership should aim for building an ecosystem which can create a virtuous cycle of education and research leading to IP creation and its subsequent commercialisation. Such aspect in turn will incentivise and inspire further innovation. Some of the policy measures that Government can use to accelerate the development of industry–academia partnership are:

- Joint ownership of IP arising out of these collaborations.
- Align the goals and annual planning processes of central research institutions with that of industries through industry associations.
- Incentivise Central/State Research institutes to create joint IPs with Industry.
- Tying up a certain percentage of their budget to the number of collaborative IPs created.
- Incentivise university and industry for forging successful partnerships in university's governance, infrastructure, course curriculum design, faculty/students development and research.
- Create cluster innovation centres at universities with the aim to foster a favourable ecosystem and enforce industry–academia linkage.
- Provide an institutional framework for active interface between funding agencies, academia and industry. Clusters (and NIMZ) Can Provide Enabling Infrastructure to Improve Technological Depth

Clusters play a critical role in propagating technological depth by facilitating technological learning and manufacturing through the presence of the entire ecosystem in the same geographical location. The National Manufacturing Policy, which outlines creation of NIMZs, was cleared by the Cabinet in November 2011. It ensures that business is provided with the ecosystem required for growth, not only in manufacturing, but also for investments in research and development. The attractiveness of NIMZs will be even higher for new high-technology industries, which will benefit from the localised presence of the entire value chain of participants. Also, the benefits of industrial clusters to MSME participants are also well understood, and the MSME Ministry is using the cluster approach to drive the growth and depth of MSME industries. Industry Improve Technical Standards, Voluntary Compliance and Conformity Assessment Standards are a form of embodied technical knowledge accessible to all types of business that enables more effective product and process development. They promote and enable the diffusion of technology in a form that is readily assimilated by firms with the complementary capabilities to take up and use the new methods. Standards, therefore, constitute one of the important foundations for the technological depth in manufacturing, and are accorded high importance by the policy planners in the developed world. During the Twelfth Five Year Plan, the focus on technical regulations should be on:

- Developing a policy on technical regulations.
- Capacity building of regulators (BIS).
- Review of technical regulations to identify the gap vis-à-vis national standards.
- Sensitising the industry regarding the need to provide scientific data to regulators to formulate effective technical regulations.
- Setting up of helpdesks in industry bodies and export promotion councils for information dissemination.

In addition, voluntary compliance initiatives must be strengthened:

- Promoting and funding a 'Standards Cell' in industry associations and Standards Developing Organisations (SDO).
- Capacity building of SDOs.
- Capacity-building programmes for the training of technical staff in the industry for writing company- and industry-level standards.

Government should also create a databasebased/software-based system to track the changes in technical standards/voluntary compliances globally and alert Indian manufacturers of development.

While the Standard-setting process sets the standards to be followed, conformity to the standards is assessed by conformity assessment agencies. While many conformity assessment agencies have sprung up in the last few years, it is important that these conformity assessment agencies are of world class and their certificates are acceptable across the world. To achieve these, the following steps are envisaged during the Twelfth Five Year Plan period:

- Promoting the acceptance of Indian conformity assessment globally
- Capacity building for inspection bodies/certification bodies
- Developing regulation on conformity assessment

Quality Council of India, set up jointly by the Government of India and the top industry associations—CII, FICCI and ASSOCHAM, has been working to:

- Establish and maintain an accreditation structure in the country
- Help representing India's interest in International forums
- Spread the quality movement through the country The Twelfth Plan will focus on strengthening the capabilities and role of the QCI. Removing Anomalies in Duty Structure
- *Remove special schemes that allow import of finished goods at concessional custom duty:* In almost all promotional schemes where import duties are reduced (nil duty project imports, certain defence purchases, SAD exemption under ITA Agreement for IT products and so on), imports get the benefit of reduced duties/nil duty. This erodes the level of protection which would have otherwise been available, thereby, creating a systemic disadvantage for local manufacturers. It is therefore recommended that import of finished goods at concessional custom duty under special schemes be discontinued.
- *Inverted duty structures (Higher duty on intermediate products vs. final products):* For specified purposes, presently there is higher duty on 66 Twelfth Five Year Plan intermediate goods (used by the domestic manufacturer for assembly/manufacture of goods), as compared to duty on finished goods. This in turn leads to higher input cost for the domestic manufacturer. It is therefore recommended that duty on intermediate goods be brought in line or set lower than applicable for final products.

The Government has corrected, as best possible, the issues related to inverted duty structures (illustrated above) raised by industry. It must review any new case that is brought to its notice and must undertake a study of effective rate of protection across sectors.

Some issues regarding CST/VAT retention and VAT/SAD were also analysed:

- *CST/VAT Retention:* Interstate movement of goods by domestic manufacturers carries added cost in the form of central sales tax (on interstate sales)/retention of input VAT credits (on interstate stock transfers). This can be avoided in case of imports by executing sales in the course of import or through directly consigning the goods to the customer's state. This creates disadvantage to domestic manufacturers. Therefore, CST on interstate sales and provisions with respect to retention of input VAT on interstate stock transfers should be abolished.

- *VAT vs. SAD*: VAT rates have been increased from 4 per cent to 5 per cent, however there has been no consequential increase in the rate of SAD on imported products which is levied in lieu of VAT. Therefore, SAD should be increased to 5 per cent to reflect the pan-India based trend of revision of the VAT/CST rate bracket of 4 per cent to 5 per cent. In order to resolve the aforesaid issues, it is necessary for the Central and State Governments to quickly build consensus on the design of a comprehensive GST and implement the same at the earliest. Encouraging FDI and Joint Ventures FDI (investments by foreign companies in Indian ventures) and Joint Ventures of Indian companies with foreign partners can provide access to technology in areas in which domestic expertise is inadequate. The Government must identify the areas, in consultation with the industry, in which FDI and Joint Ventures can help to bring technology. Several problems that are impeding FDI/JVs need to be addressed. Some these are:
- The ambiguity in the characterisation of income arising to foreign investor on transfer of technology from the perspective of direct-tax obligations. This leads to uncertainty with regards to its taxability in the hands of foreign investor thereby discouraging the flow of technology from outside India. The foreign investor is required to obtain PAN to enable the payer to withhold taxes at appropriate rates. Also, the foreign investor is required to file its annual return of income before the tax authorities in India for the purpose of claiming credit with respect to the taxes withheld by the payer in India. Such additional compliances could become quite cumbersome for the foreign investor in India especially where the foreign investor does not have any operations in India.
- The R&D cess paid by the importer cannot be adjusted against any output taxes paid by the importer, resulting in additional cost of 5 per cent for the technology importer.
- Service tax paid on import of technology cannot be adjusted against taxes paid on output, if the manufacturing is outsourced.
- Limitation on technology cost as percentage of total investment available for state tax exemptions.

Preference for Domestic Products in Government Procurement

The cost of any manufacturing activity (excluding raw materials and utilities) depends on the maturity of manufacturing technology used and the magnitude of the demand. For a matured technology, the cost of manufacturing will be relatively low, due to the learning curve effects. Similarly, due to scale effects, the unit cost of manufacturing goes down with the increasing demand. Therefore, a domestic enterprise using new indigenous technology will have a cost disadvantage Industry 67 compared to a global enterprise that has the benefits of matured technology. Unless there is some incentive provided to domestic enterprises to offset this handicap, developing indigenous technology will be difficult. Therefore, Governments in many countries, developing as well as developed, provide preference in Government procurement to domestic enterprises. However, to ensure that this policy measure does not lead to development of substandard quality products or create inefficiencies in the domestic enterprises, the preference in procurement can be made applicable with minimum quality standards; a cap on the permissible price differential between domestic and imported products, and also a sunset clause. Some ways in which the preference for indigenous products can be provided in Government purchases are:

- In sectors of strategic importance, procurement should be done only from those vendors, who have locally established manufacturing base.
- A multi-tier tax structure can be introduced, which offers concessional tax rates for products with higher local value addition.
- A certain percentage of Government procurement to be reserved for enterprises using domestic manufacturing/domestic IP; and a certain percentage of it can be reserved for firms in MSME Sector. However, as a prerequisite to implementing this procurement strategy, streamlining of procurement functions is essential. Public procurement organisations must be clear about how national policy goals should be translated into procurement practices without compromising quality. 'Least cost' is not always the right strategy and needs to be balanced by other guidelines (life-cycle costs such as service agreements, continuous improvement contracts and so on). A balanced approach should be taken to determine the weight assigned to price versus other qualifying criteria. Aligning Investment Obligations Under 'Offset Policy' Offsets as a policy tool should be encouraged for public procurement in sectors where the Indian industry does not have existing technology or capability. The obligations of investments of foreign companies under 'Offset Policy' should be targeted towards investment in industries in which the country needs to improve technological depth. Articulation of clear objectives for an offset programme, not just for defence industry but also for the economy as a whole can become an instrumental lever to further investment and growth of the country's manufacturing sector. Encouragement of Local Value Addition in Critical Natural Resources Some natural resources like good-quality coal and iron ore are becoming short in supply in the global economy with growing demand from developing economies especially China and now India. Domestic availability of some of these raw materials provides us a competitive advantage which we should leverage to build domestic industries that add value to these resources, thus creating additional jobs and improving our trade balance. Going further up the value change Government policies and duty structure should be designed in a way to incentivise value addition of steel rather than exporting steel in raw material form.

In general the trade-off between export of inputs which are in demand elsewhere in the world, and use of those inputs for improving the competitive position of domestic user industries is a tricky one, while promoting entrepreneurial freedom and free trade. These trade-offs must be understood and sensitively managed to ensure competitive and sustainable growth of domestic manufacturing. Examples of vulnerabilities that have developed for Indian industries, when longer term consequences of policies have not been foreseen, are the virtual disappearance of production of intermediaries for generic drugs which China is now dominating, and also the dwindling of Indian capital goods industries (refer to Box 13.2), where too Chinese industry is becoming a big international supplier. Chinese industrial policy 68 Twelfth Five Year Plan has evidently done far better than India's in building depth in China's industries.

HUMAN RESOURCE DEVELOPMENT, JOB CREATION AND SOCIAL PROTECTION

One of the primary objectives of the plan is to increase the competitiveness of Indian manufacturing. Human resources are of critical importance for the growth of knowledge and technology, value addition and improvement of competitiveness in manufacturing through processes of continuous improvement. In fact, the human resource is the only 'appreciating resource' in a manufacturing system. It is the only resource that has the motivation and ability to increase its value if suitable conditions are provided, whereas all other resources—machines, building, materials and so on—depreciate in value with time. The best enterprises view their people as their prime asset and the source of their competitive advantage. Nations that have achieved sustainable competitiveness in manufacturing even when they do not have raw materials required, such as Germany, Japan and South Korea, have created systems for the continuous improvement of the capabilities of their human resources. India must invest in and build its human resource capabilities to catch up with other countries that have moved ahead and thereafter sustain competitive advantages in manufacturing. Indeed the contentious debate of 'labour' versus 'capital' in the enterprise, as well as disputes between the institutions that represent the people working in the enterprise and owners of the capital could be reframed if employees were seen as assets, with value that can appreciate, rather than as labour costs. The purpose of this section is to propose a set of holistic changes in key areas that require close involvement and buy-in from various stakeholders.

Dwindling Indian Capital Goods Industry

The capital goods industry can be considered as the 'mother' of all manufacturing industry and is of strategic importance to national security and economic independence. It is in the interest of User Sectors that the capital goods industry be strengthened since it is well established that the presence of a strong domestic industry increases competition and helps in reducing the capital cost of projects. And most importantly, in economical maintenance of plant and machinery. Imported plants come at lower cost but the foreign suppliers make up for that in their high priced spares and maintenance contracts. However, Indian capital goods industry is facing severe competition from Chinese companies over the last few years. In the case of machine tools, imports account for about two-thirds of the domestic requirements and is increasing further. The import of power-generation equipment from China at much lower cost is also making the domestic industry uncompetitive. The major factors responsible for increasing Chinese competitiveness are:

- Artificially depreciated Chinese currency
- Tax advantages and Government subsidies given by the Government
- Much lower interest rates
- Simpler labour laws
- Better infrastructure leading to lower cost of power, transportation and cluster approach helping specialisation of labour and engineering skills This is further complicated by the absence of level playing field for Indian manufacturers:
- All domestic manufacturers of capital goods are rendered uncompetitive due to additional burden of sales tax, entry tax, octroi, VAT and other local duties and levies.
- For specified projects (Oil and Gas, mega nuclear/hydel power, fertiliser, refinery and so on) zero/5 per cent customs duty applies on capital goods. While it may be preferable from user industry point of view to allow the import of capital goods at lower costs in order to improve their competitiveness, this will result in over reliance of Indian industry on other countries for key strategic inputs, exposing itself to vagaries to the policies of these countries. Also, this does not help in building technological depth of the Indian industry and manufacturing ecosystem. Industry 69 Consensus about these holistic changes is more likely to be achieved if, as mentioned before, the primary challenge was reframed as the development of human assets to build India's manufacturing ecosystem and strengthen India's manufacturing enterprises, rather than merely management of costs of labour.

Challenges in meeting the objectives lie broadly in three areas:

- From a skill development perspective, there is a significant gap between the existing training capacity and people entering the workforce. A very small proportion of total manufacturing workforce is currently skilled. Moreover, less than 25 per cent of the total number of graduates are estimated to be employable 1 in manufacturing.
- The total training capacity in the country is about 4.3 million for all sectors including manufacturing. The Apprentice Training Scheme (ATS), which is supposed to provide a bridge from education to employment, has very low penetration and is suffering from significant administrative issues.
- For entrepreneurs and other employers, the perceived lack of flexibility of changing the size and nature of the workforce can act as a retardant in making investments that could lead to greater employment opportunities. Furthermore, the complexity of labour laws and the administrative mechanism of the laws make it harder to do business in the country.
- By 2025, an additional 8 million management workers (supervisors and above) are estimated to be required. Well-trained management/supervisory staff are critical for improving the productivity and industrial relations in large as well as small manufacturing enterprises. Strategy and Key Recommendations

Human resources should be managed as a source of sustainable competitive advantage. Government policy changes should induce and support such firm level strategies. The key stakeholders who will need to work together to make the necessary changes to the system in key areas mentioned above are: Government (at the Centre and State level), Industrial organisations and the unions. The strategies for meeting the objectives are in the following categories:

- Inducing job creation by reducing the cost of generating employment.
- Developing a supply of qualified human resources to meet the demand from additional job creation.
- Enhancing skill levels of current workforce to improve productivity.
- Improving the state of manufacturing management in the country.
- Providing social protection to low-income workforce.
- Improving industry–workforce relationships.

Inducing Job Creation by Reducing the Cost of Generating Employment : There are two major barriers to employment generation: limited flexibility in managing the workforce and cost of complying with labour regulations. Both these barriers must be removed in order for jobs to be created at a much faster rate.

Limited Flexibility in Managing the Workforce

The recommendations to increase the level of flexibility while ensuring fairness are:

- Companies should be allowed to retrench employees (except categories such as 'protected workmen' and so on) as long as a fair severance benefit is paid to retrenched employees. This severance benefit should be higher than what is currently mandated—and the value should be arrived at through tripartite dialogue between Government, employers' associations and employees' associations.
- In order to ensure that there is sufficient liquidity to pay the severance benefit to the retrenched employees, a mandatory loss-of-job insurance programme could be put in

place. This will especially be useful in situations where the retrenchment is due to bankruptcy or exit of the employer and will reduce the justification for requiring prior permission to shut down businesses.

Twelfth Five Year Plan

The threshold level of employment for the Chapter VB of the Industrial Disputes Act and the threshold for applicability of the Factories Act should be raised to at least 300 which was the level

- The process of engaging contract labour should be reformed—employers should be allowed freer use of contract labour while ensuring that the rights of contract workers are protected, which is not the case at present.

Cost of Complying with Labour Regulations

The traditional enforcement approach which is based on inspection—prosecution—conviction creates incentives for rent-seeking behaviour, especially if the laws are complex or have provisions that are contradictory. The complexity of compliance impacts smaller enterprises much more. They cannot Recommendations to improve compliance and also contain the cost of complying with labour regulations are:

- *Simplification of labour laws:* The implications of labour laws should be detailed through a series of ready reckoners that are easily available and regularly updated so that inspectors and employers have a common set of rules to look at.
- Improvement of administration: Higher investment should be made in the training of inspectors to ensure that they are able to efficiently identify incidences of actual non-compliance rather than harass employers.
- Facilitating easier filing: Filing of reports should be made a once a year activity with an online option. As far as possible, the interface between enterprises and Government should be computerised to increase transparency and efficiency and remove scope for rent seeking.
- Developing a self-certification model: While ensuring that regulations governing labour welfare must be complied with, a self-certification model should be developed where appropriate.
- Additionally, fiscal incentives to encourage permanent job creation should also be considered, after evaluating their implications and potential impact. For example, skill building and training costs of permanent employees can be considered for accelerated tax benefits (subject to a ceiling on percentage of salary paid to permanent employees).

The manufacturing sector may need more than 90 million people by 2022. However, the current capacity for skill development is ill equipped to meet this demand. *Role of industry*: To enable the industry to play its role in defining the requirement of manpower both in terms of quality and quantity, Sector Skills Councils envisaged in the National Skills Policy are being set up. These councils will identify skill development needs in their sector, evaluate the gaps, create plans for skill development and improve the quality of the training system. The councils are also expected to establish sector specific Labour Market Information Systems (LMIS) to assist in planning and delivery of training. *Private sector participation in skill development*: For the private sector to play a role in augmenting the skill-development capacity in the country, effective PPP models are needed. Existing ITIs should be clustered together in projects with total training capacity of at least 1,00,000 each to allow private sector service providers to leverage scale benefits leading to long term financial sustainability. For inducing the private sector to participate in creation of additional capacity, scalable and sustainable business models with direct linkages to employment should be deployed. The NSDC has created such models. They should be implemented across 20–30 projects specific to manufacturing in partnership with industry associations and from funding through NSDF. *Improving ITIs*: We need to improve privatesector involvement in upgrading existing ITIs and also improve their curriculum and content through the sector-skills councils. Industry *Attracting students*: As a long-term strategy, it is important to make acquisition and improvement of skills an aspiration for people, especially youth. This could be achieved by recognising high-skill persons at the national and State levels along with recognition of other worthy citizens. For example, an unsecured loan scheme should be created for those who aspire to undertake vocational training. Large enterprises could also provide special incentives and recognition for acquisition of high skills.

Overall coordination: A number of initiatives have already been taken by various Government ministries to tackle issues related to skill development both at the Central and the State level. Coordination between these initiatives should be improved. The role and performance of the National Skill Development Coordination Board should be assessed. To ensure that skill-development activities are aimed towards areas of maximum impact, it is important to put in place an information system that provides data on availability and requirement of skilled resources. *Enhancing Skill Levels of Current Workforce to Improve Productivity* Training and skill building of the existing workforce is an important element of the strategy for increasing productivity of manufacturing in India. Training of employees can be incentivised by allowing tax deductions for expenditure incurred on training. Currently, skill building is predominantly achieved by in-house training of workers by each enterprise. However, clusters and NIMZs provide opportunities for shared infrastructure to provide training for skilled and semi-skilled workers.

A number of existing initiatives are focused on setting up tool rooms which are necessary for SMEs. These tool rooms can be made more effective by periodic performance audits by independent agencies and also by operating them on a PPP model in collaboration with industry associations. Just as tax incentives are provided for investments in critically required infrastructure assets, fiscal measures including tax benefits on training expenditure may also be considered for investment in critical human assets. MSME Sector alone needs to skill 42 lakh persons in the Twelfth Plan period, thus, requires to increase its current training capacity from 4 lakh person per year to at least 17 lakh persons per year by 2017. Apprenticeships can be an effective way of ensuring that entry-level workers have the skills required to join the formal workforce. While there should be no obligation to employ apprentices, the current apprenticeship model needs to be reformed by simplifying workflow for engagement of apprentices by employers, inclusion of new trades and recording compliance through e-filing, removing NOC requirement for out-of-region candidates. Further, it is proposed to make all graduates eligible for apprenticeships and the duration of courses should be reduced to a minimum of three months and should be converged with MES. Outdated curriculum needs to be updated and outsourcing of classroom trainings should be allowed.

Changes in the Apprenticeship Act may have to be made. In the meantime, a new model of incompany training should be deployed. In this model, companies should be allowed to take trainees for a period of up to six months. *Improving the State of Manufacturing Management in the Country* There were a total of approximately 5 million managers in the manufacturing sector in 2008. If the manufacturing sector grows at the targeted 12–13 per cent, 8 million more managers will be needed by 2025. Well-trained managers are extremely important for improving the productivity of manufacturing enterprises and maintaining harmonious industrial relations. Currently, only a very small portion of graduates from engineering and management institutes take up careers in manufacturing. Consequently, there is a significant gap between supply and demand. 13.95. The quantity and the quality of management in the manufacturing sector can be improved by the following initiatives:

- Increasing collaboration between manufacturing companies and engineering/management institutes for joint projects in which staff and 72 Twelfth Five Year Plan students of the institutes can get some hands-on experience.
- Encouraging enterprises (especially larger ones) to run good graduate engineering programmes which can be a source of management talent for themselves as well as the manufacturing sector generally.
- Scaling up programmes such as Visionary Leadership for Manufacturing (VLFM) at the national level.
- Setting up centres of excellence for manufacturing management through MoUs between institutes, government bodies and industry partners. Business schools that focus only on manufacturing management should also be encouraged.
- Creating a PPP model for engineering and management colleges with partnership with industry associations and employers with focus on manufacturing management.
- Launching a campaign focused on attracting management talent to the manufacturing sector.
- A large source of potential managerial/supervisory staff is the current workforce. Support should be provided to enable deserving members of the workforce to be promoted to management positions.

Recent reviews with many sectors of industry reveal a crying need for better supervisors and foremen—the first and second levels of supervision— who are the backbone of productive and harmonious manufacturing enterprises. Development of supervisors and foremen, through suitable programmes, collaboratively designed and managed by industry and educational and training institutions must be ensured along with the emphasis on development of skilled workmen and good managers. *Providing Social Protection to Low-income Workforce* . Formal sector workers can leverage collective bargaining to obtain social security; however, the informal workforce is dependent on government actions to improve social protection for them. A number of social security schemes have been launched in the recent past. However, the existing coverage represents a very low percentage of the total number of workers in the manufacturing sector. For example, the New Pension Scheme (NPS) that was launched in May 2009 to increase pension coverage, particularly to the informal sector, has less than 2,00,000 voluntary subscribers—this is far less than the total intended coverage for such a scheme. Limited access to social security is exacerbated for those with low or uncertain incomes. *Unemployment benefits*: Low income workers in transitional phases of unemployment are particularly vulnerable as they are unlikely to have significant savings. To help overcome the problems associated with social protection for temporarily unemployed workers, which include contract workers at the end of their contracts, a solution could be for these workers to be part of a 'sump' as permanent employees of contract agencies that are provided with Government support to ensure skill upgradation of these workers. The focus should be on creating a pool of workers who can be available to employers and ensuring that those that are unemployed have avenues for training as well as financial assistance. For example, the Automotive Mission Plan has recommended the formation of a Supplementary Unemployment Benefits Fund to be created by automotive companies for providing compensation to laid-off workers. Such funds in other sectors too can be utilised to finance the creations and sustenance of the 'sumps' that could be the 'win-win' solution out of the 'fairness–flexibility' dilemma. *Increasing penetration of existing schemes*: To ensure that existing schemes reach the entire workforce, it is important to increase awareness of these schemes through communication programmes. The distribution channels for these schemes should be evaluated and measured regularly and private sector participation should be encouraged too. Financial literacy of the workers in the informal sector should also be improved so that they make better informed decisions about participating in social-security schemes. *Improving Industry Workforce Relationships* 13.100. Strong and effective industry relations can enable managements of enterprises and their Industry workers to collaborate in increasing the productivity and competitiveness of the manufacturing sector. Unions have a critical role to play in ensuring inclusive growth of the manufacturing sector, especially by working towards social protection for the workforce. They can also play valuable roles in other areas such as skill development. The National Skill Development Policy has recommended that trade unions contribute in areas such as developing competency standards, course design, improving awareness of and promoting participation in skill development among the workforce. To ensure that unions can play a broader and more effective role, it is important to invest in capacity development of unions through training of their leadership. The multiplicity of unions in the same enterprise for the same type of workers can lead to interunion rivalries and can weaken collective bargaining. Therefore, legislation that enables one union per enterprise is strongly recommended. The union leadership should also be held accountable for any illegal behaviour by union members during negotiations. The practice of withholding recognition of unions should be discouraged. Strong gain-sharing systems can help to improve productivity. The Government has a crucial role in enabling good industrial relations by providing platforms for the industry and the workforce to participate in policy development and implementation. Since labour figures in the concurrent list in India, both the Central and State Government's role in such platforms should be that of an impartial facilitator focused on creating consensus amongst employers and employees around solutions. In especially contentious areas such as changes in labour laws, the Government should enable the development of consensus positions between the various interested parties. The 'backbone organisation' described in the Way Forward Chapter should have the capabilities to effectively assist in such a process of consensus creation.

BUSINESS REGULATORY FRAMEWORK

Countries that have performed better than the others in terms of thriving business have, to a great extent, done so on account of the quality of the business regulatory environment, which is an important factor distinguishing better performing countries from others. The key objectives of streamlining of business activities through the regulatory framework should be:

- Low compliance cost for doing business in India
- Simple regulatory environment, saving time and energy for the businesses; and
- Ensuring fair competition country must improve regulations and implementation in many subjects to make India generally a more attractive country for doing business. These include land and environmental regulations, labour laws and their administration and so on. It should be noted that, in the context of India's federal structure, the ability to mandate specific reforms to the regulatory framework from any centralised apex body is fairly constrained. Therefore, while nodal agencies may be set up to focus attention on matters that must be attended to across the country, and this section and others mention some, it is imperative that the role of such agencies in the process of making improvements across the country fits the country's federal and decentralised political structure. Such agencies cannot and must not usurp local authority.

Status and Key Challenges

The present regulatory environment is seriously deficient for the reasons enumerated below:

- Weak institutional architecture for business regulations in the country – Despite that high priority of the business regulatory reform agenda in the country, there is no dedicated authority that can guide the whole process of reform in a structured, planned, cogent and systematic manner, which could mandate the respective departments of the Union, State and Local Governments to comply in a timely, result oriented and predictable way.
- *Ambiguous nature and vast scope of business regulations:* there are vast numbers of business regulations at different levels of Government in existence in the country. There are instances of Twelfth Five Year Plan contradictory as well as overlapping business regulations on account of these being administered by the different tiers as well as layers of Government.
- Absence of national repository of business regulations: despite the advancements in Information and Communication Technology (ICT) and its ever-growing applications and usage, there is no dedicated online repository to track all the business regulations and procedures.
- Lack of coherence in business regulatory governance across country; business facilitation is often mentioned as part of the agenda at the national as well as State levels. But there is lack of coherence in all such efforts. There are wide variations in Government-business transactions taking place in different locations of the country. It has also been found that there is a lack of predictability and standardisation in terms of timelines as well as process adopted by different State Governments when it comes to facilitating business.
- Lack of defined mechanism for consultation between Government and industry: the interface between Government and the industry is also not well defined. There are periodic consultations among various industry collectives and specific Government departments located at different levels, but such consultations are not structured enough to be guided by a well-defined and outcome- oriented process.
- Inherent limitations of regulatory system in country: lack of periodic-review clauses in regulations and Lack of Regulatory Impact Analysis (RIA).
- There have been recommendations for regulatory reforms earlier as well, but due to absence of any one dedicated agency accountable for the reforms, they could not be implemented.

Follow-up Over Previous Administrative and Regulatory Reform Endeavours

Lack of implementation of earlier recommendations on regulatory reforms has contributed to the current situation of business-regulatory framework in the country, both at the Central and State level. All these recommendations need to be reviewed and a repository of all these documents needs to be created. After this an enquiry can be taken up to check the extent to which these recommendations have been implemented or are pending by the public authority or department. There is a need for a process for responding to the existing recommendations. In such a system once a certain expert group or commission of enquiry has submitted its report, the respective departments are required to prepare a response. That response is put up in the public domain along with the original recommendations. This makes it easier for various stakeholders to understand the extent to which the recommendations have been accepted along with the reasons for non-acceptance, if any. *Establishing Enabling Institutional Architecture*

- Formulating national policy on business development and regulation – The policy should also provide the principles of optimal business regulatory governance. It is recognised that there will be a special role of the Prime Minister and Chief Ministers in the aforementioned policy making process because in the final analysis, the actual adoption of the policy will entirely be dependent on the political leadership.
- Drafting and enacting 'National Business Development and Regulation Bill'.
- Building institutional architecture for looking after the business-regulatory reforms in the country: a dedicated institution can be set up for this purpose. The institution should be set up at the national level as well as at State level.
- Enabling institutional architecture for ensuring competitiveness in manufacturing. The same is required in both, Central as well as State level. – At the Central level the National Manufacturing Competitiveness Council (NMCC) has been entrusted with this responsibility. – Similar institutions may be set up at State level; to be called State Council on Manufacturing Competitiveness and Competition Reforms.
- In June 2011, the Ministry of Corporate Affairs has set up a Committee to draft National Industry Competition Policy (NCP). In February 2012, the Drafting Committee submitted a Draft National Competition Policy and comments of all stakeholders have been invited. Once this policy is approved by the Union Cabinet, further steps are required: – Building consensus on the policy – Creating institutional framework for operationalising the policy, as recommended by the Committee – Creating incentive and disincentive mechanisms for States to implement NCP
- Operationalisation of National Manufacturing Policy and development of State manufacturing plans in line with National Manufacturing Plan. *Systematisation of Business Regulatory Governance*
- Mapping and classification of all existing business regulations and procedures and providing an online one-stop shop—'National Business Facilitation Grid' for all information related to business regulations and procedures in India. Design principles of this on line portal can be finalised through a consultative process. The Department of Industrial Policy and Promotion is the nodal agency for the NBFG repository.
- A system of mandatory reviews of existing regulations at periodic intervals should be established and operationalised. This will achieve the desired goal of making the regulatory system intrinsically strong and up to date.
- A decentralised Single Window System should be established with appropriate geographical spread. The Single Window System, governed by a common minimum standard, should, rather than being a coordination office, be endowed with access to relevant information and sufficient delegation of powers from all concerned regulators, including Central, State, Local and Sector regulators. This would help reduce the start-up time for businesses by providing all requisite approvals and licenses, if any, through the Single Window System. – Recognising the wide variations with business procedures at the country level, it is recommended to benchmark the execution timelines and processes that are undertaken by different Government entities to facilitate business requirements. – A team of Business Facilitiation Officers (BFOs), in each of the participating regulatory authorities, may be asked to aid the Single Window System, and the BFOs could be made accountable for defaults or deviations resulting in aggravated costs of compliance to businesses. The desirability and feasibility of such a Single Window System should be determined through a consultative process. *eBiz Mission Mode Project* The eBiz Mission Mode Project, under the National e-Governance Plan,

aims to create a business and investor-friendly ecosystem in India by making all business and investment related regulatory services across Central, State and Local governments available on a single portal, obviating the need for the investors or the business to visit multiple offices or a plethora of websites. It in envisaged that the services offered on eBiz will eventually cover the entire life cycle of a business—right from its establishment, through its ongoing operations, to even its possible closure. Once operational, this project will also create a platform for multiple Government agencies to cross validate their information.

The project is being implemented as a 10-year PPP with M/S Infosys. The first-year pilot includes 8 Central Departments and States (Andhra Pradesh, Haryana, Maharashtra, Tamil Nadu and Delhi) covering 29 core services. Five more states (Punjab, Uttar Pradesh, Odisha, West Bengal and Rajasthan) and 21 more services will be added during the next two years of the pilot phase. An end-to-end solution providing the services under the Andhra Pradesh Single Window Act will also be provisioned on the eBiz platform by September 2013 along with the payment solution gateway. *Adopting Regulatory Impact Assessment (RIA)*

- Tool of RIA should be developed for Indian context through a consultative process and due research reflecting upon global experiences with its adoption and usage. 76 Twelfth Five Year Plan – The parameters of RIA should be clearly spelt out for evaluation (which should gradually be expanded to include the following eight elements: policy coherence; cost of doing business; competition; innovation; SMEs; consumers; labour; environment and commons).
- Process of doing RIA should involve a wide stakeholder consultation.
- RIA has be to be mandated in the country in ex ante as well as ex post manner.
- It is recommended that Policy Coherence Units (PCUs), for conducting RIA, be established under the respective State Planning Boards and at the national level. Such policy analysis functions can be connected with the capabilities of the proposed backbone organisation. *Making Businesses More Responsible Towards Society*
- Considering the importance of the subject, 'business responsibility' should be included as a separate subject under the Government of India (Allocation of Business) Rules 1961, and Ministry of Corporate Affairs can be entrusted with the responsibility of carrying out these activities.
- Redefining the contract of business and society and developing new rules of the game for corporate conduct. – Needs to be done through a widespread consultative process.
- Stronger role of business associations in responsible business. – Business associations should be encouraged to develop and impose rules of conduct on their own members.– Business associations should be entrusted with the responsibility of overseeing the compliance to rules of corporate conduct. – Such associations should provide their members a process for debating and agreeing on voluntary imposed norms, assistance to members to develop capabilities to conform to these norms and, very necessarily for such associations to become trusted by stakeholders as effective institutions for self-governance, internal governance that disciplines errant members.
- Disclosures on the adoption of 'National Voluntary Guidelines on Social, Environmental and Economic Responsibilities on Business' (NVG) principles should be made mandatory for businesses. Adoption of NVG principles can be made mandatory for all public–private partnership projects by the relevant authority at the time of project inception. This will help in mainstreaming these principles.

- Establishing the required institutional architecture for facilitating adoption of NVG principles. Awareness and implementation of NVG principles is currently the responsibility of the Indian Institute of Corporate Affairs. The IICA's abilities in this respect should be further strengthened. *Developing an Ongoing Process of Stakeholder Consultation* For achieving the objectives of a stakeholder consultation, it is imperative to have capacity, building both ends: at the Government side as well as at the industries. A process of productive consultations, and the roles of representative institutions of employers and unions in these consultations, in improving the productivity of the country's manufacturing ecosystem, and its sustainable competitiveness, cannot be overemphasised. The competitiveness of German and Japanese manufacturing industries, in spite of high-wage costs and expensive currencies, in contrast to the relative decline of US and UK manufacturing industries, is attributed to the better collaborative processes in the former countries. The following actions must be taken to achieve this objective:
- Passing a legislation mandating stakeholder consultation and also defining the process that needs to be followed.
- Measures to strengthen industry associations and their structure to enable them to convey the view of industry in a constructive manner.
- Similar capacity building for stakeholders, such as labour unions. *Developing a Business Regulatory Governance Mechanism* to choose appropriate regulatory alternatives among self-regulation, co-regulation and public regulation.
- Currently there is no structured modality exploring various alternatives for achieving regulatory objectives.
- Detailed analysis should be undertaken to determine which alternatives to regulations are feasible as well as beneficial for Indian context.
- As each form of regulation has merits and demerits, a desirable combination of all three regulatory alternatives may be evolved gradually.
- Such mechanism will serve as a ready reference one-stop shop for the policymakers as well as the business community while arriving at the choice of appropriate mode of regulation. *Capacity Building for Carrying Out Regulatory Reforms* Since carrying out the aforementioned regulatory reforms requires a tremendous effort, capacity needs to be built in order to implement them. The capacity-building framework needs to incorporate the following:
- Developing resources such as modules, guidelines, methodologies, reference manuals, checklists, case studies and so on as reference material for regulators. – These resources should also be available through an online-knowledge portal.
- Training programmes for regulators need to be arranged.
- A review may be initiated to determine the feasibility of expanding the roles of institutions functioning under the aegis of the Ministry of Corporate Affairs, namely, Indian Institute of Corporate Affairs, Competition Commission of India, Institute of Chartered Accountants of India, Institute of Company Secretaries of India and Institute of Cost Accountants of India.

ENSURING ENVIRONMENTAL SUSTAINABILITY WITH INDUSTRIAL GROWTH

The rise in growth in the resource intensive manufacturing sector is enabled and facilitated by an ever-increasing rate of material use leading to manifold impacts to the environment. The contribution of the manufacturing sector to environmental degradation primarily occurs during the following stages:

- Procurement and use of natural resources
- Industrial processes and activities
- Product use and disposal. The air, water and land are affected through the environmental impacts created through the operations of manufacturing units. Key Objectives Rapid ecologically sustainable industrial growth with focus on
- *Mainstreaming and promoting green business:* an environment has to be created wherein being green is not viewed as just an obligatory expectation of a company, but as an area of primary focus for the company to develop further and be recognised as a leader.
- *Protecting natural resources:* natural resources have to be prolonged to their fullest use to maintain the aim for continual economic growth and lessen environmental impacts.
- *Addressing funding issues:* which act as a constraint for movement towards a more sustainable industrial model.

The Central Pollution Control Board has identified 17 highly polluting industries, the majority of which are manufacturing industries. MSMEs, in particular, can have a significant impact on the environment as they are generally liable to be equipped with obsolete, inefficient and polluting technologies and processes. Seventy per cent of the total industrial pollution load of India is attributed to MSMEs. New technologies leading to cleaner processes and operations are not being developed at a fast enough pace to address the urgent need for environmental protection. The current ecosystem does not encourage and facilitate the mainstreaming and scaling up of new technologies for widespread use, mainly due to a lack of financial support, resources and Government assistance.

The waste management and recycling industry in India is currently vast but largely unorganised. In this space, it is necessary to mainstream the industry and ensure that the livelihoods of all people dependant on this industry are supported and upgraded.

Strategy and Key Recommendations

Organised Waste Management and Recycling

- Development of a National Waste Management and Recycling Programme – This is an overarching framework to create and mainstream the organised waste management and recycling industry. Structured frameworks and guidelines for recycling industry to be developed to integrate it with the existing waste management rules and guidelines. – Development of industry and sector specific recycling standards.
- Promotion of PPP model for waste management and recycling – Establish facilities for reuse, recycling and reprocessing of wastes from various sectors should be encouraged by providing incentives and ensuring the process for setting up PPP facilities.
- R&D funding – Promoting new technologies and processes for waste management and recycling. – This should be aligned with the overall technology fund as discussed earlier.
- Building institutional capacity – Local institutional bodies must have their capacity built on recycling and waste management. *Creation of a Green Technology Fund*
- For usage in three key areas: technology upgradation, promotion of green entrepreneurs and funding for R&D.
- This could be disbursed in the form of concessional loans, grants and so on.
- This fund should be a part of the overall technology fund proposed for improving depth in manufacturing and must ensure focus on commercialisation of new technology areas. *Promotion of Green Products*

- Development of a framework and guidelines for promotion of green products – Definition of the specifications – Creation of/assignment of a new/existing entity to perform this task on a regular basis – Identification of top 100 green products (based on assessment of maximum environmental impact) and setting of standards for the same
- Promoting green public procurement through price incentives on Government tenders
- Encourage and develop voluntary rating programmes
- Creation of centres of excellence to promote green products and processes
- Incentive programmes for creation of Life Cycle Inventories
- Incentives for export of green products *Environmental Regulatory Reforms and Market Based Instruments*
- Strengthening regulatory institutions together with bringing institutional reforms – Moving towards load-based standards from concentration based regime.
- Implementing polluters-pay principle, with specific pollution loads beyond a defined benchmark should be priced and paid for by industry. – Reforming the existing environmental clearance process. – Institutionalise the concept of cumulative impact assessment of the region. – Introducing technology assessment while appraising new projects. – Process for administering the clearances needs to be streamlined—should include considerations of decentralisation, requirements and tenure of clearances.
- Establishing integrated chemical-management policy and regulatory regime – Set up a regulatory process to assess all chemicals, register and phase-out toxic chemical products and replace them with non-toxic/lesstoxic substitutes.
- Market-based instruments and emission trading – Initial pilot Emissions Trading System to limit particulate matter emissions. – Scale up the emissions market to address additional pollution problems at the State and national levels. – Monitoring technology for all types of pollutants be made as affordable as possible for industry; waiving of applicable taxes and excise duties, as well as direct subsidies to monitoring technology wherever their installation is mandated by the State pollution boards. *Sustainable Environment Management in MSMEs*
- Reconstitution of regulatory bodies – Inclusion of stakeholders/associations. – Sector-wise product sub-groups need to be formed as part of PCBs. – Grievance Redressal Mechanism should be established at each PCB.
- Creation of common infrastructure for MSMEs in clusters – Central Grant Scheme for soft infrastructure, unit level technology upgradation assistance, portion of project cost for Common Effluent Treatment Plants. – State Grant Scheme with provision for arranging land for CETPs, time-bound speedy legal clearances, provision for equity participation in SPVs by SPCBs/State agencies. *Disclosure on Performance*
- Short-term action to increase voluntary disclosure of environmental sustainability performance. – Development of reporting standard-based on several existing sustainability reporting initiatives. – Incentives for voluntary disclosure.
- Long-term steps to compare environmental sustainability performance of organisations with industry-specific benchmarks. *Development of Environment Sustainability Benchmark Index, Especially for Identified Highly Polluting Sectors Organised Waste Management and Recycling*
- As covered in the chapter on Environment, the development of a National Waste Management and Recycling Programme and the promotion of PPP model for waste management and recycling are required.

WATER ISSUES

With its increasing population and industrial activity, India is moving towards perennial water shortages. The current per-capita water availability is estimated at around 1,720.29 m3 per capita according to data from the Central Water Commission and as per the World Water Development Report— one of the United Nations, India has been ranked 133 (Out of total of 182 countries) in terms of total renewable per capita water resources. The total water demand is projected to increase by 22 per cent by 2025, and 32 per cent by 2050. A major part of the additional water demand will come from the domestic and industrial sectors. The water demands of the domestic and industrial sectors will account for 8 per cent and 11 per cent of the total water demand by 2025.

Key Objectives

- Improve the governance and management of water in order to ensure availability of water for all purposes.
- Improve the management of water by industry, in particular in terms of utilisation and pollution. Status and Key Challenges
- Inadequate storage capacity
- Governance deficit and fragmented institutional framework
- Inadequate water management by and for industry – Water intensity high as compared to global benchmarks—to the extent of ~30–50 per cent. – Recycling water in industry is not common and its proliferation is not happening at the scale as required. Strategy and Key Recommendations Strategy on improving overall governance and management of water has been covered in detail in the section on Water Resources. The proposed draft National Water Framework Bill will provide the broad overarching national legal framework of 80 Twelfth Five Year Plan general principles on water which will necessitate the requisite administrative frameworks needed for greater clarity on demand management, protection of water resources, improving efficiency of water use and so on. Specifically, strategic measures to ensure availability for and efficient utilisation of water by industry have been outlined below. *Water Management in Industry*
- Create equity-based and efficiency-based Water Pricing Regime for industries – Overcome lack of a clear policy framework based on cost-recovery principles.
- Current pricing regime is undervalued for all users.
- This would overcome wide variations in tariff structure due to current determination by various States.
- All Indian cities currently operate a mix of measured/metered or unmeasured/unmetered tariffs. – Potentially two different pricing regimes in two-tier tariff system/IBT tariff system.
- Enforce 'Water returns' – Annual return to be filed by water users on similar lines of tax returns—should include key measures like water utilisation per unit produce, effluent discharge details, rain water harvested, water reuse details, fresh water consumption and so on. – Mandatory for major water using industries and businesses.
- Promote reuse and recycle of wastewater in industry – Regulations and incentives through national frameworks and a system of water returns
- Industry specific standards – Promoting rain-water harvesting in industry, both within and beyond the fence through incentives and regulation.

LAND ISSUES

Among all the traditional factors of production for any economic activity, land being natural, immovable and non-renewable, is a distinct resource. It needs to be looked at from Industry's perspective as a tangible resource with supply and demand issues and the linkage in the form of land acquisition for industrial demand. Land in India has a special significance because it carries a huge tangible and emotional value for owners and also for those whose livelihoods depend on it. This makes it very important to consider the land acquisition process in a critical manner.

The key objectives with regard to solving the various issues and challenges related to land pertain to:

- Improving the management of land as an asset in India.
- Setting up a more transparent, fair and efficient process of land acquisition for industry development.

By achieving these key objectives, we would be able to ensure a more productive utilisation of land, and in particular, be able to spur industrial development, which has in many instances been hindered as a result of poor land management and land acquisition processes. Status and Key Challenges India has sufficient land for all uses—agriculture, industry, human dwelling, infrastructure and other uses—as long as it is used with prudence and productivity. Currently industry utilises only about 2–4 per cent of all land in India. Even at heightened industrial activity in the future, it is expected that there would be sufficient land for all users, including industry. However, there are some critical issues that need resolution in order for land to become a wellmanaged resource, especially from the point of view of Industry. Land is inherently an imperfect market, because land is an immobile asset. Hence, no two pieces of land are alike and can be differentiated. This gives rise to a monopolistic power with the Industry 81 landowners. Furthermore, the value of a piece of land effectively changes when we change its usage and due to development of surrounding areas. In addition, the owner is often emotionally attached to his land. In India, land is considered a very important asset from an emotional perspective. A major characteristic of land ownership in India is that the land holdings are typically small. Typical industrial usage requires development of large tracts of land. Consequently, industrial development has as a prerequisite need to acquire land from a large number of owners in order to develop a contiguous piece of land for industrial use.

Another problem in the land market is the incomplete, outdated and inaccurate land records, which give rise to disputes and litigation. Since industrial projects require large amounts of land and land holding in India is fragmented, industrialists have to deal with a large number of landowners and consequently face substantial risk of litigation. In addition, there are some restrictions on usage of agricultural land for non-agricultural purposes. Non-Agricultural Use Clearance (NAC) from the local/State Government is necessary before agricultural land can be considered for other uses. Strategy and Key Recommendations A three-pronged approach should be undertaken for tackling the land issues. This includes the development of an institutional framework to support the various actions, a drive to create Land Use Policies to manage land better, and a reformed process of land acquisition (Figure 13.5). *A National Land Use Policy should be developed* to take care of the growing requirements of land for sectors other than agriculture. *State Governments should formulate appropriate Land Use Policy* in alignment with the National Land Use Policy. The main features of this policy should be *Land Mapping* (record of types and quanta of land available), *Land zoning and Digitisation of Land Records*. The Land 3-part strategy to tackle land issues. Institution Framework **Land development corporations:** independent commercial entity licensed by the Regulator to acquire and develop land on behalf of the end-users (industry) **National regulator**. lay down guidelines, monitor the functioning of the sector and provide oversight Land Use Policy **National Land Use Policy to be developed:** Land mapping, zoning, digitization of land records **State Governments should formulate appropriate Land Use Policy in alignment with the National Land Use Policy** Process for Land Acquisition and R&R **Execution of land acquisition by LDCs through SPVs: based on certain considerations on** Valuation of land—open-offer price/ multiples of historical price Compensation for land R&R programme

Strategy for Land Issues Twelfth Five Year Plan

Use Policy should also look at measures to optimise utilisation of land by benchmarking current utilisation efficiency with global benchmarks, and setting standards and incentives for more efficient utilisation.

- There is a need to establish an independent and autonomous regulator which can lay down guidelines, monitor the functioning of the sector and provide oversight. The regulator should – encourage State and local Governments to define zoning of land, ear-marking them for different uses, and encourage digital land records – define guidelines for valuing various types of land for different uses – Establish norms for setting up and operating Land Development Corporations (LDCs) and monitor adherence to the norms by these institutions – lay down the guidelines for acquiring land by a corporate body–establish norms for process of land acquisition, compensation and relocation and rehabilitation of various stakeholders for different project characteristics
- *Value of land* can be determined, as per the guidelines laid down by the regulator, in the following ways: – *Open-offer price*: Land owners will be asked to submit their application for sale of land in a reverse-auction process. – *Multiple of historical price*: The regulator can set a price based on a multiple of the historical land prices, as mentioned in the land records of the government.
- The acquiring agent for land should be an independent commercial entity—Land Development Corporation— that has been licensed by the regulator to acquire land. The role of LDC would be to acquire and develop the land on behalf of its clients (end users) in exchange of the process and maintenance fee. A State can have multiple LDCs and each LDC will execute projects through SPVs. The operations of the LDC will be under the purview of the regulator.
- The *process of land acquisition* will be guided by the regulatory framework applicable for the project characteristics as defined by the LDC in its SPV. The role of local/State Government authorities in supporting the acquisition process should be laid out clearly by the regulator based on project characteristics. The acquisition process may vary depending upon – minimum per cent that the SPV needs to acquire from individual landholders before regulation mandates compulsory acquisition of land from other owners– nature of consent required from different stakeholders
- *Compensation for land* needs to factor the following: – upfront payment – annuity income stream – participation in the future appreciation due to growth as a result of land development In addition to the above factors, the land owner needs to have the flexibility to choose a compensation package – an owner can choose to take the full value in upfront compensation or take a part of it as annuity payouts (determined by prevailing financial indicators of the time) – however, every land owner will necessarily have the component of 'participation in future appreciation' as part of the compensation.
- The *LDC has to operate a rehabilitation and resettlement programme* with combination of different elements which have been defined by the regulator based on the project

characteristics; these include elements like. – alternative dwelling, if displaced – skill development – assistance in employment/income-generating opportunities – community development.

- The *Industry* must be *responsible for payment of cost of land acquisition*, including market price, share of the appreciating value and cost of the comprehensive R&R.
- There should be a *timeframe defined for land acquisition*, and the LDCs must *interface* Industry 83 *appropriately not only with the* local selfgovernance bodies, but also other grass-root level organisations in order to *build awareness about the land acquisition process*.

A description of the process of land acquisition, the role of the institutional framework and the other modalities related to land acquisition is provided Industrial clusters are increasingly recognised as an effective means of industrial development and promotion of small and medium-sized enterprises.

- For MSME participants, clusters play an important role in their inclusiveness, technology absorption, efficiency improvement and availability of common resources. Ministries dealing with MSME enterprises have been using Cluster programme as one of the key policy tools in administering Industrial Policy. There are around 7,000 clusters in traditional handloom, handicrafts and modern SME industry segments.
- The Ministry of Micro, Small and Medium Enterprises (MSMEs) adopted the cluster approach as a key strategy for enhancing the productivity and competitiveness as well as capacity building of small enterprises (including small scale industries and small scale service and business entities) and their collectives in the country. The Ministries have been administering hard and soft interventions to help the cluster participants. While hard interventions will include investments in infrastructure like common facilities, common testing centres, roads, the soft intervention will include training, capacity building, skill improvement, marketing inputs, product design and development and so on.
- In order to assess the level of intervention required, MSME Ministry has carried out a diagnostic study of about 471 clusters. However, the follow-up on these studies have been weak.
- Today, the cluster programmes are administered by various ministries (textiles, leather, food, MSME, heavy industry [auto]) under various names with different terms and conditions. This apart from putting the cluster participants through procedural hurdles also makes it very tough to learn from each other and improve the efficiency of these schemes.

www.ingramcontent.com/pod-product-compliance
Lightning Source LLC
Chambersburg PA
CBHW061323190326
41458CB00011B/3870

9789351302469